No. 8224
$19.95

THE COMPLETE HANDBOOK OF RADIO RECEIVERS & TRANSMITTERS

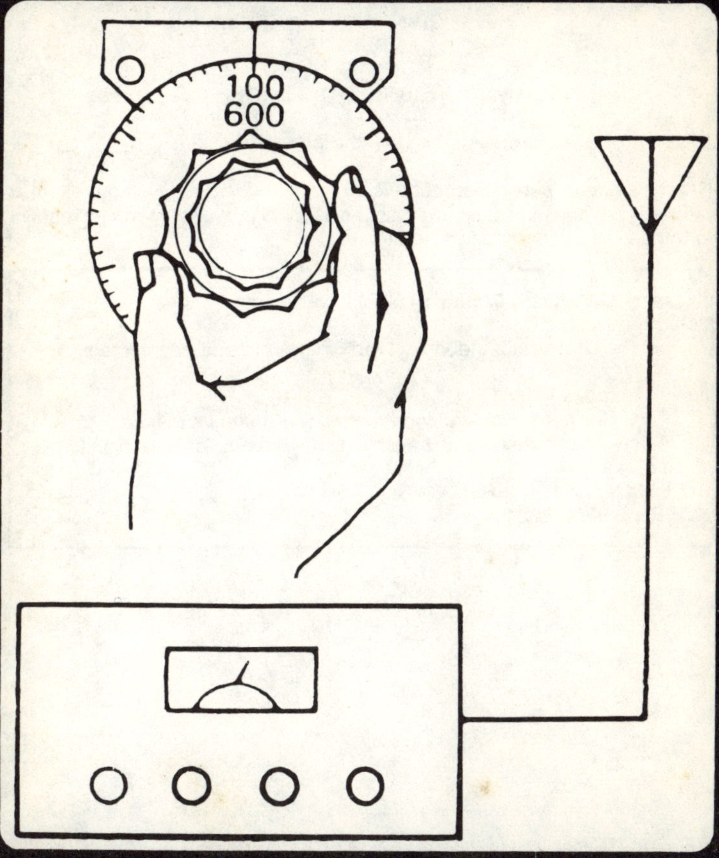

BY JOSEPH J. CARR

TAB BOOKS Inc.
BLUE RIDGE SUMMIT, PA. 17214

FIRST EDITION

FIRST PRINTING—JUNE 1980

Printed in the United States of America

Library of Congress Cataloging in Publication Data
Carr, Joseph J
The complete handbook of radio receivers and transmitters.

Includes index.
1. Radio—Receivers and reception—Amateur's manuals. 2. Radio—Transmitters and transmission—Amateur's manuals. I. Title.
TK9956 C3515 621.3841'3 80-14453
ISBN 0-8306-8224-4

THE COMPLETE HANDBOOK OF RADIO RECEIVERS

BY JOSEPH J. CARR

TAB BOOKS Inc.
BLUE RIDGE SUMMIT, PA. 17214

FIRST EDITION

FIRST PRINTING—JUNE 1980

Printed in the United States of America

Library of Congress Cataloging in Publication Data

Carr, Joseph J.
The complete handbook of radio receivers.

Includes index.
1. Radio—Receivers and reception—Handbooks, manuals, etc. I. Title.
TK6563.C36 621.3841'36 80-14491
ISBN 0-8306-1182-7 (pbk.)

Cover photo courtesy of Trio-Kenwood Communications Corp.

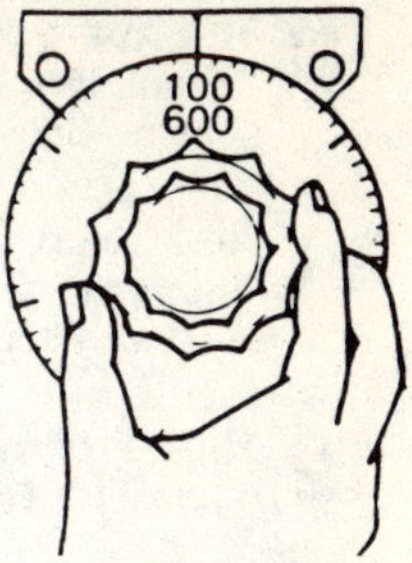

Contents

Introduction

The modern radio receiver has not been left out in the advancing technology that has affected all areas of the electronics. The radio receiver used by even low-budget amateur and CB operators is very much superior to some of the better receivers of just a decade ago. There are dozens of different types of radio receivers, and they perform as both general purpose, and highly specialized instruments.

Look at car radio receivers. In 1956, the first all transistor unit was marketed by Motorola's after market division. But it wasn't until 1962 that Delco Electronics introduced the first standard car radio for original equipment manufacturer (OEM) sale to an automobile manufacturer (General Motors). Motorola and Bendix followed in 1963 with all transistor radios for Ford and Chrysler products. Starting in 1957, however, we saw the so-called hybrid car radio receivers. These used special tubes (that allowed 12-volt operation of the anode; i.e., 12BL6) for the rf, converter, i-f and detector/1st audio stages, and a gernamium TO-3 package output transistor for the audio power amplifier. Today, most car radios are not only solid state, but use integrated circuits of advanced design, phase-locked loops for both front-end tuning and stereo decoding, and at least two models use microprocessor-controlled tuning (Delco and Chrysler all-digital tuning radios).

The amateur radio (ham) receiver and the shortwave listeners (SWL) receiver have also undergone much in the way of advancement. Even relatively low-cost receivers in this market make use of integrated circuits, ceramic crystal band-pass filters, and double conversion techniques. The average amateur/SWL receiver of today is superior to all but the best commercial receivers of a decade ago. Similarities in performance exist between these modern radios and even the best available in the early 60s. Two different categories of radio receiver are used here: the *ham band only* and *general-coverage models*. The ham band-only receivers are designed to receive only the high-frequency amateur radio bands (80 meters through 10 meters), and some will also cover the 160-meters and 6-meter bands. The general coverage receiver is designed to cover all of the high-frequency bands from 3 MHz to 30 MHz, with most also covering the AM broadcast band (0.55 MHz to 1.6 MHz) and not a few also receiving the VLF (200 kHz through 500 kHz) band. There was once a substantial difference in quality between ham band and general-coverage receivers that had the same price. It was usually the case that, dollar for dollar, the ham band receiver worked better. But today, such distinctions are only occasionally valid. Some of the general-coverage receivers boast the same selectivity and sensitivity as ham band-only (HBO) models in the same price range. Even the tuning rate of the dial (which was the really big advantage of the HBO receiver) is the same for the two models. It is now common to see 500 kHz to 1000 kHz segments on each band. We find, therefore, that the tuning dial will cover the same number of kilohertz per dial division on all bands. Formerly, when the general-coverage receiver would cover 0.55 MHz to 30 MHz in five to seven bands, the tuning rate on the high bands was very much more rapid than the tuning rate on the lower bands. Why? The VFO used as the local oscillator had a frequency coverage that is equal to the square root of the capacitance ratio in the VFO tank circuit. For most receivers, the capacitance ratio was 10:1 (they used a 365-pF variable capacitor), so the frequency ratio was approximately 3.2:1. On the low band, where the low-end frequency was 1.75 MHz, the high-end frequency would be approximately 5.5 MHz. But at the high end, the low end of the highest band is typically 15 MHz, so the 30 MHz highest frequency is found in less than the total dial sweep (sigh). Today, though, almost all of the SWL general-coverage receivers use double or triple conversion techniques for converting the high bands to some low band, in 500 kHz to 1000 kHz increments. The principal reason why amateurs don't use many of these general coverage receivers is that their receivers are part of ham band-only *transceivers* that also transmit.

A large number of VHF/UHF receivers are on the market. In the early 60s, *monitor receivers* for these bands were, in a word, *terrible*. With the VFO-tuned models, it was almost impossible to tune in a narrow band AM or FM signal, and even harder to maintain the tuning for any period of time. I have seen some models that would change frequency when somebody entered the room (their body capacitance detuned that darn VFO!). By the mid-60s, however, most of these receivers were crystal-controlled. This eliminated most of the drift problems but only at the expense of channelizing the operation of the receiver. Several manufacturers now offer scanner receivers that operate in both VHF and UHF ranges (channels programmable by the user through a keyboard), and are microprocessor-controlled.

In this book we are going to consider some of the different receiver types in block diagram analysis, discuss some of the circuits in the receivers, and develop some idea of the different applications and operation of the receivers. This book is part of a two-volume series, and its companion volume is TAB book No. 1224, *The Complete Handbook of Radio Transmitters*.

Joseph J. Carr

Chapter 1

History of the Modern Radio Receiver

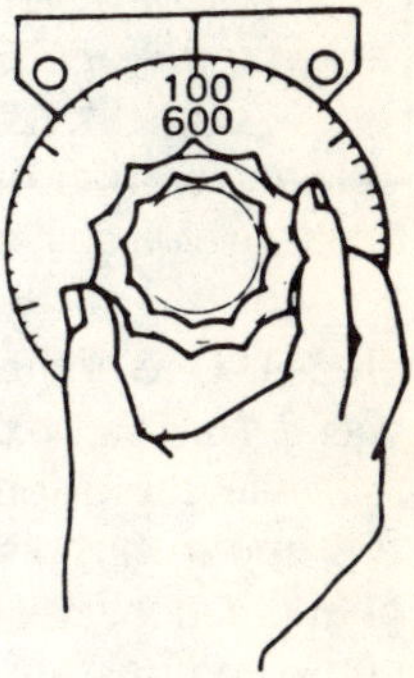

As early as the 1860s, a Virginian operated what appears to have been a wireless telegraphy unit from a tethered hot air balloon. The receiver was an induction coil coupled to an ordinary telegraph clicker. By the 1880s, the claim to wireless telegraphy was firmly established by German Heinrich Hertz, for whom our basic unit of frequency is named. His experiments used coupled induction coils to make the telegraphy work, and was able to work over only a short distance.

In 1903, Marconi established the possibility of transoceanic wireless telegraphy by transmitting morse code "S" (...) over the Atlantic ocean, to a receiver station in St. John's, Newfoundland, Canada.

At the turn of the century, radio receivers were really crude, yet worked with surprising ability. The antenna circuit was coupled to a crude detector called a *coherer*; a funny glass tube containing a set of electrodes and some loosely packed iron filings. The filings had to be shaken loose periodically in order to keep the detector working. The sensitivity would deteriorate over a period of time, and the operator would have to tap the glass envelope to shake the filings loose to resume operations. Transmitters in these days were mere spark gaps. The first commercially made radio receiver using the coherer was advertised nationally in *Scientific American* in 1906.

By World War I, it was found that certain mineral elements would detect radio signals. We know today that these crystalline minerals formed natural PN junctions, so were natural detector

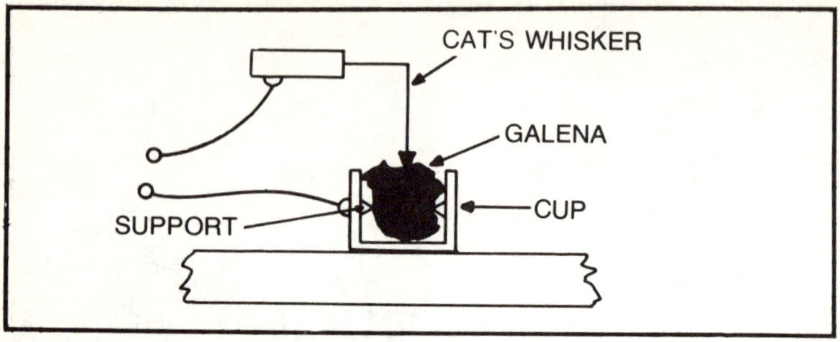

Fig. 1-1. Galena crystal detector.

diodes. One of the principal minerals used in crystal set detectors was *galena* (an oxide of lead).

But the galena crystal set was not without problems. Only a few spots on the surface of the crystal would exhibit the detection phenomenon. It was necessary to place the crystal inside a metallic cup (see Fig. 1-1), and then probe for those points with a fine wire called a *cat's whisker*. This turned out to be very tedious, and a breath of air would dislodge the probe from one of the correct spots.

In World War II, enterprising soldiers made *foxhole radios*. These were really crystal sets that used a blued razor blade as the detector. (You cannot use most blued blades, but only those which are *heat blued*; chemically blued blades do not act like PN junctions.) A cat's whisker was made to probe for the detection points on the razor blade, coils and wire antenna made from "salvaged" wire, and a pair of earphones could be liberated from a burned out enemy tank (of course, no GI would *ever* think of liberating US Army property!).

Figure 1-2 shows the basic circuit for a crystal set using modern components. The simplest form of crystal radio is simply a pair of earphones connected across a detector diode, with antenna and ground connections to bring in the signal. But this form of radio will not operate with any tuning; it will simply respond to all signals strong enough to make the diode go into rectification.

The second most simple crystal set involves an induction coil (Fig. 1-3) and the crystal detector. The induction coil is made using 100 turns or so (it didn't seem to be critical) of No. 22-No. 26 enamelled wire wound on a 1- to 3-inch diameter coil form. In most "classical" crystal sets, the coil form was the cardboard cylinder from a roll of toilet tissue, or a Quaker oats cereal box (today, the roll from the paper used in most IBM copier machines seems just about right, and is more sturdy than the Quaker oats box!). The builder would then take a file and scrape the enamel off part of the coil, making certain not to short together adjacent turns. A piece of

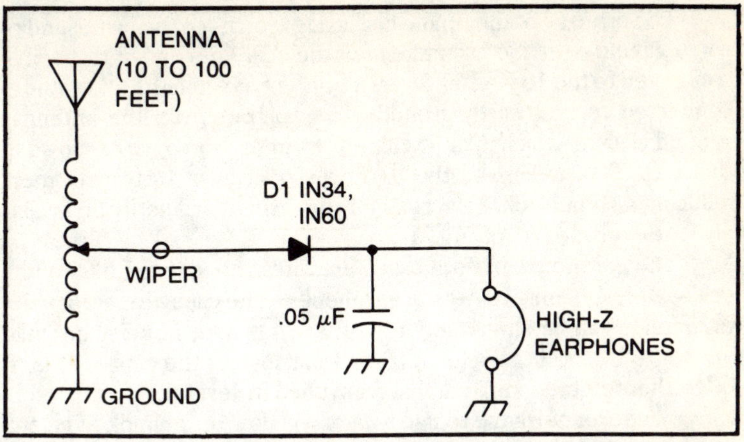

Fig. 1-2. Circuit for a simple crystal set receiver.

sheet metal was used to make a wiper to select a portion of the coil for use. Some amount of tuning is offered by this arrangement because of the stray circuit capacitances.

The fully tunable crystal set is shown in Fig. 1-3. Coil L1 could be one of the crude coils just discussed, or it can be a regulator broadcast antenna or rf coils sold by Radio Shack, Lafayette, and others. The capacitor across the coil is optional, but would make the system tunable across the AM broadcast band. Use the valve of capacitance required to resonante the coil selected.

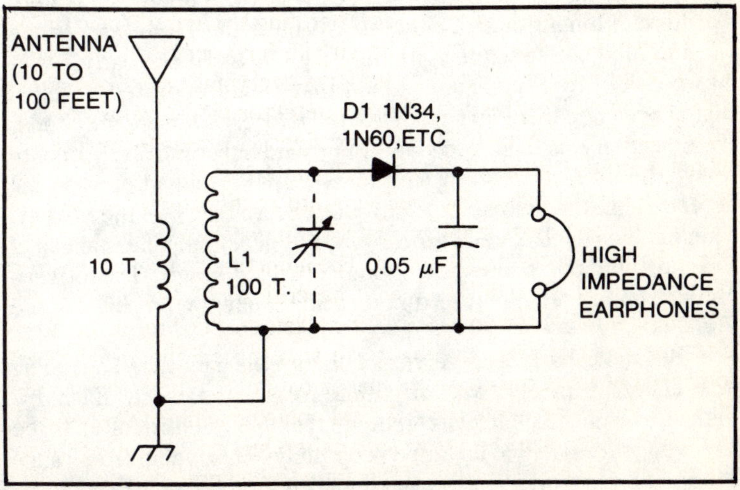

Fig. 1-3. More advanced crystal set receiver.

The primary of the tuning coil is a few turns of the same kind of wire, wound over the ground end of the coil. One end of this coil is connected to the 10- to 100-foot wire antenna, while the other end is connected to real earth ground. The current from the antenna-ground circuit, which is an rf current from the radio wave, flows in this coil. This will make the two coils operate as a transformer, inducing a higher voltage across L1 than would ordinarily be available in the crude set of Fig. 1-2.

The earphones in both cases are high-impedance (2000 ohms or greater) dynamic or crystal earphones. The capacitor across the earphones can be almost any type that you have, but disc ceramic seems to be the best selection. The purpose of the capacitor is to filter the detected audio signal from the diode.

We no longer have to use a cat's whisker on a chunk of galena or a blued razor blade in order to detect the radio signal and recover the audio. Many different signal diodes are now available. The 1N34 and 1N60 germanium diodes are specified over more modern silicon types because their junction potential is only 0.2 to 0.3 volts (as opposed to 0.6 to 0.7 volts), and this makes a more sensitive crystal set.

TRF RADIOS

The crystal set is not too efficient as a radio receiver; strong signals are needed in order to make the crystal detect. It is relatively easy to generate the large amounts of power needed ordinarily for long distance (DX) radio communication, but substantially better distance can be had if the receiver at the other end is very sensitive. Fleming and DeForest invented the diode and triode vacuum tubes, respectively, in the early 1900s. By the early 20s, it was possible to buy vacuum tubes that would operated in the rf region occupied by radio signals. With Deforest's amplifying triode, the sensitivity of radio receivers improved substantially. Undoubtedly, the first attempts placed an rf amplifier ahead of a crystal detector, but the multistage radio receiver was soon on the market. These radios used several *tuned radio frequency* amplifiers ahead of a diode (or diode-connected triode) detector stage. These were called trf receivers for the tuned radio frequency amplifiers (see Fig. 1-4).

But those early trf receivers were not the panacea. The tuning of the earlier models was frightful, to say the least. Each LC (inductor-capacitor) tank circuit in the rf amplifier chain (often three to seven stages) had to be tuned separately. This would often cause the amplifiers to break into oscillation, producing a terrible howl and scream. Some manufacturers simplified things a little bit by ganging

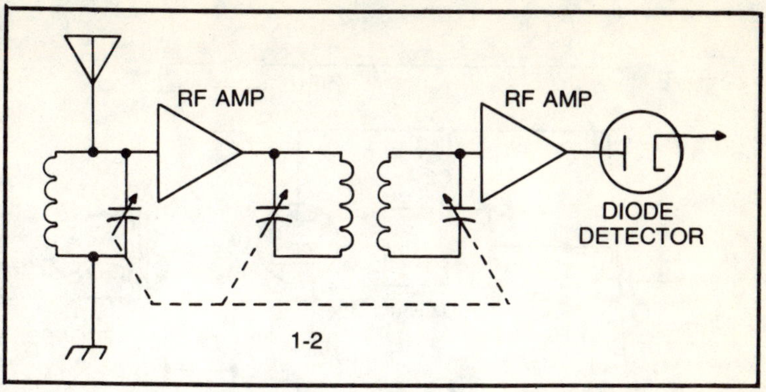

Fig. 1-4. Tuned radio frequency receiver.

the tuning controls onto the same shaft, but the trf radio still lacked something.

One day in the early 20's, so goes a radio legend, radio pioneer-inventor Major Edwin Armstrong invited David Sarnoff of RCA into a laboratory at Columbia University in New York. Armstrong had invented a new type of radio detector circuit that greatly improved the sensitivity and selectivity of radio receivers. He had constructed a radio using the detector inside a sealed wooden box, and only let General Sarnoff listen to it. Sarnoff immediately wanted it for RCA. This was the regenerative detector, an example of which is shown in Fig. 1-5A. The basic circuit very much resembles the Armstrong oscillator circuit, also named after the same inventor. A triode amplifier has a tuned LC tank circuit in the grid circuit, and produces some rf feedback through a tickler coil (Ls) in the anode circuit of the tube. The rf was decoupled to ground by a bypass capacitor (C3), but the audio component was coupled to either a set of earphones, or additional stages of af amplification, by audio transformer T1. Capacitor C4 is an audio decoupling bypass capacitor.

The idea is to adjust the position of the tickler coil viz the main tank circuit coil (L1) so that the stage would be *just short of oscillating*. This would greatly heighten the gain of the stage at the tank circuit frequency, greatly increase the apparent Q of the tank (hence the obtainable selectivity), and provide the nonlinear impedance needed for detection of the radio signal—all three in one stage. This type of detector would, at once, render multistage trf radios much more sensitive, and make possible reasonable sensitive radios with fewer stages (hence, lowered cost).

In some models, the tickler coil was mounted inside of the main inductor. Regeneration occurred because the operator could rotate

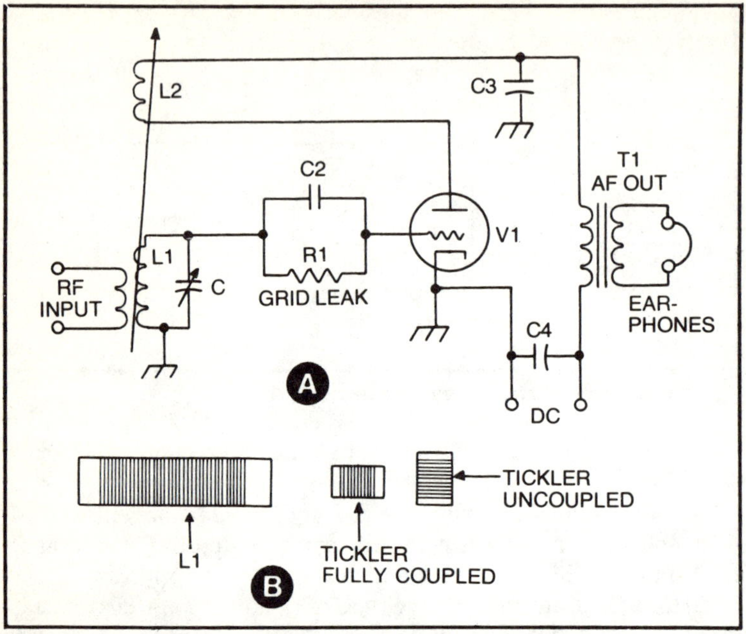

Fig. 1-5. Regenerative detector at A and coil construction at B.

the coil from fully coupled (0 degrees axis) to fully uncoupled positions (90 degrees axis), as shown in Fig. 1-5B. At some intermediate position, the regeneration would occur. In certain other cases, the position of the tickler coil was fixed, usually over one end of L1, and some other means would be used to bring the stage to the regeneration point. Popular at one time or another were filament heater potentiometers, or a series B+ potentiometer in the anode circuit.

A lot more detail on these receivers is given in an out-of-print book titled *Elements of Radio* by Marcus & Horten (later Marcus & Marcus), which was used by the Navy in the early part of World War II to train radio operators. This book is one of the best sources for those interested in early radio circuits, and should be still available in many local public and high school libraries. Try to find the pre-war editions, or the 1941/42 Navy edition (a little more difficult to locate).

SUPERHETERODYNE RECEIVERS

The trf receiver was used for several years, and represented a large jump over the crystal set. As long as the radio frequencies used were in what we today call the VLF range, maybe as high as the AM broadcast band, the receiver could not be substantially improved upon. But as the operating frequency of practical radio receivers went higher into the shortwave region, it became apparent that there were problems in the trf design. It seems that LC tank circuits become less selective, more finicky, and in general poorer at higher frequencies. At frequencies above 10 MHz, LC tank circuits become very difficult to tame in a trf radio. Clearly, some other technique was needed.

The superheterodyne answered the problems of high frequency trf designs, and is the system currently used by radio receivers at all frequencies. The basis of the superhet, as it is nicknamed, is to convert the rf signal from the antenna to another frequency. The heterodyne principle is best seen, perhaps, using an analogy from music. Suppose you are sitting at a piano console. Strike some key—say, A above middle-C. What note do you hear? Why, *A* of course. Now strike some other nearby key. What note do you hear? That note of course. But now strike *both* keys simultaneously. What do you hear? A *new* tone that is composed of both single notes, the *sum* of the two notes, and the *difference* frequency between the two notes. You might use the difference frequency to tune a guitar, or other stringed instrument.

Do you know what the tuner means by a *beat note*? He will compare the note being tuned with a standard note, either from another instrument, or some mechanical or electronic pitch source. When the difference between these two frequencies becomes small, then a wavering beat note tone is heard. The frequency of the beat note is exactly the difference in frequency between the standard and the string being tuned. When the two notes are identical, then the beat note completely disappears (the difference being zero). In the superhet radio, the rf signal is converted to some beat note called the *intermediate frequency*, or i-f.

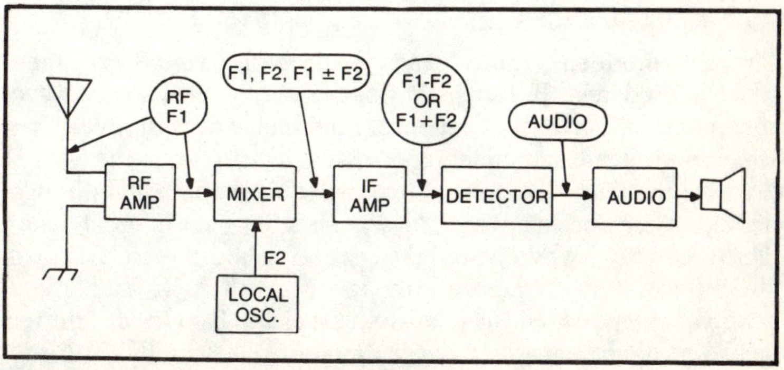

Fig. 1-6. Block diagram of a superheterodyne.

Figure 1-6 shows the block diagram of a simple superheterodyne radio receiver. The first stage is a tuned rf amplifier, and is tuned to the frequency of the radio signal being received. There are sometimes two rf stages, but in most receivers there is only one rf amplifier. The purpose is to amplify the rf signal a little bit, and provide some selectivity (called *preselection*). In some cases, there is no rf amplifier, as is typical of cheap table model AM radio receivers. The principal purpose of the rf amplifier in shortwave radios is to isolate the antenna from the mixer/oscillator stage that follows. The local oscillator signal used in this next stage can accidentally be radiated from the same antenna that we use to receive the signal. This radiation could conceivably interfere with other stations.

The mixer/oscillator stage is responsible for converting the rf signal to the i-f frequency. The output of the mixer contains four signals: the rf (F1), the local oscillator signal (F2), the sum frequency (F1 + F2F2), and the difference frequency (F1–F2). The tunable superhet would use a variable frequency oscillator (VFO) for the local oscillator stage, while a channelized receiver would use a crystal oscillator (XO) for the local oscillator stage. In many receivers, especially broadcast models, the local oscillator and mixer circuits are combined into a common stage called a *converter*.

The i-f amplifier selects either the sum frequency or the difference frequency, and rejects all others. In most classical receivers, it was the difference frequency that was used, but in modern radios it is possible to use either or both, depending upon the band in use. The bulk of the gain (for sensitivity) and selectivity of the receiver is determined by the i-f amplifier chain. In very great respect, the quality of the receiver is set by the i-f amplifier almost entirely. Following the i-f amplifier are the detector stage (which demodulates the radio signal) and audio amplifier stage or stages.

MODERN RADIO RECEIVERS

All modern radio and television receivers are superheterodynes. In fact, only a few low-cost child's shortwave receivers still use the regenerative method. But even the superheterodyne has undergone many changes over the years. Some of the best buys for radio amateurs in the late 40s through mid-60s were surplus Army/Navy models left over in great abundance following World War II. Great buys like the BC-342 and BC-348 covered frequencies up to about 18 MHz but lacked *bandspreading* needed to adequately cover the ham bands. Better still were the Hammarlund *Super Pro* series like the BC-779, and civilian counterparts HQ-129X, HQ-100, Hq-140, etc., and the venerable high-cost SP-600 series receivers. National Radio made their Model NC-183 and a series called the HRO (–5, –7, –8, –50, –60) that were the pride of many ham shacks of the 50s. Those HRO-series radios used plug-in banks of coils for bandchanging and one very good dial for main tuning.

Hallicrafters came forth with their SX-28 receiver, which was considered one of the best receivers made at the time. This receiver was made in civilian models before and after WW II, and was used as part of a division-level communications truck by the US Army. The SX-28 weighed in at over 72 pounds, as can be attested by an amateur radio friend of mine who carried one up the last 500 feet of Virginia's Bull Run Mountain for the 1958 ARRL Field Day...which means he had the hot June sun on his back.

Several modern communications receivers are shown in Fig. 1-7 through 1-9. All of these are high-frequency models and reflect several different methods of construction. The unit shown in Fig. 1-7 is a Heath SB-313 model that covers the ham bands, and selected high-frequency international shortwave bands. This type of radio uses *double conversion* with a low-frequency superhet stage and a high-frequency converter that heterodynes the selected HF bands down to the variable i-flow band. This low band would then be heterodyned down to a fixed second i-f. The main tuning dial tuned the variable intermediate frequency.

The input tank circuits in the high-frequency converter stage are not tuned by the main tuning dial, but by a separate control called

Fig. 1-7. Ham band/international broadcast receiver (courtesy of The Heath Co.).

antenna in some cases, but more commonly preselector. The latter is used in the SB-313, and is the more correct terminology. An rf attenuator control is used to control the level of rf signal reaching the rf amplifier from the antenna, while the rf gain control sets the amplification or gain of the rf amplifier. Both are important in producing proper reception of strong signals which would otherwise overload the receiver. The af gain control is merely an audio volume control.

The agc control is used to select the attack speed of an *automatic gain control* (see Chapter 11). It seems that AM, SSB and CW signals are best handled with different agc time constants, and this control is used to select the best for the given signal.

The *mode* switch is used to select the form of detection required, according to the kind of signal being received: AM, LSB, USB, or CW.

A modern general-coverage solid-state radio receiver is shown in Fig. 1-8. This is the Yaesu FRG-7. It will operate from 0.55 to 29.9 MHz, but avoids the problems of the single-conversion receivers by down converting to a low-frequency variable i-f. Note that an internal loudspeaker is provided.

Figure 1-9 is a modern digital tuning radio receiver, the Drake Model R7. A similar receiver is also part of the Drake TR-7/DR-7 amateur band transceiver. This type of radio is similar to older designs only in that it is a superhet. The main difference is in the tuning system. The local oscillator is a phase-locked loop (PLL) circuit controlled by a small microprocessor (an IC-form microcomputer). It seems that band changes can be caused by looking

Fig. 1-8. Modern general coverage radio receiver (courtesy of Yaesu).

Fig. 1-9. Digital radio receiver (courtesy of R.L. Drake & Co.).

up an appropriate code in a read-only memory. Receiver designers now have to also be competent digital computer designers as well; a phenomenon seen throughout the electronics industry as the microprocessor comes to its full maturity.

In the chapters to follow we will discuss many of the circuits found in radio receivers. We will also cover some of the different types of superhets found and many different radio receiver models.

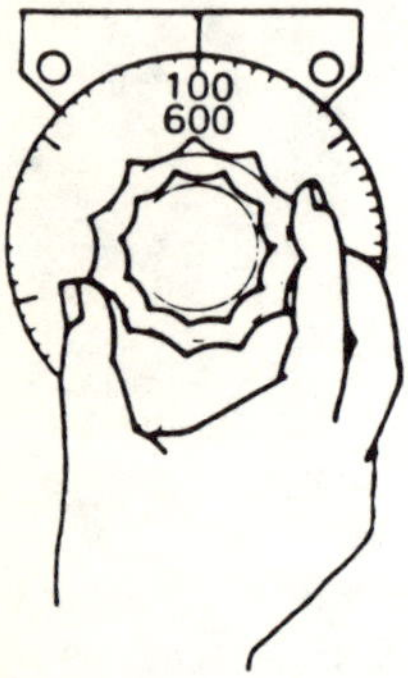

Chapter 2
Simple Radio Receivers

This topic was actually introduced in the last chapter. We stated that the simplest type of radio receiver is the *crystal set*. It uses an LC tank circuit between the antenna, and a crystal detector circuit. Germanium diodes such as the 1N34 and 1N60 are best in this application because the lower junction potential of Ge diodes makes a more sensitive radio receiver.

One of the simplest improvements on the basic crystal set is the circuit shown in Fig. 2-1. This particular circuit, and others very much like it, have been popular with beginners in the area of hobby electronics. At least one hobbyist magazine that slants toward beginners regularly publishes various versions of this circuit for newcomers. All that we have done is add a simple audio amplifier to the output of a crystal set. The audio amplifier will take the low-level signal from the output of the detector and amplify it to a point where it has sufficient power to drive a loudspeaker. In most cases, the output of the detector is sufficient only for directly driving high-impedance, high-efficiency communications (not hi-fi) earphones.

Resistor R1 is used as a load for the diode detector, so may not be eliminated. Capacitor C2 is a filter to remove the rf component of the detected signal. It has a low impedance to the rf, but is essentially a high impedance to audio signals. Capacitor C3 is used as a DC block to prevent the DC bias on the base of the audio amplifier transistor from reverse biasing the detector diode. The amplifier stage is a relatively straight forward class-A single-ended audio power amplifier. It will produce up to several hundred milliwatts of af output power.

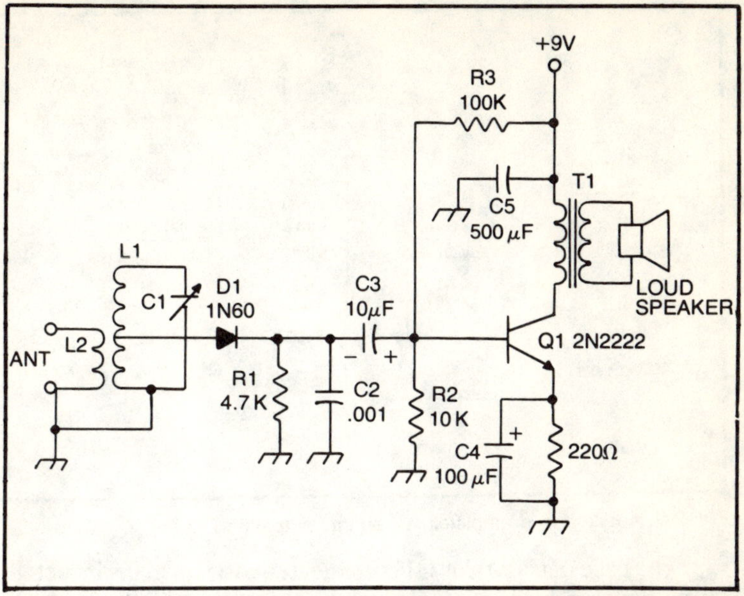

Fig. 2-1. Simple radio receiver uses an audio amplifier at the output of crystal set.

Note that most of the retail hobby electronics chains offer audio amplifier modules of up to 1 watt output that can also be used in this application. It becomes a simple matter to couple the output of the detector to the input of the premanufactured audio power amplifier.

We could also use an integrated circuit audio amplifier in place of the transistor. This would result in less distortion, but would mean a more complex connection scheme. Several semiconductor manufacturers offer audio amplifier ICs up to 4 watts in power (actually, only a couple hundred milliwatts are justified in this radio!).

We could go one step further by adding an rf amplifier ahead of the crystal detector circuit. A sample circuit is shown in Fig. 2-2. Transistor Q1 is almost any medium-beta transistor billed for use in rf/i-f/oscillator applications. Such designations are commonly used in the service replacement and experimenter lines of transistors (Motorola, HEP, GE, RCA SK-series, Sylvania ECG-series, etc). The two inductors shown are standard Japanese coils intended for *antenna* (L1) and rf (L2) use in the frequency range selected (usually the AM broadcast band). The primary of L1 is connected to the antenna-ground circuit. The antenna should be a length of wire, elevated above ground somewhere, which is at least 10 to 100 feet in length.

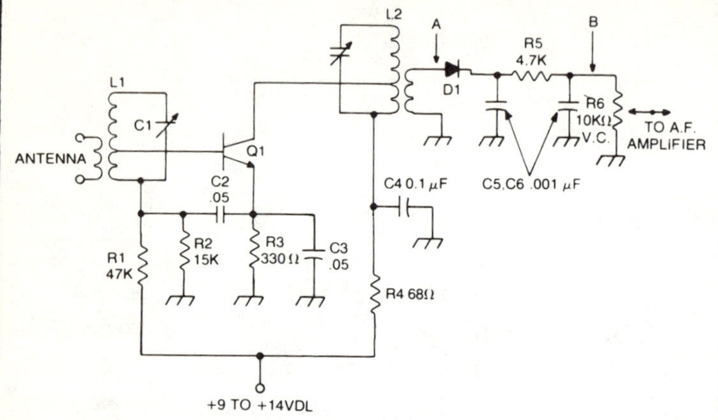

Fig. 2-2. Addition of rf amplifier makes the simplest trf radio.

The base of transistor Q1 is connected to a tap on the inductor. This is done so that the input impedance of the transistors, which is essentially low, could be matched, while simultaneously keeping the high impedance needed for a high-Q selectivity in tank circuit. Bias for the base-emitter junction of Q1 is provided by R1/R2. Capacitor C2 is a bypass that is used to keep the low end of the input tank circuit at a low impedance to ground for rf while maintaining a high DC resistance to ground for the bias network. Similarly, capacitor C3 is used to keep the emitter at a low rf impedance, while allowing us to provide thermal stability by using the emitter resistor R3.

The collector of the transistor is connected to a tap on the output tank circuit, also for reasons of impedance matching. Resistor R4 and capacitor C4 are used to decouple the collector circuit of the transistor from the base circuit. This is needed to prevent oscillation via feedback down the power supply line.

A low-impedance link to the collector coil delivers signal to the diode detector. Again, a 1N34 or 1N60 germanium rectifier are preferred for the detector. The RC low-pass filter following the detector diode is used to remove the rf component from the detected signal, leaving the audio unimpeded. This filter also performs the function of a *tweet filter* (as it sometimes called), which means that it removes the 10-kHz beat note that exists between two adjacent channel AM band stations. The standard FCC channel assignments are 10 kHz apart in this band. This radio is sufficiently sensitive to receive two adjacent channel stations, but lacks the selectivity to completely suppress the one that is not being tuned in.

This leaves a 10-kHz beat note to annoy the listener. The tweetfilter is low pass, and has a cutoff frequency that is between the 3- to 5-kHz audio frequency spectrum of the transmitter, and the 10-kHz heterodyne between adjacent stations.

The volume control selects a portion of the detected audio signal for application to an audio amplifier. A multi-stage transistor amplifier, like one of the Radio Shack modules, or an IC audio amplifier must be used. The single-ended class-A amplifier of Fig. 2-1 is also useful, but is not the preferred circuit. The radio in Fig. 2-2 is basically a single-stage tuned radio frequency (trf) radio. The transistor and tuning circuit components are easily obtainable from sources such as Radio Shack, but note that it is necessary to select capacitors and inductors that are matched to each other for the frequency range. In the AM broadcast band, for example, "standard" tuning capacitors might be either 115 pF or 365 pF, and the coil selected must match the capacitance range of the tuning capacitor selected. Most blister pack hobby components give the specifications of the components to which they match.

The radio just discussed is a good project for a beginner in the art of radio communications, and will serve to demonstrate several principles. As a result, it is frequently seen in junior and senior high school electronics shop classes.

REGENERATIVE RECEIVER

We can make a simple three-transistor regenerative receiver by adding the regenerative detector of Fig. 2-3 to the output of the rf amplifier in Fig. 2-2. The input coil to the regenerative detector is connected to the output link on the rf amplifier (see Fig. 2-2). The diode detector of the previous circuit is disconnected, or not used to begin with. The regenerative detector is basically a solid-state version of the vacuum tube circuit presented in the last chapter. It is

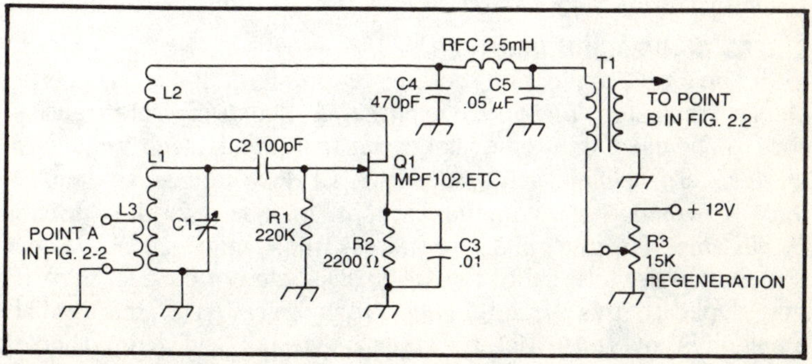

Fig. 2-3. Regenerative radio.

essentially an Armstrong oscillator that can be kept just at the verge of oscillation, so that it operates as a regenerative detector. Potentiometer R3 supplies a regeneration control, and is adjusted so that the circuit is just short of breaking into oscillation. Capacitors C4/C5 and the 2.5-mH rf choke are used to decouple the rf signal from the drain to ground. Transformer T1 is an audio transformer. Almost any high-impedance secondary model will work nicely in this application. Typical values will be 10K to 20K ohms for the primary impedance, and 500 ohms to 100 ohms for the secondary.

This circuit will demodulate AM signals when the regeneration control is kept just below the point of oscillation and the circuit does not break into oscillation even on modulation voice peaks. We obtain *autodyne* demodulation of CW signals by adjusting the regeneration control such that the circuit will not oscillate when no carrier is being sent by the transmitting station, but breaks into oscillation when the di-dah CW carrier is received. This means that the circuit breaks into oscillations when the rf carrier is received, so the output will contain an audio beat note that has a frequency equal to the difference between the (rf) carrier frequency and the frequency of oscillation. This method, incidentally, makes a surprisingly sensitive SW receiver that would work even on today's crowded bands except for the fact that it is not as selective as we might prefer. Many amateurs of a couple decades ago, however, started their novice careers using simple homebrew regenerative detectors copied out of *Popular Electronics* or *QST* magazines.

A simplified *direct conversion* receiver is shown in Fig. 2-4. This circuit is sometimes confused with the superheterodyne because it causes a local oscillator to beat against the rf signal. The principle difference between the direct conversion and superhet methods is that the local oscillator in the direct conversion receiver operates on the received frequency. You should recall that the local oscillator (LO) in the superhet operates on a frequency different from the rf frequency by a factor of ± the intermediate frequency.

DIRECT CONVERSION RECEIVER

The direct conversion receiver circuit of Fig. 2-4 is based on a bipolar transistor *differential amplifier*. Although discrete transistors can be used, it is somewhat easier to use an IC high-frequency differential amplifier such as the RCA CA3028 device (or equivalent). The pinouts shown in the circuit of Fig. 2-4 are for the CA3028 IC diff-amp. The unnumbered resistors (those in which only values are shown) are internal to the CA3028, so do not connect them if you duplicate this circuit. The actual operation of differential amplifier is covered in a later chapter, so will not be discussed here, except in the most general terms.

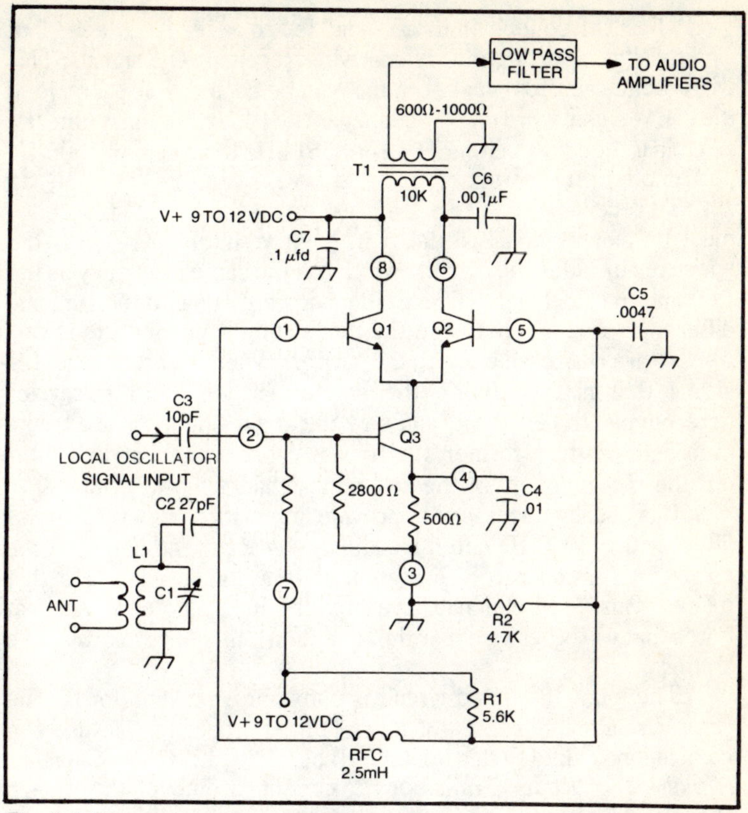

Fig. 2-4. Direct conversion radio receiver.

The local oscillator signal is derived from a VFO or XO, and is applied to the base of transistor Q3. This will vary the current to the emitters of the Q1/Q2 differential pair at a rate equal to the local oscillator frequency. This frequency, incidentally, is equal to that of the rf.

We could drive the bases of Q1/Q2 in push-pull, which is the way RCA designers intended. But in this case, we simplify the job by biasing Q2 at the same point as Q1 and placing a bypass capacitor (C5) between the base and ground so that the base of Q2 remains at a low impedance to rf.

The signal developed in the rf tank (L1/C1) is fed through a DC blocking capacitor to the base of transistor Q1. Here it is mixed with the oscillator signal in a nonlinear manner, producing heterodyne action. The difference frequency between the LO and rf frequencies appears as an audio beat note across the audio output transformer T1, and is coupled to the next stage.

The LO frequency must be adjusted to a frequency near, but slightly different from, the rf signal when receiving CW signals. This will produce an audio beat note that can be copied. For example, if the rf CW signal were on a frequency of 3550 kHz, we must tune the LO to a frequency of either 3549 or 3551 kHz to produce a 1000-Hz audio note in the output.

To receive AM and SSB signals, however, the LO frequency must be exactly equal to the rf frequency. In an AM signal, the spectrum includes the carrier (which has the same frequency as the LO) and the sidebands. These sidebands are equal to the sum and difference products between the modulating audio and the rf carrier, so are clustered close to the carrier. When we produce an LO signal that is exactly on the carrier frequency, the carrier heterodynes to zero beat, and only the audio sidebands are transmitted across transformer T1. Bypass capacitors C6 and C7 filter out the rf component of the collector signals from Q1/Q2.

It is usually necessary to provide some sort of audio low-pass filter (with a 3000-Hz cutoff frequency) between the output of the direct conversion receiver section and any audio amplifiers that follow. This is to eliminate the high-frequency heterodyne beat notes that will otherwise mar the reception if there are nearby radio stations.

The input LC tuned circuit is shown in simplified form. The actual circuit might consist of several tuned circuits in cascade. The companion volume to this book, TAB book No. 1224, *The Complete Handbook of Radio Transmitters,* contains some of the design equations for simple multiple-tuned circuit networks.

The direct-conversion receiver is capable of surprising operation at frequencies up at least the 20-meter amateur band (14 MHz). At least one manufacturer of low-cost QRP amateur radio equipment made a very simple bandswitching direct-conversion receiver for amateur radio experimenters. The Heath HW-8 QRP transceiver uses a direct-conversion receiver circuit.

SUPERHETERODYNE RECEIVER

The simple superheterodyne receiver is shown in Fig. 2-5. This receiver is an extension of the idea discussed in Chapter 1, but with all of the stages. This block diagram is typical of home, auto and portable radio receivers operating in the AM broadcast band.

The rf amplifier usually does not have great selectivity, and has a typical gain of 5 to 10. Radio engineers do not usually want to produce too much gain in this stage because they lack the selectivity to attenuate nearby signals. This means that we might amplify nearby out-of-passband signals before the filtering of the i-f

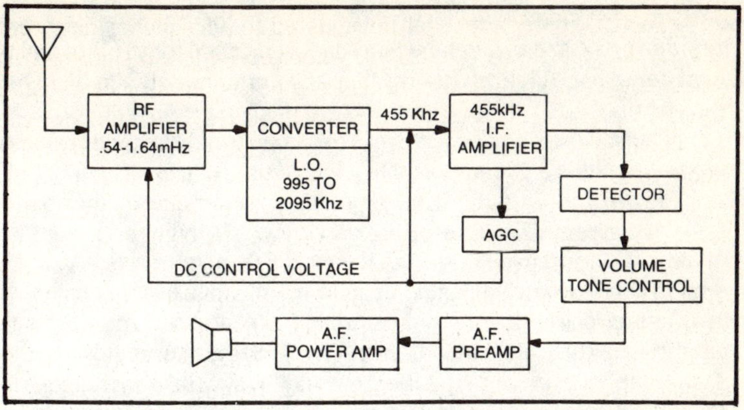

Fig. 2-5. Simple superhet uses converter in place of mixer/LO.

amplifier can get rid of them. The principle functions of this stage are to:

- ☐ isolate the antenna circuit from the local oscillator
- ☐ produce some gain
- ☐ provide some selectivity.

The converter stage is a combination mixer-local oscillator in which both functions are produced by a single active element (transistor, IC, etc). The output of the converter is tuned to the intermediate frequency. In most cases, the first i-f transformer is connected directly into the collector circuit of the converter transistor (see Chapter 6).

The i-f amplifier operates at a single frequency, and provides most of the stage gain and selectivity of the receiver. All radio frequencies are converted to the intermediate frequency be the converter (mixer-oscillator) stage(s). In older designs, the intermediate frequency was almost universally lower than the radio frequency, because it was easier to make high-gain circuits at low frequencies. Most table model AM radios made in the US used an intermediate frequency of 455 kHz (460 kHz was popular in European designs), while FM broadcast radios used a 10.7-MHz i-f. Interestingly enough, US automobile AM radios used 262.5-kHz i-f frequencies. This frequency is still used extensively today, except for one or two models in which some special reason compels the use of another i-f. One digital auto radio made since 1978, for example, uses 260-kHz because a 262.5-kHz signal tends to interfere with the operation of the digital tuning system.

The detector will be a simple diode envelope detector, of the type used in the circuit of Fig. 2-1. A tweet filter is considered part of the detector, and should always be included. In some tube radios, the detector diode (and a second diode for the agc circuit) were included in the same envelope with the triode used for the first audio amplifier stage. All three "tubes" shared a common cathode-filament structure.

The audio section consists of a preamplifier/driver and an audio power amplifier. Most portable radios use a class-B push-pull amplifier in the output, in order to take advantage of the lower average current drain (batteries, you know). It might surprise you that most car radios, up through the early 70s, used a class-A single-ended audio power amplifier, despite the obvious lack of efficiency in those circuits, In car radios where the audio transistor was mounted on the front panel of the radio, such that it was close to the dashboard of the car, the heat would sufficiently increase the dashboard temperature in the vicinity of the radio to bring the customer in for warranty service. Fortunately, many manufacturers placed the power transistor in locations where this heat was not readily transmitted to the customer's hands!

The agc section is an *automatic gain control* circuit. The typical radio receiver will encounter signals of considerably different strength as it is tuned across any band. The listener is alternately blasted out, and then faced with a low-volume signal. The automatic gain control monitors the strength of the rf signal, and then corrects the overall gain of the circuit to compensate for the differences. The gain is made very high for the low-level signals and is reduced for stronger signals.

Most agc circuits use a sample of the i-f amplifier signal. This sample is rectified and filtered to produce a control voltage (DC). In vacuum tube receivers, the agc DC control voltage was universally a negative potential, as it was applied to the grids of the rf and sometimes i-f amplifier stages. The negative agc voltage became a grid bias, which would control the gain of the stages. When the signal was large, the tubes would see a high DC negative bias and provide less amplification. On the other hand, a weak signal level in the i-f amplifier would produce little or no negative bias, so the stages operated at full gain. In modern solid-state receivers, however it is possible that the agc control voltage be either positive or negative, depending upon the polarity of the rf/i-f transistors (NPN or PNP), and whether forward or reverse agc is used in that particular set. (If this sounds confusing, Chapter 11 will confuddle you even more.)

Chapter 3

Superheterodyne Receivers

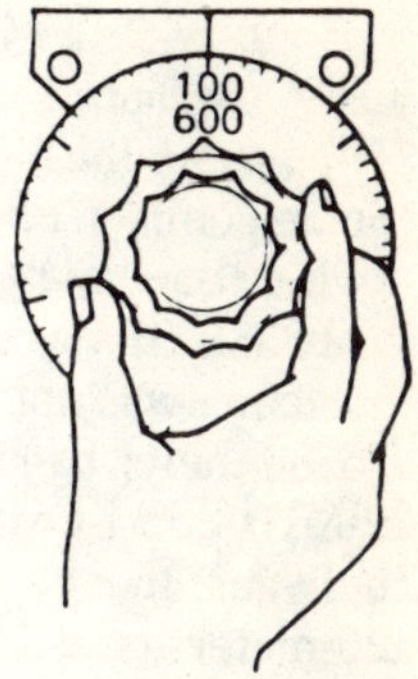

The superheterodyne receiver is *the* standard radio and television receiver. Almost every radio that you will see is some form or another of superhet design. The simple transistor AM portable radio is a superhet, and so is the fantastically complex commercial or military communications receiver. Let us briefly review the basic concept of the superheterodyne.

HETERODYNING

The first thing that we have to learn is the meaning of the phenomenon called *heterodyning*. We discussed this briefly in the first chapter. On the naive level, we can use an analogy with musical instruments. We strike another note on the piano keyboard and will hear a note of another frequency. But when we strike both notes at the same time, we hear a complex blend of the first note, the second note, a new frequency that is equal to the *sum* of the two notes, and another new frequency that is equal to the *difference* between the two notes. In actual practice, incidentally, there are also a number of cross products caused by the harmonics of the two notes. But in general, we have only the four frequencies present.

In a superheterodyne radio receiver, we will create a controlled frequency in a local oscillator circuit and mix it with the rf signal. At the output of the mixer stage, we will find the LO frequency, the rf frequency, LO + rf, and LO – rf. We convert all radio frequencies within the range of the receiver to one common, intermediate frequency, or i-f. The i-f will always be either the sum or the difference frequency (LO ± RF). The i-f amplifier contains tuned circuits or band-pass filters that select either the sum or the difference, depending upon the design.

Several common intermediate frequencies are in use. AM radio receivers typically use 455 kHz (262.5 kHz in automobile radios), while FM broadcast receivers use 10.7 MHz. Note that many (perhaps most) 2-meter FM receivers will tend to use the same i-f as the FM broadcast rig, because it's easy to get parts. Some common intermediate frequencies are used in amateur radio high-frequency receivers. Heath receivers and SSB transceivers use 3385 kHz, while many other brands use 9.0 MHz. The 9.0-MHz selection was not arbitrarily made, because it seems that both the 75/80-meter band and the 20-meter band can be generated from a 9-MHz SSB i-f with a single 5.0-5.5 MHz VFO. Note that the i-f is between the two bands, so the V.F.O. tunes up the band on 20-meters, and down the band on 75/80 meters (try to find the sum and difference frequencies, and you will see this result in the arithmetic). The typical intermediate frequency used in some older citizens band receivers was 1650 kHz, with some using 1800 kHz.

SINGLE-CONVERSION RECEIVERS

The single-conversion superhet receiver is the type introduced in the two previous chapters. A block diagram is shown in Fig. 3-1. In a single-conversion radio receiver, there is one mixer and one local oscillator, and they convert the rf signal to a single intermediate frequency. Almost all AM and FM broadcast receivers are single conversion. There was, incidentally, at least one German car radio made in the mid-50s that used double conversion, but more about that method in the next section.

In older terminology, the mixer was called the *first detector* and the demodulator at the output of the i-f amplifier was called the *second detector*. Do not let this confuse you. It does not infer that double conversion is in use. If the radio has only one heterodyne mixer, then it is a single-conversion receiver.

In an older wisdom, the single-conversion radio receiver was not considered to have very high quality. There were problems (notably image response—see chapter 4) especially when the radio frequency was high. In VHF receivers, it became almost standard practice among commercial radio receiver designers to use double conversion. They would take care of certain problems by using a 10.7-MHz i-f at the output of the first mixer, and then heterodyne the 10.7-MHz signal down to 455 kHz, where most of the receiver gain and selectivity were provided. But with phase-locked loop control of frequency generators and the availability of good quality

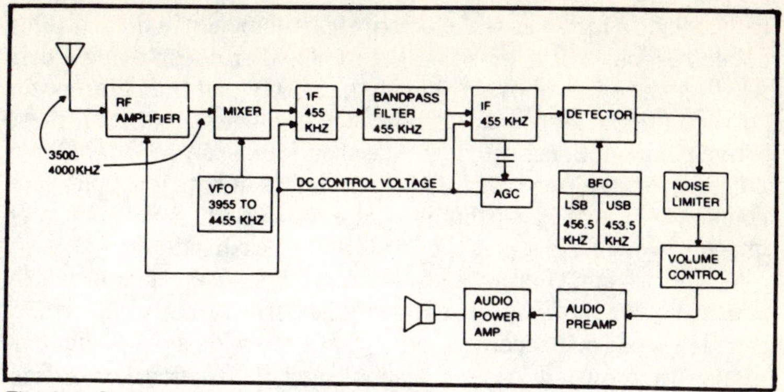

Fig. 3-1. Single-conversion communications receiver.

10.7-MHz filters, the use of single-conversion receivers has become popular again. In the past, it was simply not easy nor cheap to obtain good i-f amplifier design at 10.7 MHz.

The receiver shown in Fig. 3-1 is a 75/80-meter single-conversion ham band model. It tunes the 3500 to 4000-kHz band, and uses a 455-kHz i-f amplifier. The 3500-4000-kHz rf signal is applied to the input of the rf amplifier, where it is amplified typically 5 to 10 times. In some cases, the rf amplifier tuning is separate from the main tuning VFO and designated either *antenna* (usually incorrectly) or *preselector* (more often correct). In a few cases, amateur band receivers will gang together the rf amplifier and VFO controls.

The local oscillator is a variable frequency oscillator which tunes the range 3955 to 4455 kHz (each of these frequencies is 455 kHz above the respective frequency at the ends of the rf tuning range; i.e., 3500 + 455 is 3955, and 4000 + 455 is 4455 kHz). The VFO is the one single largest cause of drift in the radio receiver. The drift properties of the entire radio are set by this single stage, so some concern is usually given to the temperature compensation of the VFO, the regulation of its supply voltage and the ruggedness of its construction viz, vibration. All of these are factors that can alter the output frequency of a VFO circuit.

The output of the mixer stage is the difference frequency between the rf and the VFO signals. There is no design reason why we cannot use the sum frequency for the intermediate frequency. We use the difference frequency out of tradition. But some years ago, it was easier to tame low-frequency circuits, and the difference frequency was a low frequency. This is definitely true in the medium wave and long wave bands, but becomes a little less valid as the shortwaves are attacked. There is also no reason why we cannot place the local oscillator signal below, instead of above, the radio frequency. In fact, this is the principal phenomenon in the matter of image response. It seems that the i-f amplifier doesn't give a darn whether the LO is above or below the rf only that the difference is correct. If the intermediate frequency is low enough, then these image points both fall within the passband (typically broad) of the rf amplifier, so the receiver will pick up signals at both locations. The solution is to use an i-f that is high enough so that the unwanted image is very far outside the rf amplifier passband.

The i-f amplifier in a communications receiver will usually contain some sort of band-pass filter. The normal tuned transformers found in broadcast receivers simply have too wide a passband to be of use in crowded communications and international broadcast bands. The solution is to use an i-f filter with a sharp skirt selectivity characteristic. In other words, use a filter with a very rapid drop in response at frequencies outside of the specfied passband. Typical filters might include a simple crystal phasing filter in low-cost communications receivers (made 20 years ago), multiple-pole crystal lattice filters, or even a mechanical filter. These different filters are discussed in Chapter 7.

Usually, more than one i-f amplifier is in a communications receiver. The total i-f amplifier is actually several tuned amplifiers in cascade. In these receivers such as the one in Fig. 3-1, there are two stages of i-f amplification. In any given receiver, however, there might be as many as four or five i-f amplifiers.

The detector is used to demodulate the i-f signal. If the radio is for receiving an AM signal, a simple diode envelope detector is needed. But if some form of angular modulation is being received, then the detector must be a phase/frequency sensitive type. On the other hand, if a single-sideband signal is being received, a *product detector* is needed. A product detector combines the i-f signal with a local oscillator signal. This is known as a *beat frequency oscillator*, or BFO. The idea here is to replace the missing carrier with a locally generated carrier. The LSB BFO operates at 1.5 kHz above the intermediate frequency (456.5 kHz in this case), and the USB BFO operates 1.5 kHz below the intermediate frequency (453.5 kHz in this case). The BFO signal beats against the SSB signal to produce the difference frequencies—the audio signals that made up the original modulating frequency.

The noise limiter is usually a clipper circuit that will take out impulse noise from lightning, man-made machines, and other sources. All this type of noise is lumped under the rubric *static* by most people, and is characterized by short-duration, high-amplitude pulses with high-frequency content. There are also noise blankers in use, but these operate in the rf and i-f amplifiers. Chapter 11 will deal with this type of circuit, as well the clipper types.

The audio amplifier section typically provides some preamplification and some power amplification. In communications receivers, however, the power output is usually quite limited. Most communication receivers will produce only a couple hundred milliwatts of audio power when a strong signal, 100 percent modulated, is present at the antenna terminals. Some receivers have used only one tube, one IC, or two transistors as the entire radio receiver audio section.

One circuit sometimes seen in the audio section is a squelch circuit. This is a variable threshold audio switch that passes the signal to the output only when there is a signal present in the i-f amplifier. Noise will not trigger the squelch, so it remains turned off. This eliminates the harsh background noise present when there are transmissions.

The *automatic gain control* (agc) is used to keep the output of the receiver constant for a widerange of input signal levels. The agc reduces the gain of the receiver when a strong signal is being received and increases the gain when a weak signal is being received. It does this marvelous trick by controlling the bias applied to the rf amplifier, and sometimes the i-f amplifier as well. A sample of the i-f signal is taken from a later stage (usually the last stage in the sequence, although this is not universally true). This sample is then rectified and made into a DC control voltage that is proportional to the signal strength. The control voltage becomes a part of the bias applied to the rf/i-f amplifier stages.

DOUBLE-CONVERSION RECEIVERS

Several problems are apparent when high-frequency i-f amplifiers are used. These problems were especially apparent on older designs, when it was not easy to make high-gain, high-selectivity, low-noise i-f amplifiers are frequencies much above 1 MHz. A double-conversion receiver will help solve some of these problems.

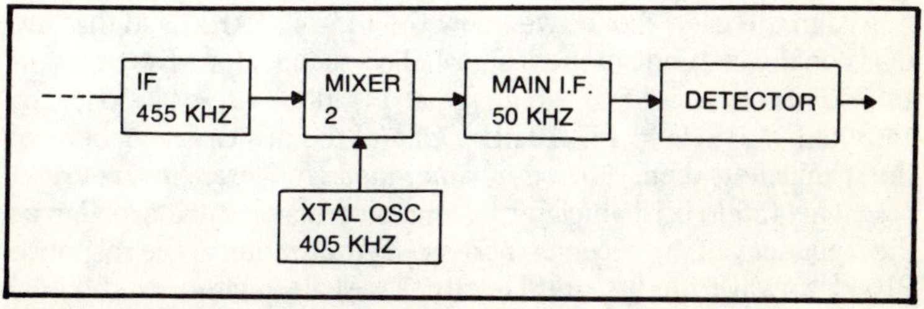

Fig. 3-2. Additional circuitry for double-conversion receiver.

The circuit in Fig. 3-2 is an extension of the single-conversion circuit in Fig. 3-1. The main high-frequency converter will down-convert the rf signal to some lower frequency. We have selected 455 kHz for our example. There is a single stage of 455-kHz i-f amplification in Fig. 3-2. The gain of this stage is minimal, because its main purpose is to insure that only the 455-kHz difference frequency is selected from the spectrum at the output of the mixer. The 455-kHz signal is fed to a mixer, where it is heterodyned against a 405-kHz signal from a crystal oscillator. The output of this mixer is 50 kHz. It is somewhat easier to make LC band-pass filters with desirable selectivity characteristics at this frequency. It is also relatively easy to obtain high gain figures because layout problems are less than at some higher frequencies. Several amateur communications receivers popular in the early to mid-60s used this system.

There are still problems with these receivers. We have solved the problems of gain, stability, and selectivity, but we still are faced with at least two problems of significant magnitude. One problem is the matter of *image response*, while the other is the *tuning ratio*, especially in the higher bands.

The mixer and i-f amplifier don't much care whether the local oscillator is above or below the rf signal frequency. The only criterion is *that* the difference between rf and LO be equal to the intermediate frequency. Consequently, it is possible for the receiver to respond to an unwanted signal on the other side of the LO frequency. For example, let us assume that a 20-meter receiver uses a 455-kHz i-f amplifier, and that the oscillator is on the high side of the radio frequency. When the receiver is tuned to 14,300 kHz, the local oscillator frequency is 14,300 + 455, or 14,755 kHz. But the mixer can respond equally well to an rf signal that is 455 kHz above the LO frequency; i.e. 14,755 + 455, or 15,210 kHz. If the rf amplifier is not selective enough to eliminate any signals at this image frequency, then they will be detected equally with the desired signal. In general, the image frequency is on the same side of the rf signal as the LO, and is found at a frequency 2 × i-f above the rf signal. In the case above, we know the i-f is 455 kHz, and that the LO is on the high side of the rf signal. This means, for a frequency of 14,300 kHz, we will find the image at 14,300 + (2 × 455 kHz), or 14,300 + 910 kHz = 15,210 kHz. One of the principle functions of the rf amplifier, then, is to *improve the image response of the receiver*.

The 455-kHz i-f amplifier becomes something of a problem as the frequency of the receiver increases. At 20 meters, the response 910 kHz away from the signal is rather well attenuated, so only real strong signals will be heard at the image frequency. But as the frequency climbs to the 25 MHz and over region, the response of the LC tank circuits becomes broad enough that the attenuation of signals 910 kHz away from the resonance point becomes less than we might desire. In these cases, we might want to adopt an intermediate frequency that is high enough to place the image far outside of the passband of the rf amplifier. For example, many amateur rigs use high-frequency i-f amplifiers. In the Heath system, the typical intermediate frequency is 3385 kHz, so the image frequencies are 6.770 kHz away from the rf signal. Even 10-meter tank circuits will attenuate signals more than 6 MHz away, so the image response of the receiver is improved. The receivers that use a 9-MHz i-f (common in receiver sections of SSB transceivers) have an image frequency 18 MHz away from the rf signal. There is at least one model on the market that uses a 43-MHz first i-f, so the image frequency for this high-frequency receiver is 86 MHz, well into the VHF range!

The problem of tuning ratio results from trying to squeeze too much frequency-coverate out of each band. Coils, and trimmer capacitors are expensive, as are multi-position bandswitches. The typical shortwave general-coverage receiver made some years ago had five bands typically:

- ☐ 0.54 to 1.6 MHz
- ☐ 1.4 to 4.0 MHz
- ☐ 3.5 to 7.5 MHz
- ☐ 7.5 to 15 MHz
- ☐ 15 to 30 MHz

But there is only a fixed, finite, resolution of the dial connected to the tuning system. Let us assume that there is 30 cm of dial available. At low frequencies, the resolution, then, is far fewer kilohertz than at high frequencies. In the AM broadcast band, for example, we must calibrate the 30-cm dial for the 1600 to 540 kHz range, or 1060 kHz. This works out to approximately 34 kHz per cm of dial. At the highest band (15 to 30 MHz), however, the same amount of dial space is calibrated with (30-15), or 15 MHz! This means a resolution of 500 kHz per cm of dial. At that rate, it would almost be impossible to tune in a narrow SSB or CW signal.

There are several ways of dealing with this problem. One is to produce a *bandspread dial*. Figure 3-3 shows a typical LC tank circuit from a communications receiver. The main tuning is accomplished by inductor L1 and capacitor C1. Capacitor C1 might typically have a maximum (plates fully meshed) value of 300 to 400 pF. A small trimmer capacitor (C2), typically only a few pF, and the tunable core inside of L1, allow alignment of the tank circuit to the

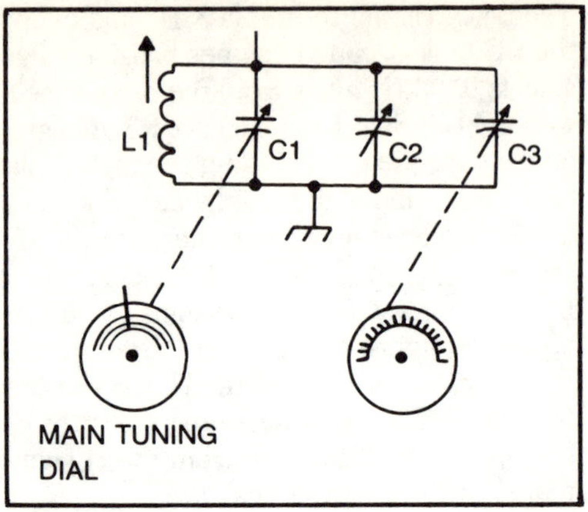

Fig. 3-3. Circuitry to add bandspread.

correct endpoints of the tuning range. In most alignment protocols, the core of L1 is used to adjust the low end of the dial, and the trimmer capacitor adjusts the high end of the band. In local oscillator tank circuits, the trimmer might be found in series with L1, in which case it is called a *padder* rather than *trimmer*. Capacitor C3 is the bandspread capacitor. It has a total capacitance that is only a fraction (typically 10 to 12 percent) of the value of the main capacitor (C1). The main tuning dial is calibrated according to the system discussed above, and its calibrations are valid only when the bandspread dial is set to a given frequency, or calibration mark. The bandspread dial might just have a simple logging scale (0 to 100 or 0 to 500), or it might have frequency calibrations for certain bands. It has been common practice to provide a logging scale for use on all bands, and then specific frequency calibrations for the 160-meter through 10-meter amateur bands. Even in this system, however, the problem of tuning ratio differences between the bands is still quite pronounced; it is merely lessened from the main dial. When using the bandspread system, it is necessary to set the main tuning dial to some specific frequency, and then tune from there on the bandspread dial. It is usually the high end of the given band to which we set the main tuning dial.

The next three receiver systems solve the problem of differential tuning rates quite nicely. In one version or another, these systems are used by most currently produced amateur and commercial shortwave receivers.

The receiver in Fig. 3-4 is a double conversion model that is very common among amateur equipment. This is the so-called *variable i-f* system. On the 75/80-meter band (3500 to 4000 kHz), the receiver operates as a single conversion model. The local

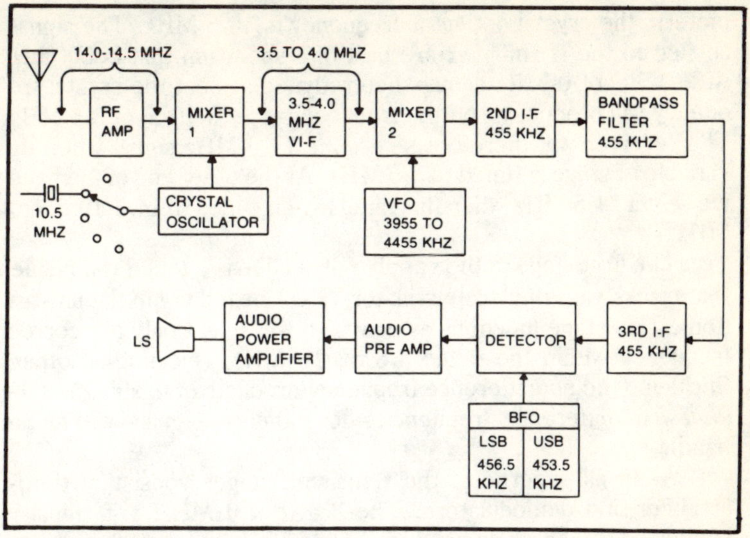

Fig. 3-4. Variable i-f double-conversion receiver.

crystal oscillator is turned off, and the mixer is either bypassed, or becomes a simple amplifier stage. It is common to find the entire high-frequency front end bypassed, and the antenna connected directly to the input of the 3.5 to 4.0 MHz variable i-f stage. This amplifier would then operate as an rf amplifier on 75/80 meters.

At frequencies high than 4 MHz, the receiver operates as a double-conversion superheterodyne, in which the front end converts the rf signals to the 3.5 to 4.0 MHz band. In this case we are converting a whole band of frequencies to a lower band, so the tuning ratio is essentially that of the lower band(3.5 to 4 MHz) for all bands on the receiver. Since our lowest intermediate frequency in the variable i-f stage is 3.5 MHz, the image frequencies will be at least 7 MHz away from the rf signal. This will help improve the image response of the receiver.

The variable i-f section is nothing more than a 3.5 to 4.0 MHz superheteorodyne receiver. But instead of receiving its signal from the antenna, it is driven by the output of the first mixer stage.

□**Example:**

The 20-meter band tuning on this receiver is 14.0 to 14.5 MHz, a 500-kHz band segment. The 14.0 to 14.5 MHz signals are amplified in the rf amplifier. This stage provides image rejection, some gain at the intermediate frequency, and isolation of the mixer from the antenna circuits. There are two signals applied to the mixer. One is the 14.0 to 14.5 MHz rf signal, while the other is the output of a crystal oscillator. When the bandswitch is set to 20 meters, the crystal oscillator frequency is 10.5 MHz. The signals applied to the rf amplifier are part of a spectrum that is 500 kHz wide. The 14.0-MHz signals heterodyned against the crystal frequency will produce an output frequency of 14.0-10.5, or 3.5 MHz. The receiver will therefore select the 14.0-MHz signal when the variable i-f stage is tuned to 3.5 MHz. At the other end of the band we select 14.5 MHz when the variable i-f is tuned to 14.5-10.5, or 4 MHz.

The first stage in the variable i-f section is a tuned rf amplifier that tracks with the local oscillator VFO labelled vi-f in the illustration). This stage functions as an i-f, so will select only the correct frequency within the range 3.5 to 4.0 MHz, rejecting all others (including the nondifference frequency products of mixing, and the *undesired* difference frequencies for other rf signals within the band).

As in all superhets, the remaining stages consist of the i-f amplifier and demodulators. The 3.5 to 4.0 MHzf vi-f signal is converted to a 455-kHz fixed i-f. The 455-kHz i-f amplifier provides most of the stage gain in the receiver, and is the principle section for shaping the bandpass of the receiver.

TRIPLE-CONVERSION RECEIVERS

Some receivers are billed as triple conversion. These models take the double-conversion process one step further, and essentially combine the schemes of Figs. 3-2 and 3-4. We see such a receiver in Fig 3-5. The output of the vi-f mixer is 455 kHz. This signal will be passed through a low to medium gain 455-kHzf stage or just a simple tuned i-f transformer that selects the difference frequency. This 455-kHz i-f signal is then applied to a third mixer, where it is heterodyned against the signal from a 405-kHz crystal oscillator. The output of mixer No. 3 is a 50-kHz i-f signal. The remainder of the receiver is the same as in Fig. 3-2. Note that some of these drawings leave out portions of the circuit for the sake of simplicity. We must assume, however, that all of the stages of the receiver in Fig. 3-1 are present in all of these radios. In fact, some triple conversion circuits use two agc stages.

CONSTANT BANDSPREAD RECEIVERS

Not one of the receivers discussed thus far will provide a constant tuning ratio over all of the bands. The bandspread system shown earlier in the chapter will reduce the tuning ratio difference between bands, but it does not make the various tuning ratios equal.

A number of newer communications receivers on the market, even in relatively low-priced models, provide equal bandspread on

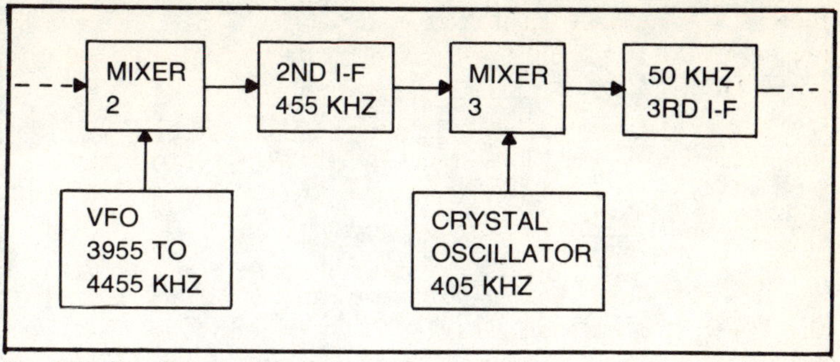

Fig. 3-5. Circuitry added to make triple-conversion.

all high-frequency bands. The system is very much like the ham band-only receiver shown in Fig. 3-4, except that it covers all bands in the high-frequency region instead of only the several ham bands. The mainstay of the receiver is a variable i-f amplifier tuning 1650 to 2150 kHz (a 500-kHz ban segment). This range is down converted to 455 kHz in the manner normal to superhet receivers. On the two lower bands, the receiver is single-conversion, so the signal is routed through a fixed 2150-kHz i-f amplifier stage.

One of the principal problems in producing a low-cost, general-coverage receiver based on the variable i-f principle is the difficulty and cost of making crystal oscillator frequencies in the entire HF region. If we want to cover the range from 3 to 30 MHz, for example, we are covering 30-3, or 27 MHz in 54 500-kHz segments. This requires 54 different crystals in the first converter section, and that is very costly. Until recently, only the most expensive communications receivers (i.e. Collins) used the vi-f system to cover all of the HF range. Today, though, a number of low-cost communications receivers do so very nicely. The secret is the HF VFO shown in Fig. 3-6. In low-cost models, the VFO is a simple tunable oscillator, exactly like the LO in the single-conversion receivers. It is calibrated in the same manner as shown in the chart earlier in this chapter.

The dial system for a receiver using this system is shown in Fig. 3-7. This receiver is the Lafayette BCR-101 high-frequency communications receiver. The main tuning dial is set by the inner knob of the concentric tuning knobs. The inner, or main, tuning knob controls the frequency of the HF VFO in Fig. 3-6. The outer knob is the bandspread knob that controls the tuning of the bandspread VFO. The outer knob is calibrated with 500 equal divisions. There are two sets of calibrations: Rotating clockwise, the markings are 1000 down to 500; The counterclockwise calibrations of the bandspread dial are 0 to 500.

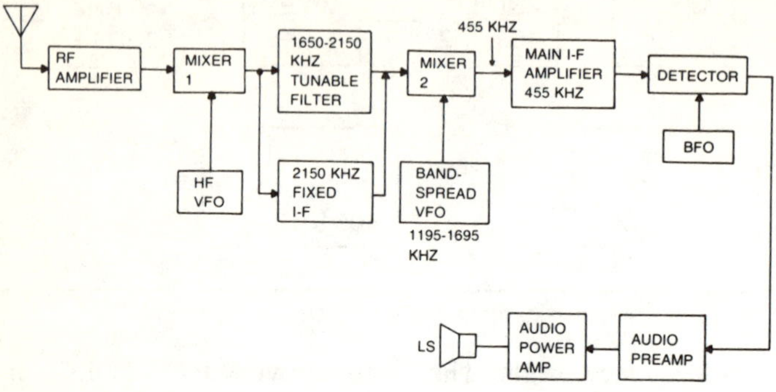

Fig. 3-6. Bandspread method using variable i-f amplifier.

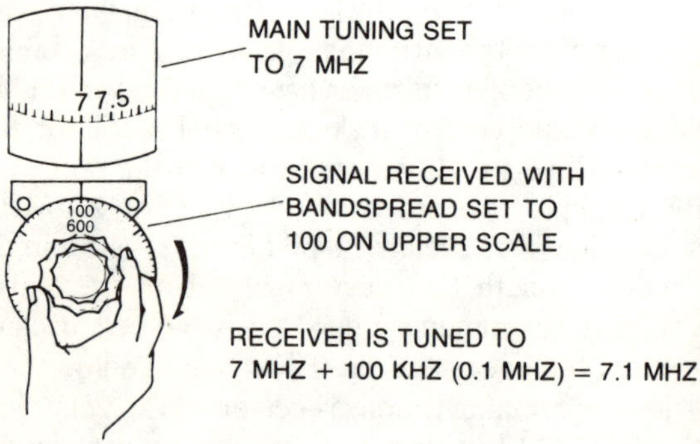

MAIN TUNING (MHZ)	MAJOR CALIBRATIONS ON BANDSPREAD DIAL (Upper Scale)	RECEIVER TUNED TO (MHz)
7.0	0	7.0
7.0	100	7.1
7.0	200	7.2
7.0	300	7.3
7.0	400	7.4
7.0	500	7.5

NOTE: EACH MINOR DIVISION ON THE BANDSPREAD DIAL REPRESENTS 5 KHZ (0.005 MHZ)

Fig. 3-7. Setting the tuning dial for Fig. 3-6.

□**Example:**

See Fig. 3-7. If we set the main dial on an even megahertz mark (i.e. 7.0 MHz), we would then read the corrected frequency off the 0 to 500 bandspread dial. To calibrate the system, we set the bandspread dial to the 0 mark and the main tuning to 7.0 (or the lower edge of whatever band we want to tune). Next, we turn on the 500-kHz crystal calibrator and readjust the main tuning dial for zero beat with the calibrator signal. We now have a calibrated dial. If we want to tune the receiver to 7100, we would leave the main dial alone and adjust the outer (bandspread) tuning knob until the 0 to 500 scale is on 100. The actual frequency, then, is 7.0 MHz + 0.100 MHz, or 7.1 MHz. Each 100 divisions on the bandspread dial represents a frequency change of 0.1 MHz, or 100 kHz.

□**Example:**

Similarly, suppose we want to tune to 9.8 MHz, and the main dial is set to 7.5 MHz. When we are set to the half-megahertz points on the main dial, we use the 1000 to 500 logging scale. To calibrate, we set the bandspread dial to 500 (same as 0 on the 0 to 500 dial). The main tuning is set to 9.5 MHz. We turn on the 500-kHz crystal calibrator and zero beat the main tuning dial to the calibrator signal. We can then read the corrected frequency from the main dial. To tune in 9.8 MHz, we would advance the bandspread tuning dial to 800. The actual frequency is then 9.5 + 0.3, or 9.8 MHz. Note that 800 is the third 100 mark on the bandspread dial.

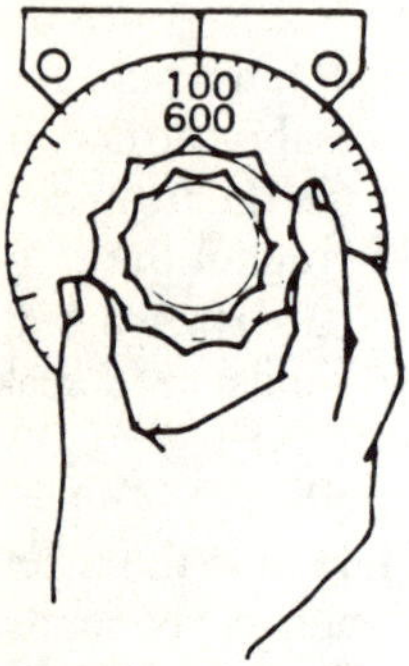

Chapter 4
Receiver Specifications and Parameters

Certain parameters and specifications are common to all superheterodyne radio receivers. Some of these include *sensitivity, selectivity, image rejection*, and *noise figure*. With FM receivers, another specification is called the *capture ratio*.

SENSITIVITY

This specification is the measure of the ability of a receiver to pick up, amplify, and demodulate weak signals. Unfortunately, several different definitions for sensitivity exist. Some of these are merely differences in engineering practice from one expert to another. Some, however, are little more than an advertising department's attempt at "improving" the appearance of a product through some creative specification writing.

One of the two most common definitions of sensitivity used for AM and SSB receivers measures the input signal level required to produce an output signal with a given amplitude. The specification will typically require a standard input waveform, such as 30 percent modulated by a 400-hertz sinewave. The engineers will then ask for a certain output power 500 or 1000 milliwatts (0.5 or 1 W), and then measure the input signal level required to produce that output level.

The other definition of sensitivity is a little more universal as to the type of receiver it measures. We know that the receiver will produce a noise output (hiss) when no signal is present, but that signal reduces this noise proportional to the signal strength. The sensitivity is the unmodulated signal level required to produce a specific reduction in the noise output amplitude, usually 10 or 20 dB

The latter method seems more reasonable, especially since it is multi-model not being limited as to modulation type. Beware, though, when two receivers have the same number of microvolts specified, but one uses 10 dB and the other 20 dB of quieting. The one using 10 dB is a lot less sensitive than the other.

SELECTIVITY

Selectivity is a measure of the ability of a receiver to reject unwanted signals removed from the receiver frequency. No radio receiver looks at the entire radio spectrum at one time, but only a small window centered about the radio frequency to which the receiver is tuned. Ideally, the receiver bandwidth exactly matches the signal bandwidth. The selectivity is measured in units of frequency—the width of the window. As with sensitivity, there seems to be several methods for specifying selectivity. We could, for example, cite the bandwidth between two points at which the response is down. The way some companies "play with" selectivity is to use different response limits. Most of the time, the bandwidth is between points on the response curve at which the output falls off −6 dB from the center band level.

But this does not tell the whole story. We must also consider the shape factor of the response. There could be two receivers with identical bandwidths at the −6 dB points, yet one is a lot more selective than the other. We find that the slope of the roll off for the better one is steeper than the other. The difference between these receivers can be seen in the shape factor, or skirt factor, of the receiver. This is a dimensionless measurement that takes the ratio of the bandwidth at −60 dB to the bandwidth at −6 dB. A perfect (and therefore unattainable) filter with ideally perpendicular slopes at the cutoff frequency will have a shape factor of unity, or 1. With real filters, values are higher than one. The best filters have a skirt factor around 2. Really poor receivers have filters with skirt factors of 10 or so.

IMAGE RESPONSE

The image is a spurious response that is unique to the superheterodyne system. It is caused by the i-f amplifier, which is tuned to a fixed frequency, but is not dependent on whether the local oscillator is above or below the rf signal. The i-f amplifier looks only to the difference between the LO and rf signals. The i-f, then, will respond equally well to signals that are on the wrong side of the LO, by the amount of the intermediate frequency.

□**Example:**

A 15-meter receiver uses a 455-kHz i-f amplifier and places the LO above the rf signal. If tuned to 21.25 MHz, the local oscillator operates on a frequency of 21,250 + 455 kHz, or 21,705 kHz. When these two signals are heterodyned together, the difference frequency will be 455 kHz, the intermediate frequency. But what happens if another signal that is 455 kHz above the local oscillator frequency, such as 22,160 kHz, also reaches the mixer? The difference frequency is still 455 kHz so the amplifier sees the mixer output as a valid i-f signal and will process it accordingly. Our only solution is to prevent this image frequency from reaching the mixer, which is a function of the receiver front end. If the receiver has no rf amplifier, the mixer sees all signals, and the image response will be almost nonexistent. This is the situation in some AM broadcast receivers, in which a single, low-Q tank circuit at the mixer input is used to select the rf and reject the image signal. When the rf amplifier is used, there will be two moderately high-Q tank circuits, one each in the input and output networks of the amplifier, to attenuate the image frequency. In fact, some designs purposely have a tunable image trap as part of the front end tuned circuitry. In other cases, the mixer is the first stage, without an rf amplifier, but the image response is reduced considerably by a very narrow band, high-Q filter circuit.

NOISE FIGURE

Seen most often in VHF and UHF receivers, noise figure is a measure of the thermal noise of the receiver. This noise can be measured when the antenna terminals are shorted together. The parameter is in either dimensionless noise figure terms or in decibels.

An interesting facet of the noise figure is that for an entire receiver it is set primarily by the noise figure of the first stage. In most cases, therefore, a low overall noise figure is obtained by using a low-noise rf amplifier in the front end of the particular receiver.

DYNAMIC RANGE

Regarding receivers, dynamic range is the ratio in decibels of the faintest to the loudest signals received without significant distortion or noise. Recently, this term has gained importance among manufacturers and critics of communications receivers and especially ham band-only receivers and transceivers. Slightly pushing aside receiver sensitivity in significance, dynamic range is the more difficult of the two specifications to measure, often requiring expensive, sophisticated equipment not normally found in the hobbyist's shop.

MEASURING RECEIVER SENSITIVITY—AUDIO OUTPUT POWER METHOD

The audio output power method of measuring the sensitivity of a radio receiver is used for AM and SSB receivers only. We input a signal that is 30 percent amplitude modulated with either a 400-Hz or 1000-Hzf tone. (Note that some individual radio service sometimes require one frequency or the other, and up to 100 percent AM modulation.)

Figure 4-1 shows the basic equipment connections needed to make this measurement. A signal generator is connected to the antenna input terminals through a dummy load. Most receiver manufacturers will specify a given dummy load for this measurement, but if none is available use the circuit shown in Fig. 4-1B for work up to the high end of the HF range.

How do you measure the audio power at the output of the receiver? Because audio power is difficult to measure accurately, measure the voltage and convert this to units of power; a justified move because the output will be a 400-Hz sine wave. The well-known relationship for power is:

$$P = E^2/R$$

From which, by solving for E, we derive:

$$E = (PR)^{1/2}$$

Let us assume that we measure the audio power across an 8-ohm standard load, and that we want 1 watt to be our reference level. The output voltage that will indicate this power level is:

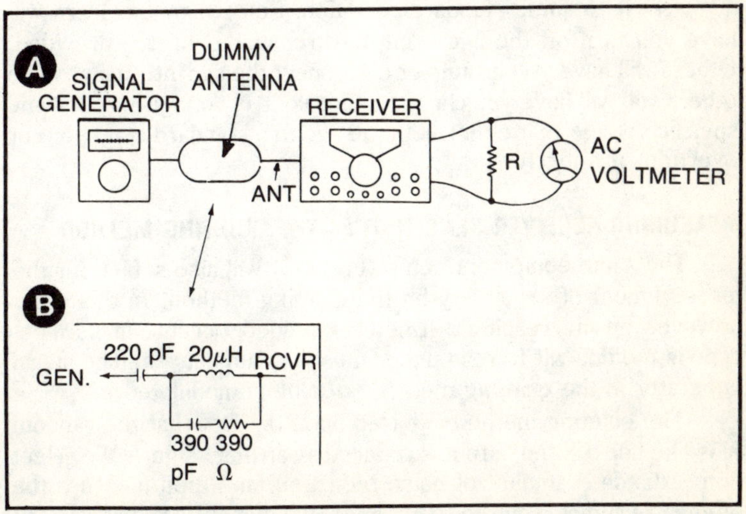

Fig. 4-1. Equipment needed to measure sensitivity by audio power method.

$$E = ((1\ W)(8\ ohms))^{1/2}$$
$$= (8)^{1/2} = 2.83 \text{ volts rms*}$$

*If an oscilloscope is used to measure the AC output voltage, it is easier to use the peak-to-peak AC value, which in this case is 8.00 volts.

The procedure is as follows:

- ☐Connect the instruments as shown in Fig. 4-1. Adjust the signal generator for 30 percent modulation by a 400-Hz sine wave.
- ☐Increase the rf output control until the reading on the AC voltmeter is 2.83 volts, or the oscilloscope shows 8.00 volts peak-to-peak.
- ☐Note the reading on the signal generator output attenuator. If given in microvolts, this is the sensitivity of the receiver. The specification should be listed in the form "__μV for 1 watt output at 400Hz, 30 percent amplitude modulation."

Different ways of producing the output reading will result in different sensitivity figures, and these are difficult to compare with any degree of confidence. In most cases, the differences between modulating frequencies of 400 Hz and 1000 Hz are insignificant, but differences in the percentage of modulation of the input signal and the output reference power level are a little more difficult to reconcile.

When making a measurement of receiver sensitivity, it is mandatory that we either turn off the automatic gain control, disable it, or set it to some standard condition. Some manufacturers will have you turn off the agc using the front panel switch provided. Others will have you ground or disconnect the agc internally, while others still will have you clamp the agc control voltage line to some specific voltage that either sets the agc to a standard condition, or overrides it entirely.

MEASURING RECEIVER SENSITIVITY—THE QUIETING METHOD

The same equipment setup (Fig. 4-1) will also suffice for the measurement of sensitivity by the quieting method. In this case, however, an rms reading voltmeter is needed, because an oscilloscope is too difficult to read under these circumstances. The signal generator in the quieting method *must* be unmodulated.

The quieting method is based upon the fact that the random hiss-like noise in the output is reduced by an input signal. We select some standard amount of noise reduction, and then measure the amount of signal required to obtain that reduction. We will use the zero-input-signal noise level as the reference level, and then find the voltage needed for the selected quieting by solving the standard decibel equation for V_f:

$$dB = 20 \ Log_{10} \frac{V_i}{V_f}$$

where dB represents the quieting level desired, usually either 10 or 20 dB; Log_{10} denotes the base-10 logarithms; V_i is the initial voltage output (when the input signal is zero); V_f is the final output voltage when the desired level of quieting is achieved.

☐**Example:**

Find the voltage ratio V_i/V_f that will exist when the quieting is 20 dB.

$$dB = 20 \ Log_{10}(V_i/V_f)$$

$$dB/10 = Log_{10}(V_i/V_f)$$

$$e^{(dB/10)} = V_i/V_f$$

$$e^{(20/10)} = V_i/V_f = 7.4$$

In this example, there will be a 7.4:1 reduction in noise level as the rf input signal is increased in amplitude from zero. For instance assume that we have 8.3 volts rms under the zero-signal condition. How much will we have to reduce the output noise to achieve the 7.4:1 ratio required for 20 dB of quieting?

First, solve the expression for V_f, when V_i is 8.3 volts:

$$\frac{V_i}{V_f} = 7.4 \qquad \frac{V_i}{7.4} = V_f \qquad \frac{8.3}{7.4} = V_f$$

$$1.12 \ V_{rms} = V_f$$

The procedure for making this measurement is as follows:

- ☐Connect the equipment as in Fig. 4-1. Set the rf output of the signal generator to zero and the modulation to *off* (or *unmod*).
- ☐Measure the noise output voltage across the load. You may use the volume control to set this noise level to some preset amount if you wish, but most authorities require that the volume control and rf gain control both be set to maximum.
- ☐Knowing the noise output voltage under zero signal condition, solve the expression $V_f = V_i/7.4$ to find the noise level that indicates 20 dB of quieting.

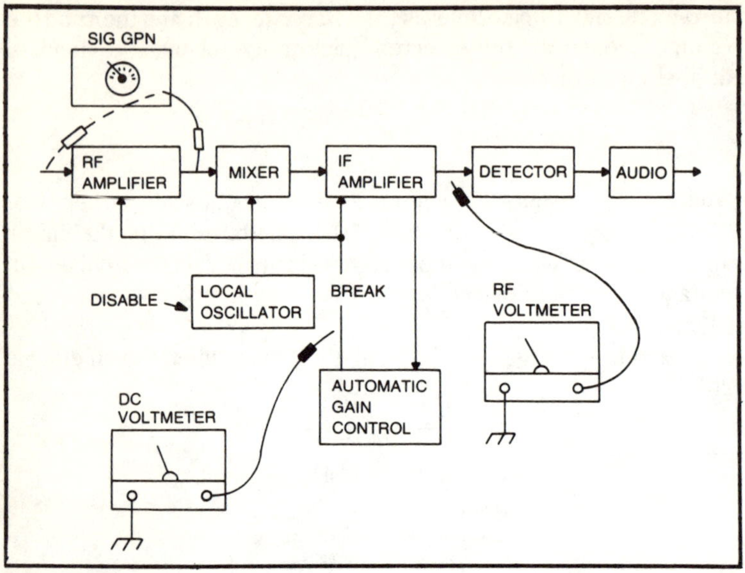

Fig. 4-2. Equipment needed to measure selectivity.

- ☐Adjust the rf output control of the signal generator until the AC voltmeter reads V_f, as calculated in step no. 3.
- ☐The signal generator output, in microvolts, required to produce this output noise level is the sensitivity of the receiver (___μV for 20 dB quieting).

An alternate quieting method uses a 30 percent amplitude modulated signal to determine the rf level required to produce an audio signal that is 10 or 20 dB above the noise level.

SELECTIVITY MEASUREMENTS

Selectivity indicates the bandwidth of the receiver and hence its ability to reject unwanted signals. Most of the selectivity in a superheterodyne receiver is from the i-f amplifier section. We may therefore measure the selectivity of the receiver by injecting a signal into the i-f amplifier chain, usually at the input of the mixer stage. In many cases, we use the rf signal port of the mixer and disable the local oscillator to reduce the possibility of spurious responses.

The equipment connections for a simple measurement of selectivity are shown in Fig. 4-2. Again, disable the agc circuit. Some method for determining relative output level is needed. If the receiver is an AM model, use an AC voltmeter across the speaker output, and a modulated signal from the signal generator. Also use an rf voltmeter (or DC voltmeter equipped with a demodulator/detector probe) at the output of the i-f amplifier. Sometimes, we can get away with using a DC voltmeter on either the agc line or (in AM models) the output of the detector, prior to any DC blocking capacitors that might be used. The idea is to obtain some indication of relative output level as the input frequency is varied. The procedure is as follows:

- ☐Adjust the signal generator frequency to the middle of the passband, as indicated by the maximum output as you tune the generator back and forth across the passband.
- ☐Set the signal generator output to some convenient level that does not overload the receiver. This level will be the 0 dB reference level, so make it a level that is easy to calculate.
- ☐Measure the output level at equally spaced frequencies above and below the middle of the passband, assuming that the input signal voltage remains constant.
- ☐Plot the data dB-down-vs.-frequency if you want a passband curve. Otherwise, measure only those frequencies above and below the center frequency at which the output level is 6

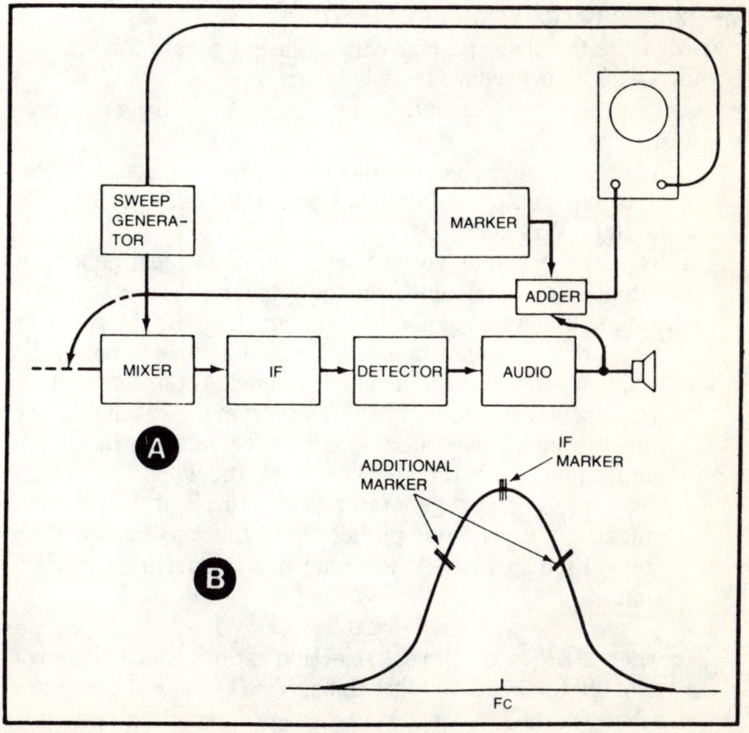

Fig. 4-3. Sweep method of measuring selectivity.

dB and 60 dB down. From these, specify the −6 dB bandwidth and the skirt factor of the receiver.

The previous technique assumes that no sweep generator is available. The sweep generator method, however, is faster and more reliable. Figure 4-3 shows the equipment setup required to make this measurement.

A sweep generator will produce an output signal that varies in frequency above and below the dial frequency by a preset amount. The sweep signal is basically an FM signal when the modulating voltage is a sine wave, but most often, it is swept by a sawtooth. This will cause the frequency to increase from its minimum towards the maximum, at which point it abruptly shifts back to the minimum point again at the falling edge of the sawtooth. By using the same sawtooth to sweep the horizontal axis of the oscilloscope, and adding together certain of the rf signals, the receiver selectivity curve is traced on the CRT of the scope.

The crystal markers are used to accurately calibrate the frequency of the system. The sweep generator calibrations are usually quite rough so these precise markers are needed.

MEASURING IMAGE RESPONSE

Measure the image response using the equipment connections of Fig. 4-1. The procedure is as follows:

- ☐Tune the receiver and signal generator to the same frequency.
- ☐Adjust the signal generator output level for some convenient audio output indication. Use a signal strength in the 25 to 100 μV range.
- ☐Record the output voltage level and the signal generator output level (obtainable from the output attenuator setting).
- ☐Retune the signal generator without changing the receiver tuning to a frequency that is twice the i-f above or below the frequency to which the receiver is tuned. At one or the other frequencies, an output will be heard from the receiver. Fine tune the signal generator—not the receiver—for maximum output level from the speaker of the receiver.
- ☐Readjust the signal generator attenuator until the receiver output is the same amplitude as in the second and third steps. Find the image rejection in decibels from the expression:

$$IRR = 20 \log_{10}(V_f/V_i)$$

where IRR is the image rejection ratio in decibels; $\log_{10}$ denotes the base-10 logarithms; V_f is the signal generator setting in the last step; V_i is the signal generator setting in the second step.

Chapter 5

RF Amplifier Circuits

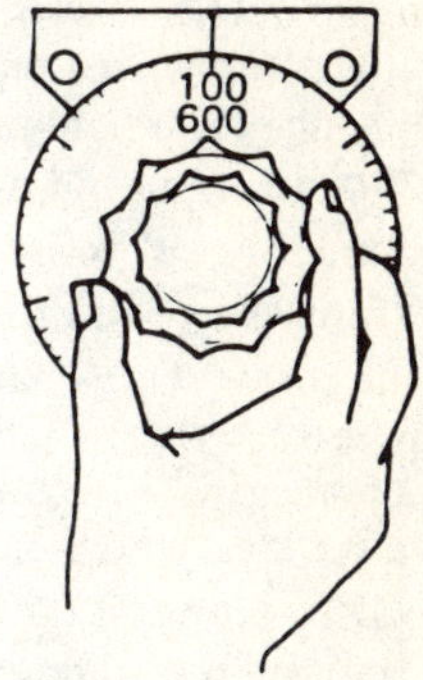

The radio frequency amplifier is the first stage in any quality radio receiver. The rf amplifier does not provide a lot of gain; its purpose is more than simple gain. The purposes of the rf amplifier are to:

- ☐provide isolation between the local oscillator/mixer stages and the antenna
- ☐improve the image rejection
- ☐provide some selectivity (but not sufficient for the overall receiver).

Isolation is needed because the mixer is not totally unilateral so some oscillator signal can be output to the antenna from the mixer. Placing an rf amplifier in the line between the mixer and the antenna limits the signal that is radiated to the atmosphere via the antenna. Radiated signals, even though weak, can interfere with other service. Military receivers typically use at least two shielded rf amplifiers in order to prevent radiation of the local oscillator. As far back as World War II, it was recognized that radiation from the LO of a receiver could provide a homing beacon for any enemy equipped with a sensitive radio direction finder.

In some receivers—notably VHF types—it is necessary to use special low-noise amplifiers in the front end. It seems that the noise figure for the entire receiver is set by the noise figure of the first rf amplifier. In dealing with weak VHF/UHF signals, it is therefore necessary to insure a very low noise figure for the rf amplifier.

This chapter will cover rf amplifier circuits primarily for the low-frequency, medium-wave and high-frequency bands. The VHF/UHF designs will be considered in a later chapter.

BIPOLAR RF AMPLIFIERS

We will not consider vacuum tube rf amplifiers at this point because they are no longer in common use. All modern receivers are solid-state, so we will start with the elementary bipolar NPN and PNP transistor rf amplifiers.

Figure 5-1 shows the circuit for a typical medium-wave (and low shortwave) bipolar rf amplifier using an NPN transistor. This particular circuit is very commonly found in automobile radios, mobile rigs, and some base station transceiver receiver sections made a few years ago. Many have since switched to field-effect transistor (FET) front ends.

When analyzing circuits, break out the components into two classes: the AC path and the DC path. This method, incidentally, is often critical to troubleshooting defects in circuits. The DC path in a circuit is where all of the DC flows (simple enough huh?). In the NPN circuit of Fig. 5-1, we have configured the transistor to operate with a negative ground power supply, making it compatible with US-made automobiles. The collector of an NPN transistor must be more positive than the emitter, so the V+ is applied to the collector. The DC path for the collector circuit consists of the DC resistance of inductor L4, resistor R3, and the collector terminal of transistor Q1. The DC path for the emitter circuit consists of the emitter terminal of the transistor, and resistor R1. Note that the emitter current also contributes to the base current ($I_c = I_e + I_b$). The current in the base circuit is approximately 1 to 5 percent of the emitter current. The DC path for the base is the base terminal of the transistor, coil L3, the agc line, and resistor R2.

The gain of the stage is set in part by the base-emitter voltage V_{be}. This will be in the 0.2 to 0.3 volt range for germanium transistors, and 0.6 to 0.7 volts for silicon transistors. The latter are more common in modern receivers, even though germanium devices were predominant in older receivers (early 60s). The silicon device is more suitable because it is somewhat more stable than the germanium, especially under changing temperatures.

The AC path in this circuit consists of any component through which the signal passes. In the base circuit of the rf amplifier, the input tank circuit C1/C2 and the input inductor L1 are part of the antenna circuitry. They are therefore part of the AC circuits of the rf amplifier. On the secondary side of the input transformer, the AC circuit consists of secondary winding L3, the base of transistor Q1, and bypass capacitors C2 and C3. Note that the base-emitter bypass capacitor does not have sufficiently low reactance to bypass the signal from the secondary coil, because of the low reactance of the coil, and the low input impedance of the transistor base. The purpose of C4 is to improve the stability of the stage.

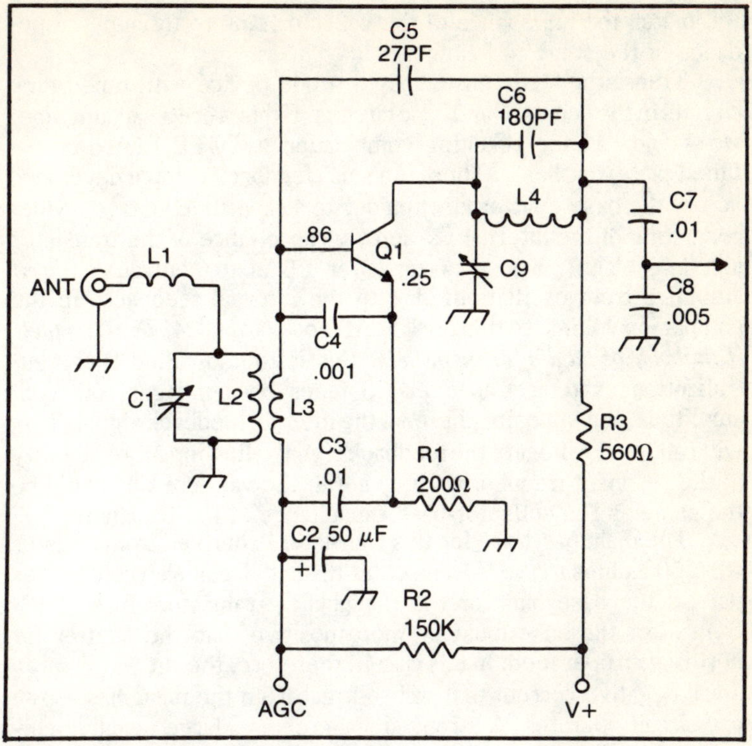

Fig. 5-1. Rf amplifier using NPN device.

The AC circuit in the collector of the transistor consists of coil L4 and capacitors C7 and C8. Inductor L4 is resonated to the rf signal by variable capacitor C9. The capacitor shunting L4 (C6) is not used to resonate L4 to the operating frequency, but to the *image frequency*. This capacitor/coil combination is used to form a parallel resonant circuit that is a high impedance to the image frequency. This particular arrangement is common in radios that use a variable inductance for tuning the front end. Inductive, or *permeability*, tuning is common in all automobile radios (except the newest digitally controlled models) and was quite common in radios produced during World War II for consumer use. The higher metal content of variable capacitors was simply too high for wartime economy, and was needed in military production.

Capacitor C9 resonates the collector inductor. This tank circuit is essentially a parallel resonant circuit *across* the signal line. The series combination of C7/C8 has a reactance that is low enough to make the capacitance of C9 the most significant in the circuit. The principal function of C7/C9 is to form a reactive voltage divider that will match the impedance of the collector tank to the input impedance of the stage to follow.

Transistor Q1 is essentially a triode device with tuned tank circuits in the collector and base circuits. If this were a vacuum tube, we would call the resulting combination a TGTP (tuned-grid—tuned-plate) oscillator. There is a small feedback voltage developed across the base-emitter junction due to a capacitive voltage divider consisting of the internal C_{bc} and C_{be} capacitance of the transistor junctions. The purpose of capacitor C5 is to neutralize the rf amplifier from oscillations due to the internal feedback. In the companion volume to this book, TAB book No. 1224, *The Complete Handbook of Radio Transmitters,* this is the so-called Rice neutralization system. Capacitor C5 feeds back a signal of equal amplitude, but opposite phase, of the internal feedback signal. This will cancel the effect of the feedback signal, allowing more stability at the resonant frequency. In general, the value of C5 should be the same as the collector-base capacitance of the transistor.

The main base bias for this transistor is derived from resistor R2 (150 Kohms to the V+ line). But the agc circuit also contributes part of the base bias for the rf amplifier transistor. In an NPN transistor, the base must be more positive than the emitter for normal gain operation. In this circuit, therefore, the agc will supply a small negative current that will subtract from the main bias as the agc circuit operates. A strong signal creates a large negative current from the agc to the base of Q1. This current will in turn reduce the positive current from bias resistor R2, thereby reducing the gain of the transistor. If the signal is very weak, on the other hand, the agc current may be as little as zero. Under that circumstance, the full positive current from the bias resistor is applied to the base of the transistor, and the gain is maximum.

A PNP version of the circuit is shown in Fig. 5-2. Again, the circuit is configured for negative ground operation and is suitable for mobile receivers. In the PNP circuit, the collector must be more negative than the emitter. Because there is no negative-to-ground power supply, we must conspire to make the collector "think" that it's negative! If we make the collector *less positive*, it is the same thing as being *more negative*. In the circuit of Fig. 5-2, we place the collector at a potential close to ground by connecting the collector DC circuit from the collector terminal of Q1, through L4, and resistor R3. The resistor forms the DC load for the collector circuit. In this circuit, the DC potential measured on the collector will depend upon the agc action, but will typically vary from 0.1 to 2.5 volts, depending upon signal strength.

Emitter resistor R2 is connected directly to the V+ power

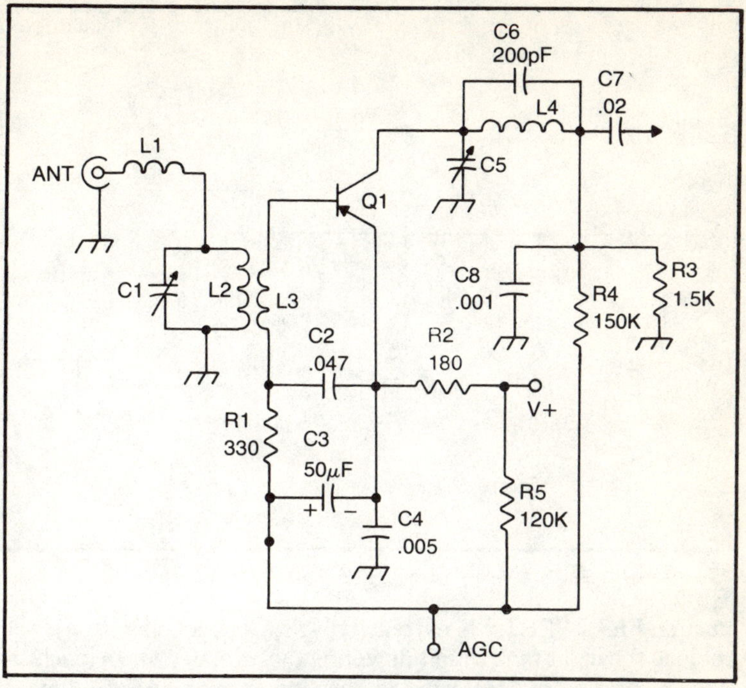

Fig. 5-2. Rf amplifier using PNP device in negative-ground configuration.

supply and is bypassed capacitor C4. The DC path for the emitter-base circuit consists of emitter resistor R2, the emitter terminal and base terminal of Q1, the DC resistance of secondary coil L3, and resistor R1. It then splits to the agc line, feedback resistor R4, and main DC bias resistor R5.

In both bipolar circuits (Figs. 5-1 and 5-2), there is a high-value electrolytic capacitor in the AC base circuit. Normally, electrolytics are not too effective at rf frequencies. The purpose of these capacitors is not to bypass rf in the base circuit (C2/C4 does that in Fig. 5-2), but to bypass the agc line. Remember that the agc is a *feedback* circuit. If this capacitor is opened, an oscillation through the rf/mixer/i-f amplifier chain produces *birdies* in the output. This capacitor, incidentally, is suspected first when troubleshooting a receiver for tunable oscillations, especially if the output is distorted on only strong stations. Such behavior, incidentally, is characteristic of most agc-related defects.

FIELD-EFFECT TRANSISTOR CIRCUITS

The field-effect transistor (FET) is capable of superior operation in rf amplifier applications. The input impedance of both types of

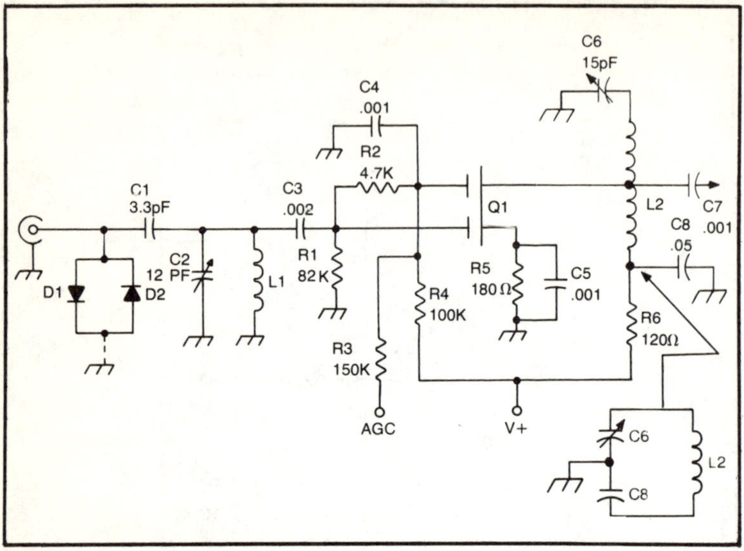

Fig. 5-3 MOSFET Rf amplifier.

junction FETs (JFETs) is extremely high. Consequently, they will not load the input tank circuit (lowering the effective Q) as much as bipolar transistors. Matching is then somewhat easier in most cases.

A MOSFET rf amplifier is shown in Fig. 5-3. This particular device is a dual-gate MOSFET. The rf signal is applied to gate No. 1, while the gate No. 2 terminal is connected to the agc control voltage. A typical MOSFET for this circuit is the RCA 40673 device. MOSFETs are sensitive to static electricity damage. The thin insulated gate structure will break down with as little as 80 volts applied. Static charges on your hands, clothing, and tools can "pop" the device rather quickly. The 40673 is preferred because it contains internal zener diodes that bypass potentials above normal operating biases across the sensitive junctions.

This circuit optionally uses another static protection scheme in the form of the back-to-back diodes across the antenna input terminals. These are low-leakage silicon diodes that will not load the input tank circuit. The silicon diode needs at least 0.6 volts to forward bias the PN junction. At normal rf signal levels (microvolts), these diodes are not forward biased. They are effectively out of the circuit. But if a strong static charge passes down the antenna line, the diodes become forward biased and shunt the potentially dangerous charge to ground. One diode will take out positive charges while the other takes out negative charges. This problem can become especially important in mobile rigs. Many years ago I was engaged in auto radio and CB repairs. I found many cases where the rf amplifier transistor popped (most were bipolar so the problem is not limited to MOSFETs) when the driver got out of the car, built up a nasty static charge from a winter coat or a vinyl car seat and adjusted the antenna. As soon as the driver touched the antenna, there was a brief spark, and the rf transistor went belly up! Of course, the customer always assumed that the problem was in the antenna; after all, didn't the radio work until he touched the antenna?

The drain tank circuit in rf amplifier of Fig. 5-3 is tapped down to match the low impedance of the drain. Oddly enough, many rf amplifiers are designed such that the tap for the drain is to the same point as the tap for the output capacitor (C7). The tank circuit is parallel resonant, not series resonant as might be suspected from the circuit configuration. The inset of Fig. 5-3 shows the parallel nature of the coil and capacitors in the tank circuit. Note that the capacitance of C8 is so large with respect to the capacitance of C6 that by the ordinary series capacitance formula, the capacitance of C6 is predominant.

The agc voltage is a bias voltage that adds to, or subtracts from, the bias applied to gate G2 of the MOSFET. This type of circuit, incidentally, has the reputation of a more significant dynamic agc range than is possible with bipolar devices. The change of gain is more linear over a broader range of signal levels.

One problem with the MOSFET amplifier is the possibility of oscillations. The same problem as we saw in the previous circuit exists in the MOSFET version. Figure 5-4 shows an example of

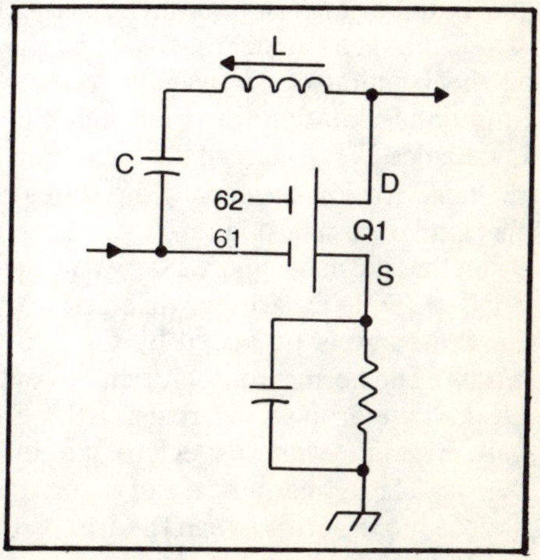

Fig. 5-4. Neutralization circuit.

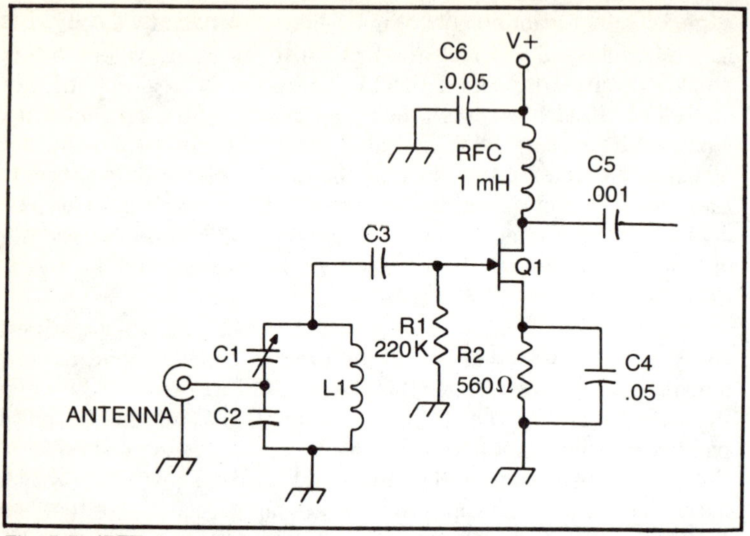

Fig. 5-5. JFET rf amplifier.

inductor neutralization. The feedback signal is applied to the signal gate through inductor L. The purpose of the capacitor (C) in this circuit is not to resonate the inductance, but to provide DC blocking of the potentials present on both G2 and the drain. This circuit is adjusted to eliminate oscillations in the circuit without unduly reducing the overall gain.

An example of a single-stage JFET rf amplifier is shown in Fig. 5-5. The JFET device can be almost any rf junction field-effect transistor, such as the popular MPF102 device. Note that the Motorola HEP and the other replacement lines offer decent rf JFETs at low cost and easy availability. The input tank circuit of this rf amplifier is a parallel resonant circuit. The input line is connected to the junction of a capacitive voltage divider, which is used to impedance match the low-impedance antenna to the high-impedance JFET gate circuit. The total capacitance that resonates the tank circuit is essentially the series combination of C1 and C2. In this kind of circuit, it is rare to find one capacitor so much higher in value that it can be ignored (as was true in Fig. 5-3). Capacitor C3 is a DC blocking capacitor, while resistor R1 is a gate load. The bias for the stage is produced by the series resistance in the source circuit. The normal source-drain current of the JFET creates a small DC voltage drop across resistor R2. Since the goal is to make the gate negative with respect to the source, the JFET is properly biased. This is another case of reversing the polarities—making the source more positive than the gate is the same as making the gate more negative than the source. The source terminal of the JFET is kept at rf ground potential by the 0.05-μF bypass capacitor across R2.

Note that the drain circuit is not tuned. The AC load for the drain is the reactance of the 1-mH rf choke. This essentially reduces somewhat the gain possible at the rf signal , but is done to eliminate the possibility of oscillation. Like the MOSFET and bipolar circuits discussed previously, a JFET will act much like a solid-state version of a TGTP oscillator if not neutralized. This circuit will perform without neutralization because the drain is not tuned to the same frequency as the gate circuit. The output signal is taken from the drain through C5, a DC blocking capacitor.

A cascade JFET rf amplifier shown in Fig. 5-6 also needs no neutralization. Input stage Q1 is essentially the same as the circuit of Fig. 5-5. The matching of the low impedance (50 to 100 ohms) of the antenna, however, is accomplished with a tapped tank inductor, rather than a tapped capacitor. The other components perform the same jobs as in other circuit of Fig. 5-5. Again, the drain circuit is untuned, eliminating the TGTP oscillator effect.

The output stage of this cascade rf amplifier is exactly the opposite from the input stage. It also needs no neutralization because we do not tune *both* input and output circuits. In this circuit, however, it is the output that is tuned; the input is left wideband. The drain signal from Q1 is impedance coupled (see TAB book No. 1224, *The Complete Handbook of Radio Transmitters*) to the gate of Q2. The drain of Q2, however, is tuned in much the same fashion as one of the previous FET circuits. The fact that neither stage has both the input and the output tuned to the same frequency results in stable operation without the need for a neutralization circuit. This

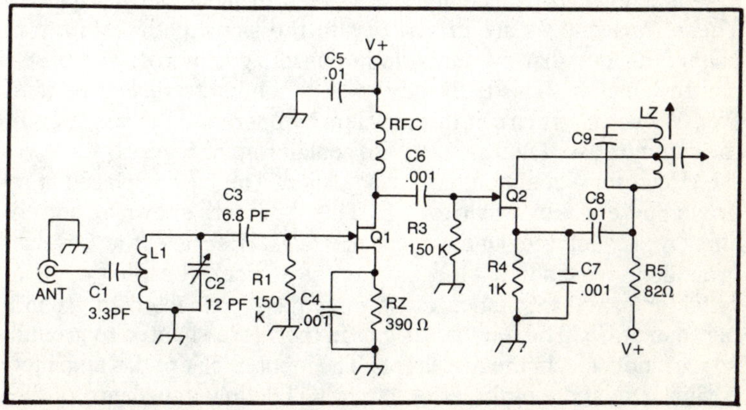

Fig. 5-6. Cascade JFET rf amplifier using common source configuration.

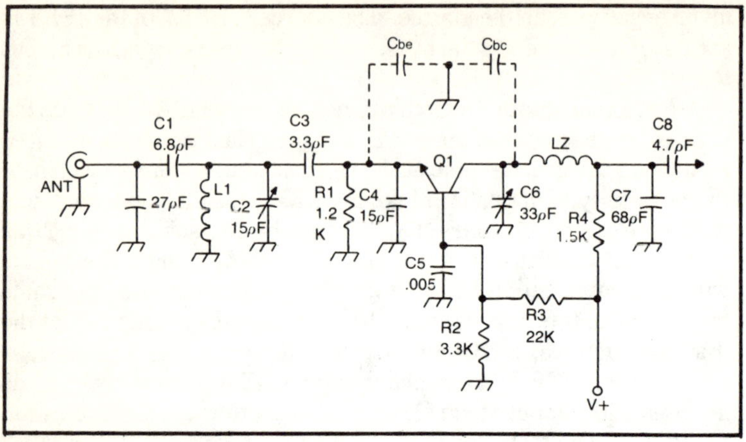

Fig. 5-7. Grounded-base rf amplifier.

configuration is often used when more gain is needed than can be provided by a single transistor, and a wide frequency range of operation is desired. Neutralization schemes sometimes fail when the frequency range is wide. This is theoretically not supposed to happen, but real circuits sometimes contain distributed parameters that are difficult to predict without a rather complex analysis. It's not that the theory doesn't work, but that there are factors to be considered that are often too difficult to predict with accuracy.

GROUNDED GATE/BASE Rf AMPLIFIERS

The principal reason why the neutralization of rf amplifiers is needed is the feedback due to the internal capacitances of the device. In a bipolar transistor, for example, there are a collector-base junction capacitance and a base-emitter junction capacitance. These capacitances are in series with the signal path and form a voltage divider that presents an in-phase signal across the base-emitter junction. This is exactly what is needed to *sustain* oscillation. We can short circuit this problem by operating the amplifier in the common base (also gate or grid) configuration. Figure 5-7 shows the circuit for a common base rf amplifier. The base is placed at rf ground potential by capacitor C5. The capacitors shown as dotted lines represent the junction capacitances. Because the base is grounded, we find that C_{be} is merely in parallel with C4. The collector base capacitance C_{bc} is effectively connected to ground because of C5. The feedback signal is therefore shunted to ground so it cannot sustain the oscillation. The input circuit to this amplifier is similar to some of the other circuits. The low impedance of the antenna is matched to the tank circuit through a capacitor voltage divider that parallel resonates the inductor coil L1. The signal is then coupled to the emitter of the rf amplifier transistor through DC blocking capacitor C3. R1 forms the emitter loading resistor. Note that the tuning of the input tank circuit is essentially quite broad because of the lowered tank Q occasioned by the low value of the emitter resistor.

The collector circuit is tuned to resonance by capacitors C6 and C7 and inductor L2. Output to the following stage is through DC blocking capacitor C8.

The circuit for a single-stage JFET grounded gate rf amplifier is shown in Fig. 5-8. Like the other circuits in this section, this circuit is intended primarily for the VHF and UHF region, but is sometimes seen also in the HF region. The gate of the JFET is connected directly to ground. No bypass capacitor is needed in this application. The bias is formed by the voltage drop across source resistor R1. This resistor is bypassed to ground, keeping the "cold" end of L1 at zero volts rf potential by capacitor C2. The input circuit is essentially nonresonant, despite the capacitor in series with the antenna lead. The purpose of this capacitor is to match the antenna impedance to the inductor. This type of circuit is seen frequently in VHF mobile and auto radio applications where the object is to obtaiin a match to a short vertical or windshield antenna which is only *close* to resonance. These are not compensation antennas, as might be used on the HF bands, but are nearly resonant antennas. The typical FM broadcast antenna, for example, is resonant at only one frequency in a 20-MHz spread (88 to 108 MHz).

The output tank circuit is an ordinary parallel resonant circuit. Capacitor C5 couples the output to the next stage. Decoupling is

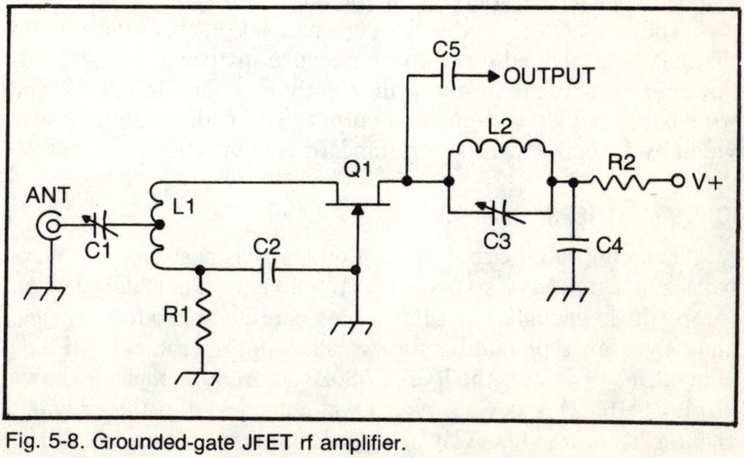

Fig. 5-8. Grounded-gate JFET rf amplifier.

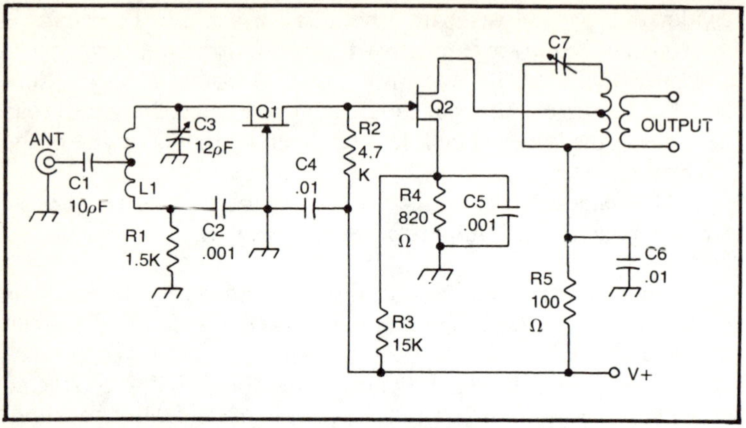

Fig. 5-9. Cascade rf amplifier using a grounded-gate input stage.

provided by capacitor C4 and resistor R2, a common combination in HF anf VHF rf amplifiers, but not always used in medium-wave units.

The circuit in Fig. 5-9 is a version of the cascade JFET rf amplifier in which the input stage is a common gate amplifier. The circuit is otherwise much similar to the other circuit presented earlier. The drain of the input stage (Q1) is direct coupled to the gate of the output stage (Q2). This means that we must also bias the source up to the level of the gate, or there would be a large positive voltage between the gate and source, forward biasing the junction. In the normal JFET operation, it is mandatory that we keep the junction reverse biased in order to prevent destruction of the device. We can then develop the bias in the normal manner; i.e. the voltage drop across source resistor R4.

The drain circuit in this case is a parallel resonant tank circuit that has been tapped down for impedance matching. The output, however, is a little different from the other circuits. In this version we have opted for line-coupled output. The rf decoupling is provided by R5/C6 which is our standard configuration.

IC Rf AMPLIFIERS

The integrated circuit revolution that has reshaped the electronics industry has also made contributions to rf amplifier design. Interestingly enough, most IC devices operate at low frequencies, and only a few type numbers are available for operation at HF and above. One of these is the RCA CA3028 differential amplifier shown in Fig. 5-10. This device, which was introduced in the previous chapter, is used extensively in rf applications to well over 50 MHz.

Transistors Q1 and Q2 form a differential pair whose emitters are connected together at the collector of Q3. If Q3 is operated as a constant current source, then true differential operation will result. The total of the emitter currents in Q1 and Q2 cannot be anything other than the collector current of Q3. If the emitter current in one of these transistors is increased, it must necessarily reduce the emitter current of the other transistor. Similarly, when the emitter current of one transistor is reduced, it necessarily increases the other transistor's emitter current.

When the CA3028 is used as an rf amplifer, the circuit can be driven as single-ended by grounding one of the differential inputs to rf with a bypass capacitor. In the circuit of Fig. 5-11, the rf signal is made to look like a differential signal by applying to input transformer L1/L2. The secondary L2 is resonated to the rf signal by capacitor C1. This tank circuit is connected across the two differential inputs, but the base of Q2 (pin No. 5) is kept at rf ground potential by capacitor C2. This point is also convenient to place the bias from the R-2R network.

Pin No. 7 of the RCA CA3028 is used to control the base bias of the constant current source transistor Q3. This point can be connected directly to V+ if the stage is to operate wide open at

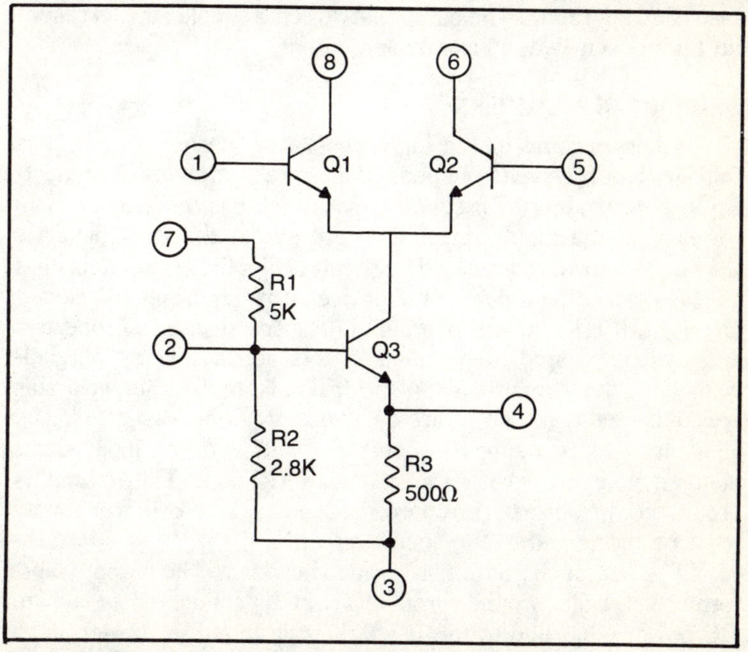

Fig. 5-10. Pinouts for the RCA CA3028 differential amplifier IC.

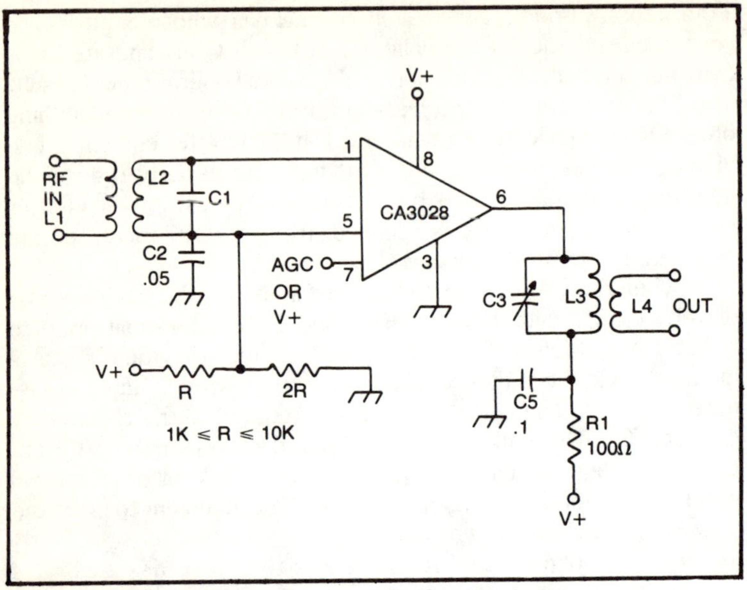

Fig. 5-11. RF amplifier based on the RCA CA3028 device.

maximum gain. Or connect pin No. 7 to the agc circuit for gain-controlled operation. The output network is a simple parallel resonant tank circuit with a link-coupled output.

WIDEBAND Rf AMPLIFIERS

It has become increasingly popular to design wideband rf amplifiers for receiver front ends. Bands are then changed connecting 50-ohm band-pass (or even low-pass/high-pass combinations) in series with the input signal. In many cases, the rf amplifier is allowed to remain wideband. The circuit of Fig. 5-12 is a wideband rf amplifier based on a dual-JFET device. It is important to select a low-noise JFET. An improvement in transconductance properties and dynamic range is obtained in this application by parallel-connecting the terminals of the dual JFET. Note that the two drains are connected together, as are the two sources and two gates. The input network is a simple wideband circuit, with the input signal being developed across the rf choke (RFC1). This circuit is grounded gate and uses no source bias, so a −12-volt power supply must be connected to the source circuit.

The output circuit is also a wideband affair. The output transformer is a bifilar wound toroid affair, cross-connected as shown. DC power is applied to the JFET through this transformer and a decoupling network consisting of a *pi* circuit (C5/C6/RFC2).

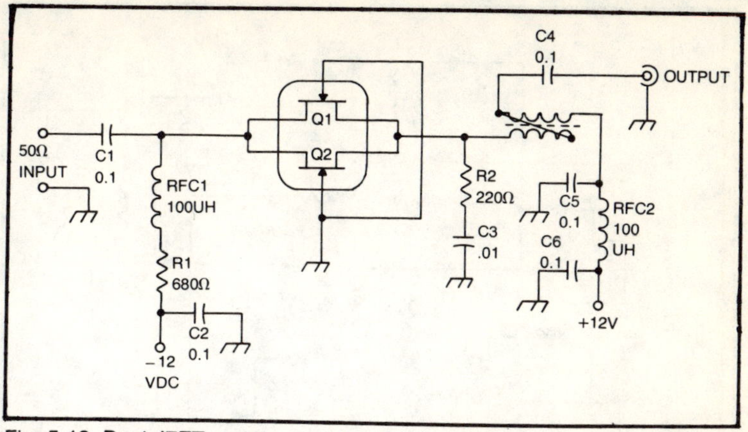

Fig. 5-12. Dual JFET used as a wideband rf amplifier.

BANDSWITCHING

The classical method of bandswitching is shown in Fig. 5-13. There will be a series of rf tank circuits. At least three sets are normally included: the rf amplifier input (often labeled *antenna*), rf amplifier output, and the local oscillator. For even the simplest receiver, this means that we have to switch into and out of the circuit not less than 15 rf tank circuits. The main tuning capacitor(s) will be permanently wired into the circuit, as will be any bandspread

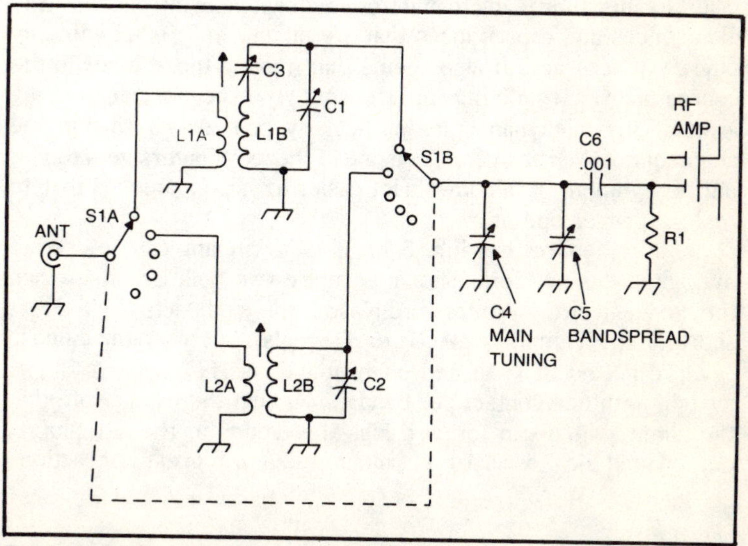

Fig. 5-13. Traditional bandswitching system.

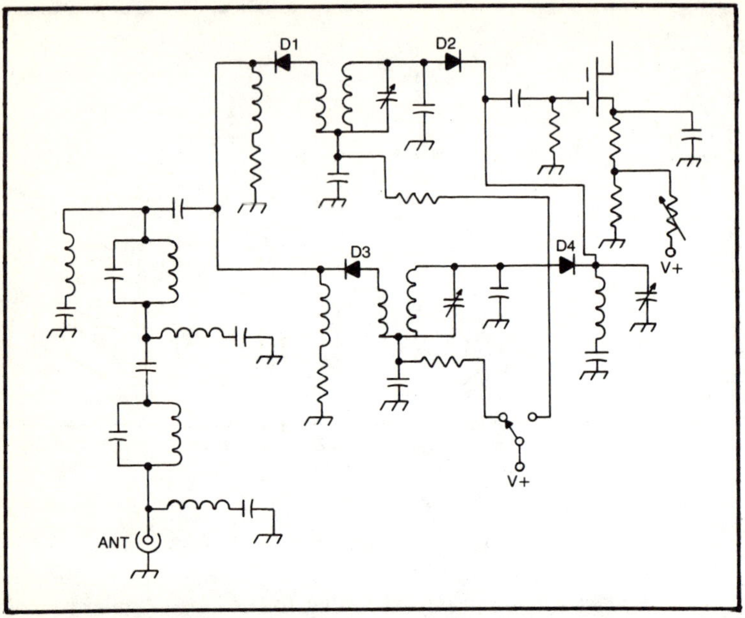

Fig. 5-14. Use of diode switching uses only DC on the bandswitch.

capacitors. This is a lot of switching, but there are other problems. One is the ordinary problem of maintenance: all of those darn switch contacts will need periodic cleaning and are potential failure points. The leads to and from the switching introduces stray inductances and capacitances that are difficult to predict with any degree of accuracy. It also seems that moving those wires in the course of troubleshooting, maintenance, modification, or messing around (to which hams are particularly prone) will change the alignment of the receiver. Once one of these receivers gets too far out of alignment, it becomes a lot easier to skin an amoeba than to restore proper operation.

As in the circuit of Fig. 5-14, diode switching can be used to overcome this problem. Shown here are two tank circuits. When the diodes in either are forward biased, the diode acts like a direct short for small amplitude AC (i.e. rf signals), but when the diode is reverse biased, it is turned off to low-level rf signals. A simple switch, with one contact per band, will select the band. Note that the same switch can forward bias the diodes in the antenna, rf output, and local oscillator circuits without any great interaction.

Chapter 6

Mixer, Oscillator, and Converter Circuits

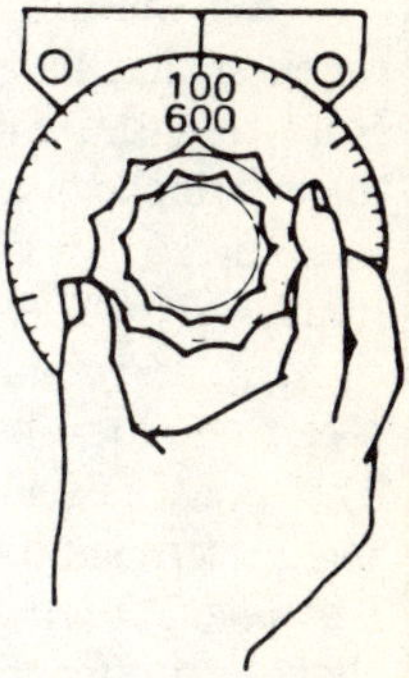

The superheterodyne radio receiver has become *the* standard type of receiver configuration. There are numerous advantages of the superhet, not the least among them is that HF, VHF, and UHF radio receivers have principal gain and selectivity properties formed at a lower frequency. Most high-frequency (and up) radio circuits are not stable and lead to many problems when used directly. We can keep the problems caused by this kind of circuit to a minimum by converting the signal to a lower frequency early in the game and then performing all of our signal processing at the easier-to-tame lower frequency. We also gain a certain amount of evenness of performance in superhet, not obtainable in trf or regenerative radios. Most radio receivers will operate nicely at VLF and even medium-wave bands. But as frequency is increased, the performance of the circuits decreases. If we made such a receiver, it might be a "dy-no-mite" performer at VLF or LF, but will exhibit steadily decreasing performance at higher frequencies. The performance above 10 MHz would more warrant a raspberry than praise. In the superheterodyne receiver, though, the first thing that we do with this signal is to down-convert it to a lower frequency where the performance of the circuits is better.

The key to superheterodyne operation is the heterodyne process. Recall from previous chapters what this term means. When we combine two signals, F1 and F2, in a *nonlinear* circuit (a necessary condition for mixing, incidentally) the output spectrum will contain at least four different frequencies. Two are the original frequencies and two are new products caused by the mixing action: F1, F2, F1+F2, and F1-F2. The last two are the *sum* and *difference* frequencies, respectively. A tuned amplifier following mixer, called the intermediate frequency (i-f) amplifier, selects either the sum or the difference frequency, rejecting the others.

We will next consider those stages of the superheterodyne receiver that are responsible for the mixing of the two frequencies. Note that the circuits given in this chapter are mostly oriented toward application in the first conversion, but are equally useful in second or third conversions in those receivers that use such systems. We will examine popular oscillator, mixer, and converter circuits. Incidentally, the converter is a stage that combines the mixer and local oscillator functions into one single active-element. The topic of oscillator circuits is also covered, but in greater detail, in the companion volume to this book, TAB book No. 1224, *The Complete Handbook of Radio Transmitters*. A lighter treatment is therefore offered.

OSCILLATOR CIRCUITS

The local oscillator in a superheterodyne receiver is used to provide a signal that is different from the radio frequency being received by an amount equal to the intermediate frequency. For example, to receive 21.35 MHz on a receiver with a 3.385-MHz i-f amplifier, and LO frequency is needed of either

$$F_{LO} = 21.35 + 3.385 = 24.735 \text{ MHz}$$

or

$$F_{LO} = 21.35 - 3.385 = 17.965 \text{ MHz}$$

There is no reason why the local oscillator frequency *must* be above or below the radio frequency. As long as the rf/LO difference is equal to the i-f, the radio will work. There are some other considerations, however, that sometimes dictate a selection of one over the other. One of these is the matter of the tuning direction when a variable LO is used. If the LO is above the rf (the usual practice in most receivers), the received radio frequency increases as the LO frequency increases. If, on the other hand, the LO is below the rf, the radio frequency being received will decrease as the oscillator frequency is increased. There might also be a matter of *mixer products*. In our simple illustrations of the superhet process, we assume that the output spectrum consists of only four frequency components: RF, LO and the sum and difference products. Additional products (remember that the mixer is nonlinear) are equal to the sum and difference products of any number of harmonic combinations of the rf and LO. Possible combinations could be expressed as:

$$F_n = N(rf) \pm M(LO)$$

where N and M are arbitrary integers (0, 1, 2, 3... n).

Receiver designers go to great lengths to study the mixer by products to see if there is a possibility of spurious signals that will fall within the range tuned by the receiver. Then this is not done properly, the result will be an annoying birdie somewhere within the tuning range.

There are two types of local oscillator circuit: *variable frequency oscillator* (VFO) and *crystal oscillator* (XO). The VFO is used in continuously tunable radio receivers, while the crystal oscillator is used in channelized applications, such as CB sets, 2-meter FM hand-helds, rigs, landmobile, etc. We should possibly claim a third class, which is the fixed-frequency LC oscillator, used more now in phase-locked loop (PLL) frequency control of channelized receivers.

Figure 6-1 shows several VFO circuits that are popular as the LO in radio receivers. The circuit in Fig. 6-1 is the Armstrong oscillator. This circuit was briefly mentioned in previous chapters in a modified form called the *regenerative detector*. But in this case, we are going to make the circuit actually oscillate to produce the LO signal. Transistor Q1 is forward biased by resistor voltage divider network R1/R2. Resistor R3 is the emitter load and is used to stabilize the circuit to temperature variations. The emitter of the transistor is kept at rf ground potential because of bypass capacitor C4. The DC collector circuit of the oscillator consists of the collector terminal of the transistor, the DC resistance of tickler coil L2, and the V+ terminal of the power supply.

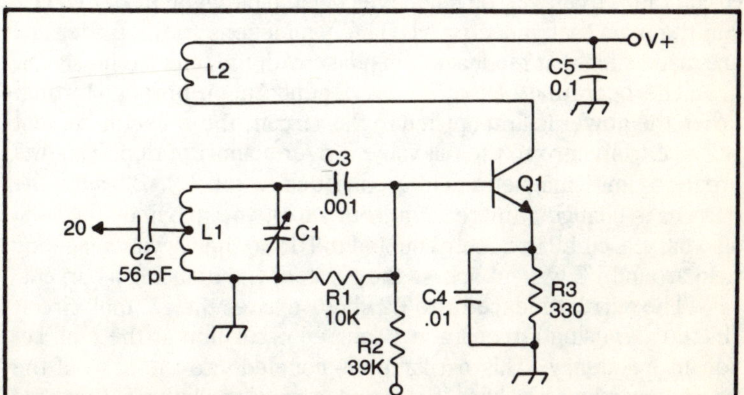

Fig. 6-1. Armstrong oscillator.

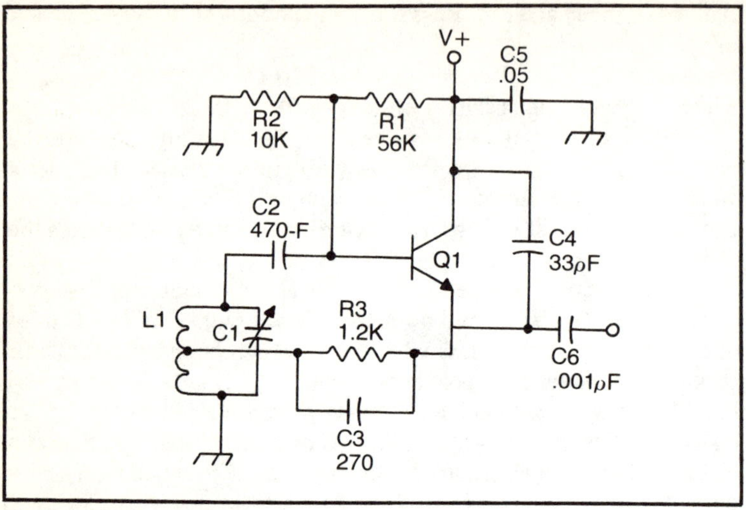

Fig. 6-2. Hartley oscillator.

The tank circuit consists of inductor L1 and variable capacitor C1. This tank circuit is coupled to the base of Q1 through DC blocking capacitor C3. Output to the mixer stage is taken through a small value capacitor (C2) connected to a tap on the tank inductor. Several other techniques could have been used for producing an output signal. Among them might be a link-coupled coil to L1, a blocking capacitor from the base of the transistor or (if C4 was reduced in value or eliminated) from the emitter of the transistor.

Feedback is a necessary condition for any oscillator to work. In this circuit we feed back some of the collector signal to the base circuit tank circuit via the inductive coupling between tickler coil L2 and the main tank inductor L1. The connections to the tickler coil are made such that feedback is in phase with the tank circuit signal.

The operation of this circuit depends upon proper starting. When the power is first applied to this circuit, the collector current will rise from zero to the full value I_c over a short period. This will create a small magnetic field around tickler coil L2. Because the current is changing (increasing from zero), the magnetic field will also vary. Coil L2 is closely coupled to L1, so the varying magnetic field around L2 will cut across the turns of L1, inducing a current.

The current induced into L2 shock excites the LC tank circuit (L1/C1), causing a momentary damped oscillation at the tank resonant frequency. This oscillation is coupled into the base of the transistor, which amplifies it, creating a large oscillation of the same frequency in the collector current. The oscillation is essentially a sinewave, so it causes a varying magnetic field of the same frequency around L2. Again, a varying current is induced into the tank circuit, sustaining the oscillations. After a few dozen cycles, as the transistor comes to full operating speed, the amplitude of the oscillations comes to equilibrium and the circuit is fully operational.

The Armstrong oscillator is not terribly popular because it produces some stability problems. We see this type of oscillator, however, in some AM broadcast receivers and VLF receivers. It is also the oscillator most commonly used as the beat frequency oscillator (BFO) in low-cost receivers which supposedly receive CW signals. The other two oscillator circuits shown in Figs. 6-2 and 6-3 are somewhat more common in radio receiver design because of their superior stability.

An example of the Hartley oscillator circuit is shown in Fig. 6-2. This circuit is closely related to the Armstrong circuit, and is probably a direct descendant of same. The principal difference is that the tickler coil is part of the tank circuit coil, connected in autotransformer fashion. Coil L1 in Fig. 6-2 is tapped for the feedback signal. A Hartley oscillator can be made several ways with both series and parallel feed of the collector current. The major identifying factor (important should you decide to take an FCC license examination, incidentally) is the tapped coil. The circuit shown here places the coil in the base circuit, but it could just as easily be placed in the collector circuit. Space does not permit us to exhibit all possibiliities, but should you see a tapped coil in a collector tank, with some form of feedback connection to the base, then suspect a Hartley circuit.

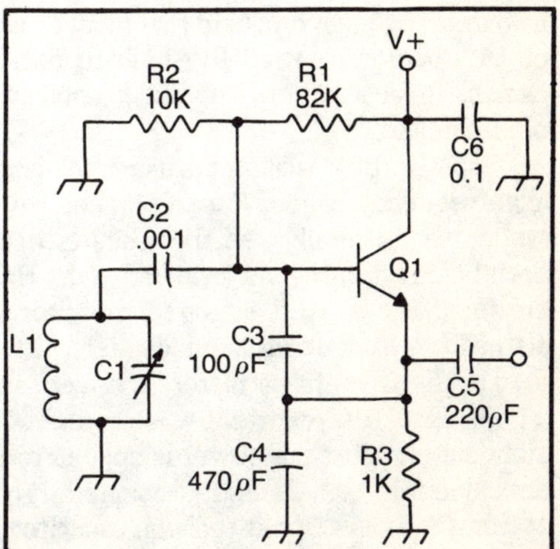

Fig. 6-3. Colpitts oscillator.

The tap on the coil in a Hartley oscillator is placed at a point that permits a trade-off between maximum oscillator output amplitude and best stability. In commercially prepared Hartley oscillator coils, of course, the tap is fixed by the manufacturer.

Like the Armstrong circuit, there are several different methods of taking the output signal. In this case, we have elected to take the signal from the emitter terminal of the transistor. We could also take the signal from a link-coupled winding to L1, or a tap (other than the feedback tap) on L1.

Note that this circuit contains a capacitance from the collector to the emitter of the transistor. This device might be seen in *any* transistor oscillator, and is used to increase the feedback signal amplitude. Some of the feedback signal is developed by the interelectrode capacitances of the transistor. At some frequencies, this internal signal is not sufficient to make the oscillation stable, so the circuit operates intermittently. This capacitor provides a little external feedback that makes the operation of the oscillator less critical.

The Hartley oscillator operates in the following manner. When power is first applied, the emitter current (almost identically equal to the collector current and different only by the amount of the tiny base current) will vary from zero to its steady-state value I_c. This varying emitter current flows in the lower portion of the tapped indictor (L1), causing a varying magnetic field that cuts across the remainder of the coil. This will shock excite the tank circuit, creating a momentary damped oscillation at the resonant frequency of the tank circuit. The tank circuit oscillation is amplified by the transistor to create a varying emitter current. This current will in turn develop an oscillating magnetic field that induces current into the remainder of L1, further exciting the tank to oscillate. After a few dozen cycles, the oscillation amplitude stabilizes and the circuit is fully operational.

The Hartley oscillator is used mostly in the lower portion of the high-frequency bands. For the higher HF range through the UHF range, we generally see the Colpitts oscillator of Fig. 6-3. This circuit is functionally very similar to the Hartley circuit, except that the feedback is from a tapped capacitor voltage divider (C3/C4) instead of a tapped inductor divider.

In the Colpitts oscillator, the feedback is initiated by the voltage drop across emitter resistor R3, which is a varying voltage immediately after the power is applied to the circuit. This signal is developed across C4 and is coupled into the base circuit, shock exciting the tank circuit through capacitor C3. A related version of this circuit is called either the Clapp oscillator or the series-tuned

Colpitts, in which the L1/C1 tank circuit is replaced with a series resonant circuit instead of the parallel resonant circuit shown. The Colpitts oscillator is used mostly in the HF and above range. It is considered more stable than some of the other oscillator circuits.

Crystal Oscillators

Variable frequency oscillators suffer from a pronounced amount of drift. The drift is tolerable, especially if the circuit construction and temperature compensation limit it, when the variable frequency capability of the VFO is needed. When channelizing the receiver, or in other cases where frequency variability is not needed, superior stability is possible with a crystal oscillator.

The crystal oscillator is based on the phenomenon called *piezoelectricity*, which is seen in certain crystalline substances, such as natural and synthetic quartz and certain ceramics. When a piezoelectric crystal slab is mechanically deformed, an electrical potential appears across the opposing surfaces. If we mechanically excite the crystal, this voltage will vary as the crystal vibrates. The vibration of the crystal—hence the frequency of the varying electrical potential—is a function of the natural resonant frequency of the crystal. There is another aspect to piezoelectricity. If an electrical potential is applied across the surfaces of the slab of quartz, the slab will then deform. We can therefore sustain the oscillation of the slab and the AC signal appearing across the surfaces by periodically shock exciting the slab with an electrical signal. This is exactly the behavior of an LC tank circuit.

One popular form of crystal oscillator is the crystallized Colpitts oscillator. If we build a circuit such as shown in Fig. 6-3 (the Colpitts VFO) and replace the LC tank circuit (L1/C1) with a parallel resonant crystal, the oscillator will generate a signal at the crystal frequency. The crystallized Colpitts oscillator is used typically from VLF through VHF ranges.

Crystal frequency is dependent upon several factors, but the seemingly most important is the physical dimensions of the slab. As the crystal slab becomes higher, the frequency of operation becomes higher. Once above 18 to 20 MHz, however, the slab becomes too thin for effective use. It requires very little power to *fracture* a thin crystal. But a crystal can operate at an *overtone* of the fundamental frequency. An overtone is not quite identical to a harmonic, although the overtone will be very close to the harmonic of the same order; i.e. the third overtone is close in frequency to the third harmonic. If we design an oscillator as if it were to operate at the overtone, and then place the crystal in the circuit, it will produce an output signal at that overtone. Overtone oscillators invariably contain a tuned tank circuit in order to select the correct overtone. Note that the overtones are all odd-order: third, fifth, seventh, etc.

The Colpitts circuit of Fig. 6-3, properly converted to crystal operation, will function at frequencies from 2 to 18 MHz in the fundamental mode only. We can operate the Miller oscillator shown in Fig. 6-4A in either the fundamental or the overtone mode. This circuit is analogous to the TGTP oscillator of vacuum tube design in that is has a tuned tank circuit in the output (drain) circuit, and a tuned tank circuit—the crystal—in the input (gate) circuit. Perhaps this could be called the TGTD, or tuned-gate-tuned-drain circuit! The LC tank circuit in the drain is tuned to a frequency very near the resonant frequency of the crystal. The tuning is supposed to be a little different from the crystal frequency, however, or the oscillation becomes unstable and may even cease altogether. Note that the tuning of the tank circuit is more critical in one direction than it is in the other. If we approach the correct resonant point from above—from a higher frequency—we find that the adjustment is faster than if we approach it from below. We know that the oscillator will stop running if the tank is tuned to the crystal frequency. We can monitor the operation by measuring the source or drain current of the JFET. As the tuning adjustment is made in the drain circuit, we find that the current rises as the oscillations start, and reach a peak right before the tank is tuned to resonance. When resonance is reached, the oscillations cease abruptly, resulting in a sudden decrease in drain current. We find that each turn of adjustment slug in coil L1 causes a larger change in drain current when the resonance point is approached from the high end, than when it is approached from the low end. The correct point for most stable operation is about one-half to three-fourths the way up the current slope from the low end of the resonance point. Start with the coil slug fully inside the coil form to insure that the tank is tuned below the resonance point. Adjust the slug until oscillations begin. Count the number of turns of the slug until the oscillations cease (resonance). We can now set the slug at a point approximately two-thirds up the slope by noting the number of turns in the range. If, for example, it took three complete turns of the slug to go through the range (note how I made the arithmetic easy!), then step back from the resonance point by one turn.

Another popular crystal oscillator is the Pierce circuit of Fig. 6-4B. This oscillator is also capable of either fundamental or overtone operation. A tank circuit in the output is tuned to approximately the resonant frequency of the crystal. The identifying feature of a Pierce oscillator is the placement of the crystal directly in the output to input path of the active device. In the bipolar transistor circuit

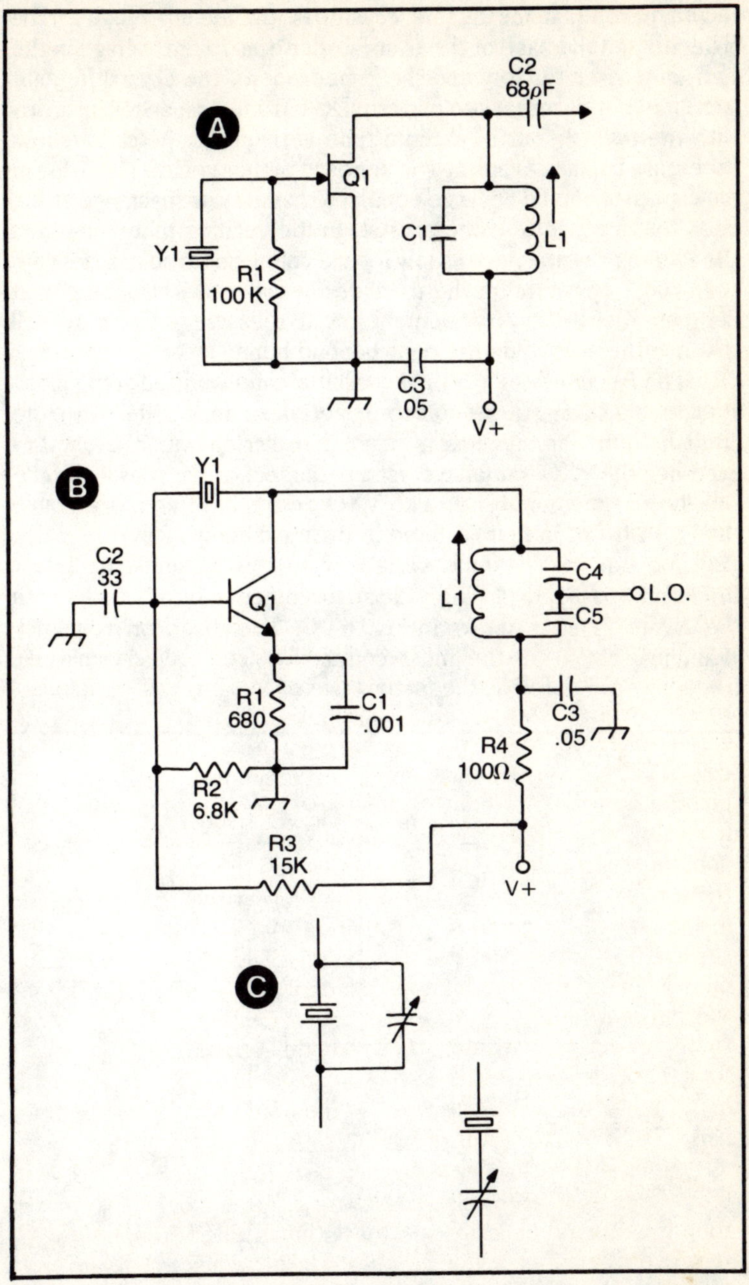

Fig. 6-4. Miller oscillator at A, pierce oscillator at B, and frequency trimming the crystal at C.

shown here, this means the crystal is connected between the collector and the base of the transistor. Because the voltage in the transistor circuit is low and the impedance of the crystal is quite high, it is not necessary to use any DC blocking capacitor in series with the crystal. But in vacuum tube versions of the circuit, it is necessary to place a capacitor in series with the "rock." The value of the capacitor should be large enough to have a low reactance at the operating frequency of the crystal. In the vacuum tube version of the Pierce oscillator, use as low a plate voltage as is consistent with the proper operation of the oscillator in order to keep the power dissipation of the crystal element low. Excessive power may well fracture the crystal, damaging it beyond hope.

The frequency of a crystal oscillator can be varied some small amount by placing a reactance in series or in parallel with the crystal. Some circuits use a inductor in series with the crystal. So-called the VXO (variable crystal oscillator) circuit, this is not the way to maximum stability. The VXO design is often more stable than a VFO but is more limited in tuning range.

We can obtain some variability (a few hundred to a few thousand cycles) by placing a capacitor in series or in parallel with the crystal. This is shown in Fig. 6-4C. The capacitor frequency changing scheme is the most commonly seen. All crystals are frequency calibrated at the factory while "looking into" a standar-

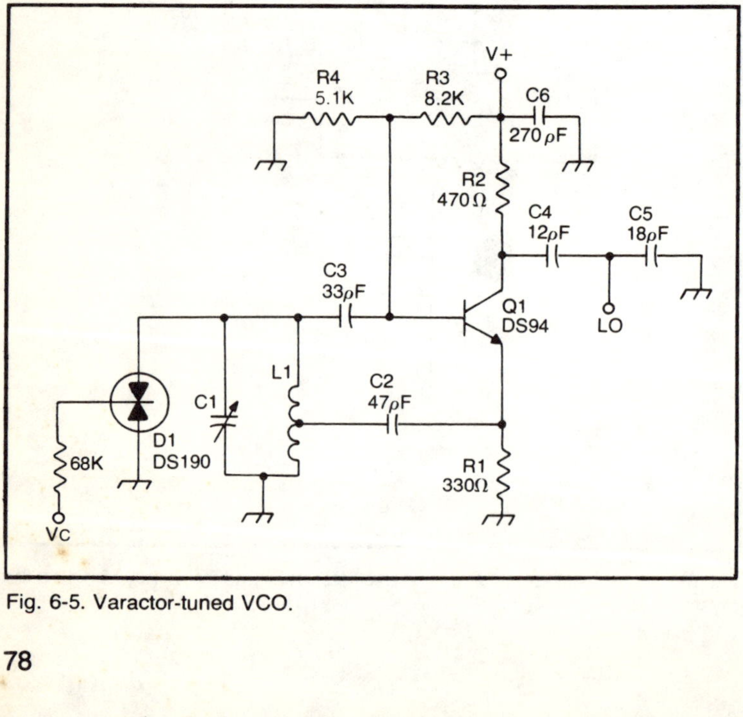

Fig. 6-5. Varactor-tuned VCO.

dized capacitive load, such as 20 pF, 32pF, etc. By changing the capacitance seen by the crystal in the actual circuit, the operating frequency can be "pulled" slightly.

One last oscillator circuit is the voltage tuned type shown in Fig. 6-5. This circuit is a standard LC-tank Hartley. Any of the oscillators, including crystal, can be used. In the case of an LC oscillator, the variable capacitor is used to trim the operating frequency and is not the main tuning capacitor. The main tuning capacitor is a dual varactor diode D1. This particular circuit is the local oscillator from the FM automobile radios made by Delco Electronics for General Motors automobiles. The DS-190 Delco dual varactor produces a capacitance that is proportional to the voltage V_c applied to the control line. The capacitance in *any* diode is controllable to some extent. This capacitance exists across the PN junction when the diode is reverse biased. A reverse biased PN junction creates a charge-free depletion zone extending a little way on both sides of the junction. Since there are charges facing each other on opposite sides of this depletion zone, the little devil thinks it's a capacitor. We change the width of the depletion zone, hence the capacitance, by varying the reverse bias potential V_c.

CONVERTER CIRCUITS

A converter is a single-stage circuit that combines the functions of the mixer and local oscillator. Converters are commonly used in lower cost radio receivers, especially those which over only one or two bands. Almost every AM broadcast receiver on the market uses a converter instead of the more costly oscillator/mixer combination.

Figure 6-6 shows the circuit for a simple converter circuit from an AM band automobile radio. Q1 is an ordinary NPN silicon transistor that functions as both the oscillator and the mixer. The rf signal is applied to the base of the transistor through an impedance matching capacitive voltage divider C1/C2. This particular circuit is used in a model that has an rf amplifier, so the low impedance of the base circuit must be matched to the higher output impedance of the rf stage. We would reverse the order of the impedance matching circuit if this converter were connected directly to an antenna, because the antenna impedance is typically lower than the base impedance of the transistor.

Bias is applied to the base of the transistor through an ordinary resistor voltage divider R1/R2. Resistor R3 forms the emitter load.

The tuned tank circuit in the primary of T1 forms the i-f signal collector load for the transistor. The oscillator inductor is coil L1, which is resonated by capacitors C3, C4, and C5. We can get away

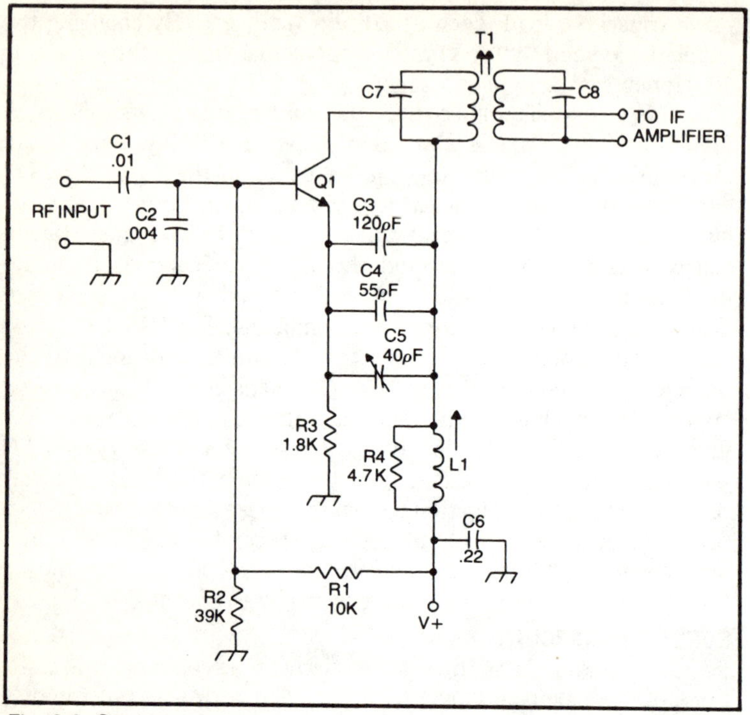

Fig. 6-6. Converter stage.

with series connecting these coils because the inductance of the T1 primary from the junction with L1 to the collector tap is very low compared with the inductance of L1. The transistor will oscillate at a frequency equal to the rf-i-f difference, and is almost universally placed above the rf signal for AM band receivers. If you view the signal on the emitter with an oscilloscope, you will see an ordinary rf carrier at the LO frequency. The base signal will be the rf modulated signal from the antenna/rf amplifier circuits. The collector signal will be a combination of the two and generally takes the form of a lightly modulated carrier at the intermediate frequency, with additional components that represent additional beat note products or the rf/LO signals. The signal transferred to the secondary of the transformer, however, will be only the modulated intermediate frequency signal. From there, it is amplified in the rf amplifier and demodulated to recover the audio.

MIXER CIRCUITS

Most of the higher quality radio receivers use separate local oscillator and mixer stages when making the heterodyne frequency conversion. Figure 6-7 shows a simple transistor mixer stage. This stage is an ordinary transistor amplifier, but the bias is not set to exactly class-A; a little nonlinearity is needed. In fact, it is necessary to insure nonlinearity by making the local oscillator signal several times greater in amplitude than the rf signal. We want the LO signal to switch transistor Q1 into and out of conduction.

The collector of transistor Q1 is tuned to the i-f signal by the primary of interstage coupling transformer T1. The primary of T1 is tapped for the collector in an attempt at matching the output impedance of the transistor.

The rf and LO signals are applied in parallel to the base of the transistor. The mixing action requires nonlinear amplification, and this is provided by the switching action of the LO signal. A series resonant trap in this particular circuit is resonant at 10.7 MHz. The model radio from which this circuit is drawn also includes an FM band, and the i-f for that band is 10.7 Mhz. This trap is needed in this particular circuit because the designer used some of the same components for the FM i-f amplifier. This intermingling is done for economic reasons, and is usually permissible because the frequencies of the two bands are considerably different.

The previous example showed the rf and LO signals applied in parallel to the same element of the transistor, i.e. the base. But this is not always necessary or, in fact, even desirable. Figure 6-8 shows several alternative methods for applying the LO signal to the mixer stage. In Fig. 6-8A, the rf signal is applied to the base of a bipolar

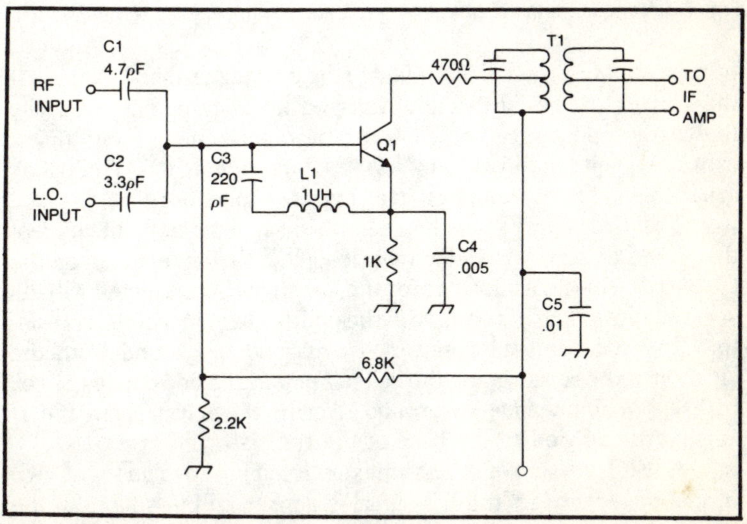

Fig. 6-7. Mixing at the base of a transistor.

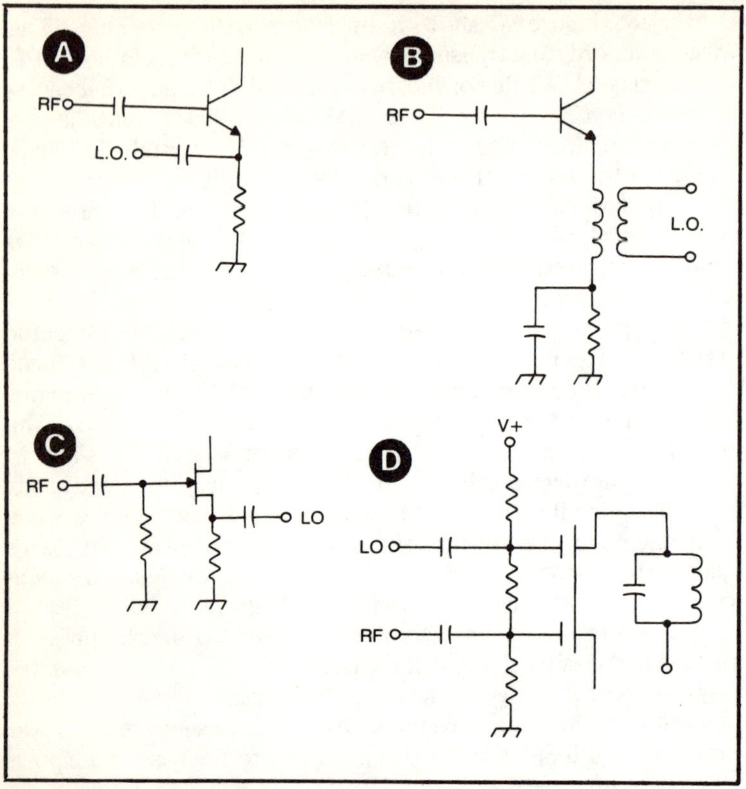

Fig. 6-8. Various methods of applying LO and rf signals to mixer stages.

transistor, while the LO is applied to the emitter terminal. Note that the emitter load resistor is not bypassed in this case. This is usually the case in most receivers, but it is also sometimes found that a capacitor voltage divider is formed between the LO coupling capacitor and a low value emitter bypass capacitor.

The circuit in Fig. 6-8B uses a similar arrangement, except that the LO signal is inductively coupled to the emitter of the transistor. An rf transformer for the LO signal is connected with its secondary winding in series with the emitter terminal of the transistor. The emitter load resistor is bypassed to ground from the junction of the resistor and the coil to keep the "cold" end of the coil with a low impedance to ground. Again, the signal from the rf amplifier is applied to the base of the transistor.

A JFET version of the first mixer circuit is shown in Fig. 6-8C. In this mixer circuit, the LO signal is applied to the source of the JFET, and the rf signal is applied to the gate.

One last transistor mixer is shown in Fig. 6-8D. This version uses a dual gate MOSFET, such as the RCA 40673 device. The drain terminal is tuned by an LC tank circuit to the intermediate frequency. We apply the rf signal to gate G1, and the LO signal to gate G2 of the MOSFET. The resistors form a bias network, which is needed depending upon whether depletion or enhancement MOSFET devices are needed.

IC differential amplifiers have become widely used as mixer stages in radio receivers, and they seem almost ideally suited for some of these applications. Figure 6-9 shows a differential amplifier used as a mixer. A similar circuit was introduced in the previous chapter as the RCA CA3028, and this circuit can be duplicated by

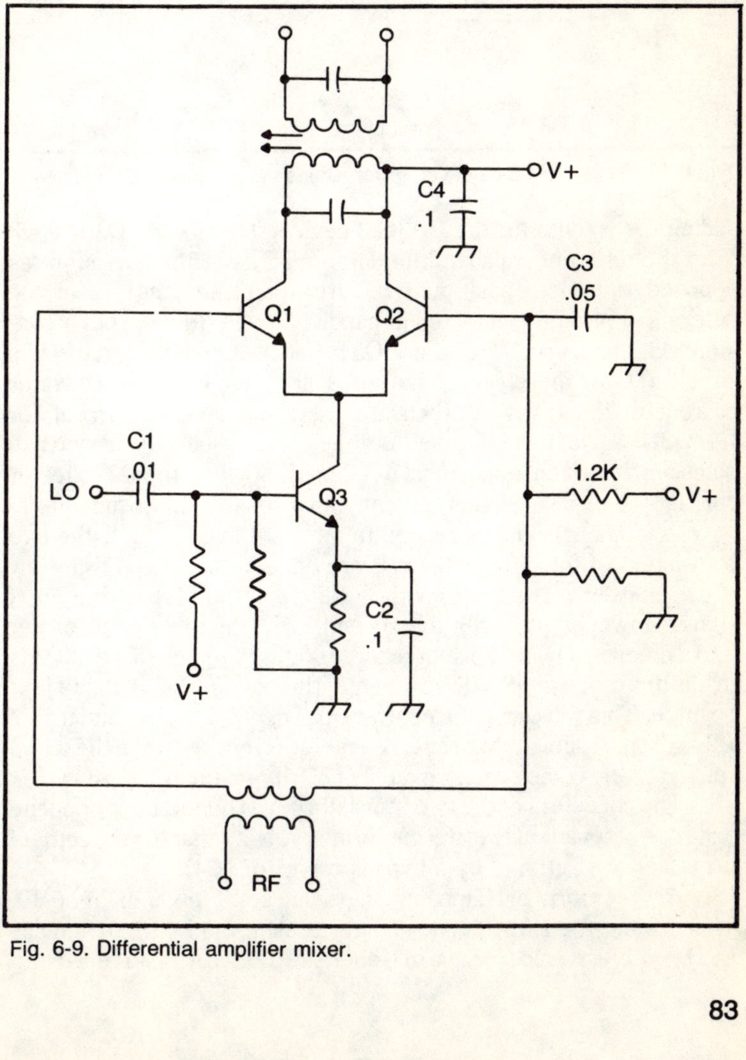

Fig. 6-9. Differential amplifier mixer.

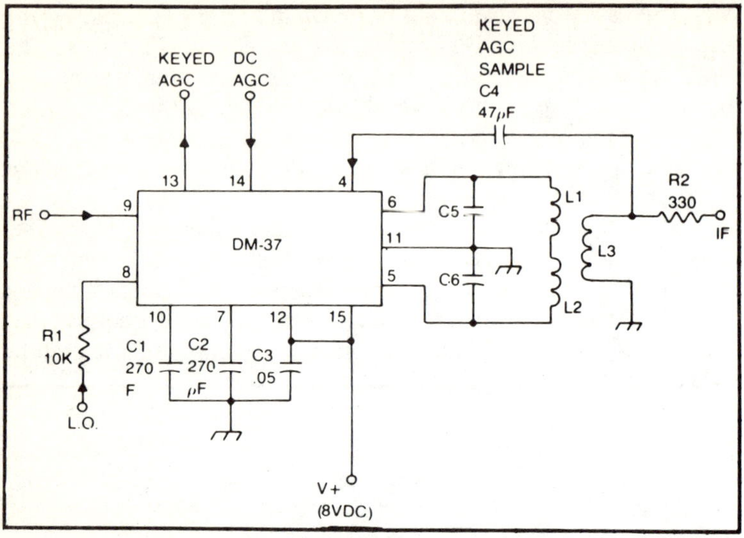

Fig. 6-10. Delco DM-38 balanced mixer IC used in FM broadcast receivers.

adding the pinouts for the CA3028 device. Transistors Q1 and Q2 form the differential pair. True differential amplifier operation (as opposed to simple push-pull) requires that the emitters of the differential pair be connected in parallel and be driven from a constant current source. Transistor Q3 forms the current source in this circuit. If Q3 is merely biased to some quiescent level, the collector current of the transistor (which is also the emitter current for Q1/Q2) is a constant DC level. But the current source is biased to a quiescent level and a sinusoidal LO signal is applied, the Q3 collector current will be a sinusoidal current with a constant amplitude peak.

When no rf signal is applied to the input transformer, the two transistors of the differential pair are equally biased and will draw equal emitter currents from the collector of Q3. Applying an rf signal, however, changes the secondary of the input transformer into a differential voltage source for Q1/Q2. When the base of Q1 is made more positive by the rf signal, the base of Q2 is made less positive. This will cause the emitter current of Q1 to rise, and that of Q2 to drop. A similar, but reverse, operation occurs when the signal applied to the transformer reverses polarity on the next half cycle.

The collectors of the two transistors are connected to opposite ends of a resonant i-f transformer primary. Note that the collector of Q2 is set to rf ground by a bypass capacitor (C4).

Another form of IC mixer is the balanced IC mixer of Fig. 6-10. This device, the DM-37 (and its later cousin, the DM-38), is made by Delco Electronics for use in General Motors car radio receivers.

This IC was first used in the FM section of the Chevette AM/FM receiver for 1976, and continues in use. The IC actually contains two sections: i-f and mixer. Only the mixer circuit connections are shown here. It is a balanced type of mixer, with the two imput signals (LO and rf) applied to pins 8 and 9, respectively. The output tuning is a balanced 10.7-MHz i-f transformer primary. The resistor in series with the secondary of the coil is used to match the impedance of the ceramic crystal band-pass filter that follows this stage (see Chapter 7).

The balanced IC mixer is theoretically superior to the single-ended (unbalanced) types, because the rf and LO signals are completely nulled in the output, leaving only the i-f difference signal.

Two types of automatic gain control (agc) are used in this circuit ADC agc signal is developed in the final i-f amplifier stage and is applied to pin No. 14 of DM-37. The keyed agc is applied eventually to gate No. 2 of a dual-gate MOSFET rf amplifier. This signal is developed from a sample of the secondary i-f difference signal picked off through a small value capacitor.

Passive Balanced Mixers

Diodes can be used to make both single- and double-balanced mixers. The double-balanced mixer (DBM) has become somewhat standard in VHF, UHF, and especially microwave applications. Many of these devices will operate from a fraction of a megahertz up to gigahertz. The material on DBMs that follows was supplied by Mini-Circuits Laboratory, a principle supplier of DBMs and other rf components.

Double-balanced mixers (DBMs) have become a standard component in communications systems, microwave links, spectrum analyzers and ECM equipment. Used properly, DMBs allow systems designers to achieve minimum levels of distortion with a high degree of isolation from interfering signals. However, used improperly, DBMs can degrade systems performance. Similarly, incor-

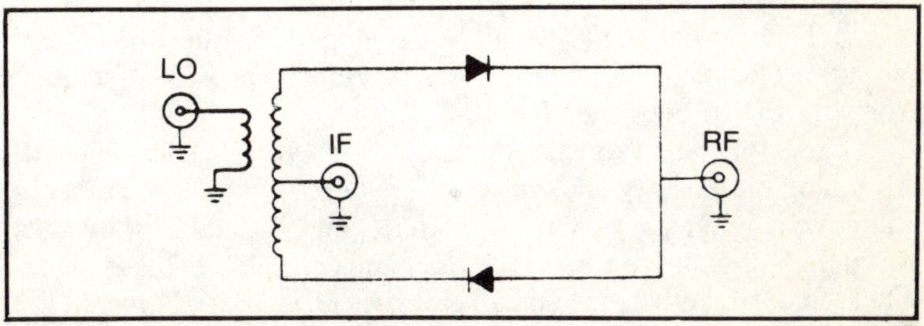

Fig. 6-11. Single balanced mixer.

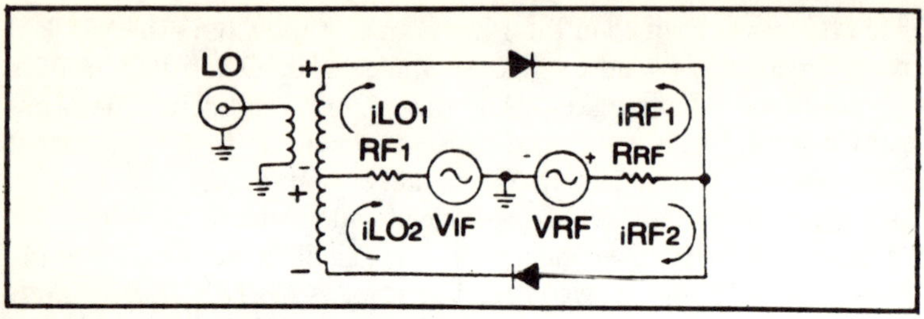

Fig. 6-12. Loop currents in single-balanced mixer.

rect interpretations of mixer specification can be costly. Underspecify and expect marginal performance; overspecify and pay for unnecessary device characteristics. To properly apply and specify a double-balanced mixer it is helpful to understand how it works, what factors influence distortion and how performance is measured.

How a Single-Balanced Mixer Works. To begin, analyze a simplfied version of a typical single-balanced mixer, shown in Fig. 6-11. The characteristic of the single-balanced mixer is that there is isolation between the local oscillator (LO) and the rf, attributable to the inherent circuit balance between the LO and rf. However, there is no balance between the rf and i-f; examination of Fig. 6-12 with its loop currents explains why.

When the LO signal is applied, assume the polarity shown on the LO input-transformer; the direction of LO currents would then appear as shown for i_{LO1} and i_{LO2}. The total current through the IF and rf source resistances consists of the sum of the currents from the LO and rf inputs. First, consider what happens when the LO signal is applied. With the polarity shown at the LO transformer secondary, current flow is clockwise for i_{LO1} and i_{LO2}. These currents flow through rf and i-f source resistances. Looking at the current flow through the rf source resistance R_{RF}, there would be complete cancellation of the LO component (assuming the i_{LO} currents were exactly identical in the amplitude and phase).

Similarly, examining the local oscillator (LO) current flow through the i-f source resistance, R_{IF}, again cancellation would occur. Thus no LO power would be developed at either the i-f or rf outputs. The single-balance mixer, therefore, provides isolation from the LO to both the rf and i-f.

Now, let's analyze what takes place when rf signal is applied. Assuming the polarity shown for the rf source, two mesh currents will flow, i_{RF1} and i_{RF2} as shown. Both i_{RF1} and i_{RF2} add as they pass through the i-f source resistance, R_{IF}, and cancellation takes place. Thus, for a single-balanced mixer, there is no isolation between the rf and i-f.

How about the balance between the rf and LO ports? Since i_{RF1} flows through the upper winding of the LO transformer in the opposite direction as i_{RF2} in the lower winding, no voltage would be developed at the LO terminal. Therefore there is isolation between the rf and LO.

All previous references to conditions of balance are based on currents being equal in amplitude. In actual practice, the following factors tend to upset the ideal conditions. Variations in transformer balance and unequal diode impedance will cause deviations in current balance. At high frequencies above 100 MHz, wiring capacitance, transformer winding capacitance and physical location of components will also upset balance. As operating frequency increases, balance will fall off and a lower isolation specification will appear on the mixer data sheet.

Double-Balanced Mixers Offer Improved Isolation. A typical schematic for a double-balanced mixer is shown in Fig. 6-13. Let's now examine how balance is achieved. If CR_1 and CR_2 and the LO transformer are symmetrical, the voltage at point A is the same as the center-tap of the transformer, or ground. Similarly, if CR_3 is equal to Cr_4, the voltage at B is the same as V_{GND}. Therefore, there is no voltage across A and B and no voltage across the rf or i-f ports. This illustrates how isolation is obtained between LO and rf and i-f ports.

Look at the rf input. If CR_4 is equal to CR_1 and CR_2 equal to CR_3, the voltage at C will be equal to that at D. There will be no voltage difference between C-D and thus no rf will appear at the LO port. From symmetry, it can be seen that the voltage at the i-f port is the same as the voltage at C, D or zero; thus there is no rf output at the i-f port. The simplified sketch for the above is shown in Fig. 6-14.

Again assumptions for balance are based on transformer symmetry and diodes being equal. As with single-balanced mixers,

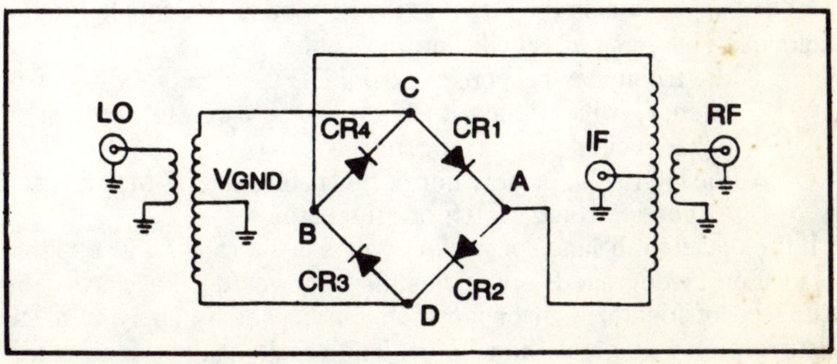

Fig. 6-13. Double-balanced mixer.

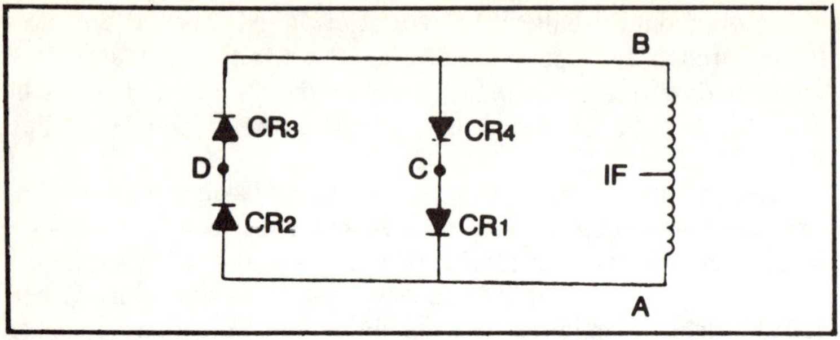

Fig. 6-14. I-f leakage to rf and LO ports are minimized in double-balanced mixers by use of balanced transformers and matched diodes.

unbalance and a subsequent drop in isolation will result from diode junction capacity differences and transformer winding variations. Here's a rule-of-thumb approximation for isolation: as frequency of operation is increased, isolation tends to fall off at the rate of 5 dB per octave. Thus, for example, if you were to measure 40 dB isolation at 300 MHz, you could predict isolation of 45 dB at 150 MHz by using the 5 dB per octabe estimate.

Whether the rf has a positive polarity or negative polarity, the inherent response of the DBM circuit is the same. For example if a sine wave is applied, there would not be a lower conversion loss if the input was positive and a higher conversion loss for a negative-going input; the response is equal and independent of RF signal polarity. Again, it is assumed the diode characteristics are symmetrical for positive and negative input signal polarities.

Basically, the three major considerations of the double-balanced mixer are the inherent isolation between the various ports, as summarized in Table 6-1. The implications of these isolation properties appear in the variety of applications to which the double-balanced mixer can be used.

Another inherent characteristic of a DBM that suggests an area of application is the fact that the i-f point is DC-coupled to the diodes. This has two significant implications:

- ☐the frequency response is from DC to some very high frequency, which lends itself to use as a phase detector in phase-locked loop arrangements
- ☐the high isolation between ports enables the DBM to be used as an electronic switch or attenuator.

If the excellent balance between ports is deliberately upset so that isolation is degraded, some feedthrough would take place. The degree of feedthrough or isolation can be set by applying a DC current or voltage at the i-f port. The resultant unbalance determines the level of attenuation.

Table 6-1. Properties of Double-Balanced Mixers.

PROPERTIES	IMPLICATIONS
• Inherent isolation between LO & RF, IF ports • Inherent isolation between RF & LO, IF ports • Inherent isolation between IF & LO, RF ports	• No filters necessary—thus broadband—can be used for suppressed-carrier modulation
• IF DC coupled	• Can be used as phase detector
• Isolation depends on symmetry of transformers and diodes	• High isolation—can be used as electronic switch/ attentuator by injecting DC current to cause unbalance
• Response to RF input same for either polarity of RF signal amplitude	• High rejection of even-order harmonics

Typically, 10 to 20 mA will change the insertion loss between LO and rf to as low as 3 dB or less. As the current through the i-f port is varied, the balance between LO and rf will be altered; a predetermined amount of current will establish a preset degree of attenuation. Since the i-f port response extends from DC to some very high frequency, very fast switching or change of attenuation characteristics is practical. Present-day switching techniques using PIN diodes as switches are not practical at lower frequencies because the PIN charactersitics are lost below 1 MHz. With a DBM, satisfactory attenuator performance can be attained down to 500 Hz.

The disadvantage of the DBM in switching/attenuator application is the generation of harmonics of the RF input. These might be undesirable in some systems designs.

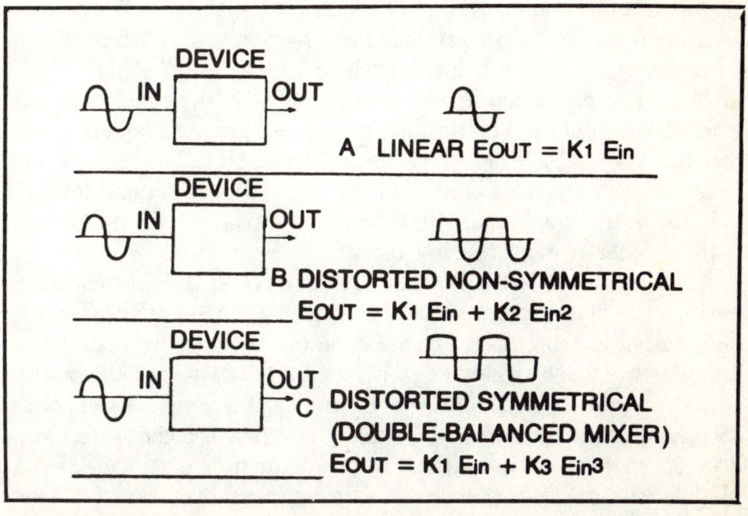

Fig. 6-15. Feed the same sine wave input into three different devices, and three different outputs will result.

Consider three different devices (not just mixers), as shown in Fig. 6-15 with a sine-wave input applied to each. In case A, the output of the linear device will be identical in waveform, although not necessarily in amplitude. In case B, unsymmetrical distortion due to unequal voltage sensitivity would develop an output that is flat-topped on one-half cycle. And finally, nonvoltage-polarity sensitive device could produce symmetrical flattening as shown in case C. Mathematically, the output of case A is a constant K_1 times the input. In case B, as a first approximation, the output is a constant K_1 times the input plus another constant K_2 times the input squared. In case C, where distortion is symmetrical, the output is in the form of a constant K_1 times the input plus another constant K_3 times the cubic of the input. Case C applies for the double-balanced mixed where all circuit elements are balanced and distortion is uniform for both voltage polarities of the signal. The important point here is that the output of a DBM can be described, on first approximation, as a cubic. In practice, of course, no device is perfect and unbalances will occur. This means that, along with the cubic term there will also appear second, fourth, etc. terms with the third-order the predominant term. In the double-balanced mixer, a third-order term will always exist because of some limiting factors that determine the saturation of the mixer, namely the diodes and transformer.

Single-tone, or harmonic intermodulation, distortion appears when only one input signal at the rf port combines with the LO signal. The interaction of the LO and its harmonics with the rf input signal and its harmonics produce higher-order distortion products as shown in Table 6-2 for models SRA-1, SRA-2, ZAD-1 and ZAD-2. The distortion levels are indicated by the number of dB below the output level of the rf input frequency ± LO frequency; LO frequency = 52 MHz and rf frequency = 50 MHz. The upper chart (a) applies where the LO level is +7 dBm and the rf input level is −10 dBm; the lower chart (b) refers to a similar +7 dBm LO level with a higher rf input level of 0 dBm.

Two-Tone, Third-Order Distortion Effects. Let's analyze the distortion product of a double-balanced mixer (DBM). The output is proportional to the cube of the input voltage. If two input signals were applied equal in amplitude ($E_A = E_B$), the output would be proportional to $(E_A)^3$. This means if the input E_A changes 10 dB, the output changes by 30 dB. This explains a confusion often arising when two-tone, third-order distortion is discussed. A guideline for calculating such distortion is that there is a two-to-one improvement in third-order response as the inputs are lowered. This merely means that for each 10 dB change in input there is a 30

Table 6-2. Typical Harmonic Intermodulation Signals on Models SRA-1, SRA-2, ZAD-1, and ZAD-2.

LO frequency = 52 MHz
Signal frequency = 50 MHz

Harmonics of the input signal	0	1	2	3	4	5	6	7	8	
7	>70	>70	>70	>70	>70	>70	>70	>70	>70	
6	>70	>70	>70	>70	>70	>70	>70	>70	>70	**(a)**
5	>70	>70	>70	>70	>70	>70	>70	>70	>70	**−10**
4	>70	62	>70	>70	>70	>70	>70	>70	>70	**dBm**
3	68	58	>70	58	>70	55	>70	52	68	**Signal level**
2	66	55	>70	66	68	67	68	66	>70	**+7 dBm**
1	30	0	40	15	46	27	52	37	52	**LO level**
0		25	42	35	46	42	49	52	54	
7	>70	>70	>70	68	>70	68	>70	60	>70	
6	>70	>70	>70	>70	>70	>70	>70	>70	>70	**(b)**
5	>70	62	>70	52	>70	48	>70	48	69	
4	>70	64	>70	>70	>70	68	>70	68	>70	**0 dBm**
3	53	42	55	42	58	52	66	47	68	**Signal level**
2	64	59	67	66	66	66	67	61	68	
1	31	0	43	15	50	27	68	48	62	**+7 dBm**
0		34	52	46	64	59	68	64	>70	**LO level**

Harmonics of the LO

dB change in third-order output for a net improvement of 20 dB or 2 to 1.

More important, how can various mixers be compared on the basis of two-tone, third-order mixer distortion? Is a −100 dBm two-tone third-order response for a −30 dBm input better than a spec of −70 dBm for an input of −20 dBm? The answer is no. Mixer distortion performance is exactly the same. So here's some advice for the design engineer selecting a mixer. First, know the input level the mixer is expected to handle in order for the intermod spec to be meaningful. As just shown, a −100 dBm intermod spec may not satisfy a −80 dBm requirement at the desired input level. The two-tone, third-order distortion is related to the symmetrical distortion characteristics of the mixer and to improve the two-tone performance of a mixer, the mixer must be able to handle a larger input signal for the same amount of distortion.

Let's take an example for Mini-Circuit's MCL SRA-IH. If the input of each tone is at 0 dBm, typically the third-order output is −60 dBm. This means the two-tone, third-order response is 54 dB below the desired i-f output (assuming a 6 dB conversion loss). Now if the level of each tone drops to −10 dBm, the two-tone, third-order product would be at −90 dBm since a change of 10 dB at the input produces a 30 dB change in the output. Now assume the

SRA-IH is a super-mixer and the two-tone input is raised to +10 dBm. The third-order output now changes from −60 dBm to −30 dB, or 30 dB worse. Extend the assumption and apply a +20 dBm two-tone input; the third-order output would again change 30 dB to +0 dBm. Now let's further extend the assumption and apply +30 dBm two tone input (of course, the SRA-IH couldn't handle this level) and now the two-tone third-order product increases to +30 dBm. Now the two-tone, third-order response is exactly equal to the two-tone input. This point is termed the intercept point where the third-order level is equal to the two-tone input level. See Fig. 6-16. In practice, mixers are not operated at this level, but the intercept point offers a figure of merit for comparison of devices. In addition, it allows comparison of mixers where specs for intermod are given at different two-tone levels. Once the intercept point is known, you can calculate the two-tone , third-order response at any input level, remembering that every 1 dB change at the two-tone input produces at 3 dB change at the third-order output.

Consequently, it is possible to predict the rf input level allowable to keep two-tone, third-order response to a given level in a systems design. To use a mixer properly, it is necessary to relate the two-tone input and third-order output levels involved to avoid generating excess distortion and compromising the final design.

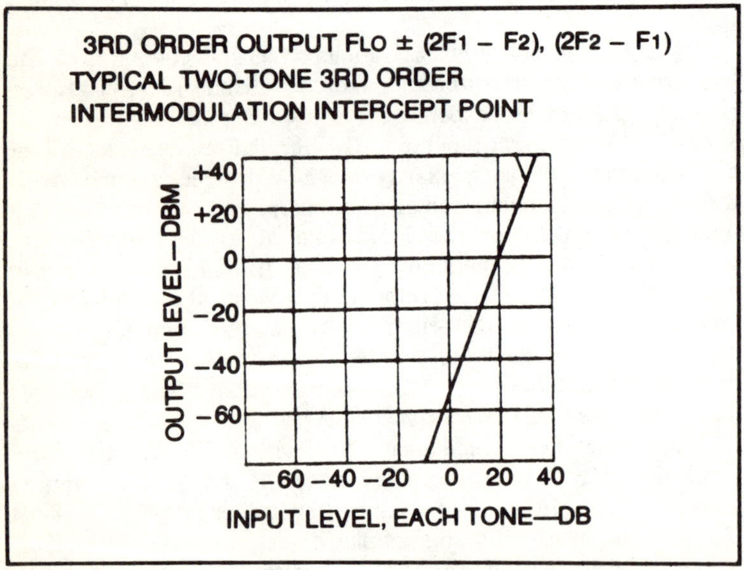

Fig. 6-16. Although the intercept point is fictitious since mixers are not operated at this level, it is a convenient figure of merit for double-balanced mixer evaluation.

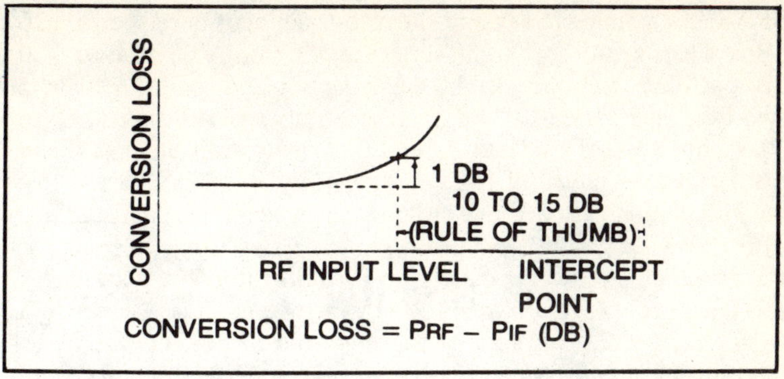

Fig. 6-17. The 1 dB compression point is an indication of the dynamic range and the maximum power capabilities.

Not as obvious, but important, is the effect of increasing operating frequency on the mixer two-tone, third-order distortion characteristics. Generally, performance is better at low frequencies and drops off as frequency is increased. For high-frequency mixers (500 MHz), fall-off starts to occur somewhere between 50 to 100 MHz.

Often a mixer data sheet does not specify intercept point, so a rule-of-thumb estimate can easily be made by examining the 1 dB compression point. As rf input level is increased, there is a point where the conversion loss will increase. A convenient point of reference is a 1 dB compression point. As rf input is increased, i-f output should follow in a linear manner. However, after a certain point, the i-f output increases at a lower rate until the mixer output becomes fairly constant. When the i-f output cannot follow the rf input linearly, and deviates by 1 dB, this point is called *1dB compression point*. Now the conversion loss is 1 dB higher than it was when the rf input signal was smaller. The importance of this figure is its utility in comparing dynamic range, maximum output and two-tone performance of various mixers.

As a rule-of-thumb, the intercept point is approximately 10 to 15 dB higher than the 1 dB compression point; at low frequencies about 15 dB, and at higher frequencies about 10 dB. See Fig. 6-17.

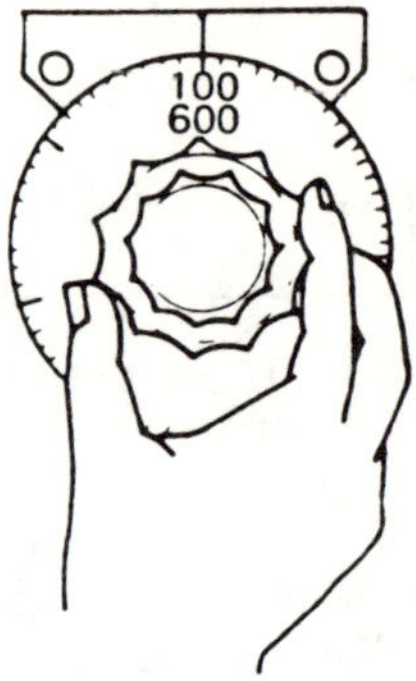

Chapter 7

I-f Amplifier and Selectivity Circuits

Much of the quality of a radio receiver is determined by the i-f amplifier section. This circuit, which might be a single stage or many, supplies the bulk of the stage gain of the receiver, so effectivly sets the sensitivity of the receiver. It also contains the sharpest of the band-pass filters and so will therefore determine the selectivity of the receiver. The frequency selected for the i-f determines such matters as the image response of the radio.

REVIEW OF THE SUPERHETERODYNE PROCESS

The superheterodyne system was invented many decades ago in an age when the best radio receivers were crude and simple compared with many radios today. It solved some of the more vexing problems of radio receiver design and operation, not the least of which was the instability and loss of sensitivity experienced in the higher frequency bands. It seems that the performance of circuits, then and now, fell rapidly as frequency was increased. In the superheterodyne process, all rf signals being received are converted to a single intermediate frequency, or i-f. The main portion of the signal processing prior to demodulator occurred in the i-f amplifier. Most i-f amplifiers built over the years have been lower in frequency than the rf stages because low-frequency amplifiers were easier to tame than high-frequency stages. The use of a single-frequency tuned i-f amplifier, and a low intermediate frequency, solved the aforementioned problems in the superheterodyne system. Radio receivers built along the superhet lines were considerably better than the trf and regenerative models of previous times.

The key to the superhet process is the mixer/local oscillator stages. The mixer will combine in a nonlinear manner the rf signal from the antenna (or rf amplifier) and the signal from a local oscillator. The output of the mixer contains at least four different frequency components: rf, LO, rf+LO and rf−LO. In most receivers, especially those designed some years ago, the desired intermediate frequency is the difference between rf and LO (rf−LO). This frequency will usually be quite low and constant. An LC-tuned tank circuit or some sort of band-pass filter (ceramic or mechanical) at the output of the mixer selects the difference frequency and passes it to the i-f amplifier.

IF AMPLIFIER CIRCUITS

The i-f amplifier provides the main gain for the radio receiver. It must also provide a certain amount of selectivity—in fact, the main selectivity for the receiver. As a result, typical i-f amplifiers will be very similar in circuit appearance to rf amplifiers, but are tuned to the same frequency all of the time. It is standard practice to house the LC tank circuits for the i-f amplifier in shielded metal "cans," a tribute to the gain provided by the amplifiers. A little stray coupling can start the i-f to oscillate in very short order. In fact, one of the principal service problems in high-gain radio receivers is the oscillating i-f, occasioned by some additional (undesired) coupling or loss of some desired decoupling somewhere in the circuit.

Intermediate frequencies in common use include 262.5 kHz for AM car radios and 455 kHz for table model AM radios. The 455-kHz frequency was also used as the second or third intermediate frequency in multiple-conversion superheterodyne receivers. Some models are designed to have a high intermediate frequency in order to improve the image rejection of the receiver. But, for the same reasons why the superhet was invented in the first place we want to place most of the gain and selectivity at the lower frequency. Many two-way communications receivers for VHF and UHF operation use an i-f of 10.7 MHz, as do almost all FM broadcast band receivers. It is probable that the communications receiver designers select 10.7 MHz because of the large number of components for this frequency available in the consumer receiver market. Amateur and many commercial/military receivers operating in the high-frequency bands use an immediate frequency of 9 MHz. This is because a single 5.0 to 5.5 MHz VFO can be used to tune both the amateur 80- and 20-meter bands when the difference frequency (i-f) is 9 MHz.

Figure 7-1 shows a typical i-f amplifier using a bipolar transistor—in this case an NPN device. The DC paths for this transistor consist of the bias and load resistors, plus portions of the tuned tank circuit. For the base circuit, the DC path consists of the base-emitter junction of Q1, resistor R4 (the emitter load is also part of the collector DC circuit) , a portion of the secondary of tuned transformer T1, and the fixed bias network R1/R2. The "cold" end of the transformer secondary is kept at a low impedance to ground for the i-f signal because of bypass capacitor C1. Similarly, C2 keeps the emitter of the transistor at a low intermediate frequency-impedance to ground, while maintaining the DC voltage on the emitter.

The DC path for the collector circuit consists of the collector-emitter path inside the transistor, emitter load resistor R4, a portion of the primary winding in tuned i-f transformer T2, collector decoupling resistor R3, and the V+ terminal of the power supply. AC decoupling for the i-f signal is supplied by an RC pi-network consisting of resistor R3 and capacitors C4 and C5. The latter, C5, will usually be in parallel with a high-value electrolytic capacitor in the V+ power supply, which is used to remove the ripple component of the rectified power supply output. This capacitor is usually not effective at the intermediate frequency, and besides it is too far away across the chassis, so a small capacitance that presents a low impedance at the intermediate frequency must be used closer to the i-f amplifier stage.

The stage is tuned by the transformers T1 and T2. Both of these transformers have tuned primaries and secondaries. Not all i-f transformers, however, meet this particular requirement. For a variety of reasons, several different types of tank circuit are used. These will be discussed soon.

Two other capacitors in the i-f amplifier of Fig. 7-1 might be present in any given receiver. Do you recall the matter of neutralization? Transistors are basically triode devices, so their internal junction capacitances can form a capacitive feedback voltage divider network that will cause oscillation when the base and collector are tuned to the same frequency. In some i-f amplifiers, a small fixed neutralization capacitance C_n is used to cancel the effect of the internal capacitances. Neutralization, which is discussed more at length in TAB book No. 1224, *The Complete Handbook of Radio Transmitters*, requires that we feed back a signal equal in magnitude, but opposite in phase, to the signal inside the transistor. The two feedback signals will then cancel each other, and the amplifier becomes stable.

The second optional capacitor is the agc take-off capacitor (C3). This capacitor will sample the i-f signal and deliver it to the automatic gain control rectifier. The signal delivered through C3 will

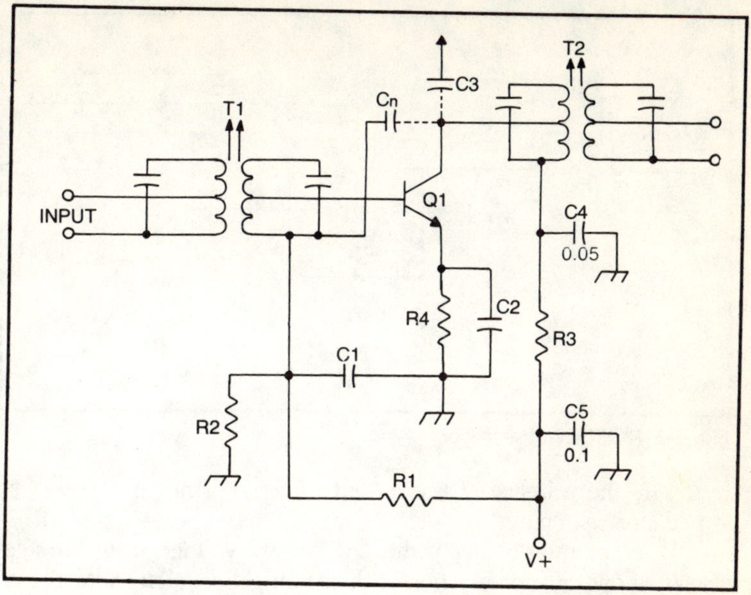

Fig. 7-1. Bipolar i-f amplifier.

be proportional to the signal in the i-f amplifier and represents the strength of the received signal. Agc systems are discussed in Chapter 11.

Different receivers have different gain requirements of the i-f amplifier. A simple table model AM radio receiver, for example, will be operated in a region where the signal strength is quite good, so the i-f amplifier will contain only one stage of moderate gain. An FM broadcast receiver or a two-way communications receiver, on the other hand, needs much more gain. In the case of FM receivers, it must be possible to quiet the receiver by driving either the detector, or a limiter stage into saturation. This will clip off the noise impulses, which tend to amplitude modulate the FM carrier. In those types of radios, there might be as many as five i-f amplifier stages. Most will contain at least three stages if bipolar transistors are used.

A MOSFET i-f amplifier stage is shown in Fig. 7-2. The MOSFET is capable of providing superior performance characteristics over the bipolar and are therefore becoming more popular in receiver design. The circuit shown is almost classical. The tuned input transformer applies signal to the gate (G1) of the MOSFET through a capacitor voltage divider made up of the resonating capacitance for the transformer secondary. The output tank is connected to the drain of the MOSFET. Like the bipolar circuit of

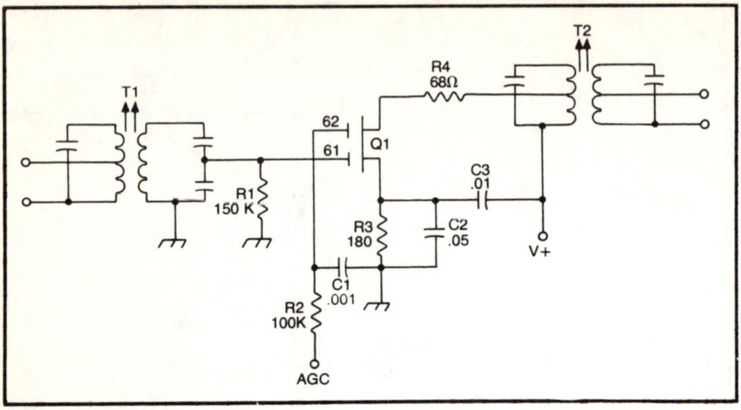

Fig. 7-2. MOSFET i-f amplifier.

Fig. 7-1, the primary of the output i-f transformer is tapped for impedance matching.

It is common to apply the DC control voltage from the agc circuit to one or more stages of the i-f amplifier, along with the rf amplifier (and appropriate decoupling). This will control the gain of the i-f amplifier inversely proportional to signal strength. The gain will increase for weak signals and will decrease for strong signals. This tactic results in a more steady volume as a listener tunes across a band.

Some receivers, notably two-way radio types, use a series resonant crystal element to shunt the emitter load resistor. The crystal will have a series resonant frequency equal to the intermediate frequency. Consequently, it will have a very low impedance to the intermediate frequency. The use of this filter (shown in Fig. 7-3) is usually in conjunction with the use of i-f band-pass ceramic filters of similar construction, in lieu of the band-pass tank transformers shown in Figs. 7-1 and 7-2.

Several different i-f transformer schemes are shown in Fig. 7-4. The circuit in Fig. 7-4A is very common in imported radios. It uses a single-tuned primary, and an untuned, low-impedance secondary. The secondary consists of only a small number of turns relative to the primary winding, and is needed to match the higher impedance of the previous stage to the low impedance of the i-f amplifier transistor. It is also useful in keeping costs down, so is especially common in low-cost radios.

The quadratuned circuit of Fig. 7-4B is used in FM broadcast receivers. The deviation of an FM broadcast station at 100 percent modulation is ± 75 kHz, so the overall bandwidth must be 150 to 200 kHz. A single-tuned or double-tuned transformer is not usually capable of providing the needed bandwidth without also producing an uneven response across the band. The bandwidth might be provided by *stagger tuning*; i.e. tuning the output transformer to a frequency high in the passband and tuning the input transformer to a frequency low in the passband. The circuit of Fig. 7-4B, however, provides the necessary bandwidth, has a nearly flat response, and produces a sharp skirt factor—steep slopes of the response outside of the passband.

The last circuit is shown in Fig. 7-4C. This circuit uses a link-coupled input, or alternatively, a tap low on the input winding. The coupling between the input and output tank circuits is through a mutual reactance. In this case, the coupling is through a small value capacitance. The signal in the input tank develops a signal across this capacitor. Because the capacitor is also in the secondary circuit, coupling is provided.

IC I-F Amplifiers

The integrated circuit is capable of providing an extremely large amount of gain in a very small package. Indeed, some operational amplifier ICs provide an open-loop gain in the over-1 million range. This fact has not been lost on the designers of i-f amplifiers, especially FM and communications receivers which require substantial amounts of i-f amplification. Although not reflected in the circuits shown here, there is a growing tendency to make a single i-c perform the functions of i-f amplifier and FM demodulation. Delco Electronics, for example, uses a two-IC i-f amplifier/detector for its FM radios. One IC, the DM-37, provides a balanced mixer stage and some of the i-f amplification, while another IC device provides the remainder of the i-f gain and the quadrature detector needed to demodulate the FM signal.

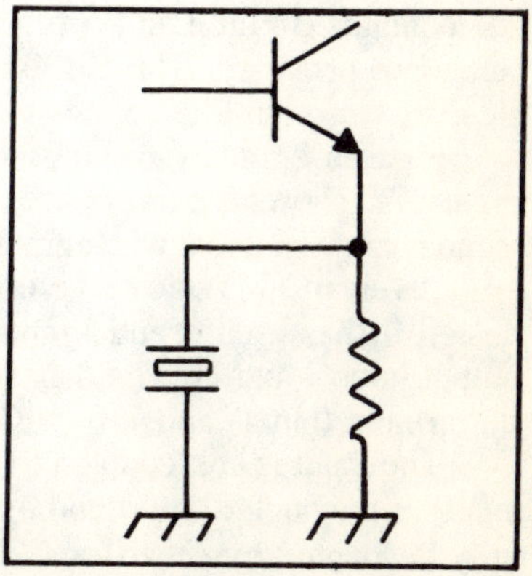

Fig. 7-3. Crystal resonator bypasses emitter resistor.

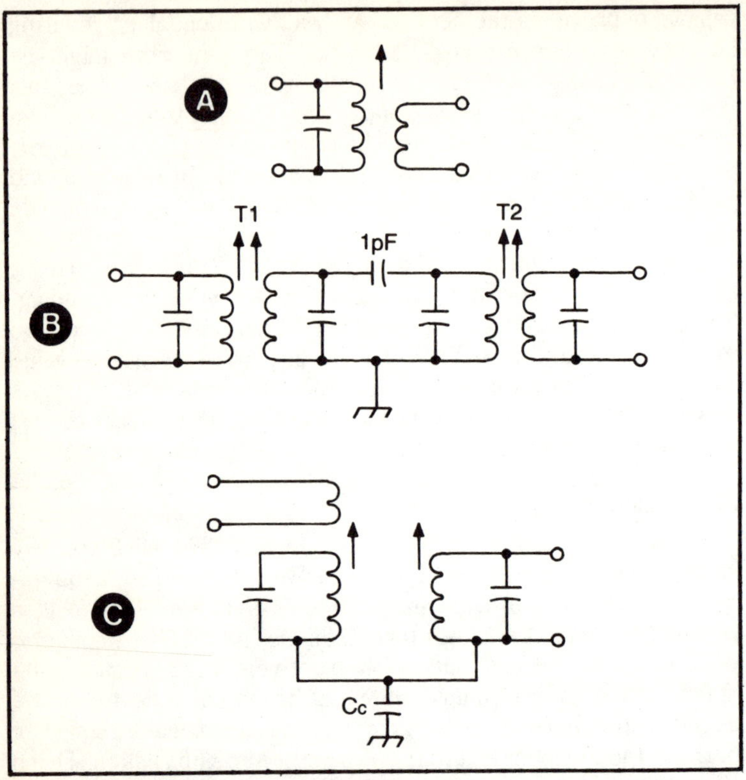

Fig. 7-4. Different i-f tuned coupling networks.

RCA and Motorola produce a series of FM i-f and detector ICs. In some cases, a quadrature detector is used inside the IC, while in other cases, the IC contains a wide-band, high-gain amplifier, and either two or four diodes that could be used in a regular FM demodulator circuit if an appropriate transformer and external circuitry are provided. The CA3014 and CA3012 devices are examples.

Figures 7-5 and 7-6 show two forms of IC i-f amplifier circuits. Figure 7-5 shows the use of the Fairchild μA703 device, one of the earliest high-frequency integrated circuits. This circuit was used extensively in FM radio receivers. The IC device is of the simplest design. It has a differential input (pins 3 and 5) and a differential output (pins 1 and 7). The only remaining pins, quite naturally, are the ground (pin 4) and V+ (pin 8).

The input is link-coupled to the i-f tank circuit. One of the two inputs is decoupled to ground by capacitor C1. The output is similarly decoupled by capacitor C2.

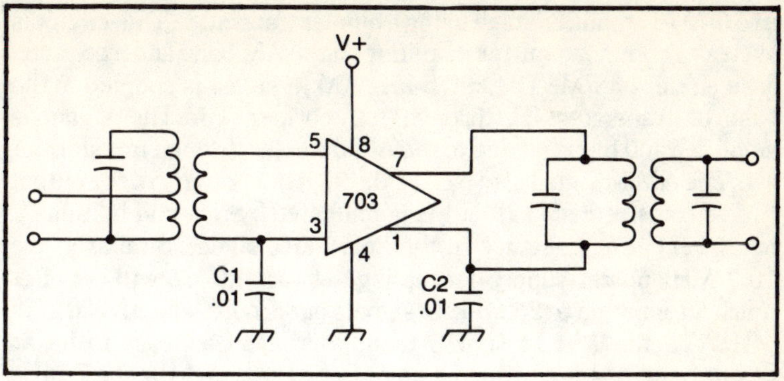

Fig. 7-5. μA703 IC i-f amplifier.

The use of the RCA CA3028 differential amplifier is shown in Fig. 7-6. Although this IC is a differential amplifier, it is single-ended in this case. The i-f input tank circuit is capacitor-coupled to the base of the constant current source transistor, which operates as a grounded emitter amplifier. The differential pair is modified by tying off Q1 (pins 1 and 8) and using Q2 as a common base amplifier. An ordinary i-f transformer tunes the i-f amplifier output.

AM/FM I-F Amplifiers

To save money, some receiver designers will use the same transistor or integrated circuit for both AM and FM bands. This is possible because of the wide difference in frequency typically used use 10.7 MHz, while the AM radio will use either 262.5 kHz (automobile) or 455 kHz (home). Figure 7-7 shows a combination

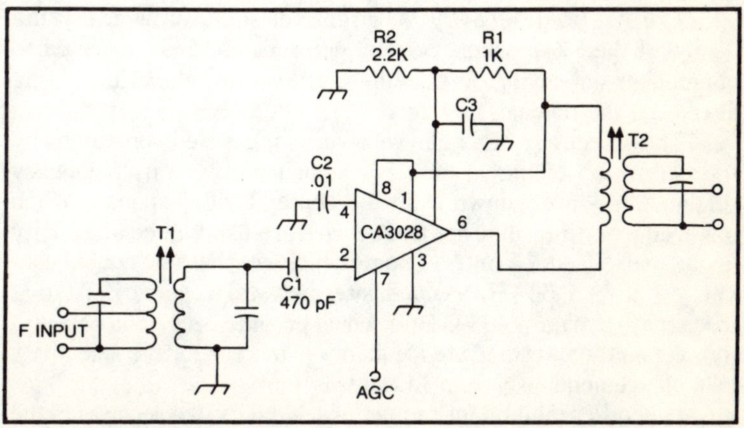

Fig. 7-6. CA3028 IC i-f amplifier.

AM/FM i-f amplifier stage using a bipolar transistor. This circuit is used as a common emitter amplifier on the AM band and a common base circuit on FM. The 262.5-kHz AM i-f signal is coupled to the base of transistor Q1 through transformer T3. The şignal is amplified and then coupled to the AM detector through transformer T4. The AM i-f signal also passes through the primary of the output FM i-f transformer (T2), but is not affected by that coil because it has a very low reactance to the 262.5-kHz signal. Similarly, the 10.7-MHz power-supply decoupling network R2/C5 will not offer much attenuation to 262.5-kHz signals but is quite effective at 10.7 MHz. To the AM i-f signal, then, this stage appears to be an ordinary common-emitter i-f amplifier stage, and the 10.7-MHz components are almost "transparent" to the AM i-f signal.

To the FM i-f signal, however, Q1 appears to be operating in the common-base mode. The 10.7-MHz FM i-f signal is coupled through the input transformer T1 to the emitter of transistor Q1. The base of Q1 is bypassed to ground for 10.7-MHz signals through 120-pF capacitor C2, but this capacitor has a high reactance at 262.5 kHz, especially compared with the impedance of the output tap on the AM i-f transformer (T3) and the input impedance of the transistor. The FM i-f output signal is coupled to the following FM i-f or the detector through transformer T2. FM power supply decoupling is offered by R2 and C5. In some circuits, notably Delco, the FM decoupling is accomplished by connecting the lower end of the FM i-f transformer to the midpoint of a capacitor voltage divider that tunes the primary of transformer T4.

SELECTIVITY METHODS

The overall selectivity of a receiver is determined by the design of the i-f amplifier. Several different methods are used to obtain high selectivity. Among these are the LC filter, the crystal filter, and the mechanical filter.

The selectivity of a radio receiver is measured sometimes by the bandwidth of the i-f amplifier at points where its frequency response is −6 dB down from the center band response. When low-frequency intermediate frequencies are used, good selectivity is obtainable from LC filters. Some receivers that use double conversion have a 50-kHz second intermediate frequency. At that frequency, a multi-pole LC filter would provide very good selectivity. But as the intermediate frequency is increased, the selectivity falls off tremendously with LC filters.

One other method for rating the selectivity of a receiver is the skirt factor of the band-pass filter. It is quite well and good to have a

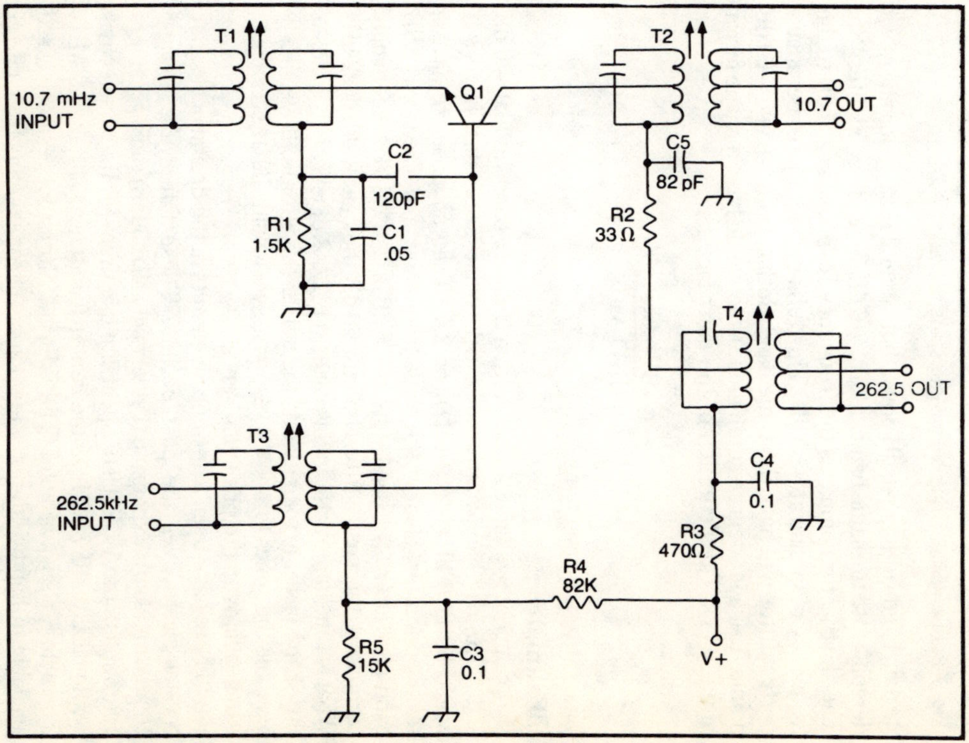

Fig. 7-7. Dual-band i-f amplifier accommodates AM and FM i-f signals.

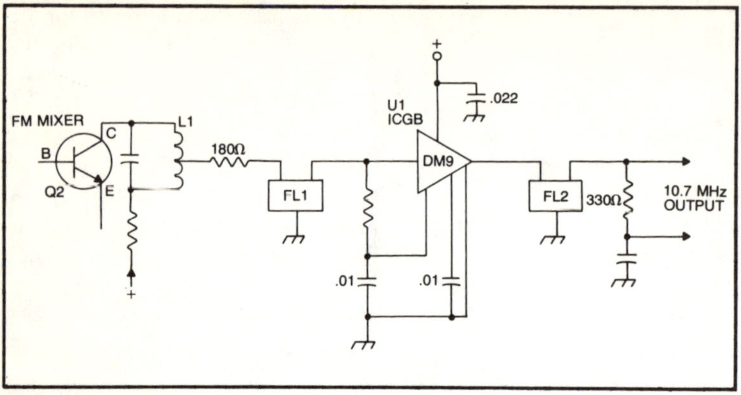

Fig. 7-8. I-f amplifier using crystal interstage filters.

certain bandwidth at −6 dB down, but several really poor filters will demonstrate the same bandwidth at −6 dB as really good filters do. The skirt factor compares the frequency response at −6 dB to the frequency response at −60 dB. This tells us a lot more, because a perfect filter will have the same bandwidth at both −6 and −60 dB points (i.e. a 90-degree cutoff slope) and would have a shape factor of unity (1). The definition of the skirt factor is:

$$\text{Skirt} = \frac{(BW)_{-60dB}}{(BW)_{-6dB}}$$

where BW is bandwidth.

☐**Example**:

Suppose we have a filter with a 2.8-kHz bandwidth at −6 dB and a 7.6-kHz bandwidth at −60 dB. What is the skirt factor? It would be (7.6 kHz)/(2.8 kHz), or 2.7. Then suppose we have a really poor filter, also for the same frequency, and billed as an SSB band-pass filter. The bandwidth at −6 dB is also 2.8 kHz, seemingly mimicking the costly filter. But at −60 dB, the bandwidth is 22 kHz. What is the shape factor in this case? It is 22/2.8, or 7.9. The closer the shape factor is to unity (1), the better the filter is. If we simply examine the specs for −6 dB response, we would still be scratching our heads over the price difference!

There have been many improvements in the design of ceramic crystal filters over the past decade. These filters are now so common and so inexpensive that virtually all modern radio receivers operating in the VHF/UHF spectrum use them for the i-f selectivity. Typically, 10.7 MHz is the center frequency, although 9-MHz filters are also available. The Murata filters for example, are available in bandwidths of a few kilohertz for communications receivers to a low-cost model that has a 200-kHz bandwidth for FM broadcast receivers. Figure 7-8 shows a typical FM i-f amplifier using these filters, while Fig. 7-9 shows the internal construction of the device.

The input to filter FL1 is connected through an impedance-matching tap on the mixer output transformer. This is the only tuned circuit in this schematic, incidentally; the filters are all fix-tuned. The filter is a simple three terminal device. There are input, output, and common terminals only. The output of the filter is connected to the input of an i-f amplifier IC gain block. This is the Delco DM-9 integrated circuit, which is used to provide a large amount of gain, and is essentially a wideband amplifier until the filter is connected. The 330-ohm resistor across the differential inputs of the IC is used to match the impedance of the filter.

The output of the DM-9 IC gain block drives the input to filter FL2 directly. A second 330-ohm resistor is used to match the output of this filter, which will be part of the IC FM demodulator circuit that follows.

The internal construction of the ceramic filter is shown in Fig. 7-9. There are two piezoelectric ceramic elements, one for each end of the band (not quite the end, because the frequency responses of two elements overlap). The two crystals are joined together by a mechanical bridge called a "horseshoe," which vibrates at the intermediate frequency to mechanically couple the two opposing crystal faces.

Examples of other forms of crystal filters are shown in Figs. 7-10 through 7-13. The simplest, and perhaps the oldest, is shown in Fig. 7-10. This circuit uses a single crystal and is found in receivers made primarily in the 50s. The capacitor adjusts the passband slightly, and will be marked on the front panel of the receiver as *crystal phasing*. Output resistor R1 is used to match the

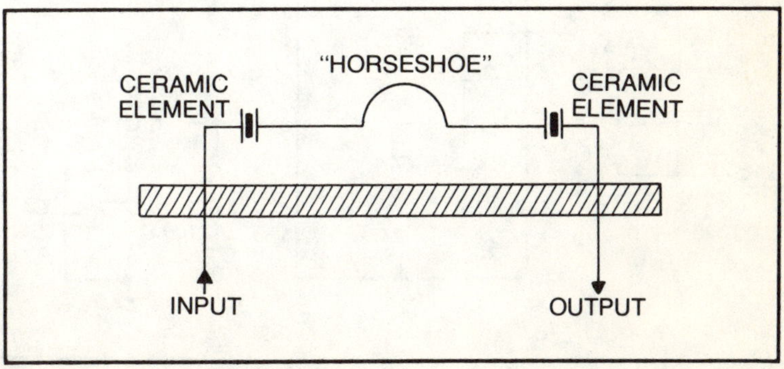

Fig. 7-9. Crystal filter construction.

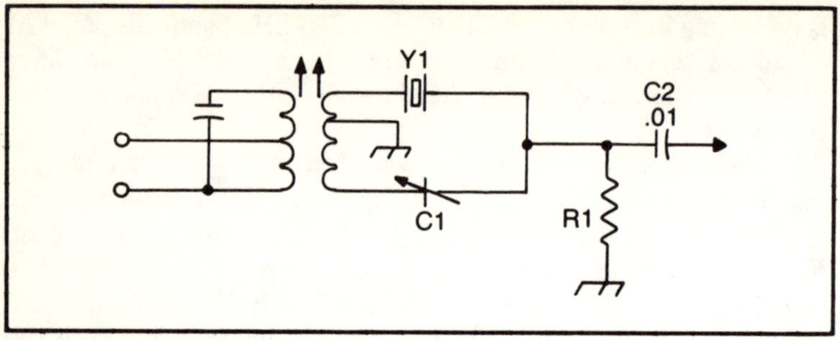

Fig. 7-10. Simple crystal phasing selectivity circuit.

impedance of the crystal at its series resonant frequency, typically a low value.

Somewhat better is the crystal phasing circuit of Fig. 7-11. This version is essentially the same, except that it uses two different crystals of slightly different frequency. While the phasing method Fig. 7-10 created a sharp spike in the resonance, this circuit is capable of a definite bandwidth. Capacitor C3 is the phasing control, and might or might not be available on the front panel of the receiver.

Figures 7-12 and 7-13 show the two main forms of crystal band-pass filters used in higher quality radio receivers. In the circuit of Fig. 7-12, two crystals have matched series resonant frequencies in each arm. Y1 and Y2 are on the same frequency close to one end of the passband, while Y3 and Y4 are on another frequency near the opposite end of the passband. Both pairs of crystals are operated in their series resonant modes. The center of the passband is adjusta-

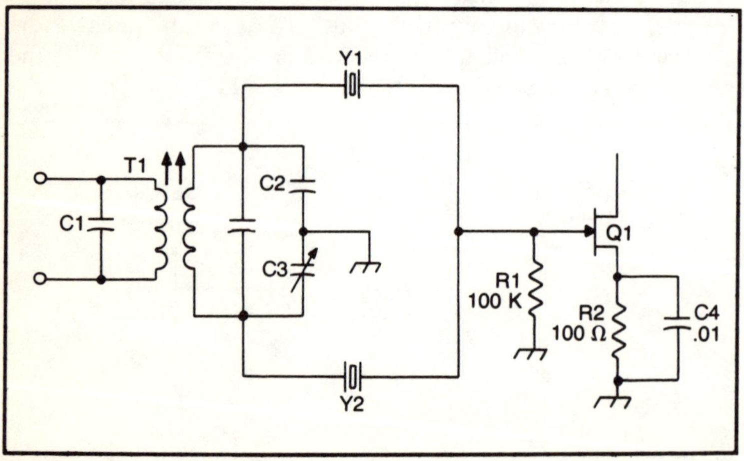

Fig. 7-11. Dual crystal circuit.

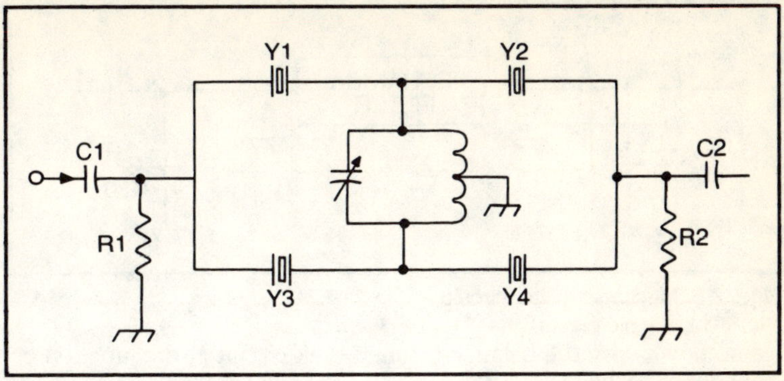

Fig. 7-12. Four-crystal filter.

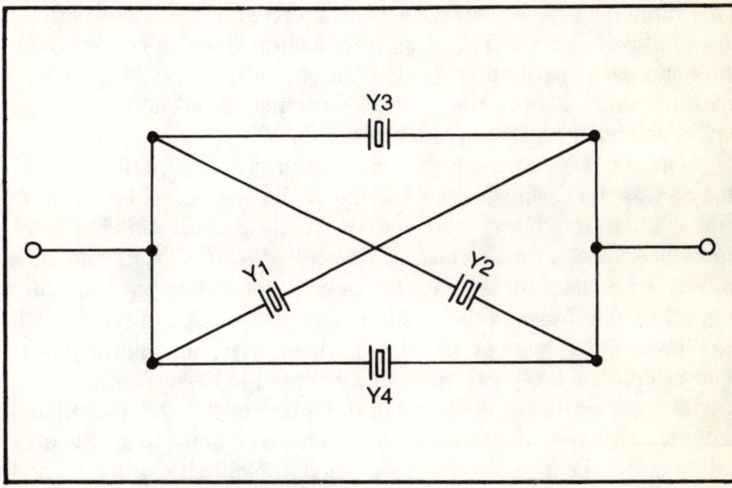

Fig. 7-13. Crystal lattice filter section.

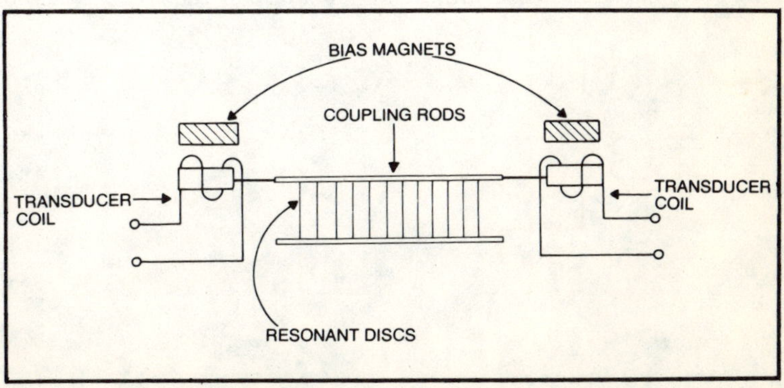

Fig. 7-14. Mechanical filter.

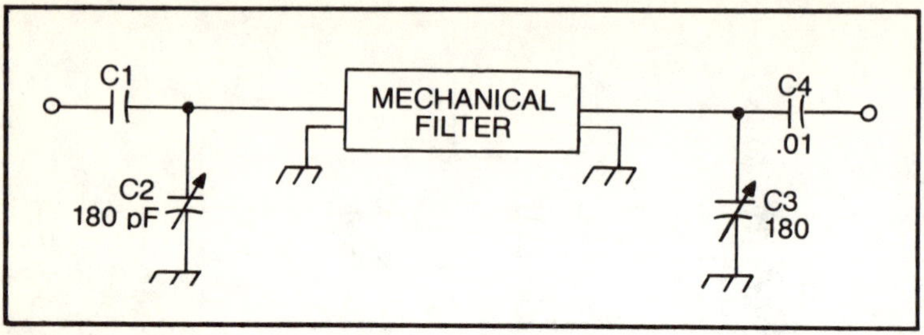

Fig. 7-15. Mechanical filter circuit.

ble somewhat by the balanced, tuned tank circuit at the junction of the two crystal pairs.

The filter in Fig. 7-13 is the well-known crystal lattice filter. This filter uses a series resonant pair (Y3 and Y4) and a parallel resonant pair (Y1 and Y2). This combination gives us control over the band-pass characteristics without any drift-prone LC tank circuits present. This is the type of filter normally encountered in single-sideband amateur transceivers.

One of the earliest filters capable of giving superior performance was the Collins mechanical filter, shown in simplified form in Fig. 7-14. In this filter, which is covered in greater detail in TAB book No. 1224, *The Complete Handbook of Radio Transmitters*, a series of resonant metal discs between input and output transducers set the bandpass of the system. The transducers operate much like permanent magnet loudspeakers or dynamic microphones. The circuit for a typical mechanical filter is shown in Fig. 7-15. Variable capacitors at each end of the filter, one each for input and output, are used to resonate the transducer coils to the center frequency of the filter (center frequencies of 250 kHz, 455 kHz, and 500 kHz are common). Collins Radio used the mechanical filter in a long series of amateur, commercial, and military radio receivers.

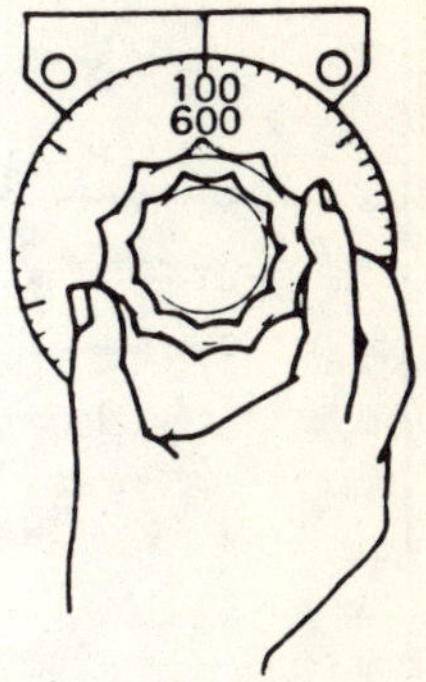

Chapter 8

AM, SSB, and CW Demodulators

The radio signal remains useless to the listener unless some means is provided to recover the information imposed upon the signal at the transmitter. In an AM or SSB system, the information is contained in the audio speech waveforms of the sidebands. In CW it is the radiotelegraph code conveyed in the dots and dashes. In the case of the CW signal, the demodulator must heterodyne a signal of close frequency against the CW appearing in the receiver passband in order to produce an audible beat note.

AM DETECTORS

An amplitude-modulated radio wave is one in which the amplitude of the rf carrier signal is varied by the modulating signal. This form of modulation is used in the 540 to 1650 kHz AM broadcast band (aptly named), in international shortwave broadcasting, and in the citizens band. It is also used in certain VHF aviation radio bands. Most other communications, however, are conducted using either single sideband or frequency modulation.

Because the information contained in an AM signal is in the rf carrier envelope, we call the detector that demodulates this type of signal an *envelope detector*. The simplest form of envelope detector is the rectifier diode, an example of which is shown in Fig. 8-1. The diode is used to rectify the i-f carrier delivered to the diode from the secondary of transformer T1. The output of the diode contains the modulating envelope and a large amount of the rf component. Tweet filter R2/C1/C2 averages the rf variations leaving only the audio

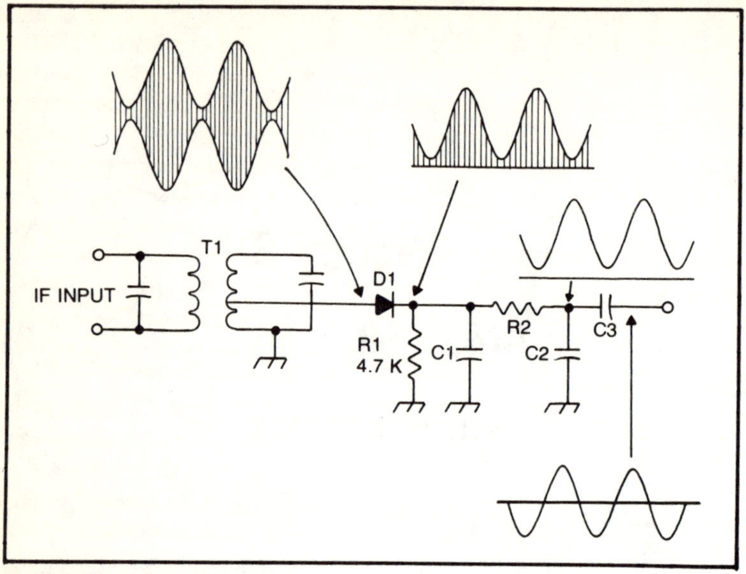

Fig. 8-1. Envelope detector.

sine wave as shown. This waveform, incidentally, contains a large DC component, which is removed by passing the signal through a coupling capacitor (C3). The tweet filter, incidentally, gets its name from the fact that it removes the beat note interference of stations on adjacent channels (10 kHz on the AM band).

The simple AM envelope detector is used extensively in radio receivers and suffers from relatively few defects. We sometimes see a full-wave version, which uses a balanced secondary on the output i-f transformer center-tapped, and two signal diodes. We also see active envelope detectors in the form of bipolar or field-effect transistor amplifiers that are easily overdriven by the signal. These circuits, which essentially operate class B, will transfer only one-half of the input signal waveform to the output circuit.

SINGLE SIDEBAND DETECTORS

A single sideband signal is formed by suppressing the rf carrier and one of the two sidebands of an ordinary AM signal at the transmitter. The carrier and one of two mirror image sidebands merely go along for the ride in AM transmissions. But in the SSB transmitter, only the upper sideband (USB) or lower sideband (LSB) can be sent. We must provide the missing carrier signal to demodulate this type of signal. This is done by heterodyning the output of the i-f amplifier against a locally generated signal from a *beat frequency oscillator* (BFO). The frequency of the BFO must have the same relationship to the LSB or USB (converted to i-f) as did the original carrier. If the original carrier is 1.5 kHz above or below (as the case may be) the center of the SSB passband, then so must the BFO frequency. An error of as little as 30 Hz can render the output almost unintelligible "Donald Duck" gibberish. The detector that performs these miracles is the *product detector*, so-called because it is a heterodyne process, and heterodyning is a *multiplication* of the two signals.

To make a product detector, we must beat the i-f and BFO signals together in some nonlinear element. Figures 8-2 through 8-5 show several methods for producing the product detection required. The circuit shown in Fig. 8-2 uses a pair of back-to-back signal diodes to perform the magic. The BFO signal is applied to the junction between the two diodes. When the BFO signal is on the positive side of its sinusoidal excursion, the diodes are reverse biased, so they will not conduct. But on the negative half cycle of the BFO signal, the diodes are heavily forward biased, so they will conduct a signal from the output of the i-f filter to the audio frequency (af) output of the product detector. The rf pi-network is used to remove residual i-f and BFO components that pass through the diodes on the forward biased half cycle.

A diode-ring product detector is shown in Fig. 8-3. This circuit suffers from a higher insertion loss than the two-diode circuit of Fig. 8-2, but it produces less of the i-f signal in the output. A half-ring product detector is shown in Fig. 8-4. This detector uses an i-f output transformer with a tapped secondary. The diodes are connected reverse from one another. The BFO signal is applied to the center tap and will be seen in-phase at each diode. Because of the reverse polarity of the diode, D1 will be forward biased by the positive half cycle, and D2 will be forward biased by the negative half

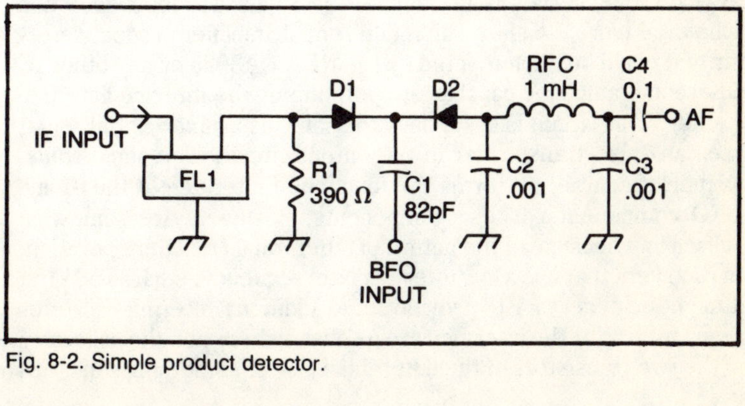

Fig. 8-2. Simple product detector.

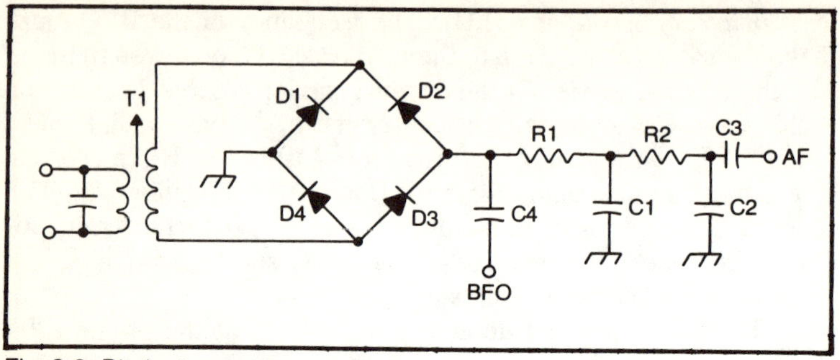

Fig. 8-3. Diode ring product detector.

cycle. On each half cycle, one diode is turned on and the other is reverse biased.

All diode product detectors suffer from a certain amount of insertion loss, typically 5 to 12 dB. This must be overcome by the amplifier ahead of the detector, or weak signals are not received well. The circuit in Fig. 8-5 is an example of an *active* product detector. This stage is a JFET amplifier in the common source configuration. The i-f signal containing the modulation is applied to the gate, while the BFO signal is applied to the drain. Note a certain family resemblance between this circuit and that of several of the mixer circuits in Chapter 6. The main difference is that there is a low-pass filter in the drain circuit to remove the residual i-f and BFO products, preventing them from getting to the audio amplifier. Some versions of this circuit use an audio interstage transformer instead of the coupling system shown. In those circuits, the natural frequency response properties of the transformer perform the rf filtering.

Integrated circuits are also useful for making product detector circuits. Almost any differential rf amplifier—again our friend, the RCA CA3028, serves as an example—will make a product detection scheme. Figure 8-6 shows an rf differential amplifier product detector that could be implemented by the RCA CA3028 or any other IC differential amplifier capable of operating at the intermediate frequency. This circuit is essentially the same as a mixer circuit, but it uses an audio transformer in the output circuit. The audio transformer, of course, performs the function of filtering out the i-f and BFO components. These components, however, are somewhat self-nulling because of the nature of differential amplifier operation. In receivers that use a low intermediate frequency, such as 50 kHz, it might be necessary to provide some additional filtering—a series resonant trap at the intermediate frequency between the collectors of the two transistors of the differential pair or something similar. If

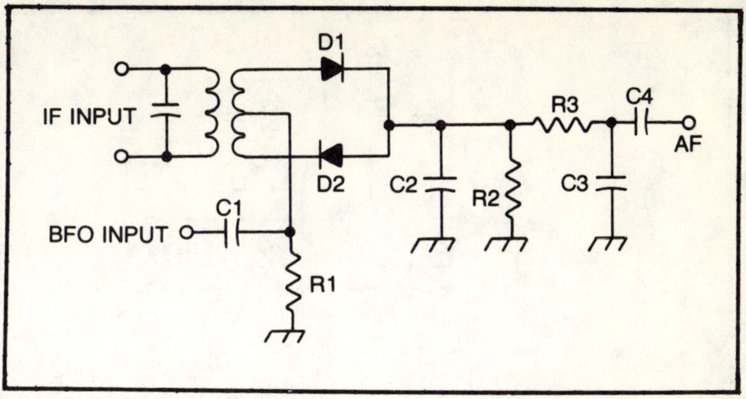

Fig. 8-4. Two-diode product detector.

the Q of the trap is low enough, it will remove both of the undesired signals.

SSB product detectors tend to use crystal-controlled BFOs for precision. Radios that use a so-called clarifier merely place a small variable capacitance across the crystals in order to pull their frequency for off-frequency stations. This is deemed better than retuning the main tuning VFO, which would also change the transmitter frequency in transceivers.

CW DETECTORS

The dits and dahs of the radiotelegraph code sent by SW stations will produce only staccato changes in the background noise when demodulated by an envelope detector. A product detector is needed to demodulate CW. In some SSB transceivers, it is common

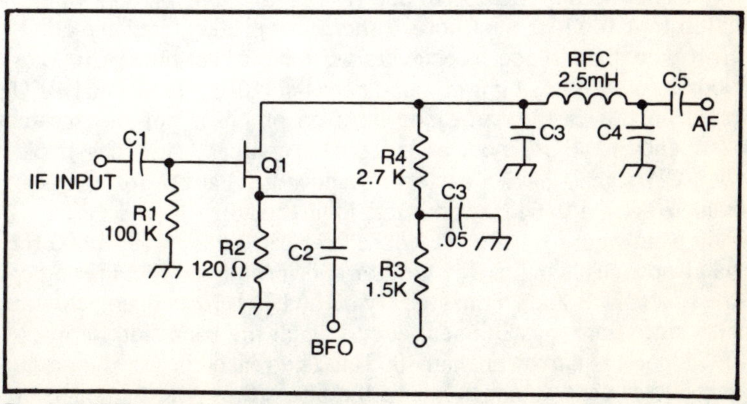

Fig. 8-5. JFET active product detector.

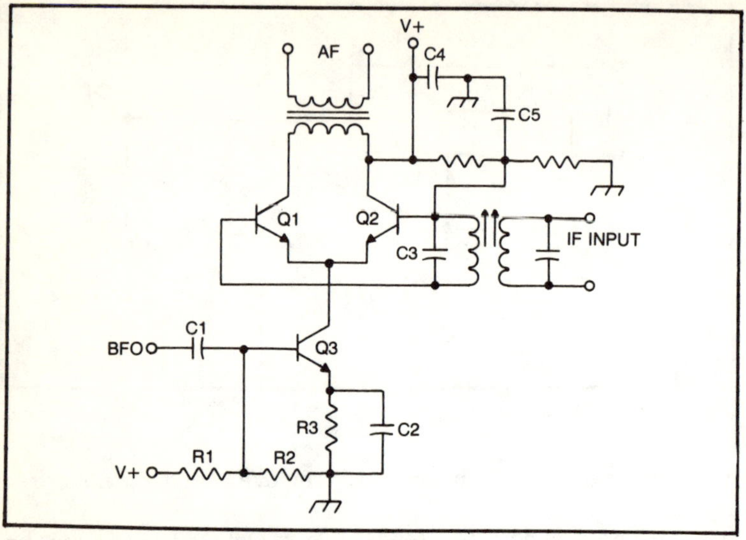

Fig. 8-6. Differential IC amplifier product detector.

practice to use the SSB product detector to demodulate the CW signal. This is assuming that CW will be a minor mode of operation and the operator is not a true CW buff.

The BFO in a CW product detector must produce a signal that has a slightly different frequency from the i-f signal containing the CW information. If the SCW signal is centered directly on 455 kHz, for example, a 456-kHz or 454-kHz BFO will produce a 1000-Hz beat note every time the CW transmitter is keyed. This makes it a lot easier to hear the signal, but some operators have different ideas of what constitutes a proper audio beat note frequency. I personally like a beat note in the 400 to 500 Hz range and find it annoying to listen to a 1000-Hz beat note. I therefore prefer a variable BFO. However, most modern receivers will have a beat note of 1000 or 500 Hz on CW, using a crystal-controlled BFO. The actual BFO used on CW is the same one as used on SSB, but the crystal frequency in the CW mode is closer to the center of the i-f passband.

CW signals have a narrower bandwidth than AM or SSB signals. While a 4 to 6 kHz bandwidth is needed for AM and a 2.5 to 3 kHz bandwidth for SSB, a CW receiver can have a 400 to 600 Hz bandwidth. This imposes a stricter specification on the filtering of the i-f amplifier and on the stability of the BFO. Note that narrower filters are not typically used, even though the bandwidth of many CW signals is narrower than 400 Hz. The reason is that the on-off keying will tend to shock excite the filter (they are LC tanks, or equivalent to LC tanks) and cause ringing.

Chapter 9

FM Demodulators

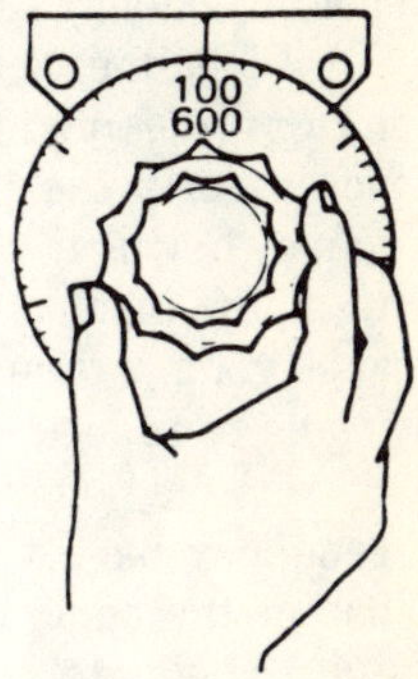

Frequency and phase modulation are examples of a general class called *angular modulation*. In any form of angular modulation, the carrier amplitude remains constant, and the speech is used to vary either the *frequency* (FM) or *phase* (PM) of the carrier. FM and PM are practically interchangeable with each other, and in fact, many so-called "FM" two-way radio transmitters are actually phase-modulated transmitters. It doesn't matter to the receiver, because the detector in each case is a phase detector.

FM/PM are capable of producing very low noise operation. The impulse noise from sparking electrical contacts, automobile ignition, and lightning will cause static in AM radios because they tend to amplitude modulate the rf carrier. The FM/PM signal, however, can be clipped so that these amplitude perturbations are eliminated, producing little or no impulse noise. Note that, despite the cackling of salesmen to the contrary, FM/PM is not guaranteed to be noise-free. We obtain the no-noise condition only when the signal amplitude is high enough to cause the clipping action just mentioned.

For many years, only two basic types of FM demodulators were used in radio receiver equipment: the *Foster-Seeley discriminator* and the *ratio detector*. More recently, however, several new designs have come to the fore and are seeing wide use. One of these discussed in a later section, uses no tuned circuits, no inductors, and is entirely *digital*.

REVIEW OF FM/PM FUNDAMENTALS

In frequency-modulated and phase-modulated systems, the carrier of a radio transmitter is varied by an audio frequency-modulating voltage. Figure 9-1 shows the relationship between the radio frequency carrier and the audio-modulating voltage.

When the value of the audio sinewave (V_a) is zero, the transmitter frequency is at the carrier frequency F_c. As the audio voltage increases in the positive direction, however, the rf carrier frequency begins to shift upwards toward F2. The rf carrier will reach F2 at the instant the audio signal reaches its positive peak. The audio voltage then begins to go back towards zero, so the rf carrier begins to reduce in frequency towards F_c.

On the negative excursion of the audio sine wave, the carrier frequency begins to decrease until it reaches a minimum at the instant the audio signal reaches its negative peak. As the audio signal begins to reduce back towards zero, the carrier frequency will once again climb towards F_c.

In the phase-modulated transmitter, the radio frequency carrier remains constant, as does the amplitude. But the phase of the rf carrier will vary with the audio-modulating signal. We can consider these as equivalent because a change in phase can be measured in the same units as frequency, and will equivalently affect the phase detector used as the demodulator.

We must be aware of several concepts in dealing with FM and PM receivers. One of these is the matter of *deviation*. The percentage of modulation of a transmitter is an arbitrarily specified deviation. We define this concept as the frequency difference between the carrier frequency F_c and *either* extreme (F1 *or* F2). We may therefore speak of positive deviation (F2-F_c) and negative deviation (F_c-F1). If the transmitter has been modulated by a pure sine wave, the positive and negative deviation are identical.

The frequency swing of the FM signal is the total frequency difference between F1 and F2, i.e. F2-F1. If the transmitter has been modulated by a pure sine wave, therefore, the frequency swing is exactly twice the deviation. This relationship holds true only when the positive and negative deviation are the same.

Neither deviation nor frequency swing is affected by the frequency of the audio modulating signal in a true FM system. The deviation and frequency swing with true FM is a function only of the *amplitude* of the modulating signal. But this is not necessarily true of a phase-modulated signal. In an uncompensated phase modulator there is a natural 6 dB/octave rising characteristic that serves as preemphasis. This means that a high audio frequency will cause

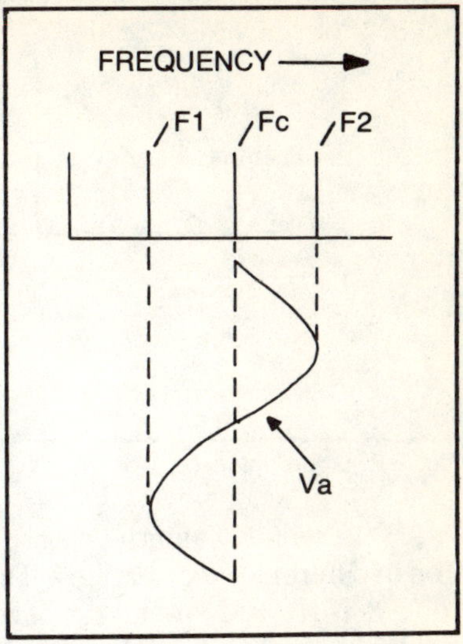

Fig. 9-1. Relationship of audio signal to FM carrier frequency.

greater deviation than a low audio frequency of the same amplitude. Of course, the amplitude is still a factor in PM circuits, but we have the added complication of the preemphasis of the audio.

Not that preemphasis is bad! In many cases, the transmitter audio section in straight FM systems will add preemphasis for purposes of improving the signal-to-noise ratio at the receiver. The receiver must then de-emphasize the audio with a mirror image frequency response characteristic in order to restore the audio balance. Until Doby ® came along, FM broadcasters typically added a substantial amount of high-frequency preemphasis to the audio, such that a 75-microsecond RC roll-off network was needed at the receiver to restore the audio balance. The audio frequency used to modulate a true FM transmitter or frequency-compensated PM transmitter does not affect the deviation; it affects the *rate* at which the rf carrier swings back and forth between F1 and F2.

In AM systems we have a nice convenient physical property of the modulated carrier on which to base our notion of "100 percent modulation." We note that the signal will double on positive audio peaks and drop to zero on negative audio peaks when modulated by a sinewave. Furthermore, in excess of 100 percent modulation, the received signal from that transmitter exhibits distortion and will *splatter* (generate spurious sidebands) over into adjacent channels. This is not true in FM and PM systems. In those systems we must arbitrarily assign a specific deviation as being "100 percent" modulation; in other words, we *define* it.

®Trademark of Doby Laboratories.

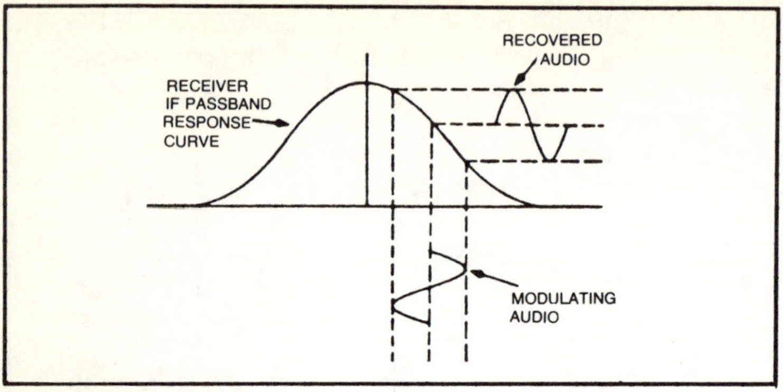

Fig. 9-2. Slope detection.

Different deviation levels are specified as 100 percent modulation in different radio services. The transmitter operating in the 88 to 108 MHz FM broadcast band, for example, is 100 percent modulated with a deviation of ± 75 kHz. The sound carrier for US television signals is FM, but it is 100 percent modulated when the deviation is ± 25 kHz. In the land mobile service, where two-way mobile and base station radios are used by taxicabs, business, and public agencies, 100 percent deviation is only ±5 kHz! Amateur radio operators tend to use the land mobile specification of + 5 kHz. But in a trade-off with older equipment that used ± 15 kHz once used extensively by amateurs (who are *adept adapters* of other people's surplus equipment), amateur transmitters are often specified as 100 percent modulated with a deviation of ± 8 kHz.

SLOPE DETECTION

A number of different detection schemes are used in ordinary practice. These include a special adaptation of the standard envelope detector scheme used in AM systems: slope detection.

Figure 9-2 shows a simple, but very crude, method of demodulating an FM signal. This system is not used much commercially anymore. It is included here to illustrate one of the primary requirements of an FM demodulator: *a frequency response which varies with the input frequency*. This method, called slope detection, requires a receiver with a bandpass that is relatively narrow compared with the FM signal bandwidth. The center of the carrier is tuned in such that it is positioned on the downslope of the i-f frequency-response curve. The FM input signal then sees a frequency response that varies with the input frequency. This system, incidentally, was once popular in mobile police/fire band converters that would directly drive the AM car radio. Once very popular, these little underdash units are still seen occasionally. For less than $30 one could modify a normal AM car radio to hear certain segments of the VHF land mobile bands, the amateur bands, etc. The frequency coverage of the AM band was roughly 1 MHz, so by changing crystals, the user could cover any 1-MHz portion of the bands. The converter performed no demodulation of its own and would merely frequency-convert a VHF band down to the AM band. The user would then tune the AM receiver to the point that corresponded to the desired VHF frequency. The car radio i-f amplifier usually had barely sufficient bandwidth properties, although it often yielded very good (if not noise-free) performance.

THE FOSTER-SEELEY DISCRIMINATOR

The Foster-Seeley discriminator (usually just called "discriminator") was one of the two main FM demodulators used over the past several decades. It still sees rather widespread application and is not any the less effective for age. An example of the discriminator is shown in Fig. 9-3, while a partial equivalent circuit is shown in Fig. 9-4. Note that rf choke L1 is common to both the primary and secondary windings of special discriminator transformer T1. In fact, it is in parallel with the primary and in series with the secondary. If you doubt this, examine the partial circuit in Fig. 9-4. This common connection of L1 allows the use of its voltage and current vectors as references. When the i-f signal is applied to the primary of T1 and is unmodulated, it will be at a frequency equal to the resonant frequency of the transformer tank circuits. This causes the voltages across L_{s1} and L_{s2} to be equal and their respective currents (I1 and I2) to also be equal. Because currents I1 and I2

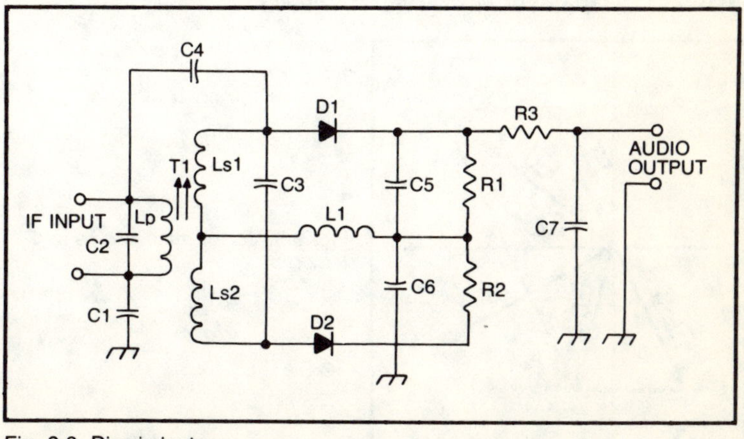

Fig. 9-3. Discriminator.

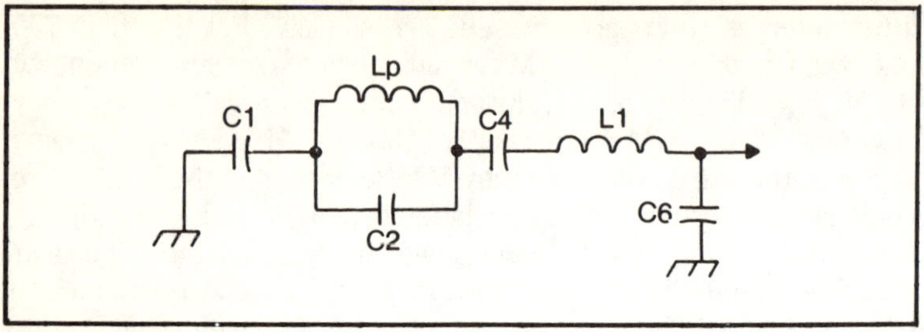

Fig. 9-4. Partial equivalent circuit.

flow in opposite directions and are of equal magnitude, they will cancel each other.

Figure 9-5 shows the voltage and current vector relationships in the discriminator when the frequency of the input signal increases above F_c. Because the secondary tank circuit takes on inductive properties when the input signal is greater than resonance, the secondary current I_s lags behind voltages E_{IS1} and E_{IS2} by 90 degrees. Since these voltages and currents are out of phase (normal for an inductive circuit), we must add them *vectorialy* to find the actual *resultant*. These are labeled E_{d1} and E_{d2} in Fig. 9-5. In this case, the voltage applied to diode D1 is greater than the voltage applied to diode D2, so the current I1 can be expected to be higher then I2. Under these circumstances, the currents no longer totally cancel, and an output voltage is thereby generated. Similarly, in Fig. 9-6, we see the vector situation existing when the carrier frequency is below F_c. The tank circuit is now capacitive, so the resultants E_{d1} and E_{d2} are reversed relative to the previous case; vector E_{d2} predominates so current I2 is greater than I1. Another output voltage is generated, one of opposite polarity.

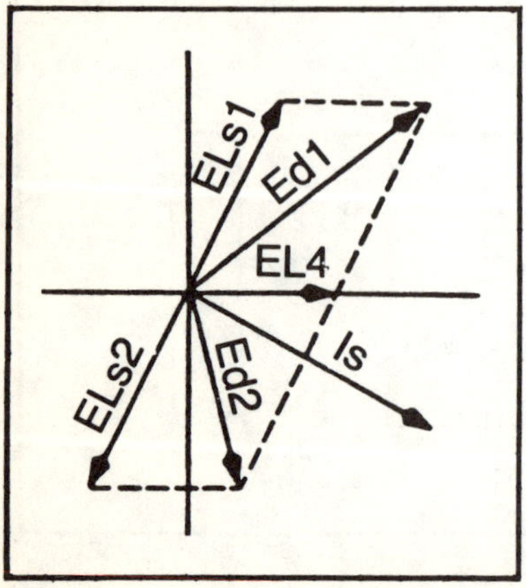

Fig. 9-5. Vector relationships in detector.

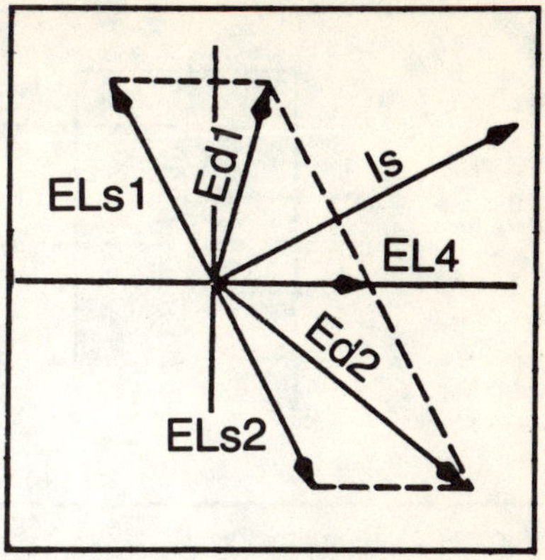

Fig. 9-6. Vector relationships in detector.

A typical *voltage output-versus-frequency* curve for a discriminator is shown in Fig. 9-7. Part of the task when aligning an FM receiver is to place F_c right as the zero crossover point on this curve. The bandwidth of this curve must be such that the anticipated deviation at 100 percent modulation (5, 25 or 75 kHz) will not drive the signal into the nonlinear extremes of this curve.

The discriminator curve also points the way to *measuring* transmitter deviation. Most commercial deviation meters are

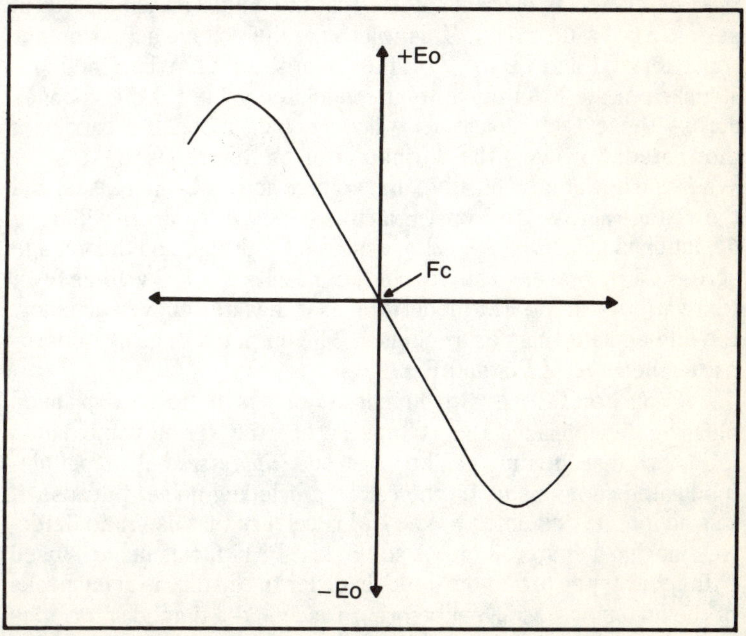

Fig. 9-7. Discriminator voltage vs. frequency curve.

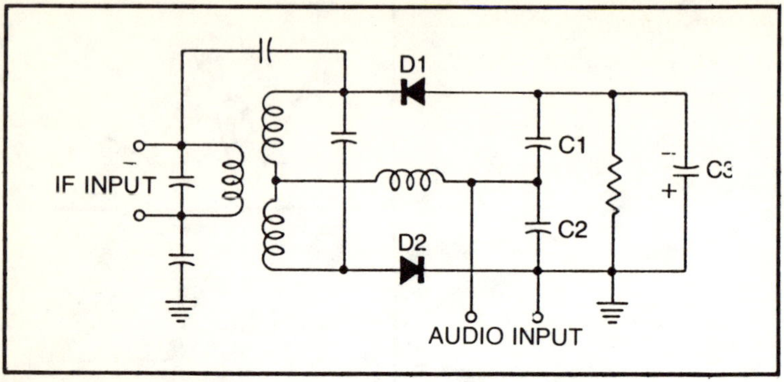

Fig. 9-8. Ratio detector.

FM receivers with a precision discriminator and a voltmeter to measure the output voltage. In any given discriminator, the output voltage is directly proportional to the deviation, hence a measurement system is at hand.

RATIO DETECTORS

Figure 9-8 shows the other classical FM demodulator: the *ratio detector*. The major identifying difference between this circuit and the Foster-Seeley discriminator is that the diodes are connected in series in the ratio circuit. This will cause the voltage across output capacitors C1 and C2 to *add* rather than subtract. When the input signal from the FM i-f is at its unmodulated value F_c, the voltages across these two capacitors will be equal. When the carrier is modulated, however, the deviation causes the frequency to shift. When the deviation is positive, the voltage across C2 increases, and the voltage across C1 drops. Exactly the opposite occurs when the deviation is negative; the voltage across C2 drops, and the voltage across C1 increases. This, of course, results in a DC voltage level that varies with the FM modulation as it deviates above and below the unmodulated carrier frequency. The output voltage is proportinal to the *ratio* of E_{c1} and E_{c2}.

Capacitor C3 nas two functions, one of them an unplanned blessing: It stabilizes the voltage across the series combination C1/C2 so that the ratio can be taken and *suppresses* any amplitude-modulated components on the carrier, including noise. Because of this action, it is common to see FM receivers that use ratio detectors lacking a limiter stage. Also the last FM i-f, this stage is used with other types of demodulators in order to clip off the noisy peaks of the incoming FM signal before it reaches the demodulator. The ratio detector, however, is almost immune to AM noise.

QUADRATURE DETECTORS

The quadrature detector is a circuit that combines two components of the FM i-f signal that are 90 degrees out of phase with each other in a manner that extracts the audio information. The quadrature detector was once popular in TV sound sections, where it used the 6BN6 gated-beam quadrature detector tube. The system went into eclipse when solid-state circuits came along until Motorola and others began to offer their integrated circuit quadrature detector (ICQD). Typical ICQDs include the Motorola MC1357P, the ULN2111 (by several manufacturers), and the RCA CA3089E i-f subsystem (which includes an ICQD). Delco Electronics also makes a number of different ICQD devices for General Motors auto radios. In fact, Delco was the first to use the designation ICQD.

Figure 9-9 shows the internal block diagram for a typical ICQD. The input stages are a wideband, high-gain, limiting preamplifier whose output is a series of square waves. These square waves are sent to two different circuits: the input of a gated synchronous detector and a 90 degree phase-shifting network. The output of the phase-shift network is fed back into the ICQD, where it is applied to the other input of the gated synchronous detector. This detector is basically a *coincidence detector*. As long as the pulses represent an unmodulated signal, they will cancel each other. But when the signal is modulated, the output of the detector is a series of constant amplitude pulses that have a *pulse width* proportional to the modulating frequency. In other words, the detector output is a string of pulses with periods that vary with the modulating frequency. If we integrate these pulses in a RC low-pass filter at the output, we will recover the audio signal.

The use of an IC in the FM detector does not automatically denote an ICQD, although that is often the case. It could mean that

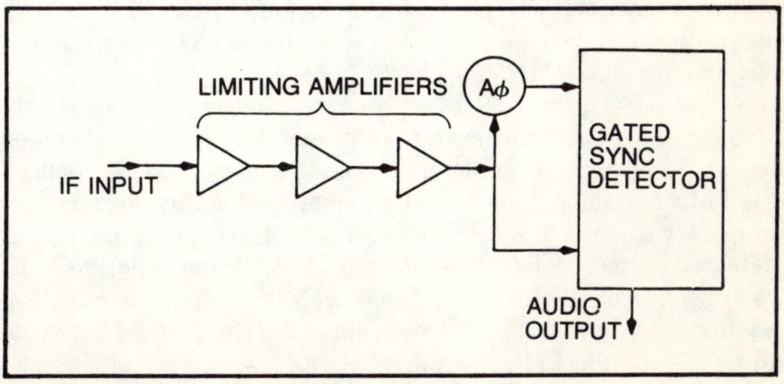

Fig. 9-9. IC quadrature detector block diagram.

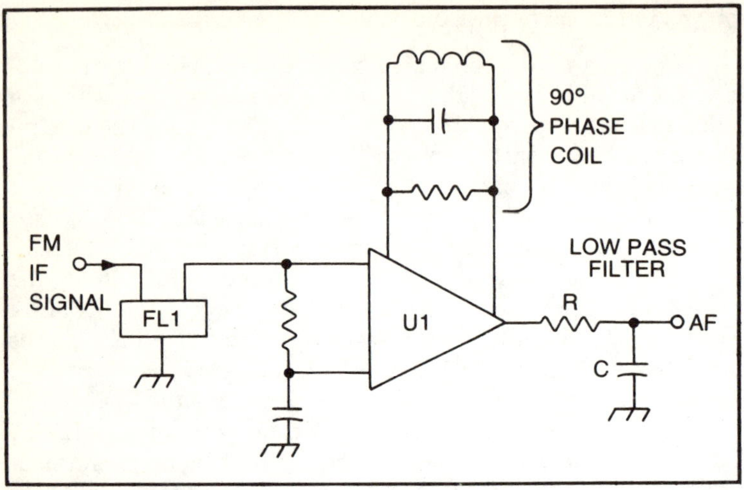

Fig. 9-10. IC quadrature detector block diagram circuit.

an IC, such as the RCA CA3012 or CA3043 devices, is used. The key to identification is the tank circuit used. If it is an ICQD, it will be a phase coil tank circuit, as in Fig. 9-10. On the other hand, though, if it is a ratio detector or discriminator, a transformer similar to those in Figs. 9-3, 9-4, and 9-8 will be connected to the IC.

PHASE-LOCKED LOOP (PLL) FM DEMODULATORS

The phase-locked loop (PLL) was developed back in the 30s when, oddly enough, it was used in an AM detector. The PLL has come into its own over the past decade or so as an FM demodulator. Several companies offer PLL FM detectors in IC form. Figure 9-11 shows the block diagram of a typical PLL circuit. Although this IC is used for purposes other than FM demodulation, we shall describe only that function. The principal components of the PLL include phase detector, voltage-controlled oscillator (VCO), low-pass filter, and an optional DC amplifier.

The voltage-controlled oscillator is designed such that it will oscillate at the carrier frequency—the center of the i-f passband—when no modulation is present. The phase detector has two inputs: the FM i-f signal and the VCO frequency. When the intermediate frequency is equal to the VCO frequency, the output of the phase detector is zero. The phase detector is a digital type, so it is basically a form of coincidence detector. It will output pulses with a width proportional to the difference in phase (frequency) between the VCO and the FM i-f signal. When the two signals are on the same frequency, the coincidence is exact and no output exists. As the FM i-f signal begins to deviate, however, the coincidence no longer exists, and an output pulse with a width proportional to the error between the VCO and FM i-f is generated. These pulses are integrated by the low-pass filter to produce a DC error voltage proportional to the difference in frequency between the FM i-f and VCO signals. This signal may be amplified in a DC amplifier or level translator circuit before being applied as the VCO control voltage. Ideally, a PLL is in equilibrium, with VCO error cancelled by the control voltage. The VCO and FM i-f signal are equal when the deviation is zero, so the DC control voltage is also zero. As the FM i-f signal deviates, though, the DC control voltage varies in a vain attempt at cancelling the error. The DC control voltage, therefore, follows the deviation and is proportional to the original modulating audio signal. This voltage is then coupled to the rest of the radio as the recovered audio signal.

Figure 9-12 shows a basic IC PLL demodulator for FM signals based on the Signetics 56OB chip. The i-f signal is coupled to the chip via capacitor C7 and pin 12. The frequency of the VCO is set to approximate range be capacitor C4. Because the internal resistances of the IC vary normally by ±20 percent due to ordinary production tolerances, this capacitor is often a trimmer. The input signal level must be at least 2 millivolts, and less than 15 millivolts. Below the lower limit, the PLL might not lock onto the FM i-f signal. Above the upper limit, the AM suppression is lost (remember that is was originally an AM detector).

RC networks R1/C1 and R2/C2 form the low-pass filter sections between the output of the phase detector and the input of the error amplifier. The output signal is obtained from pin 9 and is coupled through RC network R3/C3 to the circuits that follow. The function of capacitor C6 is de-emphasis. The value is selected to

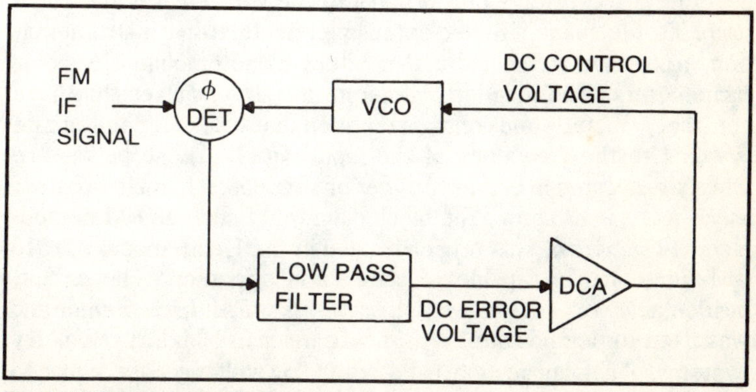

Fig. 9-11. Phase-locked loop.

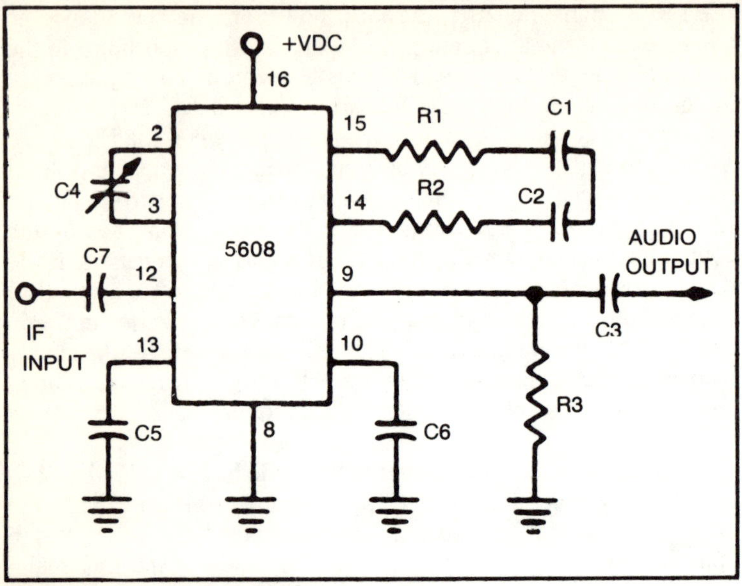

Fig. 9-12. PLL FM demodulator.

give the required de-emphasis (in this case, 75 μS) when the internal resistance of pin 10 is 8000 ohms. Since this resistance varies ± 20 percent, the usual practice to just use a 0.001-μF unit at this point.

DIGITAL FM DEMODULATORS

Digital "coilless" FM detectors are also seen occasionally. These circuits are based upon a phenomenon similar to one seen in the PLL and ICQD: Duty-cycle integration of a pulse train will generate a DC voltage proportional to the frequency of the pulse train. This technique is used extensively in electronic instrumentation, from medical heart rate monitors to automobile electronic tachometers. The idea is to generate a train of pulses that have constant amplitude and constant duration, but whose repetition rate is equal to the frequency of the input signal. These pulses are usually generated in Schmitt trigger or a monostable multivibrator.

Figure 9-13 shows the block diagram of such an FM demodulator. This circuit was originally used in the Heath model AJ1510 FM tuner and is capable of quite good operation. The circuit, incidentally, was once called a *pulse counting FM discriminator* and was used to demodulate the FM subcarrier in FM/FM telemetry systems. That application used an analog voltage data signal to frequency modulate an audio carrier (1975 Hz), which was then used to frequency modulate an *rf* carrier. The rf carrier would be demodulated in the normal manner, and the FM audio subcarrier would be demodulated in a pulse-counting detector to recover the original subaudio analog data signal.

Integrated circuit U1 in Fig. 9-13 is an FM i-f limiter amplifier. The output will be a set of square waves with a frequency deviation identical to the input i-f deviation of the signal from FM i-f filter FL1. The second stage in this circuit (U2) is a retriggerable monostable multivibrator, such as transistor-transistor logic (TTL) devices 74123 and 74123. When we use the TTL devices, we obtain a larger output voltage swing because it will have complementary Q and not-Q outputs. We could make the circuit work with either one, but the output voltage is large if both are used. Then their respective outputs are used to drive the differential inputs of operational amplifier U3.

The pulse outputs of the one-shot (U2) will have a constant amplitude and constant duration, as set by the RC network. The repetition rate of the Q and not-Q pulses, or the number of pulses per unit time (frequency), varies with the FM deviation. If we integrate the pulses, therefore, we produce an output voltage that varies with the deviation; it is the recovered audio. The RC integrators in this circuit are the networks R3/C2 and R4/C3. Note, again, that the circuit will work with either network. The outputs of the integrators are applied to the inverting and noninverting inputs

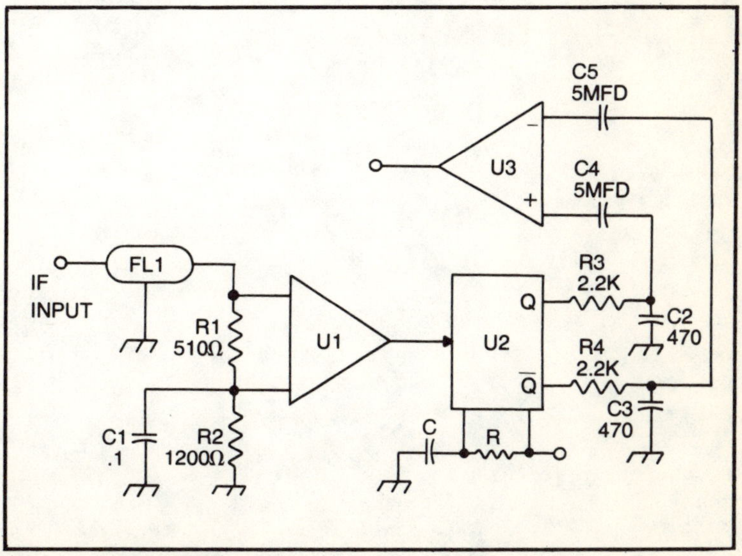

Fig. 9-13. Digital detector.

of the operational amplifier. Because these outputs from the integrators are driven 180 degrees out of phase with each other by the complementary one-shot outputs, their respective polarities are opposite each other. This means that the operational amplifier sees a differential signal. This signal will amplify the audio to produce a larger output. If the two integrators produced signals with the same amplitude and phase the operational amplifier would see a common-mode signal and produce zero output.

Chapter 10 The Audio Amplifier Section

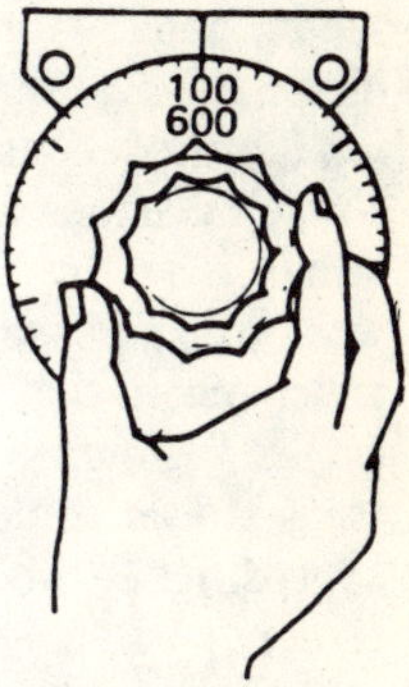

The modulated radio wave is demodulated in the detector stage to recover the audio impressed on the carrier at the transmitter. A CW signal is heterodyned against a beat frequency oscillator to produce an audio beat note that turns on and off with the dits and dahs of the international Morse code from the transmitter. The output of the detector in either case is too weak to drive a loudspeaker and may even be weak in headphones. The purpose of the audio amplifier section of the receiver is to build up this signal to a power level that will drive an efficient loudspeaker system. This means only a couple hundred milliwatts in many cases, but it might mean upward of 10 watts in some models. Hi-fi receivers typically produce 15 watts and up.

There are at least two stages of the audio section: preamplifier and power amplifier. Some circuits have a third stage called the driver, which is placed in cascade between the preamplifier and power amplifier.

Squelch circuits are also part of the audio section. This stage is used to mute the audio output (turn if off) when there is no signal in the passband of the receiver. This is a blessed relief to those who must monitor channels that are only intermittently active. The absence of a signal would produce a terribly loud hisssssss background noise.

PREAMPLIFIER/DRIVER STAGES

Preamplifiers and drivers are essentially the same, except for relative power level. The driver usually produces a somewhat larger output power than the preamplifier, because it is used to directly drive the audio output stage. In many receivers, incidentally, the preamplifier is the driver.

In this section we will consider some of the methods used to bias transistors in solid-state audio sections. Most of these circuits are for negative-ground operation, so they are compatible with the electrical systems of automobiles made in the US—and most foreign models as well.

One of the most important aspects of understanding the operation of solid-state audio amplifiers is *transistor biasing methods*. We want to bias preamplifier and driver stages for class A operation, which means that the collector current will flow over the entire input cycle. It is the usual practice in class A amplifiers to bias the transistor at some point at which the collector voltage to ground is approximately one-half of the supply voltage ($V_c = \frac{1}{2}V_{cc}$). This will require some quiescent bias current. When a sine wave audio signal is applied to the base of the transistor, then it adds to the bias current on one-half of its cycle, and subtracts from the bias current on the alternate half cycle. This will cause the collector current to rise and fall in a sinusoidal manner along with the base current. A collector load resistor means that some of the supply voltage is dropped across the collector load. When the base current causes an increase in collector current, the voltage drop across this resistor increases, thereby decreasing the collector voltage. Similarly, when the collector current decreases, the voltage drop across the load resistor also decreases, thereby increasing the voltage at the collector of the transistor (See TAB book No. 1224, *The Complete Handbook of Radio Transmitters*). The output of the common-emitter, class A transistor amplifier is therefore sinusoidal voltage that is 180 degrees out of phase with the input voltage.

Figure 10-1 shows some of the standard methods for biasing bipolar transistors. These circuits, or variations of them, are used extensively in amateur radio receivers and transmitter speech amplifiers. The method shown in Fig. 10-1A is probably the simplest of all bias arrangements. It is also one of the least practical! Bias is established by current flow from the emitter-base junction of the transistor through R1 of the supply voltage. The amount of bias is dependent upon the value of resistor R1 and the DC supply voltage. The primary disadvantage of the circuit is that it provides poor stability because no automatic means limits the collector current.

A slightly superior method is shown in Fig. 10-1B. The system uses collector feedback to self-bias the transistor. Because resistor R1 is connected to the transistor side of load resistor R2, any change in collector current will cause a proportional, but opposite change in the transistor bias. For example, if the collector current increases because of a temperature increase, the voltage at the collector becomes less positive (decreases), which in turn reduces the current through the circuit comprised of the emitter-base junction and resistor R1. Although this method of biasing does provide some stabilization, it does so at the cost of added degeneration that is caused by feedback of the AC signal voltage developed across the collector load resistor R2.

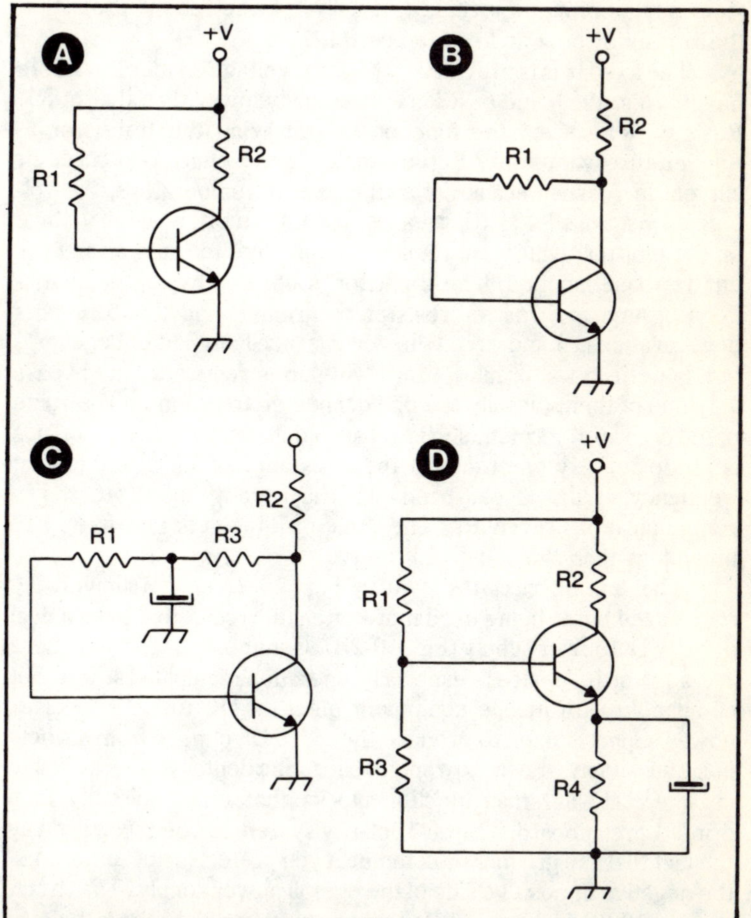

Fig. 10-1. Standard methods for biasing bipolar transistors. Emitter-base current flow biasing is shown in A, collector feedback self-bias is shown in B, collector biasing with AC bypassing is shown in C, and a combination of self-bias and fixed bias is shown in D.

We modify the basic system with the circuit of Fig. 10-1C. This is collector biasing with AC bypassing. It provides both good DC stabilization and a minimum of AC degeneration. The circuit is the same as in Fig. 10-1B, except that the resistance of R1 has been divided between two resistors in series, and an electrolytic bypass capacitor is placed between their junction and ground. This effectively removes the AC component from the feedback signal.

One of the best bias systems is shown in Fig. 10-1D. This circuit uses a combination of self-bias and fixed bias and provides superior DC stabilization and minimum degeneration. It also provides a higher input impedance, usually specified as the product of the transistor beta and resistance R4.

The fixed bias is provided by resistor voltage divider R1/R3. In most cases, the value of R3 is substantially lower than that of R1. Resistor R4 serves the function of stabilizing the transistor to temperature variations. For example, if the emitter-to-collector current increases because of an increase in temperature, the voltage drop across R4 also increases, placing a more positive voltage on the emitter, which will reduce the effective forward bias of the emitter-base junction. The capacitor is used to bypass AC variations around the emitter resistor to ground, and thus prevents degenerations. Some circuits in which a small amount of degeneration is desired will break R4 into two series resistors, and bypass only one of them (usually the one connected to ground). The value of the capacitor, incidentally, is set such that its capacitive reactance is approximately one-tenth of the resistance of R4 at the lowest frequency of the af passband. This is usually 300 Hz or so in communications receivers. The value of R4 is generally five to 10 times less than the value of R3.

The biasing circuit shown in Fig. 10-2A is not universally recognized but is being used more often in circuits that have a dual polarity DC power supply (Fig. 10-2B). Commonly used when there are a lot of IC devices, especially operational amplifiers, it is not compatible with mobile equipment unless a DC-to-DC converter power supply is used to produce the ± VDC voltages from a single positive supply. Such power supplies, incidentally, are available commercially as small function blocks that mount directly to a printed circuit board. The dual polarity system can be recognized by the fact that the ground, or common if you prefer, is not returned to the negative or positive side of the overall power supply. The circuit in Fig. 10-2B is a typical AC power supply used in this type of equipment. Instead, the ground (common) floats at the electrical midpoint of the two supplies. In most cases, the two supplies have equal voltages, although some exist in which unequal positive and negative voltages are available.

Increased output voltage swing is one of the advantages of the dual power supply system. Another claimed advantage is better thermal stability. This can mean a lot in a transceiver in which the rf power amplifier and attendant power supply circuits create a bucket full of heat, even in standby. A third advantage is that the circuit is less sensitive to hum due to power supply ripple on the V+ and V− voltages. These ripple factors tend to be equal between the two supplies, so the overall effect is that they cancel each other.

Another type of circuit frequently seen in preamplifier circuits is the Darlington amplifier (also called Darlington pair) of Fig. 10-3A. The identifying feature of this amplifier is that the emitter of the input transistor drives the base of the output transistor. The emitter for the pair is the emitter of the output transistor, and both

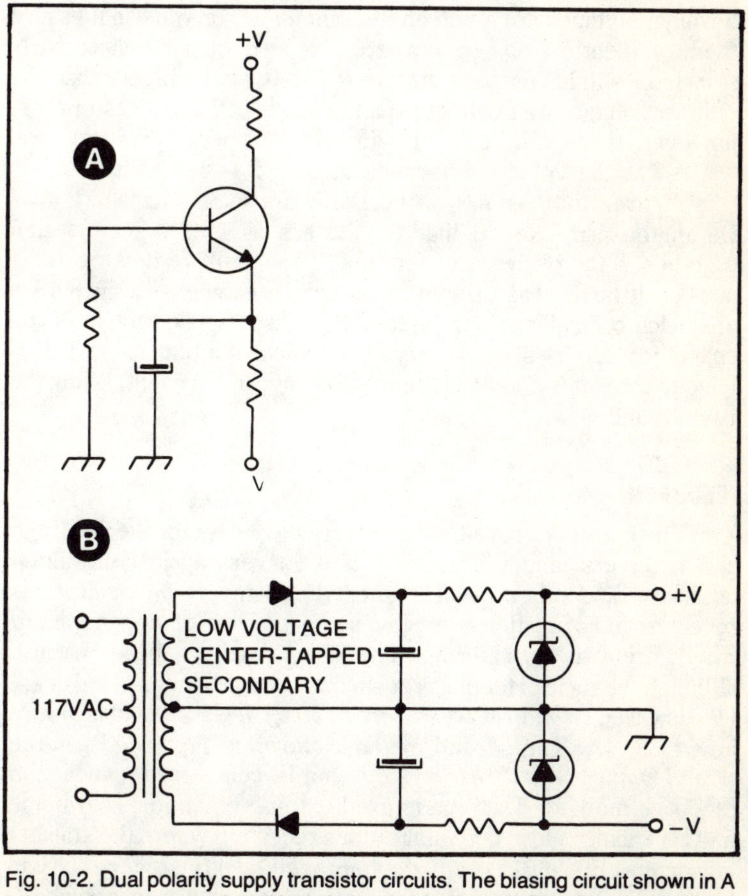

Fig. 10-2. Dual polarity supply transistor circuits. The biasing circuit shown in A is becoming more popular, although the circuit shown in B is still more typical.

collectors are tied together. This produces a much higher input impedance and a tremendous current gain for the pair. It is possible to obtain an overall current gain equal to the products of the current gains of the individual transistors. If, as is usually the case in preamplifier application the two transistors are identical, the overall current gain is $(h_{fe})^2$. Because the input impedance tends to be equal to the product of the gain and the emitter resistance, it would be $(h_{fe})^2R_e$.

Darlington pairs are available in three-terminal transistor packages and are marketed as *superbeta* transistors. One problem is that manufacturers' draftsmen tend to draw them as ordinary transistors in the schematics of service manuals. Just go to replace that device with an ordinary NPN silicon transistor, and *shesanoworkee*! You can identify a Darlington pair if the transistor terminal voltages are given on the schematic, or you can measure them on identical rigs somewhere else. An ordinary silicon NPN transistor will have a base voltage 0.6 to 0.7 volts higher than the emitter voltage. If a Darlington pair is inside of that transistor case, however, there will be *two* PN junctions between "emitter" and "base," so the voltage difference V_{be} is 1.2 to 1.4 volts.

Figure 10-3B shows an IC Darlington pair—actually a *dual* Darlington pair. Notice that the IC has all four collectors tied together. If the two emitter terminals (2 and 10) are tied together, pin 4 could be used as an input for an audio preamplifier while pin 8 is a squelch control input. If a heavy DC bias is applied to pin 8, the collector voltage will drop to zero very quickly for both pairs. This is used to turn off the audio stage when no signal is present, killing the background noise.

FEEDBACK

The use of negative feedback works wonders for the fidelity of hi-fi receivers, and will stabilize all receivers against unwanted oscillation and other major crimes. Even most communications receiver audio sections employ some degree of degeneration. There are two basic forms of feedback used in these circuits, although the topic of feedback could fill several volumes—and does! One is called *second-collector-to-first-emitter* feedback and is shown in Fig. 10-4A. The second method, shown in Fig. 10-4B, is the *second-emitter-to-first-base* system and is considered a minimum feedback method. This one is used extensively in low-cost home and car radios, and in communications receivers where the stability advantages are desired but the improved fidelity so necessary in hi-fi equipment is not desired and may, indeed, be *dis*advantageous.

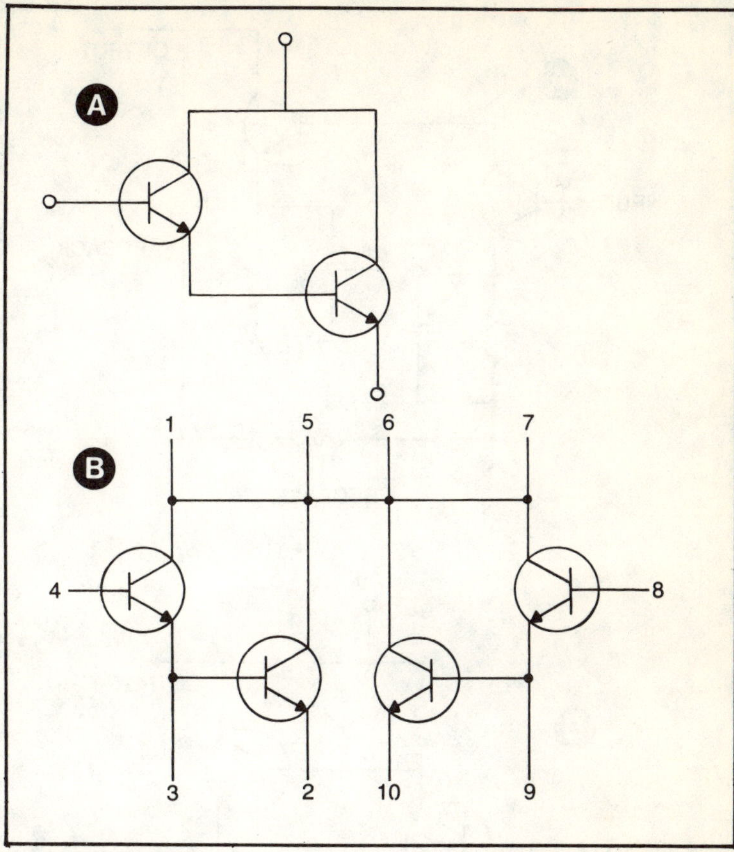

Fig. 10-3. Darlington amplifier configurations. Transistors are used in A, and an IC is used in B.

Feedback contributes a great deal to the performance of many circuits, but it is also a source of problems on two scores. One problem is that a defect in the feedback system is difficult to trace, as it presents symptoms that affect the performance of the main signal path. These symptoms vary from mildly offensive distortion to wild oscillations and very severe distortion destruction. The transistors in the power amplifier system might even be destroyed. Sometimes, only a component by component check of the feedback circuit will solve such troubleshooting problems.

The second major problem with feedback networks is seen when they are part of a hi-fi amplifier. If the feedback sample is taken from the loudspeaker output of the amplifier, it sends a signal back to an input stage. If the speaker leads are any appreciable length at all, an interfering signal from a nearby transmitter will get

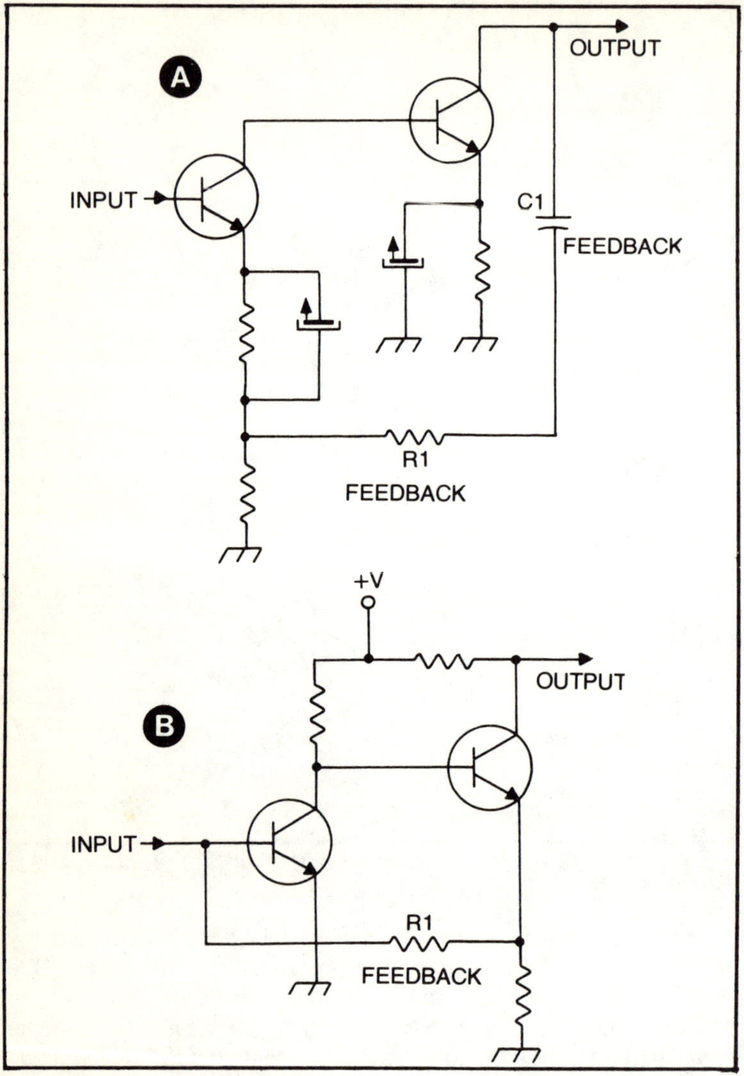

Fig. 10-4. Feedback systems. Second-collector-to-first-emitter feedback is shown in A, and second-emitter-to-first base feedback is shown in B.

into the amplifier via the speaker connections. This signal sees a low-impedance path back to an input stage, where it will be rectified, or *demodulated*, and be reproduced as your voice in the loudspeaker. The only solution is rf filtering of the speaker leads, but this will reduce the upper end of the fidelity. The owner of a hi-fi receiver usually does not understand that it is not *your* fault, but the fault of the manufacturer who failed to make a product that will reject undesired signals.

AUDIO POWER AMPLIFIERS

Because we are dealing with circuits that produce from 200 mW to no more than 10 watts of audio power, some hi-fi nuts might snicker at the title of this section. Nevertheless, the circuits *are* power amplifiers, and they are necessary if the receiver is to drive even the most efficient small loudspeaker system.

One of the oldest forms of audio power amplifier used in solid-state radio equipment is the single-ended class A design of Fig. 10-5. Even today, many car and home radios and some communications receivers or transceivers use the single-ended class A circuit (although I am not quite sure why). The preamplifier and driver circuits are direct-coupled to the audio output transistor. One interesting feature of this circuit, which is intended for negative-ground operation, is the PNP audio output stage. In the early 60s when this circuit was popular, PNP germanium transistors were common, and other types were almost unavailable. The output is taken from the collector, which must be less positive—more negative—than the emitter. It then goes, through a tapped choke, rather than a transformer.

The push-pull amplifier is widely preferred over other types where current drain economies are important. In portable receivers, for example, the battery life is very short, because a class A amplifier draws its full power all of the time. The class B amplifier, on the other hand, draws current from the power supply propor-

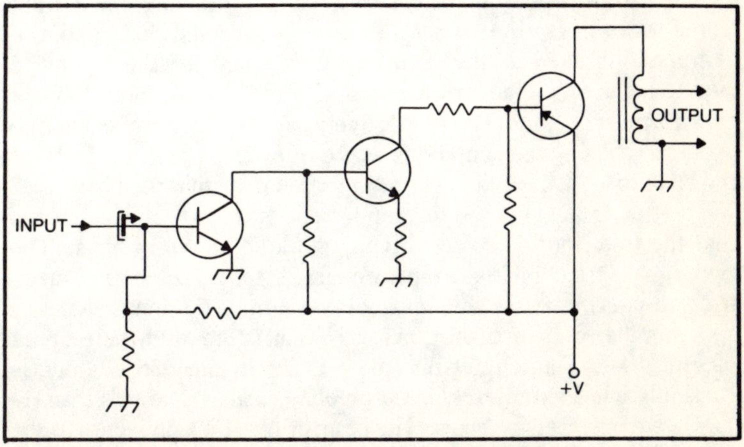

Fig. 10-5. Direct-coupled class A audio power amplifier.

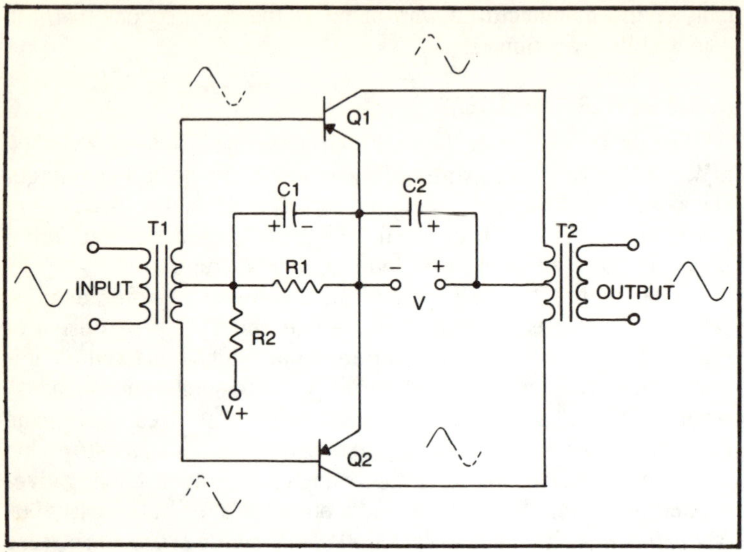

Fig. 10-6. Traditional push-pull amplifier.

tional to the input signal level; however, a class B amplifier cannot be linear if operated in single-ended circuits, because it will amplify only one one-half of the input cycle. If a pair of class-B amplifiers are operated so that they are driven by the same input signals, but 180 degrees out of phase with each other, the output signals from each class-B amplifier can be combined to form a linear power amplifier.

The classical push-pull amplifier is shown in Fig. 10-6. This circuit is merely a solid-state version of an old vacuum tube circuit. The key to the operation of this circuit lies in the input and output transformers. Both transformers are center-tapped, and the center-tap is used as the zero reference point for the input and output signals. Capacitors C1 and C2 keep the center taps of transformers T1 and T2, respectively, at a low AC impedance to ground. By using the center-tap as the reference point, we find that the base of Q1 will go positive at the same time that the base of Q2 will go negative. This will turn on Q1 and turn off Q2. This means that the two transistors are driven 180 degrees out of phase.The two signals are combined in the primary of T2, so the output taken from the secondary is the amplified version of the input signal.

Another version of the push-pull circuit is the totem pole circuit of Fig. 10-7. This circuit became popular in the mid 60s and is currently used extensively in automobile radios. The two transistors are connected in series: The emitter of Q1 is connected to the collector of Q2. These transistors form a voltage divider. When there is no input signal, the voltage at the junction of Q1/Q2 is approximately one-half of the supply voltage.

The totem pole push-pull amplifier uses a special interstage transformer to drive the bases of Q1 and Q2 out of phase with each other. The key to the operation of this circuit is that the transformers are connected with opposite *sense* so that one will be positive-going when the other is negative-going.

The circuit in Fig. 10-8 is a complementary symmetry push-pull amplifier. This circuit obtains the needed phase difference by using transistors of opposite polarities. Both of the transistors are operated in the emitter-follower configuration, with the loudspeaker as the emitter load for each transistor. The bases of the transistors are driven in parallel. But Q1 is an NPN transistor, and Q2 is a PNP transistor. The *NPN* device will turn on when the base is made more *positive* than the emitter. The *PNP* device will turn on when the base is made more *negative* than the emitter. These transistors, then, have a natural 180-degree phase difference. If we apply a sine wave to the input, Q1 will turn on when the input signal is positive, while Q1 turns on when the input signal is negative. The

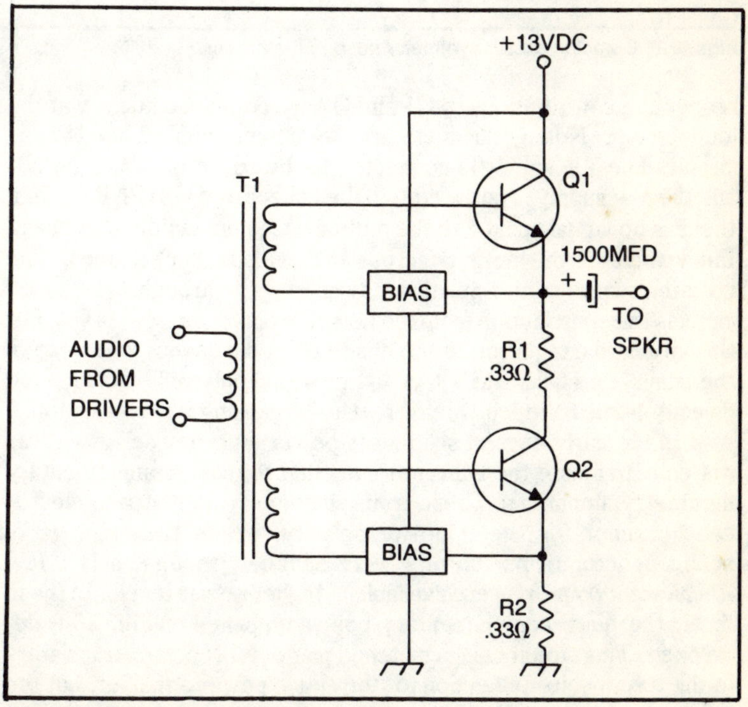

Fig. 10-7. Totem pole push-pull amplifier.

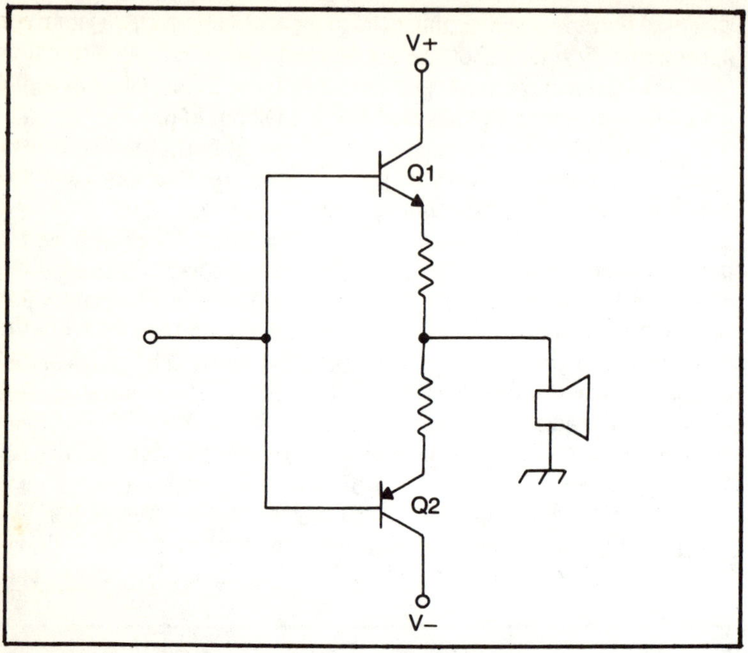

Fig. 10-8. Complementary symmetry push-pull amplifier.

respective contributions of Q1 and Q2 are combined linearly at the loudspeaker. Notice that there are two power supplies used in this circuit. The V+ supply is connected to the collector of Q1 (NPN), and the V– supply is connected to the collector of Q2 (PNP). When there is no signal applied to the mutual base connection, therefore, the voltage at the output (across the loudspeaker) is zero. The contribution from the positive power supply through Q1 exactly cancels the contribution from the negative power supply via Q2. But this circuit is dangerous to the health of loudspeakers when one of the transistors fails: the V+ or V– power supply will be connected directly across the loudspeaker, usually causing its destruction.

In the early days of solid-state power amplifier design, it was not easy to locate the transistors needed to make complementary symmetry amplifiers. These transistors must be matched electrically, except for the opposite polarity. When the catalogs of semiconductor manufacturers were slimmer, though, only a few *complementary pairs* were available at higher power levels. In those days, the quasi-complementary power amplifier of Fig. 10-9 developed. This circuit uses a matched pair of NPN power transistors in the totem pole connection for the actual power amplifier, but the *driver* transistors (Q1/Q3) are complementary symmetry. The bases of the drivers are connected in parallel. The emitter of driver transistor Q1 is used to apply signal to the base of power amplifier Q2. Similarly, the emitter of driver transistor Q3 applies the signal to power amplifier transistor Q4.

No output DC blocking capacitor is needed if bipolar power supplies are used, as shown. But if a monopolar power supply is used, the capacitor must be included because the voltage at the junction of R4 and R5 will be approximately one-half of the supply voltage. Such a voltage will easily result in a blown loudspeaker.

One last form of a totem pole push-pull amplifier is shown in Fig. 10-10. This circuit uses a pair of NPN power transistors in the final amplifier and a differential pair of driver transistors. This circuit is very simplified over what is actually found; a lot of level shifting occurs between the collector of Q1 and the base of Q4. The circuit shown here should demonstrate the phase reversal action of the differential pair, however.

Q1 and Q2 form a differential pair because they are connected with their emitter being driven together from a mutual constant

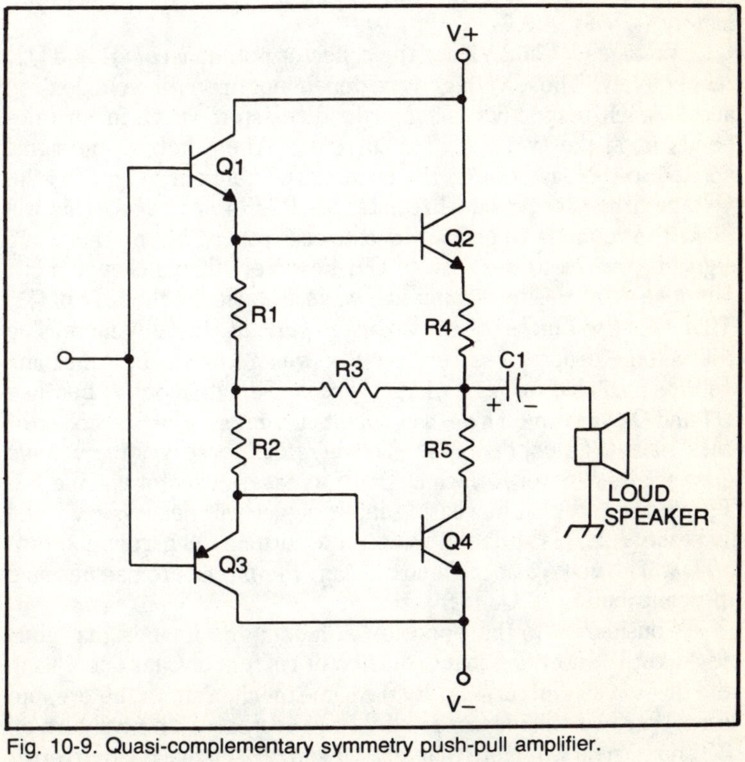

Fig. 10-9. Quasi-complementary symmetry push-pull amplifier.

current source. Under all circumstances, I3 = I1 + I2. If either I1 or I2 changes, therefore, the other transistor will change in order to maintain the equality; I3 will always remain the same. For example, if I1 is increased, I2 must decrease by the same amount. Conversely, should I1 decrease, then I2 would have to increase by the same amount.

Transistor Q2 is fixed at a given quiescent collector current by voltage divider R5/R7, and its base is kept at AC ground potential by capacitor C2. The other transistor in the differential pair, Q1, is the active signal input stage. It is also forward biased by a voltage divider network, R1 and R2. The input signal is coupled to the base of Q1 by capacitor C1.

The collector voltage of Q2 is used to drive the base of power transistor Q3, while the collector voltage of Q1 is used to drive the base of power transistor Q4. In an actual IC or discrete power amplifier built along these lines, this portion of the circuit becomes more complex due to the need for level shifting between these various transistors. After all the base of Q4, will be at a potential of approximately 0.6 to 0.7 volts, while that of Q3 will be approximately (½V+) + 0.7.

Voltages V1 and V2 are the collector potentials of Q1 and Q2, respectively. These voltages are dependent upon the voltage drop across each respective collector load resistor, which in turn depends upon the two collector currents. When there is no signal applied to the base of Q1, the collector currents are equal, so the voltage drops across equal resistances R3/R4 are equal. This will make the voltages V1 and V2 equal to each other. When a sine wave signal is applied to the base of Q1, however, things change a bit. Suppose that a positive-going sine wave is applied to the base of Q1. This signal will increase the collector current, thereby increasing the voltage drop across resistor R3, which lowers the remaining voltage available for V1; V1 then drops. Simultaneously, because Q1 and Q2 are driven from a constant current source, the collector current of Q2 must decrease. This in effect decreases the voltage drop across resistor R4, which then increases collector voltage V2. To recapitulate, an increasing input signal will decrease V1 and increase V2. This situation will tend to turn on Q3 harder and turn off Q4. The voltage at the loudspeaker, then, tends to rise because the contribution of Q3 increases.

Consider now the opposite situation: the input signal goes negative. We now find that the collector current of Q1 is decreasing and that of Q2 is increasing. By the same mechanisms as before, but in reverse, V1 will increase and V2 will decrease. This will turn off Q3 and turn on Q4. Note that *turn on* and *turn off* are relative terms

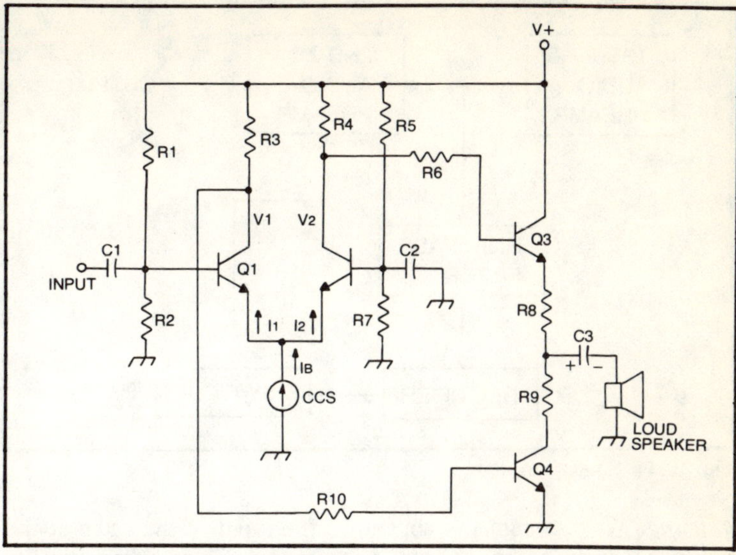

Fig. 10-10. Differential amplifier provides phase inversion.

that do not indicate switch-like action. They rather mean turned on less hard and turned on harder.

This circuit is seen in some hi-fi receivers, with many more transistors to improve the stability and the level-shifting noted earlier. In most previous cases, the circuit was seen in only hi-fi sets. Today, several integrated circuit manufacturers offer IC audio power amplifiers that include all necessary preamplifier and power amplifier stages to make an audio section. These ICs tend to use the differential amplifier method of phase inversion because it becomes easier to implement in IC form. Devices are available with output power rating from 250 mW to 4 to 5 watts. Some hybrid ICs offer power levels up to 100 watts! A hybrid is a device containing ICs and transistors in *chip form* (no packages) on a ceramic multi-layer substrate that bears printed circuit tracks. Most so-called analog function modules are based on this type of construction. It is probably true that most solid-state communications receivers and transceivers use such an IC (no the hybrid kind) for the entire audio amplifier section. This is popular with the manufacturer because it reduces the size, lowers the component count and the assembly costs, and is somewhat easier to tame than discrete circuits.

SQUELCH CIRCUITS

The background noise in a radio receiver becomes very much greater when no signal is applied. We note a phenomenon called

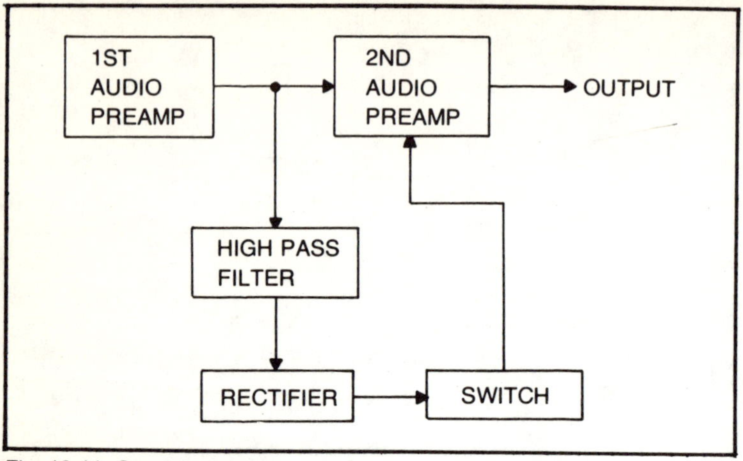

Fig. 10-11. Squelch system.

quieting. When a signal is applied to the input of the radio at the antenna terminals, the apparent noise level goes down dramatically. This is especially true of VHF and UHF receivers. When the signal reaches maximum strength, the radio becomes fully quieted, and little or no noise is heard. Consider what happens in a two-way radio system in which transmissions are intermittent. If you turn up the volume control high enough to hear the other station when a transmission arrives, the background noise between transmissions will drive you clean out of the place! The *squelch* is an appropriately named circuit that squelches the output of the radio when no signal is being received.

There are two methods of deriving the squelch signal. One of the simplest is the use of the receiver agc voltage to turn on a switch when a signal arrives. An electronic switch, sometimes called a squelch gate, allows the signal to pass when the agc voltage is greater than a certain level. But most squelch systems use the circuit shown in Fig. 10-11. This is the block diagram of a communications receiver audio preamplifier system in which squelch is used. The signal at the output of the first audio preamplifier is split into two portions. One goes directly to the controlled second audio preamplifier. This stage will not pass a signal unless the line from the electronic switch is high.

The electronic switch signal is derived from a high-pass filter that is often combined with a noise amplifier (a wideband amplifier), and rectifier circuit. Recall that the audio bandwidth of the communications signal will be 3000 Hz, but the noise signal contains a wideband spectrum with very strong components above 3000 hz.

This noise signal, because of the quieting effect, reduces almost to zero when a strong signal is applied to the input of the first audio preamplifier. So by placing a high-pass filter with a 3000-Hz cutoff frequency in the line, an output will exist only when no signal is being received. This filter output signal is rectified and used to turn on the electronic switch (Q1 in Fig. 10-12). Transistor Q1 is turned on when the base voltage is 0.6 to 0.7 volts greater than the emitter potential. The strength of the signal that will accomplish this is set by the squelch control (R1). When the switch is turned on, its emitter voltage rises, and this causes the emitter voltage of the PNP transistor used as the second audio amplifier to rise also. Because the PNP transistor conducts when the emitter voltage is more positive than the base voltage, the second audio stage turns on and passes signals to the power amplifier stage.

A hard on and off characteristic for the squelch circuit is not always desired. The squelch should hang just a little bit when the other operator stops speaking. Capacitor C2, which forms the audio emitter bypass for Q2, will also cause a slight RC delay in the squelch turnoff.

Figure 10-13 shows a diode squelch circuit. A diode will pass small AC signals undistorted when the diode is forward biased heavily. This places the diode in the linear portion of its operating characteristic curve, so distortion will not be evident unless the AC signal is strong enough to drive the diode either into saturation or into the nonlinear region close to its 0.6 to 0.7 volt junction potential. Diodes D1 and D2 are forward biased and will pass audio when points A and B in Fig. 10-13 are high. This will occur when transistor Q2 is turned off. Normally, when there is no output from the rectifier (see Fig. 10-11), transistor Q1 is turned off, so its collector

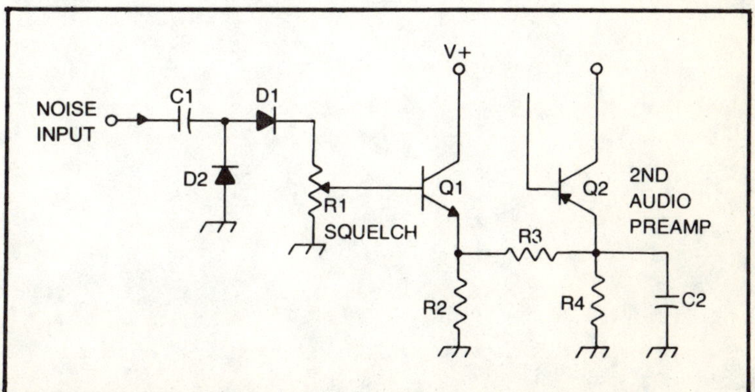

Fig. 10-12. Squelch circuit.

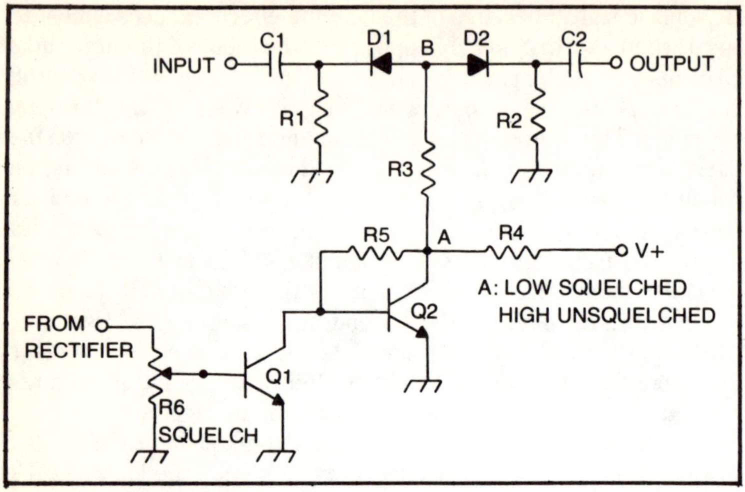

Fig. 10-13. Squelch circuit.

voltage is high. This will turn on the base of Q2, which is direct-coupled to the collector of Q1, keeping the collector of Q1 low. When a signal produces an output from the rectifier, however, Q1 is turned on, so its collector voltage (the base voltage of Q2) drops to zero. This turns off Q2 and forces its collector voltage (point A) high. When point A is high, the diodes are turned on and the audio signal is gated into the audio amplifier stage.

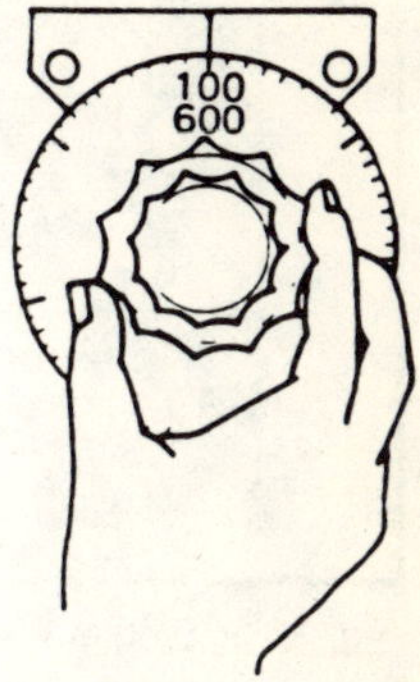

Chapter 11
Additional Circuits

Several common circuits in commercial receivers do not fit neatly into any of the other chapters, even though they were alluded to previously. Among these are the automatic gain control,the noise eliminators, and the S-meter circuits.

AUTOMATIC GAIN CONTROL

As you tune across any radio band you will encounter a widely diverse collection of signal strengths. These will vary from a mighty blast from the station down the street to a breath of warm (rf) air from a weak, distant station with a poor antenna. The listener with the gain controls set to receive the strong local signal at a reasonable volume would miss the weak station altogether. Similarly, the receiver set to receive the weak station at a comfortable volume level will be blasted out when the dial is tuned across the strong local signal.

The signal strength will also vary in long distance shortwave transmissions. The properties of the transmission path change, so the signal strength will go up and down over the course of several minutes. In an hour's time, there might be many different signal levels at the antenna terminals.

It would be nice to be able to vary the gain of the receiver to account for these signal strength variations. "Riding the gain control is not too practical. It is better to use an *automatic gain control* (agc) circuit. These are also known as *automatic volume control* (avc) circuits in older radios. The two are, however, functionally the same, and the two terms can be regarded as synonyms.

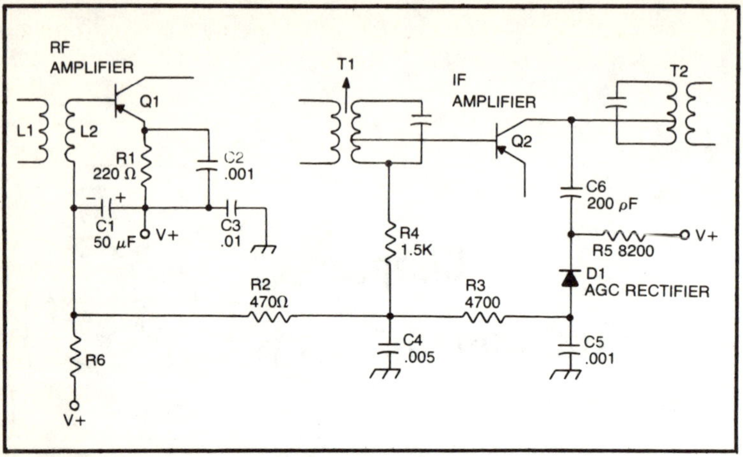

Fig. 11-1. Single-diode agc circuit for PNP stages.

The idea of the agc circuit is to sample the signal strength and then create a DC control voltage proportional to the strength that will vary the gain of the amplifiers. We want the gain of the receiver to increase when the signal strength is weak and decrease when the signal strength is strong. This will tend to even out changes in signal strength. In most receivers, the agc sample is taken in one of the last i-f amplifier stages. Usually it is *the* last i-f amplifier, except in FM radios where the last i-f might be a limiter stage. Even the secondary of the output i-f transformer might be used.

Figure 11-1 shows a simple agc system that is in common use. Transistor Q2 is the last i-f amplifier. Transistor Q1 is the rf amplifier. In most agc systems it is common practice to provide the gain control in at least the rf amplifier and very often in one or more i-f amplifiers.

Bias to the rf and i-f amplifier stages is supplied from two sources. The normal, fixed bias is through resistor R6 and L2 in the rf amplifier and through R6, R2, and a portion of the secondary of T1 in the i-f amplifier. The second source of bias for these stages is the agc rectifier, diode D1. The bias from D1 will counteract the fixed bias and is proportional to the signal strength.

The signal level is sampled from the collector of the last i-f amplifier (Q2) through capacitor C6. This signal is rectified in diode D1 to form a negative bias voltage across capacitor C5. A reverse bias is also applied to D1, so that a certain minimum signal strength is needed to cause the diode to go into operation. When the signal in the i-f amplifier is strong enough to cause rectification in D1, the negative voltage across C5 will tend to reduce the bias applied to the i-f and rf amplifier transistor bases. When the signal strength is high, the DC voltage at the output of the rectifier is high and counteracts most of the fixed bias. This will reduce the gain of the rf and i-f amplifier transistors. Similarly, a weaker i-f signal will produce a small output voltage, so less of the forward bias is counteracted. The gain is proportionally higher.

The simple circuit of Fig. 11-1 is for radios that use a PNP transistor in the rf and i-f amplifier stages. A version intended to work in radios that have NPN transistors in these functions is shown in Fig. 11-2. In both cases, we are assuming negative-ground operation common to mobile rigs. Again, the signal level at the collector of the last i-f amplifier is sampled by a small value capacitor. A rectifier diode D1 creates a negative DC potential—proportional to the signal strength—from the sample produced by C3. This negative potential is applied to the agc line in order to reduce the gain of the rf and i-f amplifier stages. In a negative-ground system, the voltages on this line are closer to ground for NPN transistors, so the rectifier diode is electrically closer to ground.

Note in both circuits that there is a 50-μF decoupling capacitor across the output of the agc system, usually placed very near the rf amplifier. The purpose of this capacitor is to prevent oscillation of the circuit. An agc system is a feedback circuit, so oscillation at some frequency or another is possible if this cpacitor is not used.

A dual-diode, voltage-divider agc system is shown in Fig. 11-3. This particular circuit is intended for radios that use PNP transis-

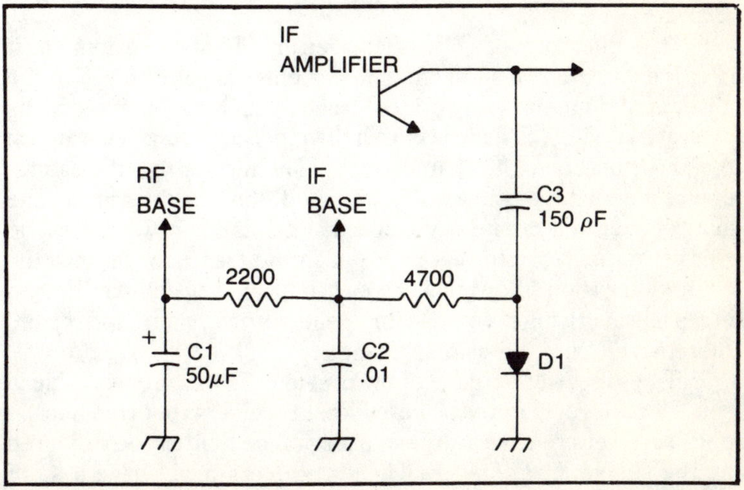

Fig. 11-2. Single-diode agc circuit for NPN stages.

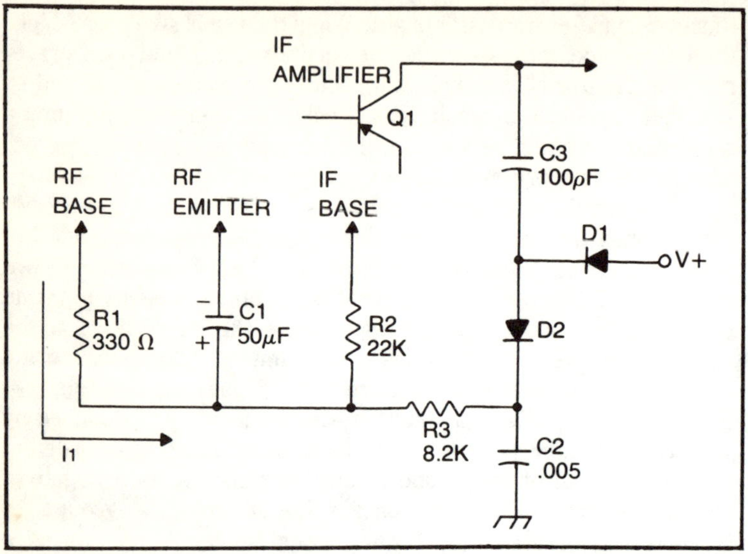

Fig. 11-3. Two-diode agc system.

tors. These diodes are directly in the DC bias path between the base of the rf amplifier transistor and the V+ power supply. Current I1 is the bias current. When the signal strength is low, the current is smaller, so the voltage drop across resistor R1 is less. This will make the bias higher, therefore increasing the gain of the stage. Similarly, a high signal strength reduces the bias by increasing the voltage drop across R1. A comparable scheme also applies to the i-f amplifier.

One other form of agc is often seen in solid-state equipment. It is called *forward agc* and sometimes seems less like an agc to the uninitiated. Forward agc works because a transistor operated in the saturated region will have less gain than the same stage operated in the linear portion of its characteristic. The further into the saturation region that the transistor is operated, the less the gain is. The agc rectifier will create a voltage that tends to forward bias the transistor. A weak signal will not push the transistor far into the saturation region, allowing a higher gain. A strong station, though, is capable of driving the transistor deeper into the saturation region, thereby reducing the gain substantially.

The simple agc circuits of the previous figures are workable in simpler receivers, especially broadcast receivers, but communications receivers require a large dynamic range than can be delivered by the simple diode stages. These receivers might have a more complex agc system, such as shown in Fig. 11-4. The overall concept is the same, except that amplifiers are used to create appropriate signal levels within the agc system. An agc amplifier that operates at the intermediate frequency receives the signal sample and amplifies it to a point where it is well into the linear range of the diode rectifier. In fact, the rectifier itself might not be a simple diode, but instead an active "precision" rectifier circuit (see TAB book No. 787, *OP AMP Circuit Design & Applications*) that produces a nearly linear transfer function. Finally, there will be a DC amplifier or level translator that actually drives the rf amplifier transistor. This becomes especially needed when certain types of MOSFET or IC are used, and a larger DC control voltage is needed at the agc input.

The time constant of the agc system determines how fast the circuit operates. In AM circuits, a relatively slow agc is needed, so the time constant will be long. This is done to make sure that the control voltage is averaged out and that none of the modulation appears on the agc line. With CW and SSB receivers, the agc should be faster. In fact, its properties should be shaped to permit a fast attack and a slow decay.

Agc Defects

One might think that the principal symptom of agc circuit defects would be loss of the gain control action. This is true to some extent, but it is overshadowed by other defects that become so

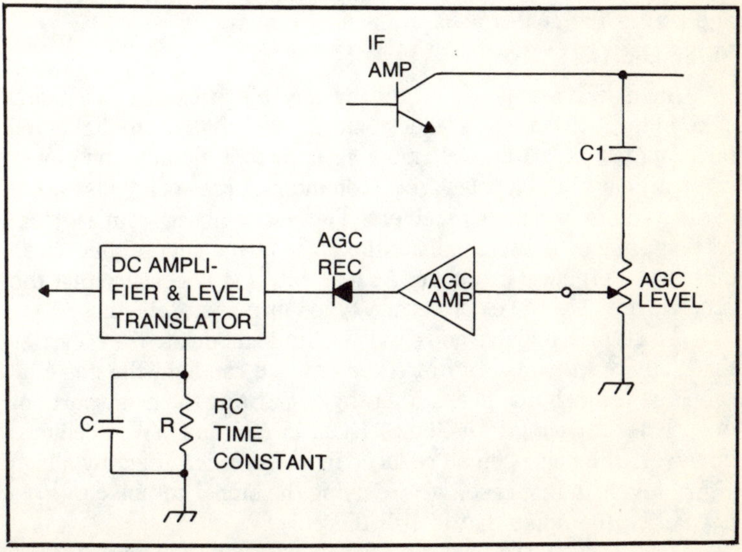

Fig. 11-4. More complex agc.

annoying as to mask the loss of gain control. One of the main defects noted will be oscillations. The oscillations are *tunable*, meaning that the oscillation can be heterodyned when the radio is tuned. The tunable type of oscillation is a dead giveaway that the problem is in the rf, mixer, or i-f amplifier, or at least some stage prior to demodulation. The other symptom is distortion on the strong stations, but not weaker stations. This particular symptom is almost 100 percent traceable to agc problems. (There *is* at least one exception, but more of that later.) Whenever you notice distortion on only strong stations, suspect the agc. This occurs because the defect tends to make the agc operate close to cutoff all of the time. A weak signal will not drive the rf or i-f amplifiers hard into cutoff, but a strong signal has no such trouble. If the distortion problem seems to be without significant oscillations, look to the agc rectifiers first. If there is a substantial oscillation present at the same time, look first to the large electrolytic capacitor at the output of the agc network. These are only rules of thumb that help make the troubleshooting faster; they are not absolutes. In some cases, the opposite defect will cause the symptoms specified.

The case where there is distortion on strong signals, but not on weak, in which the agc is *not* involved, is a partial short between the collector and base of the transistor used in the controlled stage (rf or i-f , mostly the rf). This problem was particularly common some years ago when most rf transistors were made of germanium and would increase in severity with temperature, which made the leakage across the collector-base junction increase.

NOISE LIMITERS

Impulse noise is generated by any electrical arc, including lightning. Electric motors, electronic dimmer controls for lights, oil furnace ignition, automobile ignition, and loose tie lines on power company distribution poles are all common sources of impulse noise in radio communications receivers. The *best* solution, from a technical point of view, is to eliminate the noise at the source, but this is not always practical or politic. As a result, it is necessary that the receiver be able to reject certain types of noise problems.

The typical impulse noise will amplitude modulate the rf carrier that you are tuned to. In FM receivers, we can fight the noise by clipping the amplitude peaks, thereby eliminating the noisy portion, or at least reducing its volume. Even in AM and CW receivers, however, the noise can be reduced in its *apparent* effect by a high signal level. In that case, we are using the signal-to-noise ratio to not notice the noise that is still there!

There are two classes of noise reduction system: *noise limiters* and *noise blankers*. Of these, the blanker is generally considered the most effective, but the noise limiter is a lot simpler. This means, of course, that low-cost receivers tend to use noise limiters, while complex, expensive radios use noise blanker circuits.

The noise limiter is sometimes called the automatic noise limiter (ANL). There are two basic configurations for post detection ANL circuits: series and shunt. These are shown in Fig. 11-5. The diode in Fig. 11-5A operates as a series gate. It is biased by a reverse voltage V_r that permits only signals below a certain level to pass. If a high-amplitude signal comes along, it will reverse bias the diode, causing it to momentarily cut off. The bias is set such that ordinary signals will not turn off the diode, however, noise signals have a very high amplitude. This will drive the diode into its nonconducting region, knocking a small hole in the signal at the point where the noise existed.

The series noise gate is placed in the line between the output of the detector and the input of the audio amplifier. We can also use a shunt noise limiter in this line, an example of which is shown in Fig. 11-5B. In this circuit, the noise gate is shunted across the line. Bias levels V1 and V2 are set such that diodes D1 and D2 will not conduct when ordinary signal levels are present. When there is no noise on the signal, therefore, the diodes are effectively reverse biased. A high amplitude noise peak, however, will drive one of the two diodes hard into conduction, causing it to short out the line for the duration of the noise spike. Again, a hole has been poked in the signal for the period of the noise signal. Because two diodes are used, noise spikes of both polarities can be accommodated.

Neither of these noise limiters are considered very good, especially because of the need to place them in the audio section. Somewhat better, however, are versions of these circuits that

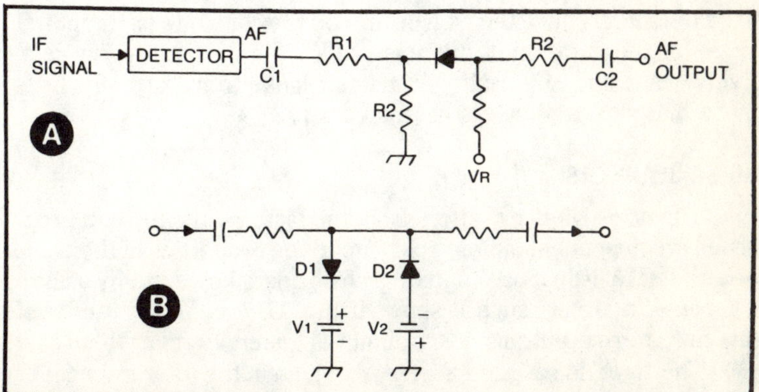

Fig. 11-5. Simple noise limiters.

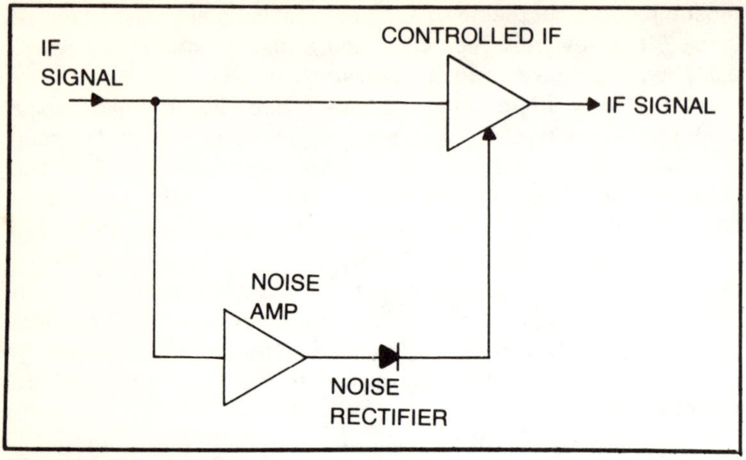

Fig. 11-6. Noise silencer.

operate in the i-f amplifier or even the rf amplifier. A shunt noise limiter placed across the primary of the output i-f amplifier transformer will operate better than the audio version. Part of the problem is that the noise spike tends to be broadened in time in the high-Q tank circuits, and in the detector. For this reason, it is wise to place the noise limiter earlier in the chain.

Figure 11-6 shows the block diagram for an i-f noise silencer. This circuit was once very popular in communications receivers, but has since been surpassed in popularity by noise blanker circuits. In this circuit, either a gain controlled or a gated (on/off) i-f amplifier can be used. A noise amplifier will build up the noise signal, but this is filtered to prevent the i-f signal from being amplified an equal amount. The output of the noise amplifier is rectified and is then used to either reduce the gain or turn off the controlled i-f amplifier stage for the duration of the noise spike. This circuit works relatively well, and some alleged noise blankers are actually little more than noise silencers of this variety.

NOISE BLANKERS

The noise silencer suffers from the fact that the turn-off properties are dependent almost solely upon the properties of the noise pulse itself. A little more control can be gained, however, by using a *noise blanker* circuit, such as shown in Fig. 11-7. Several varieties of this circuit exist, but this description will generally cover them all.

The signal is sampled at an early stage in the i-f or rf amplifier chain. In one early noise blanker, made by Collins in the late 50s, a separate sense antenna and transmission line was mounted near the receiver antenna, but in modern receivers, a sample is taken early in the chain. The earlier the sample is taken, the better, because the LC tank circuits that follow will tend to broaden the noise pulse, making the problem even worse.

The signal at the sample point is split into two paths: One passes through the regular circuitry, and the other goes to the noise blanker section. It is in the first path that the various circuits differ from each other. In some cases, the signal to the main section is passed through several stages of i-f amplification and possibly the mixer/converter stage. The normal propagation time through these stages allows the time delay needed for the noise blanker circuits to operate. If the signal were applied to the gated i-f immediately, the blanking pulse would come along too late, and the pulse would have passed. In some other circuits, the main path signal does not go through additional i-f amplifier stages, but through a *delay line*—an RLC network that provides the needed phase delay. This system is particularly common in add-on noise blankers, where all of the activity must take place inside a box added to the receiver after manufacture.

The noise blanker path for the signal contains a high-pass filter that has the effect of stripping off the signal and leaving only the noise signal. This is done because the signal will tend to have lower frequency components, while the high-amplitude, short-duration noise pulse is primarily a high-frequency signal. The output of the high-pass filter is a noise signal and is amplified in a wideband *noise*

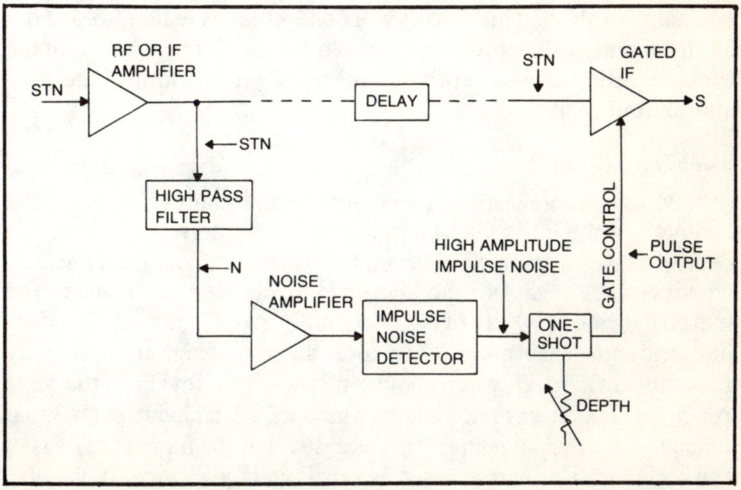

Fig. 11-7. Noise blanker.

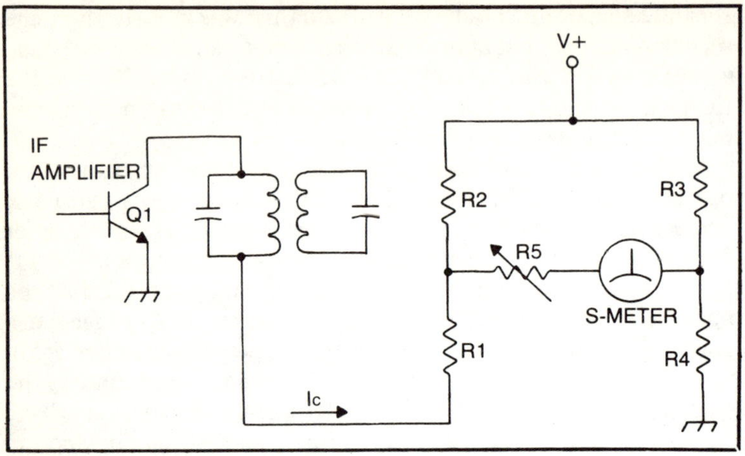

Fig. 11-8. Bridge-type S-meter.

amplifier stage. The output of the amplifier is applied to a circuit that detects the noise pulse. This is usually something like a high-pass filter or differentiator that will depend upon the fast rising leading edge of the noise pulse in order to generate an output. The signal produced by the noise detector is similar to the control signal produced at the output of the noise rectifier in the example of Fig. 11-6. But—and this is one of the principal differences between the signals—this signal is used to fire a monostable multivibrator, or *one-shot*. The output of the one-shot is a pulse of constant amplitude (sufficient to turn off the gated i-f amplifier stage) and constant duration. This circuit will poke a much cleaner hole in the i-f signal during the noise impulse. We can adjust the duration of the turn-off pulse using a depth control to vary the width of the one-shot output pulse.

S-METER CIRCUITS

Most communications receivers use an S-meter to indicate the strength of the input signal. An S-meter is a meter calibrated in a decibel scale, with so-called S-units 0 through 9 and then decibels above S-9. A really strong signal, for example, might make the meter indicate +30 dB over S-9. Unfortunately, not all S-meters have the same calibration. Therefore, these readings are practically meaningless, except when comparing two signal levels on the *same* receiver. There was once a semi-standard which made each S-unit equal to a 3 dB change in signal *power* level at the input terminals of the receiver. This is translated as a 6 dB voltage change. It was also stipulated that S-9 was equal to a signal level of 100 microvolts at the antenna terminals of the receiver. Do not make any assumptions regarding any given receiver, however, until you read the manual specifications.

One of the simplest S-meter circuits is the Wheatstone bridge circuit of Fig. 11-8. The S-meter movement is placed across the output terminals of the bridge. Three of the bridge arms are made of fixed resistors: R2, R3, and R4. The fourth arm is made of a combination or resistor R1 in series with the collector resistance of i-f amplifier transistor Q1. When the signal strength is high, collector current I_c is also high, so the voltage drop across R1 is high. This will unbalance the bridge a larger amount than we would find when a weak signal made I_c smaller. The bridge is designed to be balanced (resulting in a zero current in the meter) when there is no signal applied to the input of the i-f amplifier. In fact, most S-meter adjustment instructions ask us to zero the meter to S-0 when the antenna input terminals are shorted together, guaranteeing zero signal.

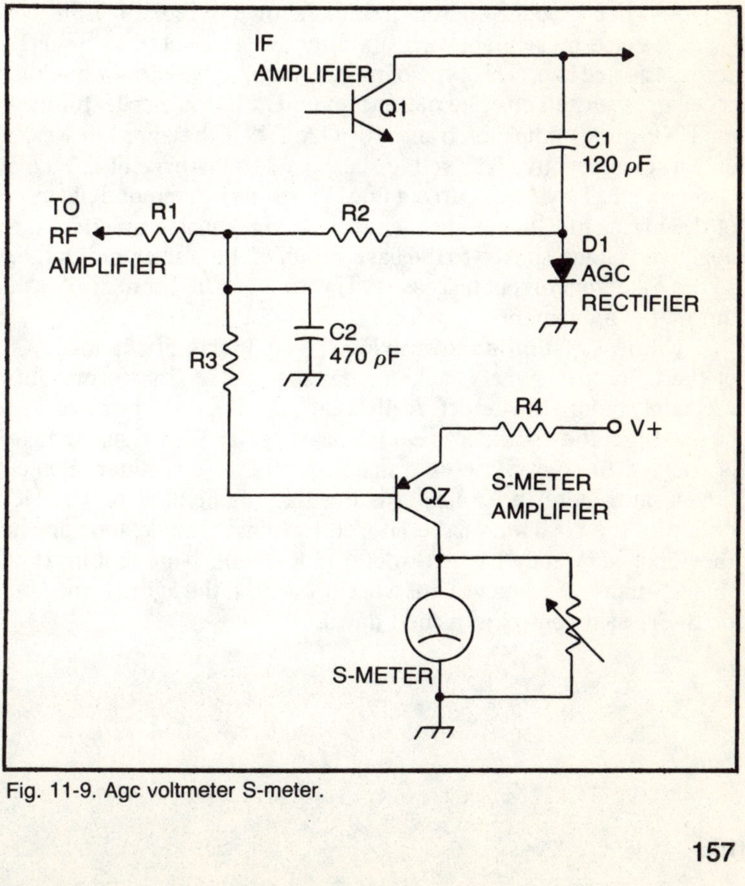

Fig. 11-9. Agc voltmeter S-meter.

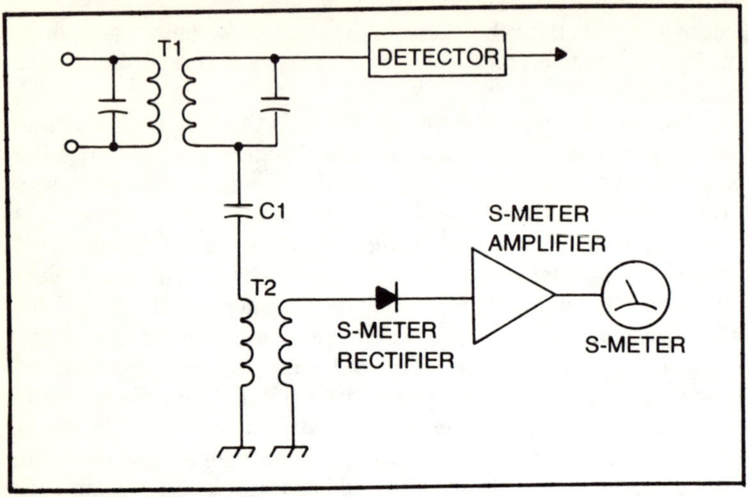

Fig. 11-10. Amplifier-type S-meter.

Another simple S-meter circuit is shown in Fig. 11-9. In this case, we are merely applying a voltmeter, calibrated in S-units, across the agc line. This type of meter, then, only works when the receiver is turned on. The meter movement forms a collector load for PNP meter amplifier transistor Q2. When the signal is weak, only a small negative bias voltage is applied to the base of Q2. This causes a small collector current flow, resulting in a minor deflection of the meter movement. But when a strong signal is tuned in, the negative voltage applied to the base circuit of the transistor is large, so the collector current increases. This causes a larger deflection of the meter movement.

One last system is shown in Fig. 11-10. In this circuit, a sample of the i-f output signal is taken from the last i-f transformer immediately before the detector. Alternatively, the primary could also be used, but the secondary seems more popular. The signal sample is coupled through S-meter transformer T2 to a rectifier. Sometimes, an S-meter preamplifier is used at this point also. The DC output of the rectifier will be filtered to remove modulation and is then applied directly to the S-meter movement. Note that most of these S-meter circuits will not work properly if the agc is turned off or the rf gain control is turned down.

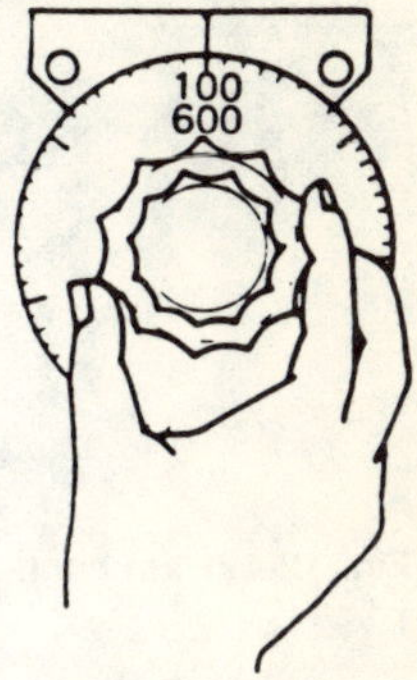

Chapter 12 High-Frequency Shortwave Receiver Systems

The modern shortwave receiver will operate over the entire shortwave spectrum, from 3 to 30 MHz, and many will also operate in the AM broadcast band and the VLF band below the AM broadcast band (150 to 450 kHz). Some receivers on the market today are traditional designs in the sense that they use an ordinary VFO or variable i-f design and an analog frequency readout dial. Some traditional designs now boast digital frequency counter dials in place of the analog dial. Although the existence of a digital dial does not guarantee better frequency readout accuracy, it is possible if the counter is correctly designed. The newest designs use phase-locked loop (PLL) circuitry in order to precisely control the local oscillator signal. One of the first of the new types was the National HRO-500. Almost all of the major receiver manufacturers now offer PLL designs, which have become greatly simplified as to front panel operation than the receivers of a decade ago.

TYPICAL HF RECEIVER CONTROLS

Despite the variety of designs used in HF communications receivers, there are certain similarities which are noted in the functioning (if not labeling) of the front panel controls. Figures 12-1 and 12-2 show two typical, medium-grade shortwave receivers. The receiver shown in Fig. 12-1 is the Heath HR-1680, which is a ham band-only model. Shown in Fig. 12-2 is the Yaesu FRG-7000 general-coverage receiver. Both of these receivers are all solid state and have been introduced within the past few years.

Fig. 12-1. Ham band-only receiver (courtesy of Heath Co.).

Band. The bandswitch selects the frequency range which will be tuned. In an amateur band only receiver, this control might be labeled with the meter band, such as 80, 40, 20, 15, and 10 meters. In some shortwave general-coverage receivers this same method of labeling is used, but in the terms of international shortwave broadcasting: 49, 31, 25, 17, and 13 meters. Many general-coverage receivers tune broader segments than limited ham band-only or international broadcasting models. These will be labeled with a frequency *range* for each band, such as 1.6 to 4 MHfz, 4 to 12 MHz, etc. Some wide-range high-frequency receivers cover the entire spectrum from 3 to 30 MHz, but do so in 500 or 1000 KHz segments. The Collins 51S3 and 51J4 models are examples. These receivers often have an unlabeled bandswitch, depending upon the dial, to indicate the frequency range. The dial will typically change with the bandswitch.

In single-conversion receivers the bandswitch typically changes all of the LC tank circuits in the antenna, rf amplifier, and local oscillator circuits. These bandswitches are usually quite complex and impressive. Many of the double-conversion systems, however, change only the crystal oscillator in the front end and possible some rf amplifier tank circuit components. This simplified bandswitching is seen on most of the amateur band-only receivers.

There has been a swing away from rf switching in the past few years. When all of the rf components are routed through the bandswitch, a large amount of stray capacitance and inductance makes it difficult to correctly calibrate the receiver through all of its ranges. Alignment and servicing of these receivers frequently changes the calibration of the dial. Recently, though, it has been DC switching of the various circuits has been used. Each of the tank circuits is selected by switching diodes. The bandswitch will forward bias the diodes in the desired tank circuit (or crystal) and reverse bias the diodes in the undesired circuits. This is allowable because a heavily forward biased diode will pass small AC (rf) signals *undistorted*. The main criterion is that the rf signal amplitude be small with respect to the bias suppled from the bandswitch. This requirement is easily met by most rf signals.

Tuning. The tuning control will select the actual frequency to which the receiver is tuned. This control generally varies the frequency of the local oscillator. The tuning control will be ganged to the main dial in receivers which have an analog dial. A system of pulleys and a dial cord is usually used for this purpose, so that the tuning ratio is reduced. Generally, the tuning rate when direct drive is used is too fast for use in the shortwave bands. In fact, only the cheapest AM radios use direct drive, because even there the tuning rate is too fast. The pulleys will make it possible to cover fewer kilohertz per turn of the dial.

Digital counters can make the most accurate tuning dials, if done properly. The main problems are the resolution and the accuracy of the time base. If a dial reads out only to 100 Hz, it is no more accurate than an analog dial that reads out to 100 Hz. In most cases, however, the digital dial is more accurate because only a few analog dial receivers will readout to the 100-Hz level, while it is a relatively easy trick in digital dials. The time base can also greatly affect the readout accuracy. A digital dial is no more than a digital frequency counter. These circuits work by opening a gate to local oscillator pulses for some decade fraction of a second, allowing them into a decade digital counter. The counter circuit will accumulate

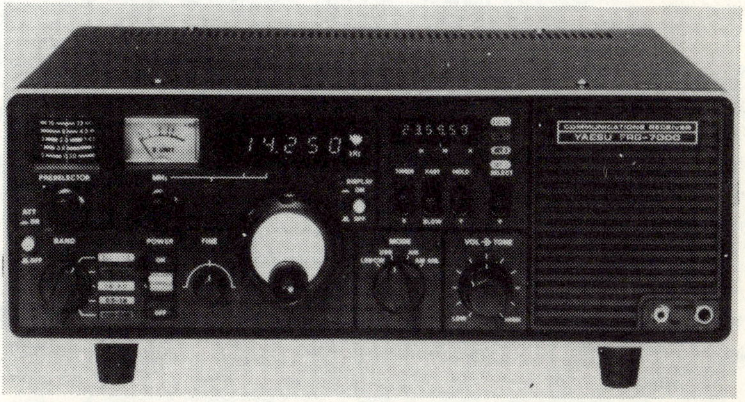

Fig. 12-2. General-coverage receiver (courtesy of Yaesu).

pulses for the duration of the gate-open period and display the result on a set of seven-segment readouts.

The digital counter used in receiver dials must be biased to account for the intermediate frequency. Recall that the local oscillator is different from the rf signal by the amount of the intermediate frequency. It is the radio frequency that we wish to display on the dial, yet it is the LO that is counted. Various digital circuits have been developed for this purpose. One source of error here is that the intermediate frequency may not be exact in lower-cost receivers and is set by the actual, not nominal, center frequency of the crystal or LC filter used in the i-f amplifier. This center frequency must be tightly spaced or there will be an error. Whether the receiver manufacturer does it in-house, or orders such filters from another vendor, the tight specification costs money. As a result, only higher cost digital readout receivers will have this error reduced to the minimum.

Preselector. The preselector control is found on receivers in which the rf amplifier is tuned separately from the local oscillator. This is particularly common in models that use a crystal-controlled or PLL-controlled front end to drive a variable i-f amplifer. The preslector control is merely the rf amplifier and antenna tuning controls ganged to one knob.

Antenna. Some receivers which gang the rf amplifier and local oscillator controls to one tuning knob will have a separate tuning knob for the antenna. This will allow the operator to match the antenna to the receiver, permitting a wider range of antenna configurations. An unfortunate tendency among some manufacturers is to confuse the *antenna* and *preselector* designations. I have seen some controls marked *antenna* which were in actual function *preselector* controls.

Rf Gain. This control will vary the gain of the rf amplifier stage. It should be reduced when very strong signals are being received to prevent front-end overloading of the receiver. The S-meter, however, will normally not work or will be inaccurate unless the rf gain control is set to a certain point (usually maximum). In most cases, keep the rf gain control at or near maximum and use the af gain control to control the output volume. However, it is easier to tune in some single sideband signals if the rf gain is reduced a little bit.

Af Gain. The af gain control varies the amplitude of the audio signal that is fed to the audio amplifier section. As such, it functions essentially as an audio gain control. This control is sometimes labeled *volume*.

Mode. This switch selects the type of demodulation that is recognized by the detector. There will be at least two positions: AM and BFO (also labeled SSB, CW or SSB/CW). The AM position selects an envelope detector for amplitude modulation reception, while the other position selects a product detector for demodulation of SSB or CW signals. In the simple systems, there will be a BFO control on the front panel that adjusts the frequency of the beat frequency oscillator in the product detector. The BFO frequency must be set differently for LSB and USB reception to receive alternate sides of the nominal center frequency. For CW reception, the BFO frequency is set to produce a pleasing output tone. This a distinctly personal preference, with most people selecting some frequency in the 300 to 1200 Hz range.

In most genuine communications receivers, including some which are actually in the low to medium price range, four positions are on the mode switch: AM, SW, LSB, and USB. These allow individual crystal-controlled BFO frequencies for the three modes recognized by the product detector. The CW BFO frequency is usually 1 kHz away from the i-f center frequency, so that a 1000-Hz beat note is produced. The LSB and USB BFO frequencies will be 1.5 kHz above and below the i-f center frequency, respectively.

RIT. The *receiver incremental tuning* (RIT) changes the received frequency a few hundred hertz from the indicated frequency. This control is often found on amateur radio SSB transceivers where it allows an operator to independently tune in different stations of a net or roundtable, even though all are ostensibly on the same frequency. No change in the transmitted frequency is made.

Attenuator on/off. Some receivers have an input attenuator between the antenna connector and the input of the rf amplifier. This circuit will reduce the level of signal applied to the rf amplifier, preventing overload. Because the attentuator factor is fixed and usually precise, the S-meter reading can be biased to account for the difference. The S-meter will then read correctly, eliminating the problem we faced with the rf gain control.

Agc. It is not always desirable to have an agc (automatic gain control) circuit operational. We may want to turn the agc off or vary its time constant. AM signals have different agc attack time requirements than do CW or SSB signals.

Clarifier. The clarifier is usually a small trimmer capacitor across the crystal in the BFO of the product detector. This control will allow the operator to adjust the BFO frequency when listening to single sideband transmitters. It is helpful when listening to different stations in a network, and some people use it for the same purpose as receiver incremental tuning. It is not, however, quite as

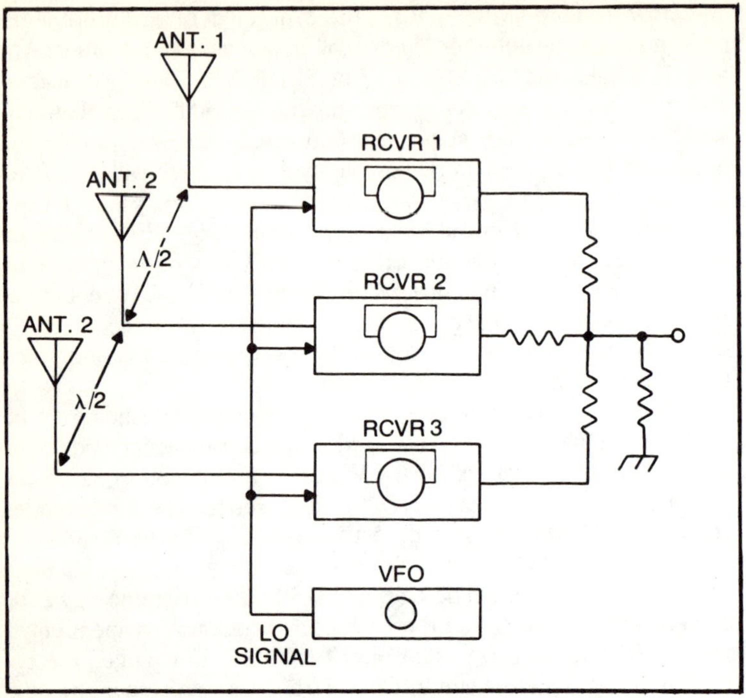

Fig. 12-3. Diversity reception system.

versatile as RIT, which allows stations close to the passband edges to be tuned out.

DIVERSITY RECEPTION

Commerical, broadcast, and military stations frequently use a multiple antenna and receiver system in order to reduce the effects of certain problems in shortwave reception. Many shortwave broadcasters operate as relay stations for programs that were originally transmitted from the mother country. The British Broadcasting Corp. and Voice of America (VOA), for example, transmit programs worldwide from a network of local, regional, and international transmitters. In the case of the VOA, stateside transmitters, such as at the complex at Greenville, NC, will make the original transmission, and then station in countries like Monrovia and Liberia will pickup that broadcast and retransmit it over their own transmitter.

But shortwave reception is a long way from being reliable. There are several causes of fading. Shifts in the ionosphere produce several different problems for the receiving station. The signal level might go up and down, or it might change either in the up or the down direction over a period of time. When the ionosphere is in an unstable condition, shortwave reception suffers greatly. Diversity reception is an attempt at reducing the effects of this problem. Figure 12-3 shows one form of diversity setup. Three receivers are used, each tuned to the same frequency. A common VFO supplies the same LO frequency to all three receivers, insuring that all three are, indeed, tuned to the same frequency. Some commercial and military receivers that are intended for diversity operation have LO output/input jacks so that one of the three can be used to supply an LO signal to the others.

Three antennas are used to feed signal to the three receivers. Each receiver has its own antenna system. These antennas will be spaced either a half-wavelength quarter-wavelength apart on the theory that the fading signal will move from one antenna to the other (not a bad assumption, as it turns out).

The audio signals from the three receivers are combined in a linear network and will be amplified in a common power amplifier (or line amplifier, if fed to the modulator of another transmitter). Some diversity systems will have a switching system that detects which signal is the strongest and then switches to the audio output of that receiver.

Another form of diversity reception is shown in Fig. 12-4. This system uses two antennas of opposite polarity to feed the two

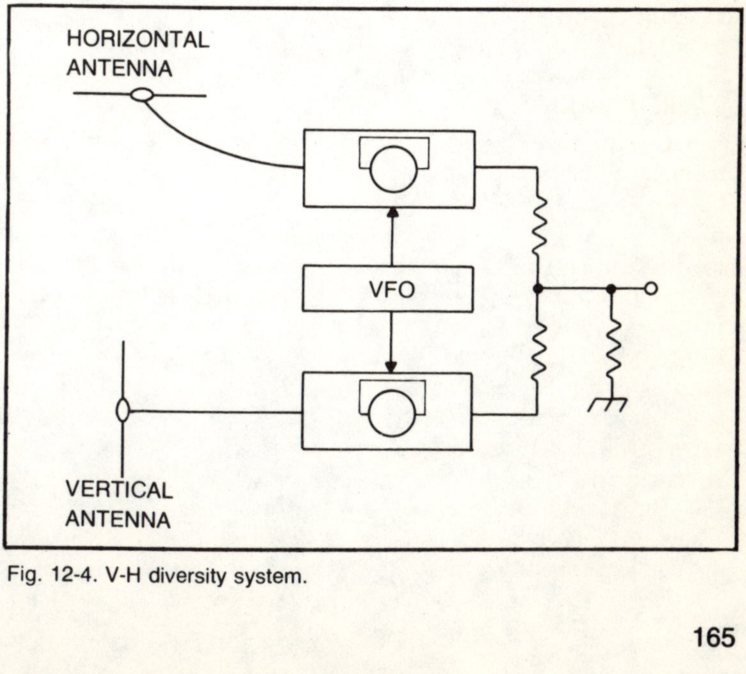

Fig. 12-4. V-H diversity system.

receivers. The polarity of a radio signal is usually expressed as either vertical or horizontal, which denotes the direction of the *electrical field*. The magnetic field will be at right angles to the electrical field, but it is the electrical field that determines the polarization of the signal. A vertical antenna produces a vertically polarized signal, and a horizontal antenna produces a horizontally polarized signal. The best reception occurs when the receiver antenna polarization matches the transmitter antenna polarization. A substantial loss of signal strength is caused by cross-polarization between the transmitter and receiver. In addition, reflections and certain ionospheric problems will reverse the polarization of the transmitted signal. Some signals arrive in such condition as to have multiple polarization components, although the original was either vertical or horizontal. The diversity reception system of Fig. 12-4 combines the two components in two different receivers to produce maximum, almost fade-free output.

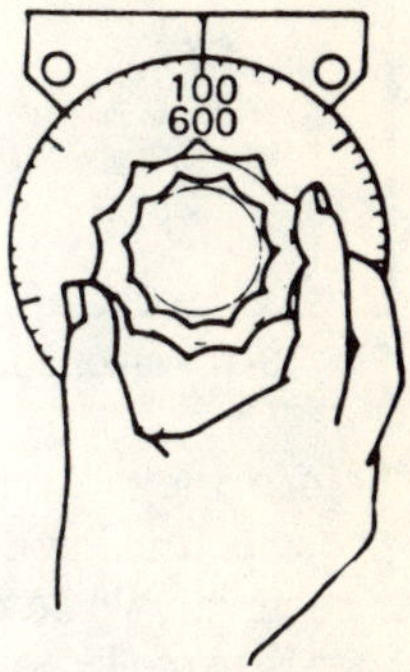

Chapter 13

Receiver Accessories

A number of popular accessories are commonly found on communications receivers, especially high-frequency shortwave receivers. Some accessories are intended merely to improve the convenience to the operator. Other accessories are designed to correct real or perceived difficulties in the design of the receiver. Low-cost receivers, for example, frequently lack much in the way of both selectivity and sensitivity. Owners of these receivers often find that a *preselector* will improve the sensiticity and an *audio filter* will improve the selectivity somewhat. In still other cases, it's found that an antenna tuner will act to improve the performance. This is especially true when the receiver is designed for a 50-ohm antenna, and we are actually using an antenna of some other impedance. Another use for the antenna tuner is to limit the signals applied to the antenna terminals to those in the range of interest. Strong local, out-of-band signals tend to desensitize the receiver or cause spurious in-band responses. An antenna tuner will attenuate these out of band signals considerably, allowing the receiver to operate on in-band signals at full sensitivity.

A crystal calibrator is a useful adjunct to most receivers that have an analog dial. The calibration points are rarely accurate, yet can be made accurate over a limited range by comparison with a calibrator. Most crystal calibrators will output a frequency of 100, 500 or 1000 Hz, which is rich enough in harmonics to produce signals in the shortwave bands.

PRESELECTORS

A preselector is a tuned rf amplifier placed ahead of the receiver. Its purpose is to amplify the rf signals from the antenna

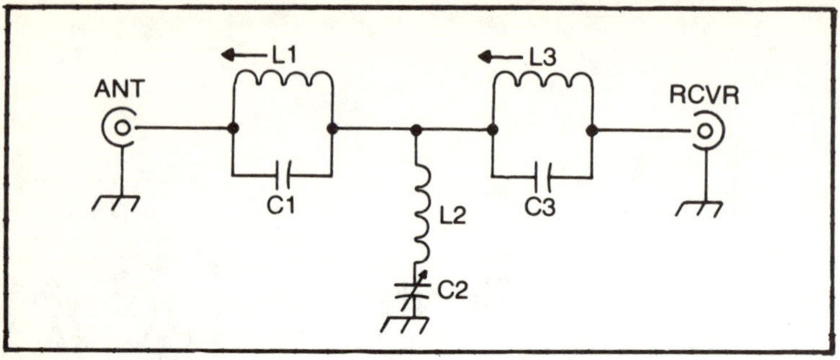

Fig. 13-1. Wavetrap.

before they reach the antenna terminals of the receiver. Some preselectors have a tuning dial that looks much like the tuning dial of a receiver, while others are marked with only a logging scale. The idea is to peak the signal strength with the tuning dial, while listening to the receiver output or watching the S-meter.

Figure 13-1 shows the block diagram to a typical rf preselector. The tuned rf amplifier portion can be any of the standard rf amplifier circuits of Chapter 5, but most people seem to prefer the MOSFET or JFET variety, with a distinct subpreference for the cascade types of circuits. A double-pole-double-throw switch (S1A) is used as an in-out switch. When the switch is connected to the *out* position, the antenna will be connected directly to the antenna terminals of the receiver. But when the switch is connected to the *in* position, the antenna is connected to the input of the preselector amplifier, while the antenna terminals of the receiver are connected to the output of the preselector rf amplifier.

The preselector is also used to improve the image rejection of some low-cost superheterodyne receivers. Recall that an image is a signal response at a frequency two times the i-f removed from the rf signal. This response is due to the fact that the mixer is not affected by whether the LO is above or below the rf signal. If the LO frequency is above the radio frequency, the signals above the LO by the same amount (the i-f) will also cause a response in the mixer, and that response is seen as a valid i-f signal by the receiver. If the preselector has sufficient Q in the two tank circuits, the signals at the image frequency will be attenuated, while the signals at the radio frequency are amplified. This apparent difference tends to improve the image response of the receiver.

WAVETRAPS

A wavetrap is a filter circuit designed to attenuate certain frequencies and pass all others. These are especially handy devices to use when the receiver is located close to a strong local broadcasting station. AM broadcast stations are the worst offenders for most amateurs and SWLs. VHF users, of course, have to worry about FM broadcasters. When a strong signal is applied to the input of an rf amplifier, it tends|to form a bias voltage that will cut down the gain of the stage. This we don't need. The only cure is to rid the receiver of the unwanted signal prior to the rf amplifier input. The usual prescription is to make a *wavetrap* to take out the signal. Figure 13-2 shows an elementary wavetrap. Recall the characteristics of series and parallel resonant LC tank circuits. The parallel resonant circuit exhibits a high impedance to the resonant frequency and a low impedance to all other frequencies. The series resonant circuits is just the inverse; it presents a low impedance to signals at the frequency of resonance, and a high impedance to all others. Both types of tank circuits are used in a wavetrap. Two parallel resonant tank circuits, L1/C1 and L3/C3, are placed in series with the signal path. This means that the resonant frequency of tank circuits will be attenuated. The series resonant tank circuit is connected in shunt with the path at the junction of the two parallel resonant tank circuits. The combination of series and parallel resonant tank circuits, all on the frequency of the interfering signal, will attenuate the signal a great deal, but offers only minimal loss to the signals within the desired passband.

The specific trap shown in Fig. 13-2 is intended to remove a specific frequency. This is the usual case for most locations, where nearby a single radio station operates on only one frequency that is removed from the band of interest. The tank circuit components are adjusted to null the undesired frequency.

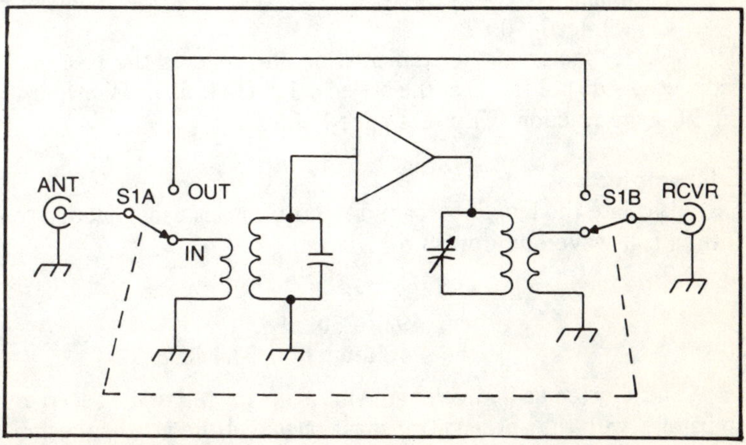

Fig. 13-2. Preselector.

If several stations are to be removed, some other approach might be needed. One might be to build a high-pass rf filter that will pass the high-frequency shortwave signals but attentuate the AM broadcast signals.

Note that FM interference is a problem in many television systems, so antenna manufacturers offer 75-ohm and 300-ohm FM traps that offer attenuation levels difficult to attain in homebrew products. These could be pressed into service for the 6-meter and 2-meter amateur bands, where such interference may also be found. This reverse TVI can also be taken out by using a half-wave stub across the antenna terminals of the receiver. The impedance at the end of a half-wavelength piece of transmission line is repeated at the feedpoint. This same principle is why many low-cost antenna instruments work only when the transmission line is exactly a half wavelength long. The end of the stub is shorted out, producing an effective short circuit at the feedpoint at the frequency for which the cable is one-half wavelength long. The correct length is given by:

$$L = \frac{492V}{F}$$

where L is the length in feet, V is the velocity factor* of the transmission line, and F is the frequency of the interfering signal in megahertz.

* *V* will be different for various types of transmission line. The most common are:

- ☐Foam coaxial cable—0.80
- ☐Regular coax—0.66
- ☐Twin-lead—0.82

The velocity factor of the transmission line reflects the fact that radio waves do not travel at the speed of light in transmission lines, but at some fraction (*V*) of the speed of light.

☐**Example:**

Calculate the length of foam dielectric coaxial cable needed to form a half-wave stub for 88.5 MHz.

$$\begin{aligned} L &= 492\ V/F \\ &= (492)(0.80)/(88.5) \\ &= 4.45 \text{ feet} = 53.4 \text{ inches} \end{aligned}$$

The shorted half-wavelength stub is connected to the receiver in parallel with the antenna transmission line at the antenna terminals. If coaxial cable is used, a tee-connector can be used to interconnect the two cables. Note that some error may exist in the calculation caused by differences in the actual and approximate velocity factors, and in measurement tolerances. It is therefore standard practice to cut the stub slightly longer than needed and then trim to the correct length by watching the attenuation of the interference signal increase.

ANTENNA MATCHING NETWORKS

There are two reasons to use an antenna matching network. One is to match the impedance of the antenna to the input impedance of the receiver. Maximum power transfer occurs when these impedances are matched. Since the input circuitry of an antenna responds to power phenomena, it can be an improvement if these impedances are matched to each other.

The other reason for using the antenna tuner has nothing to do with power transfer. Indeed power transfer is almost a moot point when the receiver is of low grade. These same receivers will exhibit poor image response and poor overload properties when confronted with the same strong local interference just discussed. This problem can be reduced considerably by using a matching network—an antenna tuner—between the antenna and the receiver input terminals. The antenna matching circuit is tuned so that it passes the desired signals and rejects all others, including the interfering signal. Antenna tuners made with high-Q components can be quite sharp, providing a large degree of isolation of the receiver from the interfering signal.

Figure 13-3 shows two popular forms of antenna tuners. The circuit shown in Fig. 13-3A is a simple L-section coupler using a simple coil and capacitor combination. This circuit assumes that the impedance of the antenna at resonance, R_a, is greater than the input impedance of the receiver. This is the usual case in systems where the receiver has a standard 50-ohm or 75-ohm antenna input impedance, and a high-impedance long-wire antenna is used. The component values for this circuit are found from:

$$X_c = R_a \sqrt{\frac{R_r}{R_a - R_r}}$$

and

$$X_L = \frac{R_a R_r}{X_c}$$

where X_c is the reactance of capacitor C, X_L is the reactance of inductor L, R_r is the input impedance of the receiver, and R_a is the impedance of the antenna at resonance.

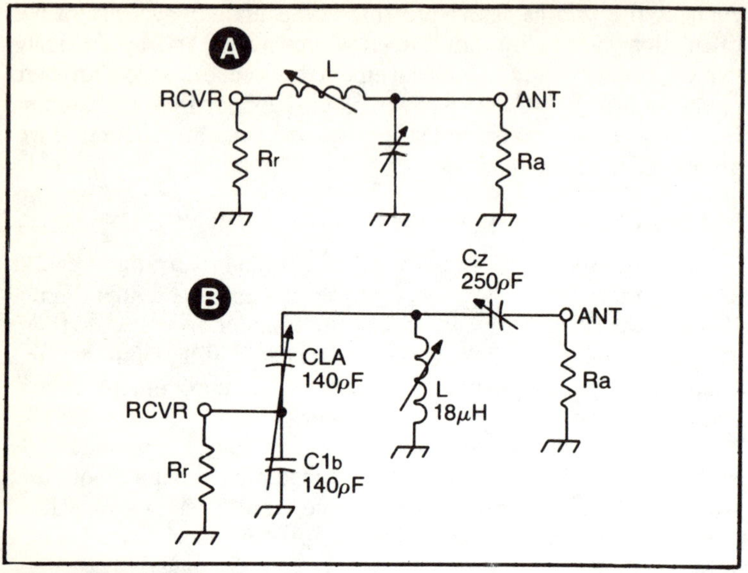

Fig. 13-3. Antenna tuners.

The antenna tuner shown in Fig. 13-3B has a narrower range of impedances, but is capable of matching most common antennas to the 50 to 100 ohm input impedance of most receivers. This type of tuner is well-behaved and will provide a large degree of signal attenuation away from the signal frequency. This circuit has become very popular with amateurs, who use it to match the SWR-sensitive outputs of solid-state final amplifiers to antennas. It also tends to keep their harmonics "at home," reducing TVI.

The input capacitor C1 is a split-stator type. This capacitor can be made by linking two 140-pF variable capacitors together with a shaft coupling. A lot of the smaller types used in receivers have both forward and rear shafts for this very purpose. A coupling (¼ in. to ¼ in.) can be purchased for pennies at a local parts wholesaler. The inductor should be a high-Q type, meaning Illumitronics or B&W air-wound coils. If a rotary inductor is not available, use a piece of Air Dux or Miniductor and tap it every turn or so. Then mount a selector switch to short out the unwanted turns.

The output capacitor must be insulated from ground. This can be done with an ordinary capacitor, provided that it is mounted on a piece of Bakelite or Lucite, which is in turn mounted to the chassis.

AUDIO FILTERS FOR CW RECEPTION

One of the first things you will notice on a low-cost receiver is the lack of selectivity. Try tuning across a busy CW band and copy through all of the interference. Novice class licenses especially have difficulty because their bands are more crowded, and they usually lack the practice required to filter out the unwanted signals "in their head." Figure 13-4 shows a type of filter that is commonly used in the line between the receiver output and the headphones. Two parallel resonant tank circuits select some frequency in the 1000-Hz range and reject all others. If the Q of the filter is high enough, the filter will tend to reject all signals that are not close to 1000-Hz. A well-known surplus filter from World War II provided a 100-Hz bandwidth around 1020 hertz, so it had a Q of roughly 10. These filters greatly reduced the interference, but because they were passive devices, they also attenuated the main signal. The audio power output of the receiver had to be great enough to drive a signal through the filter that would be strong enough to drive the earphones of the operator. We can overcome this problem by using amplifiers in association with the filter. The input amplifier in Fig. 13-4 is used mostly for isolation (buffering), while the output amplifier increases the power of the signal.

ACTIVE FILTERS

If we are going to the trouble of providing power supplies and the like for the amplifiers in Fig. 13-4, we might as well consider instead an operational amplifier active filter. The operational amplifier is a universal gain block with the following properties:

- ☐Infinite open-loop (no feedback) gain
- ☐Zero output impedance
- ☐Infinite input impedance
- ☐Infinite (?) bandwidth

These properties make it possible to set the transfer function of the entire amplifier entirely by manipulating the feedback loop. If the components of the feedback loop are resistors, a wideband amplifier is created that has a gain determined entirely by the values of the

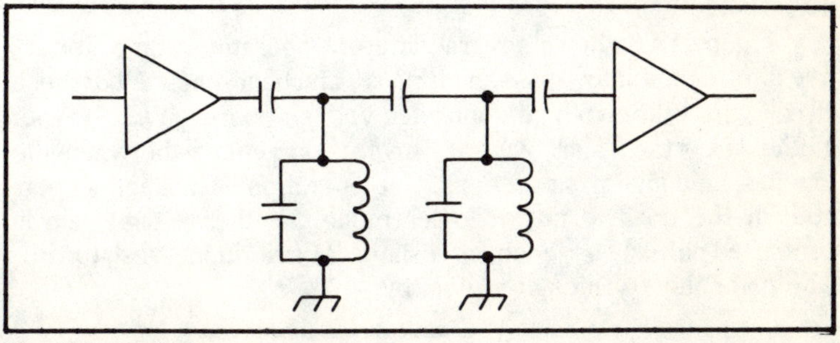

Fig. 13-4. Audio filter.

resistors. If the feedback network contains a frequency selective network, though, then the transfer function will have a frequency term, and it becomes possible to make a filter.

Figure 13-5 shows the basic frequency response properties of the different major classes of filter. The response shown in Fig. 13-5A is for a low-pass filter. This type of filter will pass all frequencies from DC to some cutoff frequency F_c. The response at frequencies greater than F_c falls off at a slope that is usually specified in terms of dB/octave or dB/decade. An octave is a 2:1 change in frequency, while a decade is a 10:1 frequency change. At some frequency higher than F_c, the frequency response becomes asymptotic to zero, or some specific level. We could use the low-pass filter to limit the bandwidth of the audio stages to some upper limit, such as 3000 Hz.

The high-pass filter response is shown in Fig 13-5B. This is the inverse of the previous response: It passes all frequencies above some cutoff frequency F_c and attenuates those frequencies between DC and F_c.

The response for a band-pass filter is shown in Fig. 13-5C. This response can be considered to be a combination of a low-pass responses are the same, there is a peaking response as shown. This type of response could be used to pass only the frequency of interest and attenuate nearby frequencies. This is useful in a CW receiver because the other signals in the passband of the receiver may well have a different audio beat note than the desired signal. We can, therefore, use the peaking filter to pass only the beat note of the desired signal.

A notch filter response is shown in Fig. 13-5D. This is the inverse of the peak filter: It passes all frequencies except a narrow band of frequencies close to the resonant frequency F_o. This response can also be made from custom-tailored high-pass and low-pass filters, with cutoff frequencies slightly amended to place the resonant frequency deep into attenuation slopes for the two sections.

Figure 13-6 shows several different operational amplifier active filter circuits: low-pass, high-pass , and band-pass. All of these circuits are of the voltage-controlled voltage-source (VCVS) type. Notice a certain, shall we say, inverse symmetry between the low-pass and high-pass sections. The operational amplifier is operated in the noninverting follower mode, so the voltage gain is controlled partially by feedback resistor R4 and "input" resistor R3. The component values are given by:

$$F = \frac{1}{(6.28)(R_1R_2C_1C_2)^{1/2}}$$

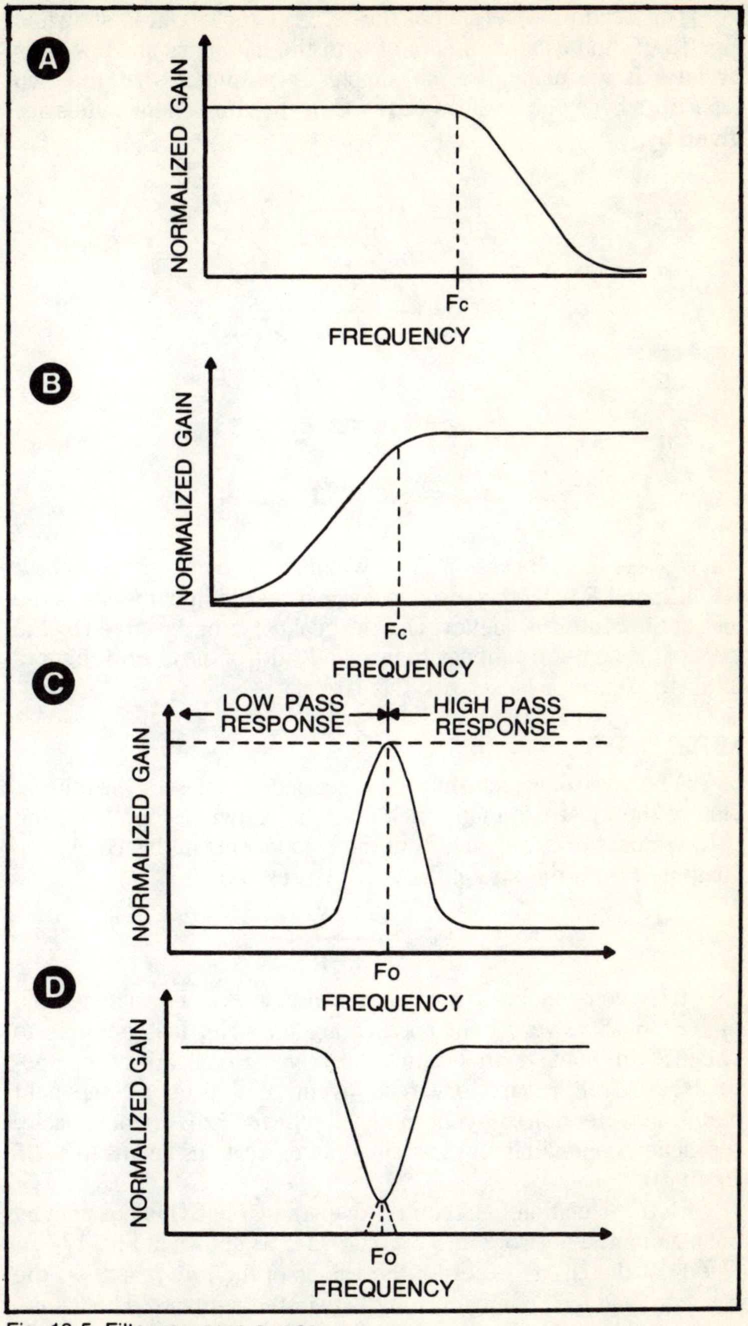

Fig. 13-5. Filter response curves.

The band-pass version of the VCVS filter circuit is shown in Fig. 13-6C and uses elements of both the high-pass and low-pass circuits. If we make the job simple by assuming that the two capacitors have equal values (C1 = C2), the component values are given by:

$$R1 = \frac{Q}{A_v C \omega_c}$$

$$R2 = \frac{Q}{(2Q^2 - A_v) C \omega_o}$$

$$R3 = \frac{2Q}{C \omega_o}$$

$$A_v = R3/2R1$$

where ω_o is the expression $2\pi F$, when F is the center frequency; R1, R2, and R3 are the resistances, expressed in ohms; A_v is the gain at the center frequency; Q is the quality factor described by the ratio of the center frequency to the bandwidth of the filter. Q figures of up to 20 are possible with this circuit.

NOTCH FILTERS

A notch filter rejects only one frequency and passes the others. One of the most common notch filters is shown in Fig. 13-7: the twin-tee network. The values of the components in this circuit are computed from the straightforward expression:

$$F_o = \frac{1}{2\pi R C}$$

If the components in the twin-tee filter are closely matched and of good quality, very tight notches are possible. It is possible to obtain 25 dB notch depths using relatively "garden variety" components, with little regard for circuit layout. With proper component match and attention to layout, notch depths to 55 dB are obtainable. A practical upper limit for most amateur applications seems to be 35 to 40 dB.

Better isolation, and control over the circuit Q is possible by connecting the network into a circuit such as shown in Fig. 13-7B. The Q of the circuit is set by the values of the two resistors, the capacitor and the components in the twin-tee network. The following expresses the proper relationships:

$$Q = \frac{C}{C1} = \frac{R_2}{2R}$$

The Q and the other components are usually selected first, and then the values for R2 and C1 which will offer the proper Q values.

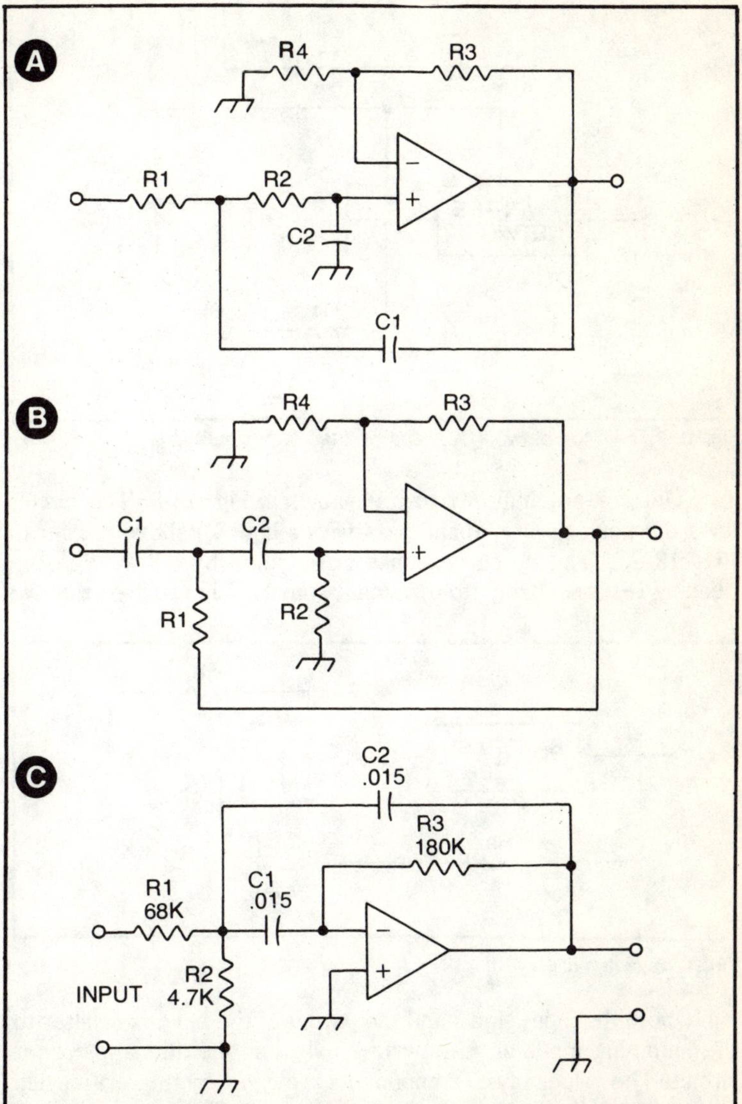

Fig. 13-6. Low-pass filter at A, high-pass filter at B, and band-pass filter at C.

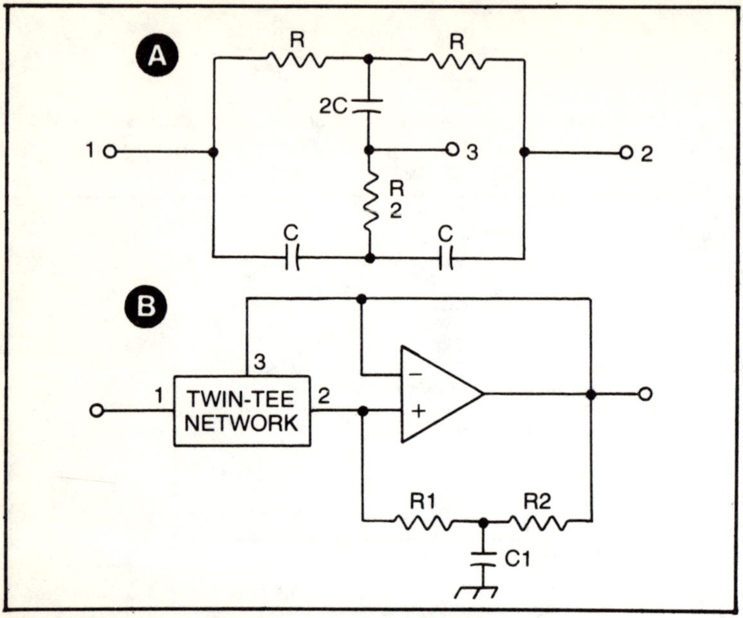

Fig. 13-7. Twin-tee network at A and twin-tee filter circuit at B.

One last notch filter circuit is shown in Fig. 13-8. This circuit uses the peaking filter (band-pass with a high Q) shown earlier in Fig. 13-6C. The notch response is obtained by subtracting the peaking response from its own input signal. This requires that we

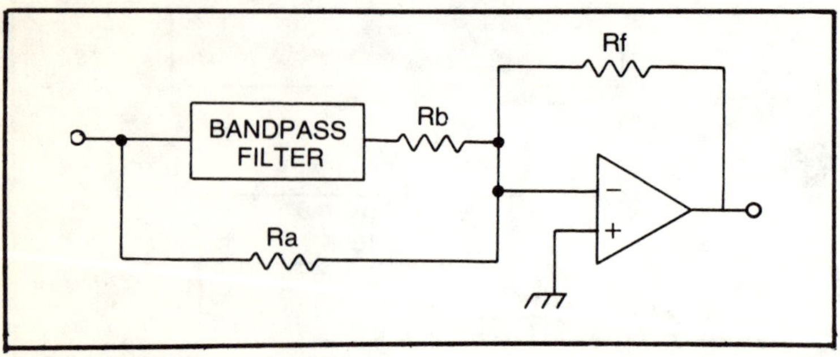

Fig. 13-8. Notch filter.

apply both the input signal and the output of the band-pass filter to the summing input of an inverting follower operational amplifier circuit. The values of the components are given by the relationship below, in which R1 and R3 are from Fig. 13-6C:

$$R_b / R_a = R_3/2R_1$$

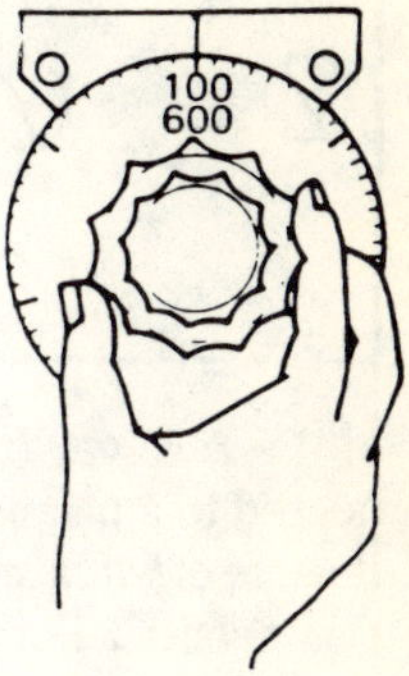

Chapter 14

Shortwave Receiver Antennas

The antenna is probably the single most important part of the receiving system. The best, highest cost commercial and military receivers will not work properly if an insufficient antenna is provided. Conversely, the performance of low-cost, marginal radio receiving equipment can often be improved substantially by the antenna system. The purpose of the antenna is to pick up the electromagnetic wave that is the radio signal, and then convert it to an equivalent electrical current that the receiver can process. In most cases, this current passes through the antenna wire to the earth ground via the primary coil of the input transformer of the receiver. When the antenna current passes through the receiver coil, it creates an induced voltage at the secondary of the antenna coil, which becomes the input signal to the rf amplifier. In this chapter we will cover in detail the construction of the most elementary types of receiver antennas. However, assume that almost any transmitter antenna will also work properly for receivers.

The principal difference between a receiver antenna and a transmitter antenna is that the transmitter is more critical and therefore requires a resonant antenna before it will work properly. A receiver will work *better* if the antenna is resonant and impedance-matched to the receiver, but it is possible to obtain very good performance with nonresonant models as well.

Figure 14-1 shows the typical form of receiver antenna terminals. In some models, there are only two screw terminals, one for the antenna and another for the earth ground. The antenna input coil

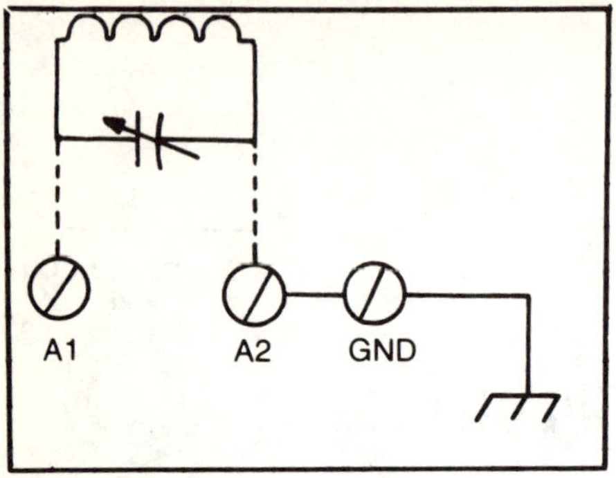

Fig. 14-1. Antenna terminals.

of the receiver (the primary of the rf amplifier input tank) is connected to a pair of terminals, often marked A1 and A2. This allows the use of *balanced antennas* such as the folded dipole shown later in this section. For balanced operation, the ground terminal is connected to the ground wire, and the two antenna wires from the transmission line are connected to A1 and A2. But in the more popular unbalanced operation, it is necessary to short one of the antenna terminals to ground, usually the one immediately next to the ground terminal. Many receivers are shipped from the factory with A2/GND shorted with a small metal link. Most users never have any reason to remove the shorting strap!

One of the first modifications made to a receiver by many amateurs is the addition of a coaxial connector for the antenna input. In some cases, one can simply remove the terminal strip (see Fig. 14-1) and ream out the hole(s) to take an SO-239 coaxial chassis receptacle. In other cases, it is necessary to punch a 5/8-in. hole for the SO-239 adjacent to the adjacent to the antenna terminal strip. In that case, it is wise to disconnect the antenna terminal strip from the antenna input circuitry and then reconnect it to the coaxial connector. Also, provide some ground connector on the chassis, if only by leaving the ground on the terminal strip in place when the modifications are made. An ordinary machine screw can form a ground connection. Use two nuts and a lockwasher. Cinch one of the nuts down onto the lockwasher in order to hold the screw (No. 6 to No. 10) in place. Then use the remaining nut to hold the wire. Even better is a five-way ground binding post, such as a banana jack. If you cannot come up with a grounding—a noninsulated binding post—use an ordinary insulated type and ground the terminal.

LONG-WIRE ANTENNAS

The principal form of nonresonant long-wire antennas is shown in Fig. 14-2. This antenna is probably used by more shortwave listeners than any other form, and is detailed in the operator's manual for almost every commercially produced shortwave receiver produced in the past 50 years. It is popular because it is so utterly simple! The antenna consists of a single piece of No. 16 through No. 10 wire mounted high off the ground between two insulators. The down lead (also called lead-in wire, or transmission line) is another piece of the same, or thinner, wire. For most high-frequency receivers, the length of the long-wire is specified as 25 to 100 feet. The lower the frequency is that you wish to "specialize" in, the longer the wire needs to be. For example, if you plan to DX the 49-meter band almost exclusively, then it would be advisable to use an antenna in the over-50 feet category. True, the receiver will work, and probably work well with a 25 foot antenna, but it will work even better with a longer wire. If, on the other hand, the frequencies above 15 MHz are your preference, longer antennas will work better than short antennas, but only marginally better. In those bands, the length is less critical. These rules of thumb are based upon the idea of wavelength. The wavelength of a radio signal is given by:

$$L = \frac{300}{F}$$

Where F is the frequency in megahertz and L is the wavelength in meters. The term "300" is derived from the velocity of travel for a radio wave, or the speed of light (300,000,000 meters/second).

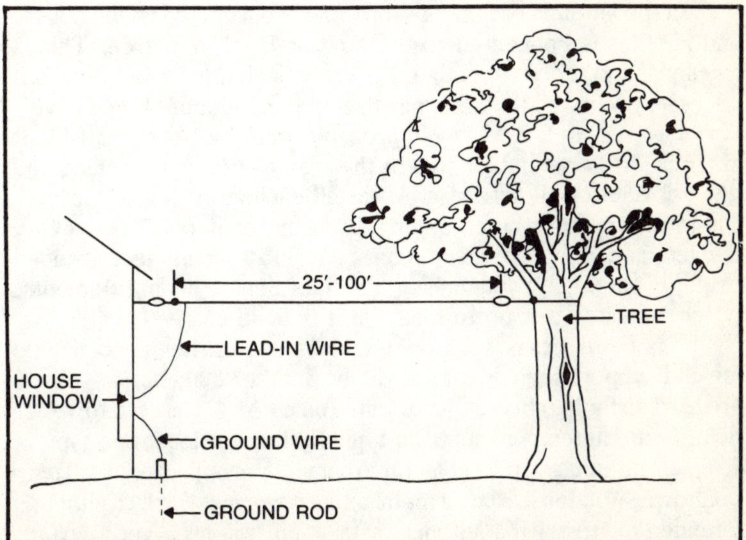

Fig. 14-2. Simple long-wire antenna.

☐**Example:**

What is the wavelength in meters of a radio signal that has a frequency of 9.85 MHz?

$$\begin{aligned} L &= 300/F \\ &= 300/9.85 \\ &= 30.45 \text{ meters} \end{aligned}$$

In shortwave parlance, incidentally, the band designations are based upon this relationship. We often refer to a band by the approximate wavelength of the lowest frequency in the band. In the 9.5 to 10.0 MHz band, for example, we designate the entire band as the 31-meter band.

Equation 14-1 also tells us that there is a direct relationship between the operating frequency and the wavelength of the signal. In the United States, it has been the custom to calibrate radio receiver dials in units of *frequency*. But in much of the rest of the world, which was more attuned to the use of the international Metric system, the radio receiver dials were calibrated in units of wavelength. In the years before World War II, until well into the 50s, European radios were often calibrated in meters (wavelength). German auto radio manufacturers of the 50s who supplied automobile radios to Volkswagen and Mercedes-Benz would pack a spare dial scale with the "export" model (that was the only difference!) so that the installation mechanic in the US could change the clip-in dial to something that was more familiar to the company's North American customers.

In the antenna of Fig. 14-1 one end is connected to the house, and another is connected to the limb or trunk of a tree. This is usually the practice, although it is not chiseled in stone. Nor is it necessary to make the antenna absolutely horizontal. If you have a very tall tree, in fact, it would probably work better to install that antenna with one end as high in the tree as possible. In fact, the general rule for all antennas is "install each as high as possible." There is a rule of thumb among "transmitter people" that every element added to a beam antenna is equal to so many feet of added height. Conversely, adding height is equivalent to adding elements, which means better performance if it is done correctly.

It is not necessary to use a tree as one of the supports; the far-end support could just as easily be another building, or—if you are hard up for height—a fence post. You could also erect a tower or pole. Some antenna handbooks typically have at least one type of wooden antenna support structure that you can duplicate or adapt to your own situation. Or buy a metal telescoping mast, which although intended for television antenna installation, will also work well for other types of antenna. A few simple guy wires will make the antenna support itself. Of course, if you want to go "all out," order and install a three-sided antenna tower as used by amateur and commerical radio stations to support large antenna arrays.

The ground rod is needed for two purposes: proper operation of the receiver and safety. The ideal long-wire antenna should eventually send its signal to ground, so connect the ground wire of the receiver to earth ground. Some people use the center screw that holds on the plate of an AC power outlet as the antenna ground. I have done this myself, but I do not recommend it; the ground wire might fray and get into the socket. In addition, although the ground screw is "ground" to 60 Hz AC, it might not be a good rf ground. The main rule for good rf grounds is a heavy wire, as short as possible, directly to earth. If you cannot, for some reason, provide the proper ground, use a ground clamp to a metal cold water pipe. Do not use a plastic cold water pipe, a hot water pipe, or any gas pipe!

Ground rods can be purchased in any amateur radio retail store, most general electronic hobbyist stores, and almost all electronic wholesale houses. These are intended for use with television antennas. The minimum length is 36 inches, and this length is *really* minimum! Use a longer ground rod if available. Try to obtain one at least 5 feet long and preferably 8 to 10 feet long. Of course, the longer ground rods are difficult to obtain and difficult to install (wait until you try hammering an 8-foot ground rod into the hard earth). Some amateurs, mindful that a proper grounding system is capable of reducing TVI, will rent a well-point tool, which is a pointed drilling tool that allows you to drill a 2-in. hole for water exploration, and dig the hole for an eight footer. After the ground rod is installed, the hole is then refilled with a mixture of mud and rock salt or copper sulphate.

The other reason for a good ground is safety from lightning. It is not necessarily true that an antenna attracts lightning, but lightning will strike an antenna occasionally, especially if it is the highest object around. It is the height of the object, incidentally, that makes it more prone to lightning strikes, and not the simple fact that it is an antenna. Some antennas will actually act more like a lightning rod, drawing the lightning that would otherwise have struck the house, even though they pose no additional threat of their own. Use an approved *lightning arrester* on any antenna system. In fact, almost all local electrical codes *require* the installation of the arrester. These devices are available from the same sources that supply antenna materials.

Figure 14-3 shows the installation of a lightning arrester at the point where the antenna enters the house. It is the usual practice to

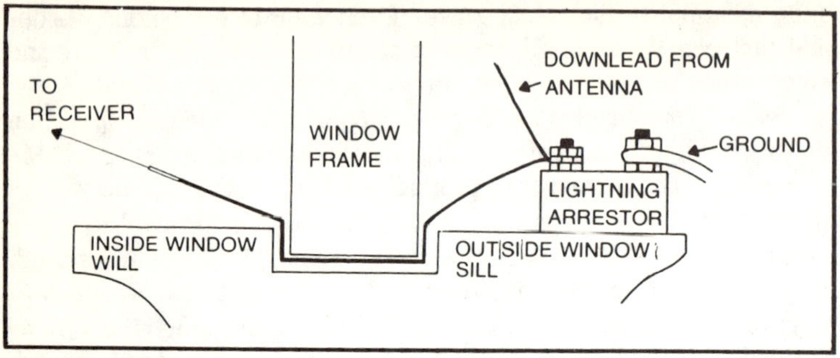

Fig. 14-3. Bringing the transmission line inside of the house.

mount it on a window sill or metal bracket outside the house. The lightning arrester is usually little more than a spark gap encapsulated in a ceramic housing. The antenna line is connected to one terminal (one point of the spark gap), and the ground is connected to the other terminal. The tiny electrical signals from the antenna will not jump the gap, but the high potential from a lightning strike will bridge the gap and dissipate most of its force in the ground.

Some people prefer to install a knife switch near the lightning arrester for supposedly added protection (see Fig. 14-4). When an electrical storm is threatening, the switch is set to ground the antenna. When they wish to operate the receiver and no storm is threatening, the switch is set to apply the antenna to the receiver input.

Note that the ground wire in both cases should be as large as you can make work with the components used; there is no such thing as *too big*! Your local electrical code may well have some specifications regarding the size of the wire, and it will be different for aluminum and copper wires.

Bringing the lead-in into the house can present some problems, especially if you live in the higher latitudes where cold weather will make you want to keep the window tightly closed. One simple method is to use an antenna *window strap*. These are often sold in hobby stores, or as part of an SWL antenna kit. Basically, it is a strap with a foil wire buried inside. In fact, there might be two wires with one of them for ground. You can simulate these straps crudely with a piece of 300-ohm TV twin-lead. This wire is slim enough to fit underneath the window, as shown in Fig. 14-3. One of the conductors can be used for the antenna, and the other for the ground wire.

It is also possible to replace one pane of the window with either a piece of quarter-inch plywood or plexiglass. Drill a hole through the plexiglass or plywood pane for screws that will act as the antenna lead-in. Figure 14-5 shows several different types of connectors attached to the replacement pane. The simplest is the machine screw method. Solder lugs on either side of the pane are used to attach the wires. Spray the assembly, especially on the outside, with some clear plastic spray such as *Datacoat* or *Krylon*. This will prevent corrosion of the connections. Also, use only brass or plated brass machine screws and nuts. Steel ones will rust and aluminum is not sufficiently strong to maintain a stable connection. Besides, the wire may tend to react with steel or aluminum to form a nonlinear, rectifying PN junction.

Another of the methods shown in Fig. 14-5 is the five-way (banana) binding post. Mount it so that the solder connection end is outside. Then coat the finished connection with plastic spray as before. Then either attach the inside antenna lead-in with a banana plug as shown, or use the attachment hole that will be revealed when the binding post cap is unscrewed a few turns.

The most expensive of these methods is the use of a coaxial barrel connector that will mate with the standard PL-259 "UHF" connector used on most amateur radio transmitters for the antenna connector. An Amphenol dealer will be able to supply the barrel connector (2-in. size is best), but make sure you also get the nuts that go with the connector. Some blister packaged barrel connectors do not have the nuts, without which the connector is useless. If you use an antenna that requires coaxial cable, then look for a coaxial lightning arrester. One brand of such is the *Blitzbug*. It is a barrel-like connector with the lightning points built inside. A machine-screw terminal is provided for the ground wire.

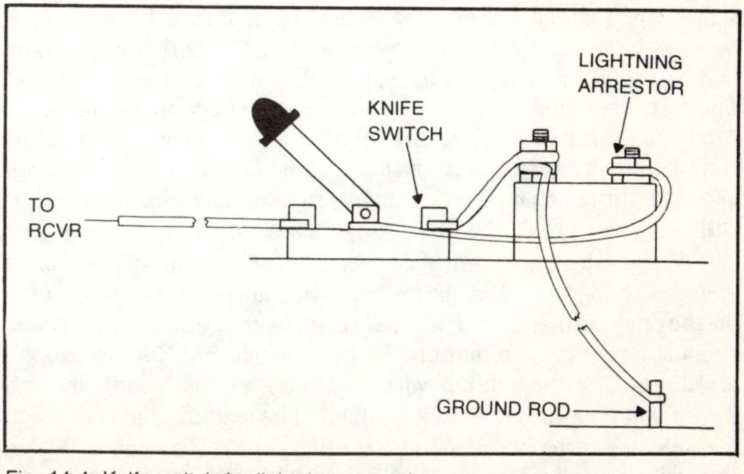

Fig. 14-4. Knife switch for lightning protection.

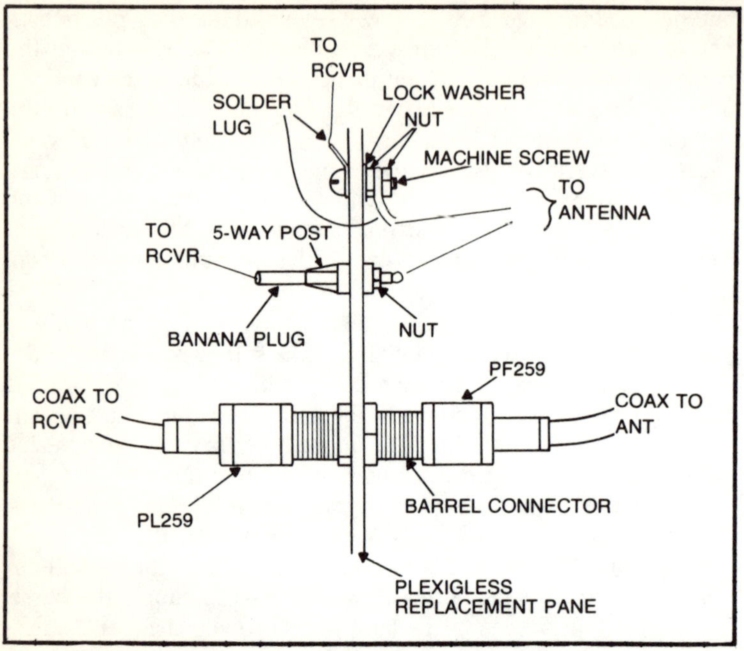

Fig. 14-5. Using replacement window pane for bringing in the transmission line.

Figure 14-6 shows how the lead-in wire, the antenna wire, and the support insulators are fastened together. The support for a short antenna (up to about 50 feet can be a simple 2-in. screw eye fastened into the wood eaves of the building or mast. For antennas that are longer, or if the wood seems a little weak, use a toggle bolt screw eye instead of the woodscrew type shown here. A rope fastens the end insulator to the screw eye. Use a thin nylon rope that has sufficient strength to support the antenna in a windstorm. The forces applied to the rope and the screw eye in a gale force wind can be terrific, especially in 100-foot antennas. Tie a sufficient knot at either end, not just a couple of turns; use something like a multiple square knot. If nylon rope is used, burn the ends with a flame to melt the fibers together to prevent unraveling.

The end insulators are glass, ceramic, or porcelin affairs. Some specifically made by Hygain for long-wire antennas are made of a plastic or nylon material. The idea is to insulate the antenna from the support structure at either end of the antenna run. Note the good-sized splice for the antenna wire. Make a good splice with at least seven twists of the wire back on itself. The lead-in splice need not be quite so strong because there is little force on this splice. Make the splice strong mechanically and then solder the joint. A 200-watt or larger soldering iron is needed; the antenna wire and outdoor environment sink too much heat for smaller irons. Some people use a propane torch instead. The solder joint is not to add strength, for ordinary lead-tin radio solder has none, but it does prevent corrosion of the splice in the weather. Failure to so protect the splice will result in taking the antenna down for repairs or replacement in a few months.

Some people like to use pulleys at one end of the antenna wire. These allow them to run the antenna up or down at will, making repairs easier. An ordinary brass marine pulley and rope system works wonders in this case.

The wire used for a long-wire antenna is important too. The wire size should be at least No. 16, with No. 14 or No. 12 preferable. Even more important, however is the *type* of wire used. Stranded wire is definitely preferred over solid wire. Also, at the very least, use hard-drawn copper wire. Soft-drawn wire will tend to stretch and eventually break of its own weight. The best wire to use is a product called *Copperweld* wire. This is a steel-core, copper-clad wire. Because rf currents flow only on the surface of an antenna wire (AC *skin effect*), this type of wire has no higher AC resistance than a solid wire. But the steel core adds strength not available in pure copper wire. Unfortunately, Copperweld is expensive and difficult to find, although once the mainstay of the SWL antenna business. Also, one word of warning: Don't ever let Copperweld wire kink—it will remain kinked!

DIPOLE ANTENNAS

A very simple form of resonant antenna is the dipole shown in Fig. 14-7. This is a center-fed long-wire antenna that is cut to a specific frequency of operation. The cutting is critical in transmitter antennas but becomes less so in receiver systems. A resonant

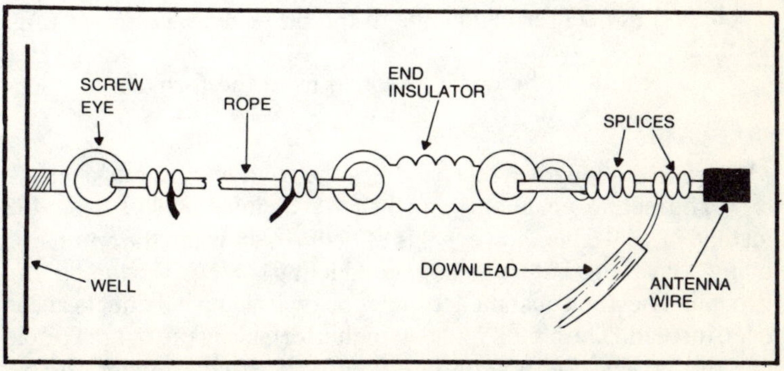

Fig. 14-6. Proper methods for tying to the end insulators.

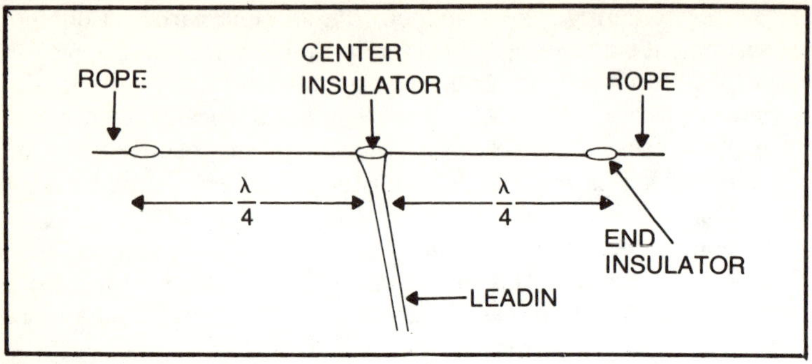

Fig. 14-7. Simple half-wavelength dipole.

antenna will perform better with most receivers. The performance off resonance becomes that of the equivalent length long-wire antenna.

The two sections of the dipole are separated with a center insulator, and the transmission line is attached at this point. Theoretically, the impedance of a dipole at the center-feed point is around 70 ohms. In actual practice, this impedance is rarely realized except in outer space. Two type of feedlines provide a good match to the simple dipole. One is 75-ohm twin-lead. This line is approximated well by ordinary household lamp cord (also called zip cord). Figure 14-8 shows how the zip cord is attached to the center insulator. In this example, an ordinary end insulator is used in the center, although you can buy actual center insulators too. The solder connections will not provide enough strength to hold the zip cord to the antenna wires, so loop the cord over the insulator, back on itself, and secure it with several turns of fishing line or thin twine. Alternatively, a cable clamp will secure the wire.

Also, 75-ohm coaxial cable is usable. Dress back the outer insulator of the cable and connect the shield to one element of the dipole and the center conductor to the other dipole element (Fig. 14-9).

The length of the dipole is found from the formula:

$$L = 468/F$$

where L is the length in feet, and F is the frequency in megahertz.

The actual equation for a half-wave, incidentally, uses 492 instead of 468. This, however, is effective only when the antenna is many wavelengths from any conducting object, including the ground. The 468 constant accounts for typical ground effects and is used instead. The length found with this formula is the overall dipole length. To find the appropriate length for each element, simply divide by two.

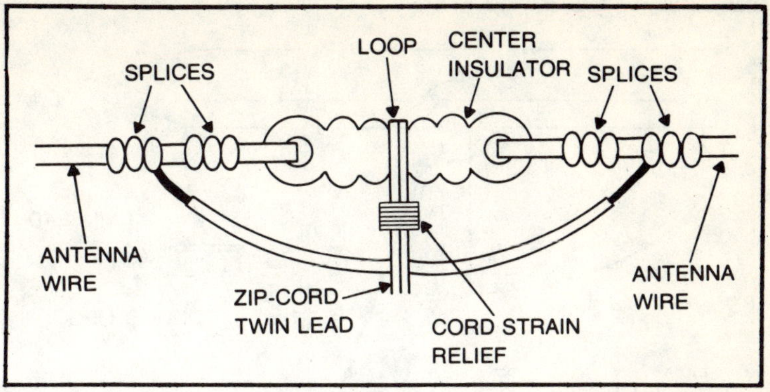

Fig. 14-8. Connecting the center insulator.

A *folded dipole* antenna is shown in Fig. 14-10. This antenna provides a better impedance match for most receivers but is a real "dog" to build with any strength. As a result, it is best used only on the higher frequency bands above 25 MHz. The insets show how to connect the ends together and the feedline to the radiator when ordinary 300-ohm TV-type twin lead is used to form the radiator and transmission line.

Compensation dipoles are useful where there is insufficient space to make a full dipole. The antenna shown in Fig. 14-11A is an inverted-vee dipole. If the angle between the elements is greater than 90 degrees, make the overall length approximately 5 percent longer than derived from the equation for ordinary dipole antennas. The center of the antenna can be supported at the roof line of the house or on a pole.

The antenna shown in Fig. 14-11B is for really tight locations. The elements can be bent at angles exceeding 90 degrees. Interest-

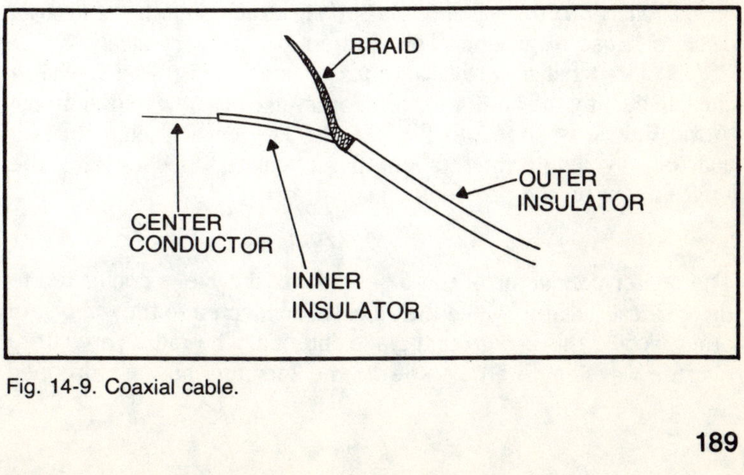

Fig. 14-9. Coaxial cable.

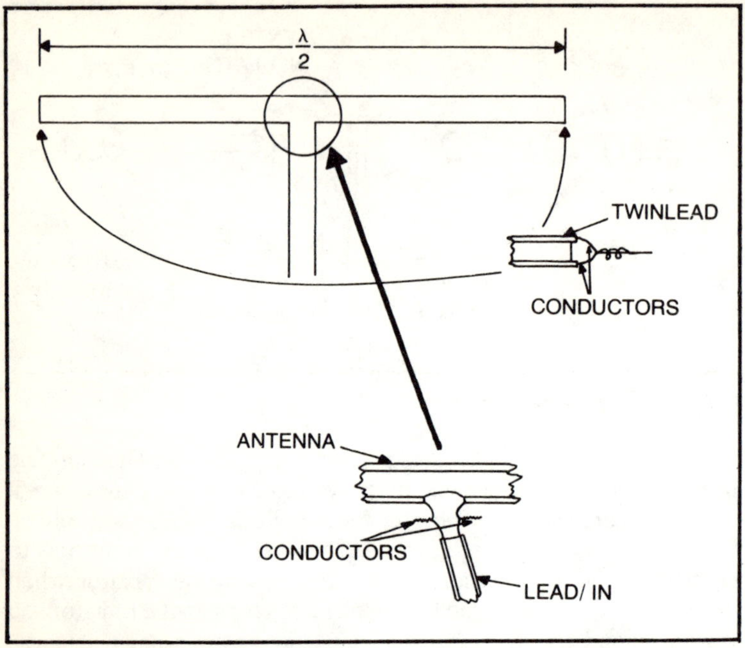

Fig. 14-10. Folded dipole.

ingly enough, this illustration could be either a top view or side view. The bending isn't too critical in receiving antennas. In transmitting antennas, however, try to make the individual sections of each element symmetrical with the same section in the other element.

VERTICAL ANTENNAS

A simple vertical antenna can be made by mounting the dipole in a vertical position. Place one end of the antenna at the roof peak, or to a mast. Stretch the antenna out and attach the other end to an insulator close to ground. This system works very nicely.

A more familiar vertical antenna is shown in Fig. 14-12, while a cheapie built by the author for temporary use in a new (old, but new to me) house is shown in Fig. 14-13. The vertical element is a quarter wavelength, and its length is calculated as one-half of the length given by the previous equation, or:

$$L_{ft} = 234/F_{MHz}$$

The center conductor of the 52-ohm coaxial cable is connected to the vertical radiator, while the shield is connected to the *radials*; in other words, this is a ground-plane antenna. The radials are cut to quarter wavelength (the same as the radiator) and are drooped

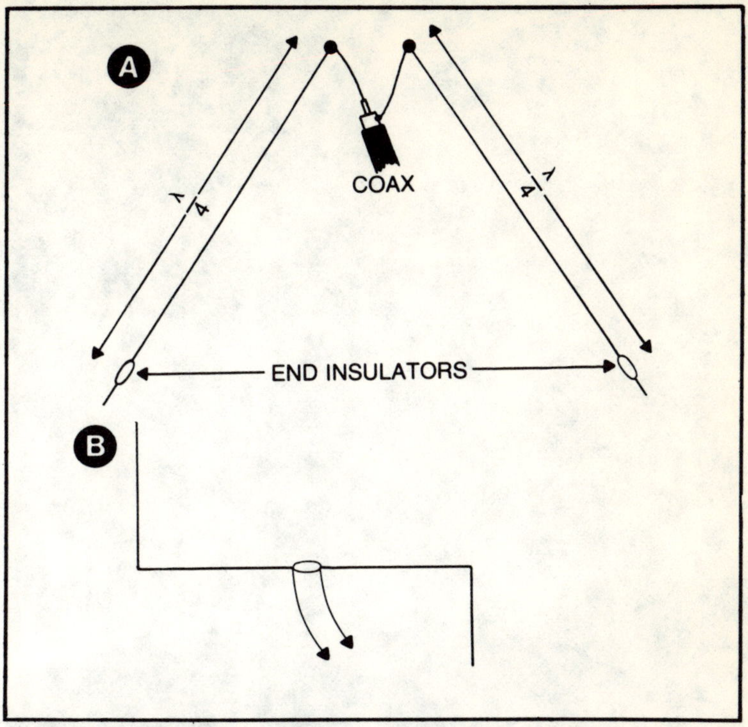

Fig. 14-11. Inverted-vee antenna at A and limited space antenna at B.

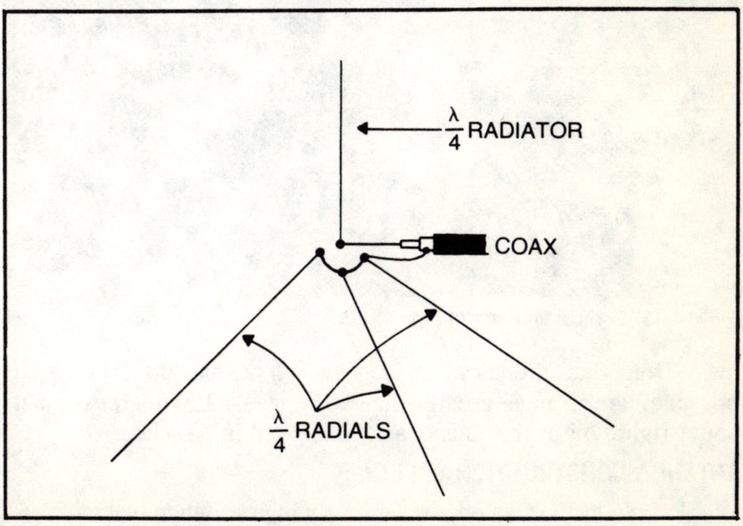

Fig. 14-12. Vertical antenna.

Fig. 14-13. Vertical antenna of Fig. 14-12.

away from the antenna at an angle *not less* than 90 degrees. If possible, use an angle greater than 90 degrees; 120 degrees is just about right. All of the radials are connected in parallel.

ANTENNA CONSTRUCTION SAFETY

People can get killed and seriously injured when making mistakes during antenna installations. Of course, the obvious hazards are that they can fall off of ladders, tumble from roofs, and topple out of trees. But just as important is the matter of electrical safety. Follow two rules:

- □Never work on any antenna when it is connected to a piece of radio equipment.
- □Never work on an antenna where it can come into contact with the AC power mains wiring. No exceptions to this rule exist, and any that you try to dream up will all be lethal.

Every year radio and electronics magazines report cases of hobbyists building antennas near AC wires and being *electrocuted*. In most cases, either a pipe antenna (beam or vertical) fell against the wires, or the installers tried to toss a long-wire *over* the power wires to erect the antenna above the wires. There is nothing more stupid than this last maneuver. The insulation on the power wires is often in precarious shape—it is rotted. This poses no problem until you stress the insulation mechanically by attaching the wire to a weight and tossing it over the power lines.

Also, if you have a low-cost radio marked "AC/DC" do not connect either an antenna or a ground until it is professionally checked. Because most transistor radios use a power transformer to isolate the power supply from the AC mains, this is not much of a problem. However, lower cost radios of a decade ago could kill if a slight defect occurred. These radios can be safely operated if an external *isolation transformer* (115 VAC to 115 VAC) is used between the radio and the house wiring. If the radio is known to be in good condition, the danger is not quite so pronounced. The defect that can cause the problem is hidden, though, and will not be noticed in the operation of the radio.

Chapter 15
Shortwave Listening

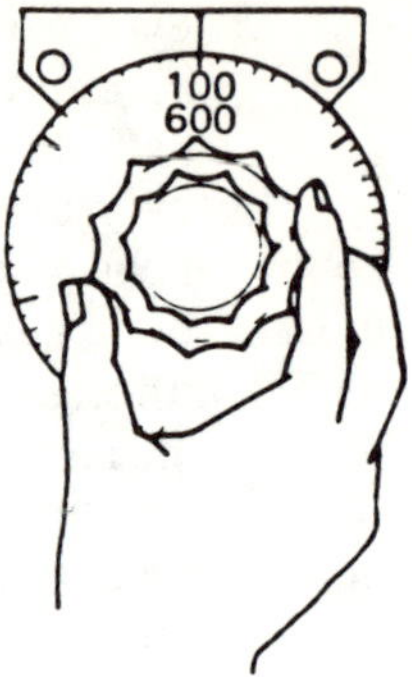

When commercial interests began to threaten the frequencies used by amateur radio operators during the early years of wireless, the main frequencies used for communications were in the very long-wave region below the present AM broadcast band. These long-wave frequencies had wavelengths longer than 200 meters. The cynical comment for amateur stations, often quoted, was "put 'em on 200 meters and below—they will never get out of their backyards with those frequencies!" But those frequencies were the *shortwaves*, now recognized as the prime frequencies for international communications and broadcasting. I once heard an amateur give a talk at a local ham club. He was present in amateur radio 60 years ago. His younger brother obtained a license just about the time the speaker left for studies in electrical engineering at the University of Virginia in 1921. When he returned home, his 200-meter multi-wire "flat-top" antenna array was torn down, and a much shorter wire was in it's place. Somewhat dismayed, he approached his kid brother and asked the reason why. "Well," came the reply, "we're not using 200 meters anymore, but 40 meters."

"40-meters! You can't work anybody more than across town on 40 meters!" After supper, however, the guy sat down to the key and called CQ. A station with the call sign 8EG replied. Thinking the fellow was in either neighboring West Virginia or Ohio, he asked if he would take a message for relay to an interested classmate from Cincinnati. The other station replied: "Sure, but you're in a better position to relay it than me—I'm *French* 8EG, not US 8EG!"

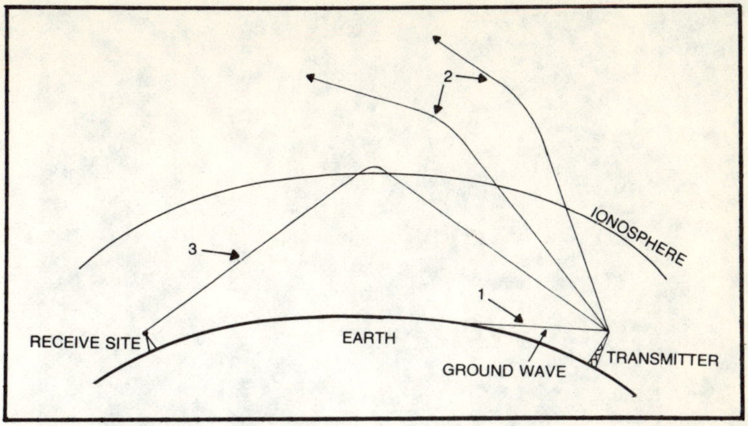

Fig. 15-1. Shortwave propagation.

(Amateur call signs in those days did not have a prefixes to identify each country.)

SKYWAVE PROPAGATION

Obviously the shortwaves worked well. The reason why is illustrated in Fig. 15-1: sky-wave propagation. There are actually two waves from an antenna. The ground wave travels close to the surface of the earth and is good only up to a few hundred miles on the medium-wave and high-frequency bands. In the VHF and UHF region, the ground wave is good only to a short distance beyond the horizon, or what is called *line of sight*. The sky wave travels from the transmitter antenna to the ionosphere, a layer of the atmosphere a few dozen miles in altitude. The ionosphere acts like a radio mirror and reflects some signals back to Earth. Signals above a *maximum usable frequency* (MUF) are not reflected back to Earth, but disappear into space. A reflected signal will bounce back to Earth some distance away from where it began. The dead zone between the end of the ground wave and the point where the sky wave comes down can be thousands of miles. It is also possible for the wave to reflect from the Earth back to the ionosphere, and then back to Earth at an even still more distant point. This double-hop propagation. There are examples of multihop propagation where the signal travels all the way back around the Earth to the transmitter site. This is rare, but it does happen, especially with the multimegawatt transmissions of international broadcasters.

SHORTWAVE SERVICES

Most major countries of the word employ high-power shortwave broadcast transmitters in order to promulgate their own

Fig. 15-2. Shortwave transmitter (courtesy of HCJB, of Quito, Ecuador).

version of the news and opinion of international events. There are also many religious broadcasters operating in the shortwave bands. One of the earliest is The Voice of the Andes, radio station HCJB (see Fig. 15-2).

There are also several amateur radio bands in the shortwave region, and more were agreed upon for the 80s at the 1979 General World Administrative Radio Conference in Geneva. US amateur radio operators use the following bands:

75/80 meters	3500 to 4000 kHz
40 meters	7000 to 7300 kHz
20 meters	14,000 to 14, 400 kHz
15 meters	21,000 to 21,450 kHz
10 meters	28,500 to 29,900 kHz

The US citizens band is in the 11-meter band. Popular international shortwave bands include:

49 meters	6000 to 6500 kHz
31 meters	9500 to 10,000 kHz
25 meters	11,500 to 12,000 kHz
19 meters	15,500 to 16,000 kHz
17 meters	17500 to 18500 kHz
13 meters	23,000 to 24,000 kHz

Fig. 15-3. Shortwave antenna farm (courtesy of The Voice of America).

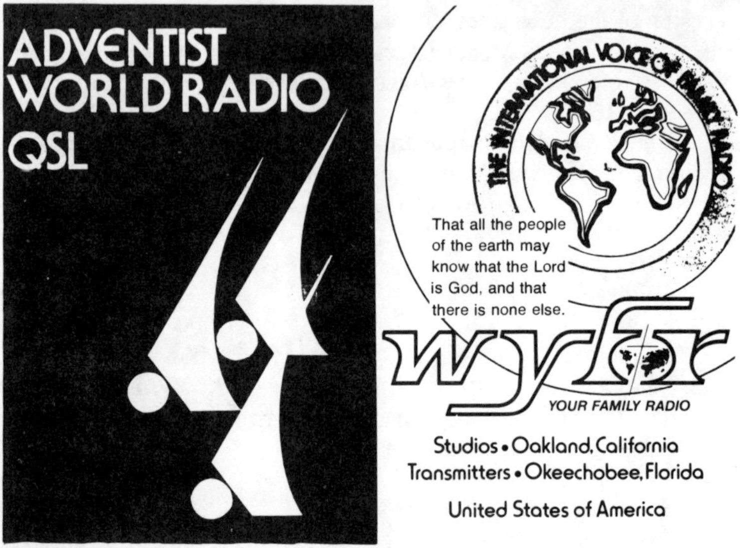

Fig. 15-4. QSL cards (courtesy of Adventist World Radio and WYFR).

There are also numerous regional broadcasters in many parts of the world that operate in the bands between the 75/80-meter amateur radio bands and the 49-meter shortwave band. US listeners can hear many Spanish-speaking stations from Central and South America on these frequencies.

The antennas required for high-power shortwave transmitter do not resemble the small receiver antennas that you build (see Fig. 15-3). These antenna systems might include an array of towers or a large wire curtain antenna. One Voice of America antenna array covers 6700 acres in North Carolina.

In the early days of wireless, amateurs began to exchange written confirmation of contacts, especially long-distance contacts. These became the basis for a "pecking order" among operators, and soon became formalized into QSL awknowledgement cards. International broadcasters also provide QSL cards for their listeners who send in adequate reception reports (see Fig. 15-4). These reports are valuable to the station at budget time and when making engineering decisions about antenna patterns or power levels transmitted in any given direction. Reports should include the date, program, and reception matters such as signal strength and interference. A mention of your receiving system will aid in evaluating the report in the engineering department. Some serious SWLs tap record a few moments of the transmission and send it along with the report.

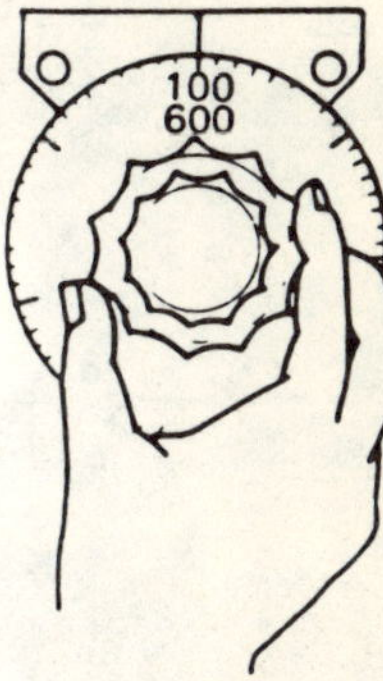

Chapter 16

CB Receiver Circuits

The typical citizens band (CB) transceiver contains a single-conversion or double-conversion receiver circuit that uses the same frequency synthesizer as the transmitter to select the channel of operation. Today there are 40 CB channels in the 11-meter band, but the original arrangement was 23 channels.

THE EARLY DAYS

In the earliest CB sets, a crystal oscillator required a separate crystal for each channel. The output signal from the oscillator was heterodyned with the rf signal to form an i-f signal. In most cases, the intermediate frequencies were in the 1650 to 2500 kHz range, with the double-conversion jobs using 455 kHz as the second intermediate frequency. Many CB receivers use a crystal filter for selectivity, needed because of the closeness of channel spacing (10 kHz) and a tendency of some transmitters to "bleed over" into adjacent channels.

When the transceiver used transmit and receiver oscillators that required a separate crystal for each channel—in fact two with one each for receive and transmit—it required a total of or 46 crystals (two times 23) to make the thing work. At an average cost of approximately $6 each for crystal (in those days), the owner would lay out $276 for crystals alone. It was easily possible to double the cost of a CB rig just by adding all of the "rocks" necessary to do the job.

Most transceivers of that era only allowed 5, 6 or 12 channels of operation, so one would have to change a crystal pair every time

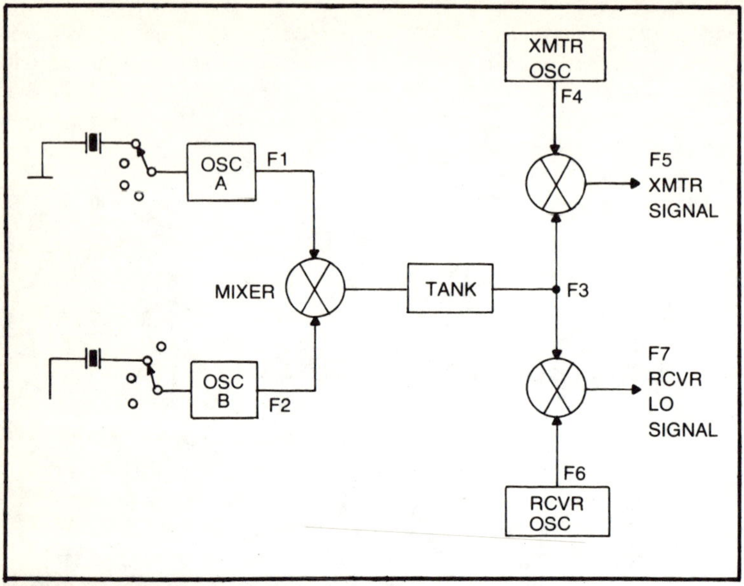

Fig. 16-1. Crystalplexer.

the channel selector switch was to be reprogrammed to another channel. And then, it could be done only if he sacrificed one of the existing channels.

Just before the mid-60s, a circuit called a *frequency synthesizer* was introduced, which needed only 11 crystals to cover the entire 23 channels for both transmit and receive. The type of frequency synthesis used in these CB sets is called *crystalplexing*. This type of circuit requires several crystal oscillators which are heterodyned together to form the correct frequency. A simplified example of a crystalplexer is shown in Fig. 16-1. There are four oscillators in this circuit. Two are master oscillators (A and B), one is the transmit oscillator, and the last is the receiver oscillator.

Master oscillators A and B produce several selected frequencies. In the CB receiver, a switching system is used that selects particular combinations of the crystals in the A and B cystal banks. The outputs of these oscillators are frequencies F1 and F2, and these are mixed together to produce frequency F3. An LC tank circuit selects the mixer product frequency needed for F3, which is usually the difference frequency. The output of the master mixer is fed to two additional mixers. One of these mixers accepts F3 and the signal from a transmit oscillator, F4, to produce channel frequency F5 to drive the transmitter stages. The other mixer accepts F3 and the output of the receiver oscillator, F6, to produce F7, the receiver local oscillator frequency. In both cases, tank circuits are used to select the correct mixer product. A large number of CB crystalplexers used only one mixer past the master mixer, but would select either transmit or receive crystals as the input against which F3 is heterodyned.

MODERN TIMES

Modern CB receivers use phase-locked loop frequency synthesizers to directly derive the local oscillator and transmitter signals. Several semiconductor manufacturers offer complete PLL synthesizers in chip form. Figure 16-2 shows the block diagram of a CB receiver section. This particular design is from the Delso CBD-20 series, which is intended to operate with the General Motors car radios. Delco produces some CB units that are an add-on underdash model controlled from the front panel of the radio, and some others that contain the CB transceiver as an integral part of the radio chassis. Circuitry is provided to automatically silence the AM/FM radio when the CB radio is operated in transmit and when the CB receiver squelch breaks, indicating a strong signal in the passband.

The circuit of Fig. 16-2 is a double-conversion CB radio. Let's follow a specific conversion through the sequence of operations from rf amplifier to the detector. Assume that a channel 1 signal is applied to the input of the CB receiver. The frequency of a channel 1 signal is 26.965 MHz. This signal is amplified by the rf amplifier and then mixed with a signal from the PLL frequency synthesizer. When channel 1 is selected, the PLL frequency will be 37.66 MHz, producing an i-f resultant of the difference frequency (37.66 - 26.965), or 10.695 MHz. This signal is then heterodyned against the signal from a 10.24-MHz oscillator (also part of the PLL syn-

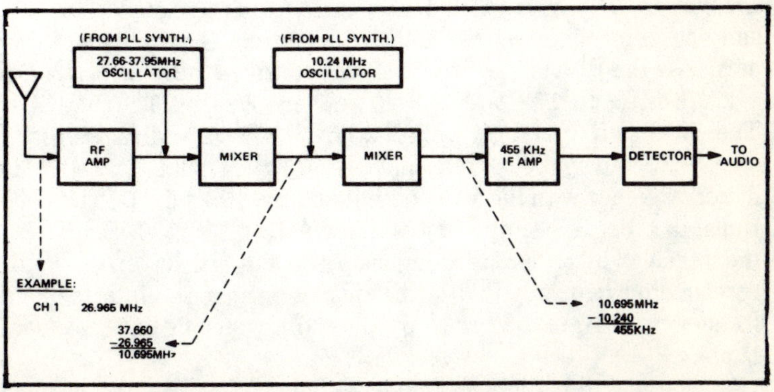

Fig. 16-2. Synthesized double-conversion receiver (courtesy of Delco Electronics).

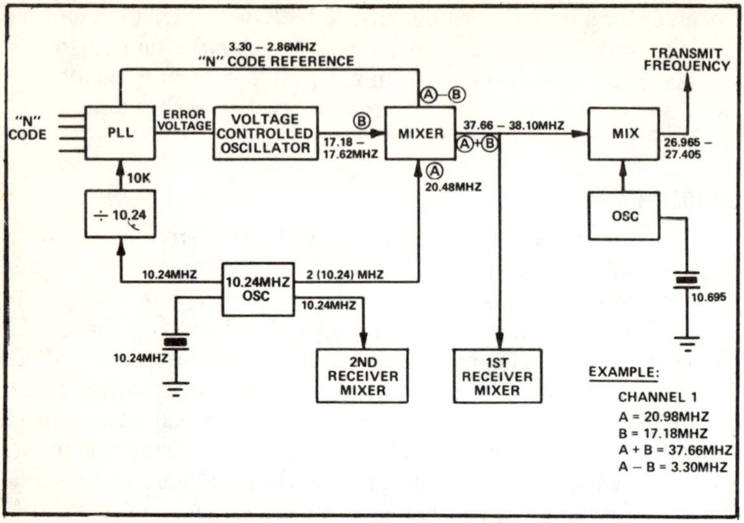

Fig. 16-3. PLL synthesizer (courtesy of Delco Electronics).

thesizer) to produce the difference frequency of 455 kHz. The 455-kHz signal is processed, bandwidth limited, amplified, and then detected to recover the audio.

Figure 16-3 shows the block diagram of the phase-locked loop frequency synthesizer section of the radio. The 10.24-MHz oscillator section is crystal controlled and forms the reference signal for the PLL, as well as the second signal used in the receiver. One output of the 10.24-MHz oscillator is divided in a binary counter by a factor of 1024 (2^{10}) to produce a 10-kHz reference signal for application to a classic phase-locked loop section. An N-code is used to control the division ratio of the divide-by-N counter in the PLL (see Chapter 6). The N-code is a binary code from the channel selector and will determine which channel is selected. The output of this portion of the PLL is an error voltage that drives the control input of a 17.18-MHz to 17.62-MHz voltage-controlled oscillator (VCO). The signal output of the PLL is a 2.86-MHz to 3.3-MHz reference signal selected by the N-code applied to the PLL input. This signal is mixed together with the VCO output and the 10.24-MHz reference signal to produce a sum signal of 37.66 MHz to 38.10 MHz. The is the receiver local oscillator signal. An additional mixer/oscillator combination heterodynes the LO signal against the intermediate frequency of 10.695 MHz to produce the transmitter output frequency.

The N-code is a binary code. Table 16-1 shows the channel numbers (1 to 40), the channel (transmit) frequency, the VCO output frequency (B), and the sum (A+B) and difference (A×B) frequencies, in which frequency A is the doubled output of the 10.24-MHz oscillator. Also shown are the VCO control voltage for each channel and the N-code.

A little explanation is in order for the *program word* and how it relates to the N-code. These program words are given in decimal, where the N-code is best visualized by using the hexadecimal number system. Let's consider the shaded example, channel 19. The N-code is the binary word 0110100, and the program word is "52." If we split the binary N-code into two portions, we can rewrite it in the form of a hexadecimal number that would produce the same bit pattern:

0011/0100 (leading zero added)

Table 16-1. PLL Troubleshooting Chart (courtesy of Delco Electronics).

CHAN	IC PIN NO. 9 10 11 12 13 14 15	PROG WORD	VCO VOLT	B	A − B	A + B	TRANSMIT FREQ
1	1 0 0 1 0 1 0	74	3.85	17.18	3.30	37.66	26.965
2	1 0 0 1 0 0 1	73	3.82	17.19	3.29	37.67	26.975
3	1 0 0 1 0 0 0	72	3.82	17.20	3.28	37.68	26.985
4	1 0 0 0 1 1 0	70	3.76	17.22	3.26	37.70	27.005
5	1 0 0 0 1 1 0	69	3.74	17.23	3.25	37.71	27.015
6	1 0 0 0 1 0 0	68	3.73	17.24	3.24	37.72	27.025
7	1 0 0 0 1 0 0	67	3.69	17.25	3.23	37.73	27.035
8	1 0 0 0 0 1 0	65	3.65	17.27	3.21	37.75	27.055
9	1 0 0 0 0 0 0	64	3.63	17.28	3.20	37.76	27.065
10	0 1 1 1 1 1 1	63	3.55	17.29	3.19	37.77	27.075
11	0 1 1 1 1 1 0	62	3.54	17.30	3.18	37.78	27.085
12	0 1 1 1 1 0 0	60	3.50	17.32	3.16	37.80	27.105
13	0 1 1 1 0 1 1	59	3.47	17.33	3.15	37.81	27.115
14	0 1 1 1 0 1 0	58	3.45	17.34	3.14	38.82	27.125
15	0 1 1 1 0 0 1	57	3.42	17.35	3.13	37.83	27.135
16	0 1 1 0 1 1 1	55	3.36	17.37	3.11	37.85	27.155
17	0 1 1 0 1 1 0	54	3.34	17.38	3.10	37.86	27.165
18	0 1 1 0 1 0 1	53	3.39	17.39	3.09	37.87	27.175
19	**0 1 1 0 1 0 0**	**52**	**3.29**	**17.40**	**3.08**	**37.88**	**27.185**
20	0 1 1 0 0 1 1	50	3.23	17.42	3.06	37.90	27.205
21	0 1 1 0 0 0 1	49	3.20	17.43	3.05	37.91	27.215
22	0 1 1 0 0 0 0	48	3.18	17.44	3.06	37.92	27.225
23	0 1 0 1 1 0 1	45	3.07	17.47	3.01	37.95	27.255
24	0 1 0 1 1 1 1	47	3.12	17.45	3.03	37.93	27.235
25	0 1 0 1 1 1 0	46	3.10	17.44	3.02	37.94	27.245
26	0 1 0 1 1 0 0	44	3.05	17.48	3.00	37.96	27.265
27	0 1 0 1 0 1 1	43	3.01	17.49	2.99	37.97	27.275
28	0 1 0 1 0 1 0	42	2.98	17.50	2.98	37.98	27.285
29	0 1 0 1 0 0 1	41	2.95	17.51	2.97	37.99	27.295
30	0 1 0 1 0 0 0	40	2.92	17.52	2.96	38.00	27.305
31	0 1 0 0 1 1 1	39	2.87	17.53	2.87	38.01	27.315
32	0 1 0 0 1 1 0	38	2.85	17.54	2.94	38.02	27.325
33	0 1 0 0 1 0 1	37	2.82	17.55	2.93	38.03	27.335
34	0 1 0 0 1 0 0	36	2.79	17.56	2.92	38.04	27.345
35	0 1 0 0 0 1 1	35	2.75	17.57	2.91	38.05	27.355
36	0 1 0 0 0 1 0	34	2.72	17.58	2.90	38.06	27.365
37	0 1 0 0 0 0 1	33	2.68	17.59	2.89	38.07	27.375
38	0 1 0 0 0 0 0	32	2.62	17.60	3.85	38.08	27.385
39	0 0 1 1 1 1 1	31	2.54	17.61	3.95	38.09	27.395
40	0 0 1 1 1 1 0	30	2.52	17.62	4.05	38.10	27.405

Binary 0011 is the same as hexadecimal "3." And binary 0100 is the same as hexadecimal "4." We may therefore write the binary number for the channel 19 N-code in the form "34H," in which the "H" indicates that hexadecimal notation in use. But why the use of program word "52"? The decimal equivalent of 34H is 52. Table 16-1 lists the program words in decimal.

The N-code generator is a simple binary counter that will count in either up or down directions. A low-frequency clock oscillator is turned on when the operator selects either up or down directions to change the channel. The output of the counter becomes the N-code. There is a problem, though. A CB set must not operate on frequencies outside the channels allowed by FCC regulation. The N-code counter and PLL circuit are capable of producing many channels outside this range. The PLL synthesizer, however, examines each N-code and compares it with the codes for legal channels. If the criterion for an illegal channel is met, the circuit refuses to permit it and will continue until it finds a legal channel.

Chapter 17
FM Receivers

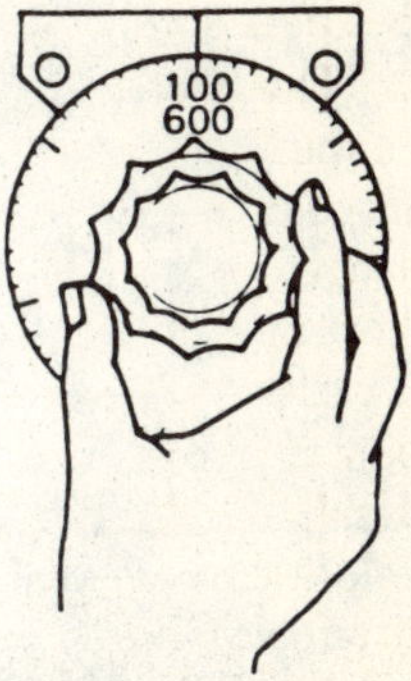

Frequency modulation was invented in the 1930s by Major Edwin Armstrong, but it was not until after World War II that FM systems began the long road to commercial acceptance. Some land mobile use of this mode was noted in the late 40s and 50s, but FM broadcasting was considered an orphan until the late 50s and early 60s. In the decade of the 50s, some radio broadcasters took out FM licenses just to hold a frequency for possible future use, should the FM band become commercially viable. Coupled with the high fidelity possible in the FM band, it seemed almost a sure thing that FM broadcasting would appeal to a limited market of hi-fi enthusiasts who preferred classical music and jazz. The FM band of the 50s, then, was known as a highbrow broadcast medium. But the equipment used on the FM band in those days was often quite crude by today's standards, and reception was limited to the larger metropolitan areas. Commercial FM broadcasting was so poor in revenue in those days that most broadcasters merely simulcast the AM program onto the FM affiliate. The background music, or SCA, signal, was authorized partially to offer ailing FM stations some means of obtaining revenue other than begging.

The land mobile services took to FM in a big way. There are certain very definite advantages to FM in the two-way radio communications world. One is the fact that the FM signal can be made almost noise-free, given the right circumstances. In addition, the problems of the FM broadcast receiver could be overcome with money, and most land mobile users could afford receivers that would overcome these problems.

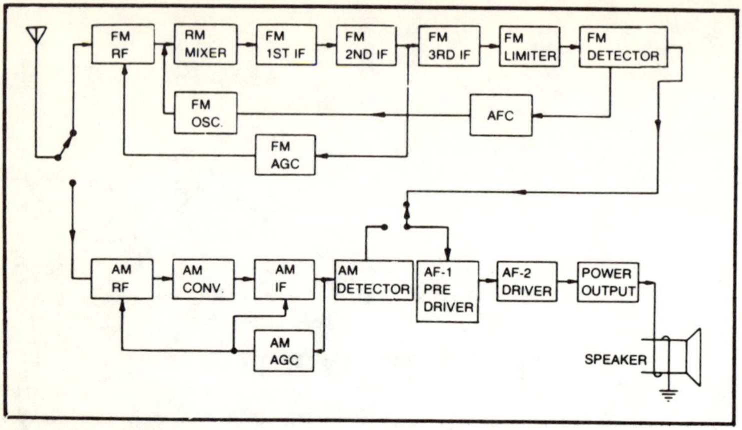

Fig. 17-1. AM/FM broadcast receiver.

Today, FM is *the* way to go for local communications and for high-fidelity broadcasting Although there is some talk of AM hi-fi stereo broadcasting, FM still carries the load in that area.

FM BROADCAST RECEIVERS

The FM broadcast receiver has always been more complex than almost any form of AM receiver. A more complicated form of demodulator is required, as well a lot more stages. Figure 17-1 shows the block diagram for a commercial AM/FM broadcast radio receiver. Let's consider only the FM section. The radio is a superheterodyne receiver. The signal from the antenna is amplified in the FM rf amplifier and then converted to the 10.7-Mhz intermediate frequency in the mixer/oscillator stages. There are three FM i-f signals. This is done for at least two reasons. One of the reasons is to gain sensitivity, but the other is to make sure that the signal is strong enough to drive the limiter into full clipping. Because the FM broadcast signal is 100 percent modulated with a deviation of ± 75 kHz, the bandwidth of the FM i-f must be at least 150 kHz, with 200 kHz being common.

The key to the claims of noise-free operation comes mostly from the fact that there is a limiter stage. Impulse noise, which is high-amplitude, short-duration noise pulses, is generated from a variety of sources. Among these are automobile ignition systems, fluorescent lights, and anything that produces an electrical arc. The static produced by lightning is an example of impulse noise. Impulse noise tends to *amplitude modulate* radio signals. The FM signal does not use amplitude variations to convey information, so it is possible to pass the FM signal through a clipper that chops off the amplitude peaks bearing the noise. If we tried that on an AM signal, the result would be noise-free *distortion*!

The FM signal is not inherently noise-free, however. There must be i-f signal capable of driving the limiter stage into hard clipping. A sample limiter circuit is shown in Fig. 17-2. The circuit works by virtue of some DC feedback to the third FM i-f amplifier stage. Note that the i-f amplifier and the limiter are direct-coupled, with the collector voltage of the i-f transistor being the base voltage of the limiter circuit. The base bias of the third FM i-f amplifier is derived from the emitter current in the limiter stage. In other words, the voltage drop across limiter emitter resistor R75 forms the base bias for the third FM i-f amplifier transistor. When the limiter transistor conducts harder, the voltage drop across R75 increases. This in turn increases the voltage applied to the base of the i-f amplifier. The FM signal is applied to the base of the third FM i-f. The collector current of this stage depends upon the signal strength. If the signal strength is low, the collector current is low. This means that the voltage drop across collector load resistor R64 is low, making the base voltage applied to the limiter stage high. All this makes the limiter act as one more amplifier. But as the signal level increases, the base voltage of the limiter decreases caused by an increase in the voltage drop across FM i-f collector load resistor R64. A point is reached at which the output signal from the limiter stage is constant, despite changes in the input signal level. As long as the input signal level remains above a certain threshold, the output of the limiter is constant.

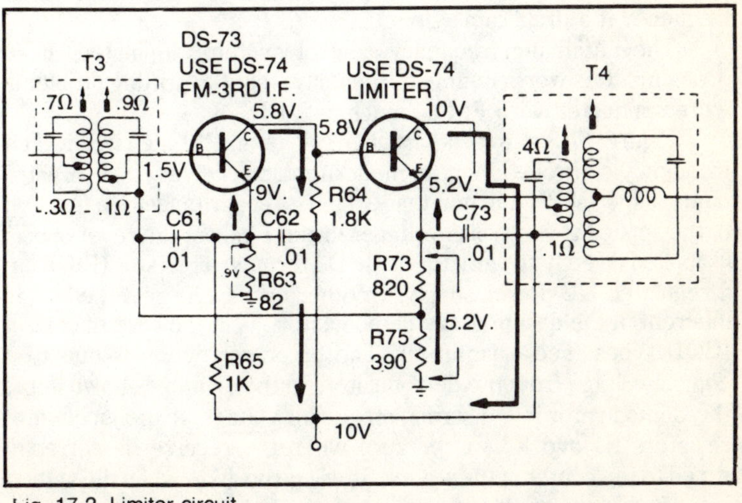

Fig. 17-2. Limiter circuit.

The signal from the limiter is applied to an FM detector stage, which could be almost any of the circuits discussed in Chapter 9. The audio section is a three-stage circuit that uses a preamplifier, driver, and power amplifier.

The automatic gain control (agc) circuit is pretty much the same as in any radio receiver and serves the same function. Note in this case, however, that the agc control voltage is applied to the rf amplifier alone, where in most AM receivers (including the AM section of this receiver) it is applied to both the i-f and rf amplifier stages.

One stage in this radio is unique to FM radios: *automatic frequency control* (afc). This circuit is a feedback control system that keeps the station tuned in correctly. One of the principal defects with early radios used in FM broadcast reception was the instability of the local oscillator. If the LO frequency changes, of course, the station drifts out of the passband of the receiver, and reception is lost. One popular kit-form FM tuner was so prone to drift that the station would change if the user tried to get too close to the cabinet! Imagine trying to tune a radio when the capacitance of your body (the radio did not have a metal, shielding cabinet) would affect the tuning. If someone walked into the room or walked across the floor, the vibrations would even cause the radio to drift off station. Even if everything else was perfect—no hand-body capacitance and no vibration from heavy footfalls on the carpet—the darn radio would still drift off frequency from thermal problems. As the radio heated up, the local oscillator frequency would change. It required a 30-minute warm-up period to settle down, and then would still change frequency if a draft came along.

The automatic frequency control system eliminated these problems. It is worth noting, incidentally, that the drifting problems were connected with FM as much as VHF.

Figure 17-3 shows a partial circuit of an FM afc circuit. The frequency control element in modern radio receivers is a varactor diode, a special PN junction that exhibits a capacitance as a function of the applied reverse bias voltage. In some cases, the reverse bias is applied directly to varactor diode D3 through a resistor (R6) from a regulated DC power supply. In other cases, the reverse bias is inherent in the design of the demodulator, as in the case of certain ICQD types (see Chapter 9). The error correction is due to a control voltage from the demodulator. In the example shown here, the demodulator is a discriminator. The voltage at the junction of resistors R1 and R2 will be zero when the receiver is correctly tuned to the center of the station. It will go positive when the station is too high in the passband or negative when the tuning error is in the

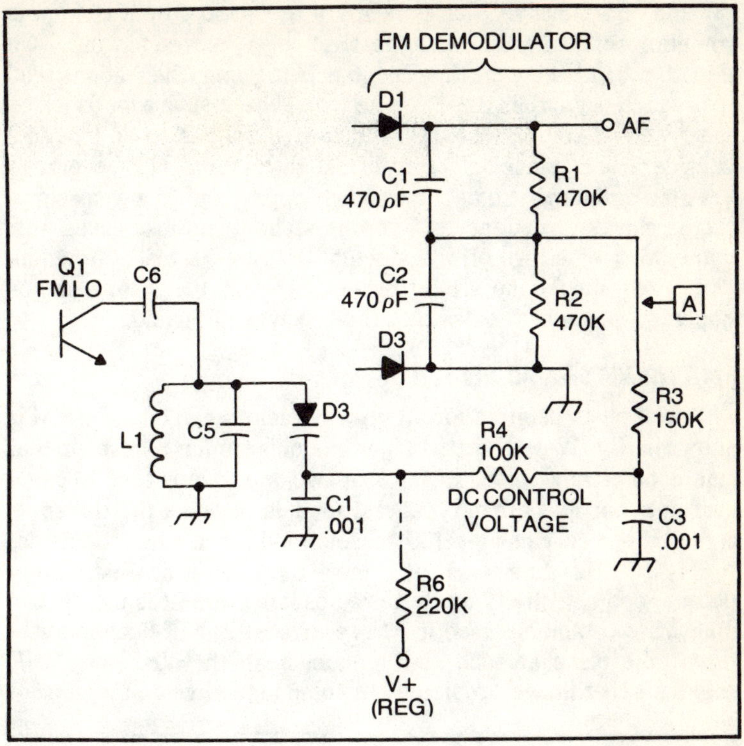

Fig. 17-3. Afc circuit.

opposite direction. We can use that voltage to cancel the drift of the FM local oscillator by applying it to the varactor diode in the LO. When the station drifts off in one direction, the detector generates an error voltage that will tend to pull the local oscillator frequency back to the correct point. The opposite voltage is generated when the LO drifts off in the other direction. The result, however, is the same: the LO frequency is pulled back to the correct point.

When tuning an FM receiver, you might notice that it is a little different from an AM, CW, or SSB model. The stations seem to *jump* into the receiver passband as the correct point on the dial is neared. If the time constant of the afc system is very short, the stations will actually "thump" into the passband, much to the chagrin of the loudspeaker designer. The afc time constant is set by the resistors (R3/R4) and capacitors (C3/C4) in the afc feedback loop. There must be a trade-off in tunable FM receivers between short and fast time constants. If the time constant is too short, the afc may lose its lock on weak stations. This also occurs when the signal strength varies over a range wide enough to drop the limiter out of clipping occasionally. What is heard in the audio output is what is sometimes called (somewhat descriptively) "picket fencing." On the other hand, if the afc time constant is too long, the stations tend to broaden out across the dial. The afc will have such a hard lock on the received station that the main tuning dial has to be moved a considerable distance before it will lose the station. This becomes a very big problem to someone who is trying to tune in a weak signal that is close in frequency to a strong station. In those cases, it is sometimes wise turn off the afc until the weak signal is tuned in. Only then should the afc be turned on, with the strong station hopefully far enough outside of the receiver passband.

FM TWO-WAY RADIO RECEIVERS

The block diagram for a two-way radio receiver using FM is shown in Fig. 17-4. This particular model is similar to many in that it is double-conversion. Recall that two-way radios tend to use a narrower bandpass than broadcast models because the transmitters in those services are 100 percent modulated when the deviation is ±5 kHz. As a result, it is more difficult (or at least *was*) to obtain good selectivity on these radios when the i-f is the normal 10.7 MHz commonly used in FM receivers. But FM communications take place at such high frequencies—the VHF and UHF regions—that image problems are compounded when the needed

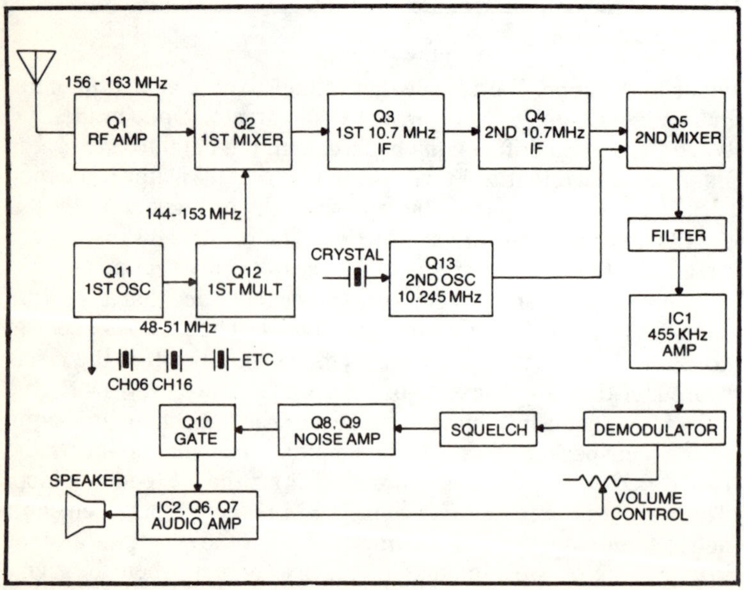

Fig. 17-4. Marine band VHF FM receiver.

low-frequency i-f is used to insure proper selectivity. A 455-kHz intermediate frequency at 2 meters, for example, would be disastrous, especially in today's crowded bands. The double-conversion design of Fig. 17-4 takes advantage of both worlds. The first mixer stage converts the rf signal from the rf amplifier to the high i-f of 10.7 MHz. This will insure the needed image rejection. The 10.7-MHz high i-f is heterodyned against the signal from 10.245-MHz crystal oscillator to produce a 455-kHz low intermediate frequency. It is in the low i-f that the main selectivity filter and often the bulk of the gain are located.

The demodulator could be any of the circuits discussed in Chapter 9. In most applications, however, it is common to find either one of three: ratio detector, Foster-Seeley discriminator, or the integrated circuit quadrature detector. The latter is most commonly found in modern amateur and commercial two-way radios. The radio shown in Fig. 17-4, incidentally, is a marine two-way radio operating in the 153 to 163 MHz band, but is similar in all respects other than radio frequency to amateur radio and land mobile equipment. The FCC forced all marine users to abandon the 2 to 3 MHz AM ship-to-shore and ship-to-ship band in favor of the VHF FM band. The results are somewhat better performance over the short ranges normally encountered in this work and less interference from "skip" stations.

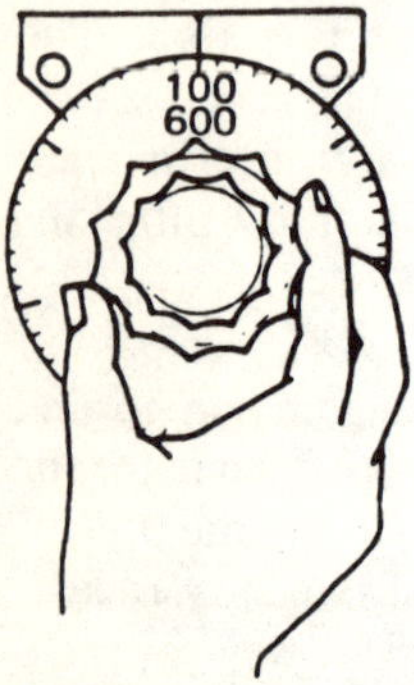

Chapter 18

VHF/UHF Receivers

The VHF spectrum consists of those frequencies between 30 MHz and 300 MHz. The UHF frequencies are 300 MHz to 3000 MHz, although some people now call everything above 1000 MHz "microwaves." But UHF they are.

Special techniques are needed when dealing with VHF and UHF receiver circuits. The values of the inductors and capacitors become so small, that a piece of wire can be used uncoiled as the inductor, and the capacitances can be the strays found in any circuit. The schematic diagrams for some television receivers, which are VHF/UHF devices, show no capacitors in the rf amplifier and local oscillator circuits; only coils are shown. The resonating capacitance are the stray, interelectrode capacitances normally found in any electronic circuit. These capacitances are ignorable at VLF and become only an annoyance to be accounted for in the HF region. At VHF and UHF, though, these capacitances become the total capacitance of the circuit and are in fact often a limiting factor. These circuits are so sensitive that moving a small component in a TV UHF tuner—a necessary act when making repairs—will shift the dial calibration by several *dozen* megahertz!

The six rf amplifiers used in VHF/UHF receivers can be almost any of the circuits shown in Chapter 5, but there seems to be a predominance of grounded-base, grounded-gate or grounded grid circuits because these are able to function without any external neutralization. Most of the triode circuits—those using bipolar and field-effect transistors—will require such neutralization if used in the normal common emitter or common source circuits.

VHF and UHF receivers also need low-noise rf amplifiers. The noise figure of the receiver becomes the limiting factor of the smallest signal that can be handled. While the noise figure of any rf amplifier is important, it becomes crucial in many VHF and UHF applications.

COUPLING NETWORKS

One of the principal physical differences between VHF/UHF receivers and receivers used at lower frequencies is the coupling networks. In the low VHF region—under 100-MHz for example—the coils and capacitors seem like those of the lower frequencies, except in physical dimensions. Only a few turns are needed at these frequencies. Many FM broadcast receiver turners (88 to 108 MHz), for example, use only three to five turns of wire on a 3/16-in. form. At these frequencies, we see some inductors made of printed-circuit material cemented to the cylindrical form. As the frequencies increase further, the printed-circuit inductors are printed right on the main PC board with the rf amplifier circuitry. At 200 MHz, the typical PC indictor is merly a short section of PC track, less than ¼-in. wide. As the frequency increases above 300 MHz, it becomes practical to make the coupling circuits in the form of cavity resonators. These are chambers cut to a resonant frequency with input and output coupling in the form of hairpin loops of wire.

Figures 18-1 through 18-5 show several different coupling networks used in VHF and UHF circuits. The helical resonator is shown in Fig. 18-1. This network consists of two parallel resonant LC tank circuits shielded from the outside world and each other, except for a coupling slot between the two chambers. The helical

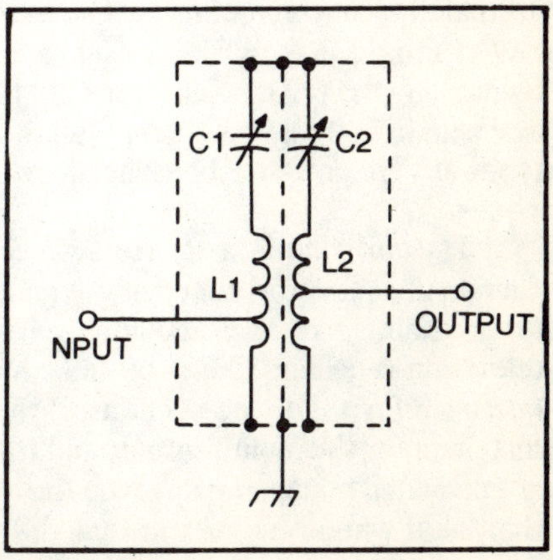

Fig. 18-1. VHF/UHF coupling arrangement.

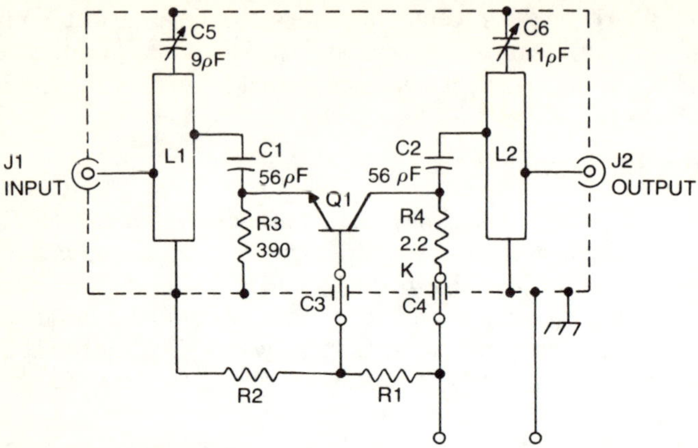

Fig. 18-2. VHF/UHF coupling arrangement.

resonator operates not so much as an LC tank circuit, but as a resonant section of transmission line. The Q of the tank is dependent upon the chamber dimensions. Some high-quality VHF communications receivers will cascade two or more helical resonators of the type shown above to form a very sharp resonant tank circuit for incoming rf signals. This becomes a very necessary property in circuits that have to contend with a large number of high-power signals in the same frequency range, which is common in the land mobile ranges.

Stripline resonators become useful in the VHF/UHF region. Figure 18-2 shows an example of such coupling networks. This is a circuit for a UHF rf amplifier, with the transistor operated in grounded-base configuration. The base of the transistor is held at a low rf impedance, while retaining a high DC resistance by the feedthrough coupling capacitor C3. These capacitors are typically available in 500 pF and 1000 pF units and are preferred over other types at VHF and UHF because of lower lead inductances and series losses.

The stripline is a quarter-wavelength (or multiples thereof) section of conductor, tuned at the ends with a trimmer capacitor. A quick example of such inductances is seen in UHF tuners from television receivers. Most of these have three stripline inductors, one each for the three sections of the converter. The impedance matching for the input, output, and transistor emitter and collector are provided by tapping down on the stripline in a manner similar to that used when real coils are used.

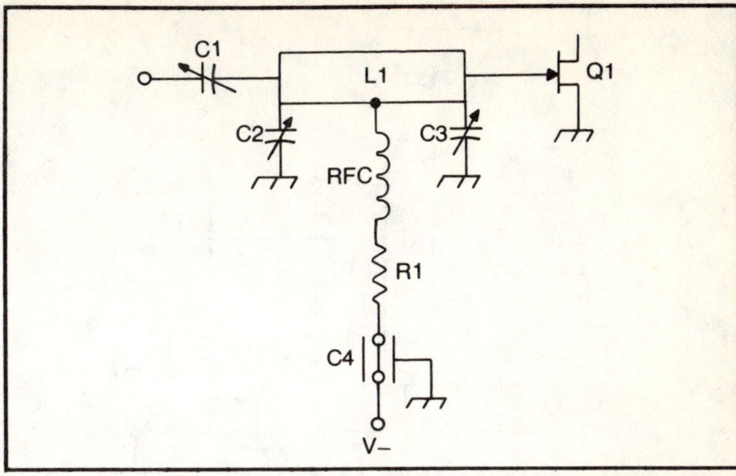

Fig. 18-3. VHF/UHF coupling arrangement.

Another form of stripline resonator is shown in Fig. 18-3. In this case, we are coupling the antenna to the input of a junction field-effect transistor. The input impedance is matched to the stripline with capacitor voltage divider C1/C2. The input network is resonated to L1 (the stripline) with capacitors C1, C2, and C3. C1 and L1 form a series resonant circuit, while C2/C3/L1 form a pi-network. Bias is applied to the JFET through a tap on the body of the stripline, usually made at the exact center. Rf isolation to the bias supply is provided by an rf choke, decoupling resistor, and feedthrough bypass capacitor C4.

A special form of "rf choke" is used in VHF and UHF circuits: ferrite beads placed over a length of wire (Fig. 18-4). These small cylindrical beads will act as a short circuit to DC and radio frequencies into the low HF region, but offer substantial inductive reactance to frequencies in the VHF and UHF region. They tend to have less losses than coils wound for these frequencies. Ordinary coils become something of a problem to wind precisely; in fact, coils often

Fig. 18-4. VHF/UHF coupling arrangement.

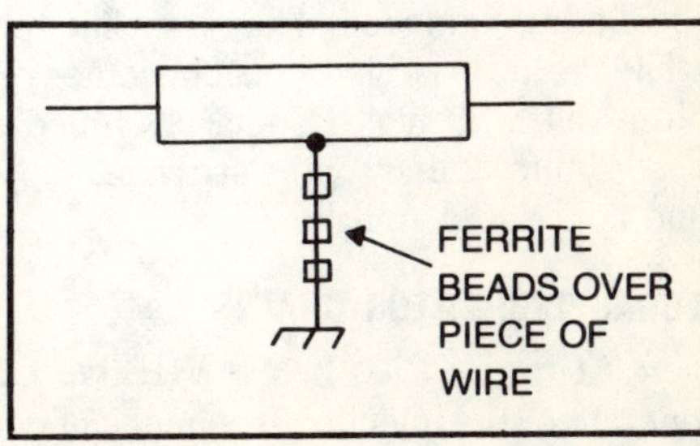

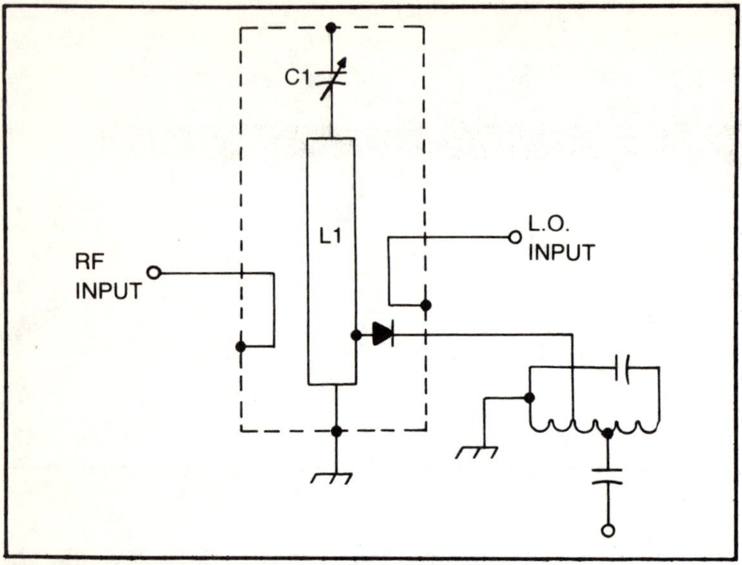

Fig. 18-5. VHF/UHF coupling arrangement.

have a brass core so that *more* turns can be used to obtain the needed low inductance. Where ferrite cores will increase the inductance of a coil, brass and other nonferrous cores will reduce the inductance of the same coil, allowing more manageable proportions of the coil.

A stripline mixer is shown in Fig. 18-5. The interior of the shielded chamber is resonated by the parallel resonant stripline/capacitor combination. A hairpin loop of wire comes through a hole in the side and runs alongside the stripline indictor to couple rf energy into the tank. A similar hairpin wire is used to couple the local oscillator into the stripline. A UHF mixer diode is connected to an impedance matching point on the stripline and is then brought to the outside where it connects to a tank circuit resonant at the intermediate frequency.

The same arrangement is useful for asking a stripline filter if the diode is removed. The rf amplifier port then becomes the input, while the local oscillator port becomes redesignated the output. The filter will be a high-Q, sharply tuned affair that passes only the resonant frequency of the stripline, suppressing all of the harmonics and nonrelated signals.

LOCAL OSCILLATOR CHAINS

At frequencies below VHF we can make the local oscillator operate at the injection frequency needed by the superheterodyne principle (rf ± i-f). Even at VHF, such as in the FM broadcast band, some receivers use local oscillators operating at the right frequency. But in many cases the LO operates at some lower frequency and is then frequency multiplied to the correct frequency. Using doublers, triplers and quadruplers is common in VHF and UHF receivers.

Figures 18-6 through 18-8 show several different types of frequency multiplier circuits. The primary goal in all of them is to pass the signal from a crystal oscillator through a nonlinear element which distorts the waveform. Any distorted waveform will produce harmonics of the fundamental sine wave frequency. A tank circuit in the output of the multiplier will then pick off the harmonic desired.

The nonlinear element in Fig. 18-6 is a simple unbiased bipolar transistor. A bipolar transistor requires a forward bias if it is to be operated linearly. If zero bias is used, the transistor is operating class B, which produces an output waveform resembling the half-wave rectifier output waveform. In fact, we might well characterize the class B, single-ended stage as an active half-wave rectifier. In the case shown, a small resistor is placed in series with the signal applied to the transistor base. If the driving signal is strong enough, there will be a small voltage drop across the resistor that tends to reverse bias the transistor, making the stage operate class C. Either class B or class C produces large amounts of harmonic energy from a pure sine wave input. The parallel resonant tank circuit (L1/C1) in the collector of the transistor is used to select which of the harmonics will be the output. If *N* is an integer and the

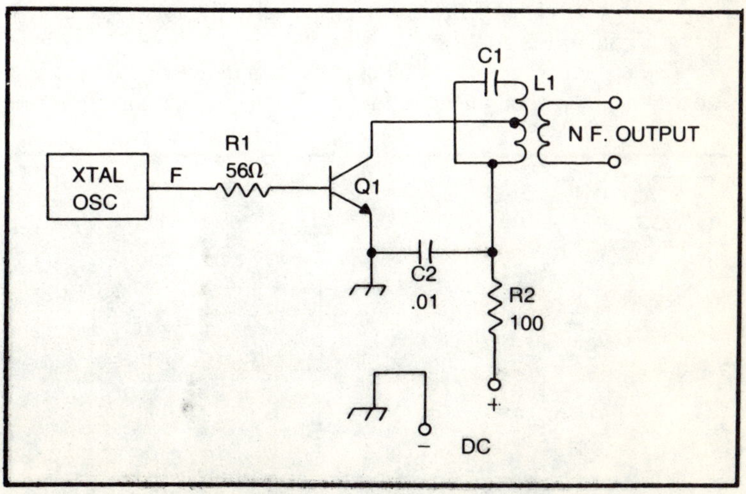

Fig. 18-6. Frequency multiplier.

input frequency is F, then the output frequency will be the produce NF.

A simple diode frequency multipler is shown in Fig. 18-7. This circuit uses a diode half-wave rectifier to produce the needed nonlinearity. The only problem with this circuit is that it is passive, so no amplification is provided. This means that the driving signal must have sufficient power to produce a harmonic at the desired frequency with the needed amplitude for injection into the mixer. Sometimes, a tuned or untuned buffer amplifier will follow the output of the multiplier and will build up the signal level to the required amplitude.

A full-wave frequency doubler is shown in Fig. 18-8. Recall from your power supply theory that the ripple frequency of a full-wave rectifier is twice the input, or line, frequency. This can be used to our advantage in the construction of a simple passive frequency doubler. Transformer T1 is "trifilar" wound, meaning that all three windings are parallel to each other at all points. This is done by laying the wires onto the core (usually toroidal) side-by-side. They will then form three identical, matched windings. The output of the crystal oscillator is applied to one of the windings, while the diodes are connected to the other windings in opposite sense from each other. One diode, then, must be connected to the dotted end of winding B, while the other diode is connected to the undotted end of winding C. The cathodes of the diodes are then joined together at the input of a tank circuit that is tuned to the desired output frequency—twice the input frequency. Like the other circuit, the output amplitude is less than the input amplitude. There is, however, an amplitude advantage to the full-wave circuit simply because it *is* full wave.

The output waveform of all of these frequency multipliers is a sine wave, even though the input sine wave was distorted in the

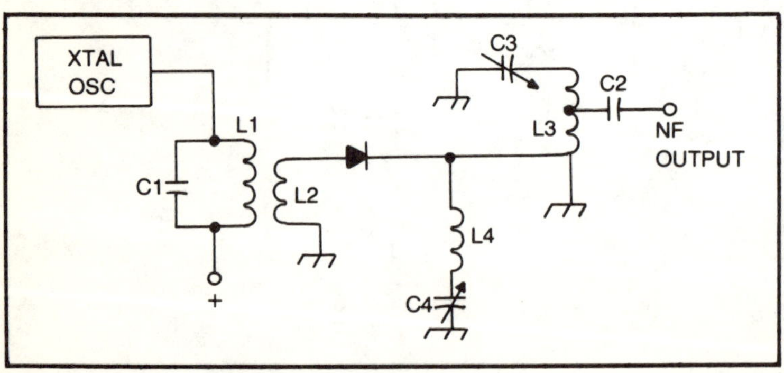

Fig. 18-7. Frequency multiplier.

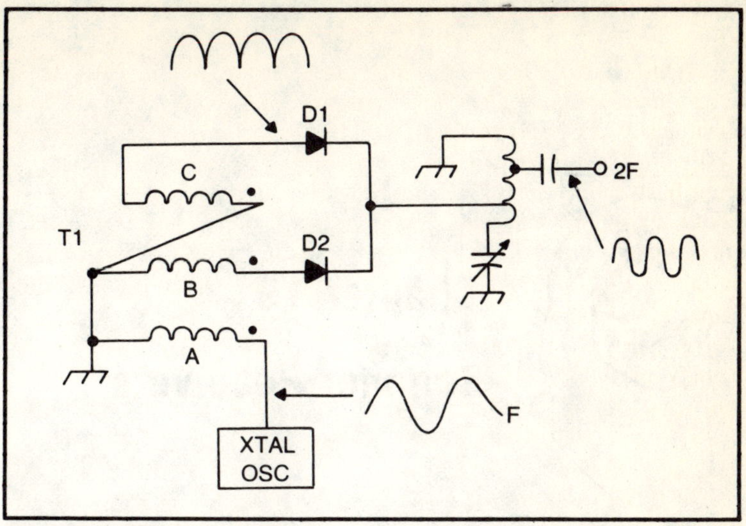

Fig. 18-8. Frequency multiplier.

multiplication process. The sine shape is restored by the flywheel effect of the output tank circuit. Any shock excited LC tank circuit will produce a damped oscillation at its natural resonant frequency, and these oscillations have sinusoidal shape. The pulses from the diode or transistor used as the nonlinear shock excite the tank circuit.

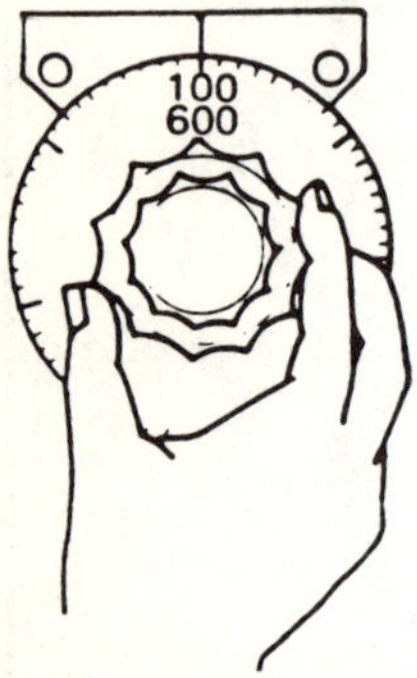

Chapter 19

Scanner Receivers

A scanner receiver, or just scanner, is designed to sequentially examine two or more channels in the VHF or UHF land mobile, amateur radio or marine two-way radio bands. The idea is to continuously monitor these channels for activity and then stop scanning and start listening when activity is present on one of the channels. The scanner is typically squelched until one of the scanned channels becomes active. The radio then stops scanning, unsquelches the output, and begins listening to the active channels.

The crystal-controlled scanner radio made monitoring police, fire, and other VHF/UHF radio services a lot more popular than in the past. At one time, these bands were monitored only by a few dyed-in-the-wood buffs, off-duty police and fire officers, newspaper reporters, and so forth. The earliest VHF monitors were ordinary VFO-tuned superhets that were subject to a lot of drift, and the slightest vibration could make them jump off channel. The typical radio signal in these ranges had a bandwidth of a few kilohertz, and the typical radio dial covered 10 to 30 *megahertz*. The tuning rate, therefore, was extremely fast, and no mean skill was required to successfully tune in the narrow-band signals. In the early 60s several crystal-controlled models were introduced that allowed the user to select six to 12 channels, buy the crystals for them, and use the radio only for those channels. This was terribly limiting but produced superior radio performance. The earliest scanners were also crystal-controlled, and would allow the selection of eight to 10 channels. A separate crystal had to be bought for each channel selected, which was an expensive proposition. But modern scanners use digital phase-locked loop techniques that permit direct register entry of the desired channel from a front panel keyboard. Some will allow dozens of channels to be programmed, and the user is no longer limited to the few channels he has on hand.

SIMPLE SCANNER EXAMPLE

Probably the simplest form of scanner is the dual-receive citizens band transceiver. This feature allows the operator to monitor a favorite channel while simultaneously keeping an electronic eye on channel 9, the road emergency channel. Amateur radio, marine, and certain other forms of two-way radios are now offering this same feature, even though it originated on the dual-receive CB rigs.

Figure 19-1 shows a simplified block diagram of a dual-receive CB rig. There are two local oscillators in this superheterodyne. One is the 40-channel synthesized LO that is the main tuning for the rig. The operator uses this LO to select the transmit/receive channel for the radio. The second LO is a channel 9 crystal-controlled oscillator. A J-K flip-flop selects these two LO circuits alternately. A low-frequency clock signal causes the flip-flop to alternate between SET and RESET states. In the SET state, the Q output is HIGH and the not-Q is LOW. In the RESET state, the Q output is LOW and the not-Q is HIGH (exactly the opposite of the SET). The local oscillator circuits are designed such that they are enabled with either a HIGH or a LOW. But both must use the same level—HIGH or LOW. One cannot be active-HIGH and the other active-LOW or they will both turn on at the same time. An inhibit line on the J-K

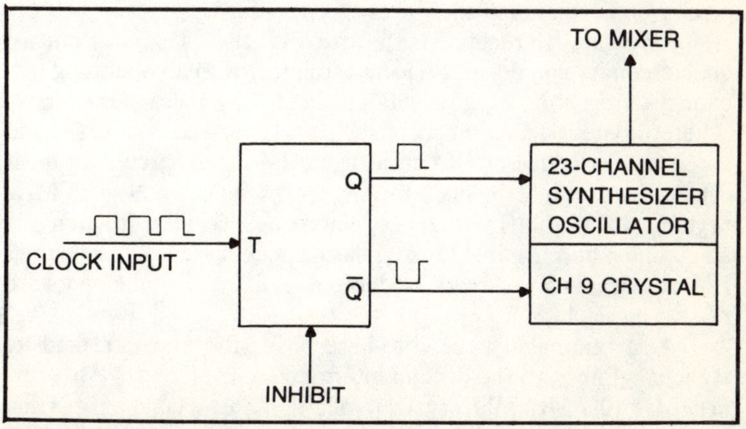

Fig. 19-1. Simple two-channel CB scanner.

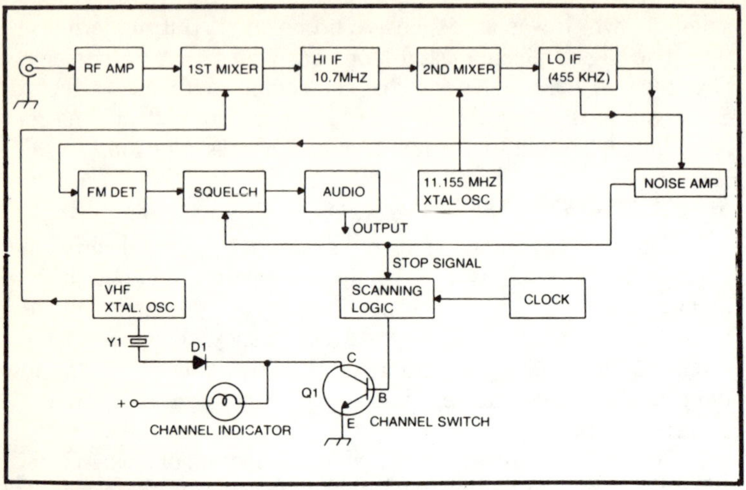

Fig. 19-2. VHF/UHF scanner receiver.

flip-flop is used to latch the outputs to their present state when an agc signal indicates that a strong signal is present in the passband of the CB radio when the channel-9 LO is active.

VHF/UHF SCANNERS

Most of the scanners on the market today receive either VHF, UHF, or a combination of VHF and UHF bands. In fact, the original market for these device was the law enforcement group that had to monitor several VHF or UHF communications channels in order to do their jobs. A federal or state agent, for example, might need to monitor his own communications channels, plus those of city and county departments in their area of operation.

A typical VHF receiver is shown in Fig. 19-2. This is a standard double-conversion design. A local oscillator, with an operating frequency determined by the crystals selected by the scanner circuits, is heterodyned against the incoming signal from the rf amplifier and antenna to produce an i-f output from the mixer circuit. In most cases, the first intermediate frequency will be in the 9 to 15 MHz region, with 10.7 MHz being very common. This latter frequency is so common because it is the standard frequency used in consumer FM broadcasting receivers, so large quantities of components are readily available.

A second mixer is used to heterodyne the first intermediate frequency down to the second intermediate frequency. When the first i-f is 10.7 MHz, the crystal oscillator in the second mixer must operate at either 10.245 MHz or 11.155 MHz, as shown. The output of the second mixer will be the standard 455-kHz frequency, which is then handled in the usual manner for superhets.

The squelch circuitry provides more than mere silence in the output. It also provides the control signal for the scanner stop cycle. The clock will continue incrementing the counter in the scanner logic section until it sees an *open squelch* command. At this instant, the counting ceases, and the listener hears the chatter present on the active channel.

Channel selection is performed in most cases with a simple NPN transistor switch in the crystal circuits. Only one channel is shown here for sake of simplicity. Crystal Y1 is one of the crystals in the first local oscillator. It is connected to ground when the associated transistor switch (Q1) is turned on. A HIGH on the base of Q1 selects the channel. This will also turn on the channel indicator lamp or LED. The purpose of the isolation diode is to separate the DC from the lamp circuit and the crystal.

Figure 19-3 shows the partial circuitry of a scanner. Some form of low-frequency oscillator is needed to form a clock signal. A 555 timer is sometimes used, but in this case, the clock is unijunction transistor (UJT). A UJT, you might recall, will not conduct between the emitter and the base channel unless the emitter/base voltage exceeds a certain level. A capacitor connected across this junction is charged from a constant current source. When the capacitor voltage builds up to a preset point, the emitter will breakover and dump the capacitor charge through the base/emitter junction path. The

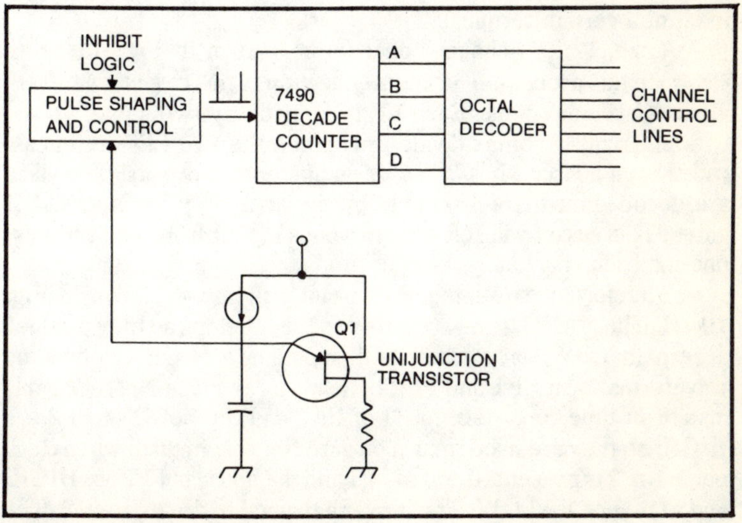

Fig. 19-3. Counter circuit.

waveform across the capacitor is a sawtooth with a period set by the constant current level, the value of the capacitor, and the breakover point of the UJT. If values are selected correctly, the UHT output (the capacitor signal) can be connected directly to the TTL counter. In most cases, however, some form of waveshaping is provided. This could be a Schmitt trigger or some other circuit that will output TTL compatible pulses.

The control section will shape the pulses and provide them to the counter. When the inhibit signal is received, indicating that a station is in the passband, the control section will interrupt the flow of pulses to the counter. The output state of the counter will then remain on the active channel until the inhibit signal goes away. In most circuits, a time delay is provided so that the channel is not lost by resumed scanning when the two operators break to let the other one talk.

The counter will be some form of binary or BCD version. In the example shown, we have a BCD (type 7490) counter, which will allow us to scan up to 10 different channels. A 7492 octal counter would allow eight channels, while a 7493 would allow 16. A decoder at the output of the counter is used to provide a unique output indication for each count. If we wanted to select 10 channels, for example, we could use a common Nixie tube driver from the TTL line. These devices are called 1-of-10 decoders and provide a single active-LOW output for each of the 10 counts permitted by 7490 device. A 74154 will allow up to 16 unique outputs, and can be made either active-HIGH or active-LOW depending upon the programming of a certain terminal.

A simple four-channel decoder is shown in Fig. 19-4A to illustrate this process. The counter section of this circuit consists of a pair of J-K flip-flops. Each FF is a divide-by-two stage, so the overall circuit becomes divide-by-4 when the two FFs are in cascade. Such a circuit has 2^2, or 4, unique output possibilities which are decoded into 1-of-4 outputs by the array of two-input NAND gates. This circuit will select from a bank of four front-end crystals, one for each channel.

Waveforms explaining the circuit action are shown in Fig. 19-4B. The NAND gates are wired to the flip-flops so they produce a zero (a LOW) only when both inputs are HIGH. Notice the waveforms form FF1 and FF2 beneath clock pulse No. 1. At this instant in time, only the not-Q of FF1 and the not-Q of FF2 are HIGH, so they are used to turn on gate No. G1. In turn, when clock pulse No. 2 is present, the Q of FF1 and the not-Q of FF2 are HIGH, and all others are LOW. These are then used to drive gate G2. The sequence continues through all four possible combinations.

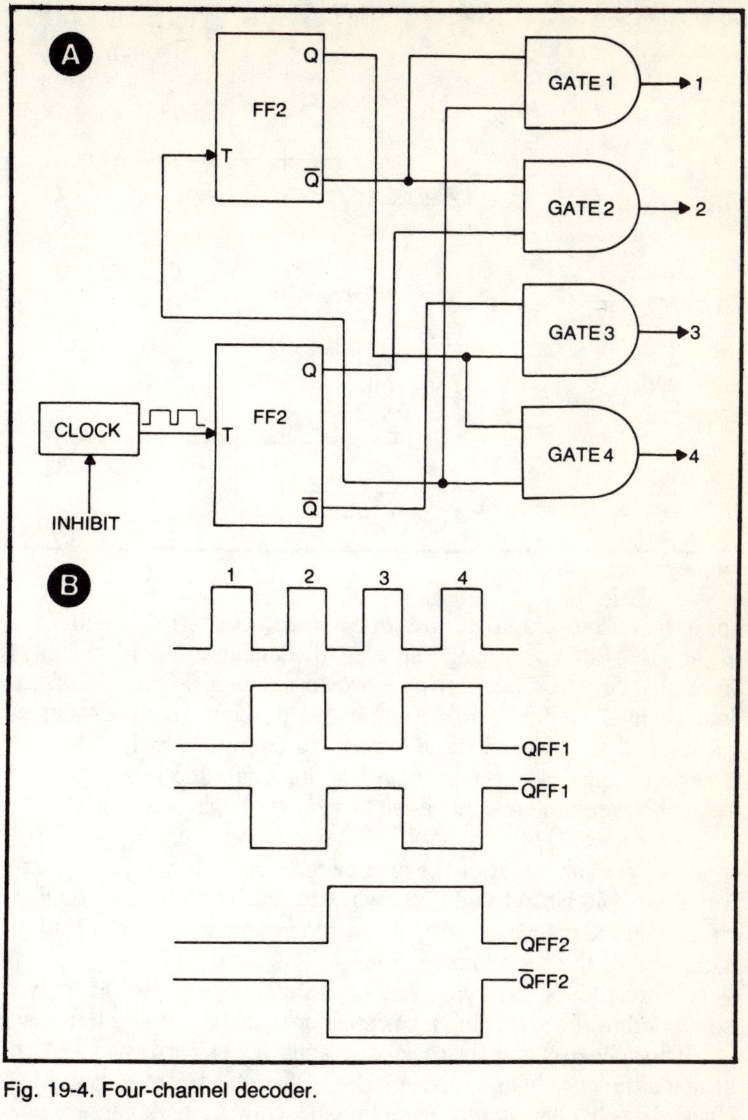

Fig. 19-4. Four-channel decoder.

Most scanners offer more than four channels. Only a few pocket, hand-held models are limited to the four channels accommodated by the above circuit. In fact, the most common arrangement in the crystal-controlled (non-PLL) models is eight channels. Because these devices operate using base-2 arithmetic, we can double the number of channels with just one more flip-flop. If a third flip-flop is added, then the channels available are 2^3, or 8. Similarly, if

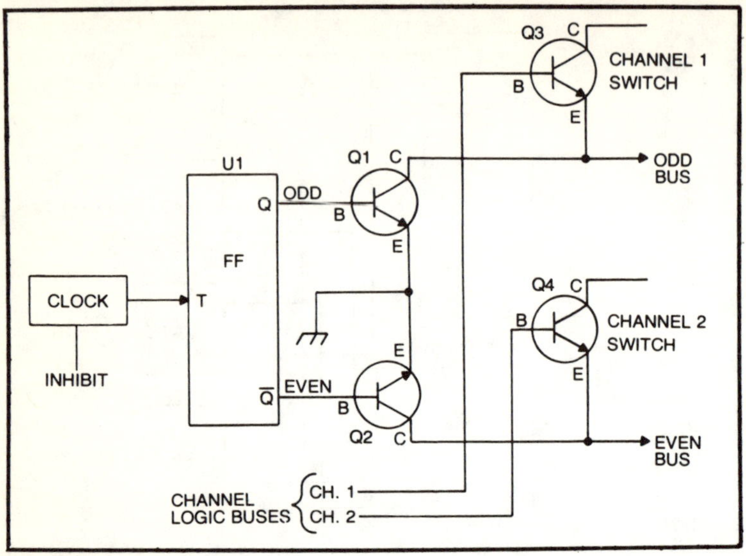

Fig. 19-5. Odd-even system.

there were four flip-flops, the total number of possible channels is 2^4, or 16. But the gating required to decode 8 or 16 channels becomes very complex, so one would not use the circuit of the four-channel model. To overcome this problem, some manufacturer used an odd and even system to simultaneously scan two different banks of crystals. One additional flip-flop is required to select between the odd and even banks of crystals (see Fig. 19-5).

Transistor Q1 in Fig. 19-6 is the regular VHF crystal oscillator which drives one input of the first mixer stage. Although only one complete switching circuit is shown here, each crystal has a similar circuit and will be grounded by its switching transistor like Q2 at the command of the switching circuit.

When the logic circuit selects a channel, a positive voltage will be applied to the switching transistor for that channel—in this case Q2. The collector-emitter resistance approaches zero ohms, which grounds the crystal and turns on the LED indicator to show which channel has been selected. Sometimes, lock-in or lock-out switches are provided. These will either ground the crystal manually or prevent it from being addressed by the counter. In most receivers, each crystal is also provided with a small value trimmer capacitor so that its frequency can be netted exactly on the correct frequency for the selected channel. Even precision crystals, which most of these are not, require some netting due to manufacturing tolerances in both the crystal itself and in the receiver LO circuit.

The case markings on crystals sometimes go out to the very hertz, but this is rarely if ever the true situation. When the crystal is placed in the circuit, the actual oscillating frequency is a little different from the marked frequency. The frequency changes slightly with changes in frequency. It is therefore necessary to provide the crystal vendor with certain pertinent information when ordering replacement "rocks." Of course, one piece of information required is the operating frequency. It is also necessary to know whether this is the fundamental frequency, or which of several possible overtones (third, fifth, seventh or ninth).

Low-band VHF receivers (30 to 50 MHz) typically use fundamental mode crystals below the radio frequency or third overtone crystals above the radio frequency. The proper frequency to order, then, will be:

$$\text{Crystal Freq.} = \text{Radio Freq.} - \text{intermediate freq. (fundamental)}$$

If an overtone crystal is used, then the required frequency will be:

$$\text{Crystal Frequency} = \frac{\text{Radio freq.} \pm \text{Intermediate freq.}}{N}$$

where ± depends upon whether the LO is above (+) or below (–) the radio frequency, and N is the overtone used (3, 5, 7 or 9).

□**Example:**

A receiver operates in the fifth overtone made (N=5) with the LO above the radio frequency. Find the frequency of the first LO crystal if the frequency to be monitored is 146.52 MHz. The i-f is 10.7 MHz.

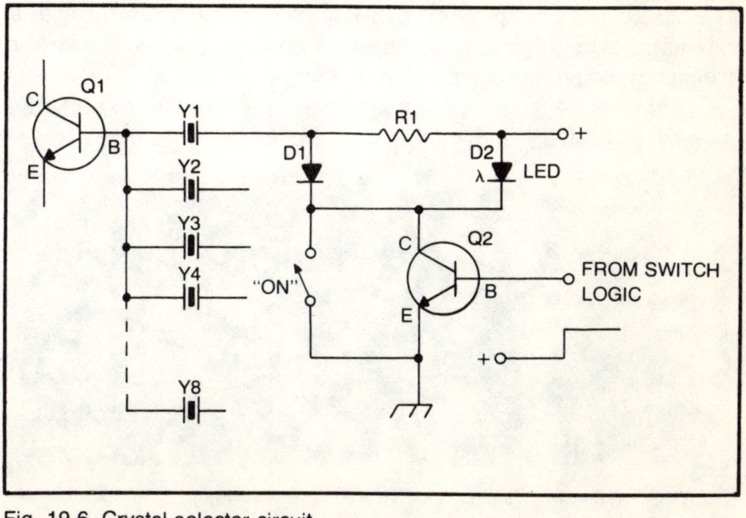

Fig. 19-6. Crystal selector circuit.

$$\text{Crystal Freq.} = \frac{\text{rf} + \text{i-f}}{5}$$

$$= \frac{(146.52) + (1p.0)}{5}$$

$$= 157.22 \text{ MHz}/5 = 31.444 \text{ MHz}$$

This, then, is the fundamental frequency required if a fifth harmonic is selected. The crystal manufacturer can often select the right crystal for you if given the make and model number (sometimes also the serial number, if there has been a major design change in the circuit). Provide this information first. If the firm does not have the file data required for your mode, it will need the following.

- □Crystal frequency desired.
- □Style of crystal holder (match style in the catalog for correct designation).
- □Mode of oscillation (fundamental, 3rd overtone, etc.).
- □Operating capacitance.
- □Drive level in milliwatts.
- □Maximum allowable series resistance.

If you lack this information, contact the manufacturer of the radio.

A PROGRAMMABLE SCANNER

A new programmable scanning monitor receiver shown in Fig. 19-7 just introduced by Radio Shack lets you listen in on police, fire, emergency, government, business, and amateur radio communications in your area. The Realistic PRO-2008 direct entry programmable scanner provides direct access to 18,600 frequencies with no crystals to buy. Simply enter the desired frequencies through a calculator-type keyboard. A large fluorescent display shows the frequency being monitored.

Up to eight frequencies may be entered from 30 to 50 MHz, 144 to 174 MHz, and 410 to 512 MHz. Any channel may be

Fig. 19-7. Radio Shack's Realistic PRO-2008 direct entry programmable scanner.

programmed with a scan delay to prevent missed replies, or locked-out and skipped when desired. Channels can be scanned automatically, or selected manually.

Features include a built-in telescoping antenna, a jack for an external antenna, a headphone jack for private listening, and a reset button that instantly clears all frequencies from the scanner. An optional 9-volt battery prevents loss of programmed frequencies if AC power is disconnected.

WARNING

Many people use scanners for amusement. This is legal in many jurisdictions—perhaps most. Be aware, though, that some areas regulate the use of police-band monitor receivers, especially in cars. Some cities and counties require a police permit for such receivers and limit these permits to bonafide news media personnel, tow truck operators licensed by the city or county and a few others. In some other cases they will license anyone who applies, except perhaps felons, but simply want to know *who* might be listening.

In any event, the Communications Act of 1934 prohibits the use for personal gain of any information transmitted by radio stations, other than CB amateur radio, and broadcasters. If you own a tow truck and respond to an accident scene because of a police transmission, you are in violation of the law unless you have prior permission to do so. A taxicab driver in my hometown bought a scanner to pick up the transmissions of a competing company. He would then respond to their calls, grabbing the fare ahead of the other company. I am not sure who was harder on him—the FCC who nailed him or the other drivers who hung him out to dry when they found out!

Chapter 20

Auto Radio Receivers

In a scene from the old TV program, *The Untouchables,* agent Eliot Ness learned too late some vital information that had been broadcast over a Chicago radio station. He turned to his companion, another rackets-busting Federal agent, and said (paraphrased), "Someday, someone's going to invent an automobile radio, and I'm going to be the first to buy one!" The real agent Eliot Ness (oh, yes, he *was* a real agent, not merely a figment of some Hollywood writer's imagination) could have bought a car radio. Inventor Bill Lear, who also brought us the Learjet, eight-track tape cartridges (which are known officially as *Learjetpaks*) and the two-way airplane radio, is usually credited with the first auto radio to be produced in the US. The Chicago-area firm now known as *Motorola* could have sold agent Ness his longed-for car radio. They, in fact, pioneered commercial automobile radio production. The very name *Motorola* takes its meaning from motor, for motor car, and the *ola* commonly used to brand radios of the late 20s.

HISTORY

The auto radio of the late 20s and early 30s was a far cry from the products made today. Those oldsters used wire aerials under the running board of the car, and some models required an AC alternator be mounted under the hood so that it could be driven by the engine to produce AC for the radio. It seems that the earliest car radios were little more than adapted home radios. The alternator was dropped early in the game, however, in factor of an electronic switching device called a *vibrator*. This was an SPDT switch driven by an electrical buzzer and was used to chop the DC from the car battery so that it could be applied to the primary of a power transformer. Car radios from the early 30s to the mid-50s used vibrator power supplies. I can recall customers who owned the new solid-state radios bringing in filter capacitors ripped from the PC board, asking us to "replace this vibrator." It seems that the vibrator looked much like an electrolytic capacitor can, and the customer was unaware that a quiet revolution had taken place in the design of car radios.

Radios made from 1957 until 1962/63 were known as hybrids. The development of transistor technology had not progressed to the point where it was economical and reliable to use transistors in all stages of the radio receiver. Rf transistors were in short supply and were costly. Motorola, once again a major leader in car radio design, introduced an all-transistor model in 1956, but it was a limited production model intended for the so-called "after market" (people who do not have a factory radio in the car or want to upgrade to a fancier model). The automakers use radios intended for the OEM (original equipment manufacturer) market, so they kept with hybrid designs for several years.

The hybrid radios used special vacuum tubes in the rf, converter, i-f, detector, and first audio stages, and a power transistor in the output stage. The tubes (12BL6, 12AD6, etc.) were specially designed for use in automobile radios to operate with as little as 12 volts on the anode.

The power transistors were germanium PNP types. Motorola used its 2N176 (made by another division of the Motorola company) for many years. In fact, it used technical descendents of the 2N176 into the 70s. Delco Electronics used its DS-501 and DS-503 devices, also well into the 70s. They have recently switched over almost entirely to a special integrated circuit bridge audio system for the power amplifier. As you might guess, many of the repairs to car radios of that era revolved around the transistor output stage. Those early germanium devices had much more leakage than do modern transistors, so failures were somewhat more frequent.

In 1962, Delco produced the first OEM car radio manufactured on a large scale for a major car producer, Delco's parent company, General Motors. The other car makers followed suit in 1963 using Motorola OEM models in Chrysler and Ford. Delco introduced the first AM/FM car radio in 1963.

Many car buyers opt to purchase a new car without the radio on the theory that they are going to an after-market shop and buy a "better quality" car radio. This is not generally true. Although some

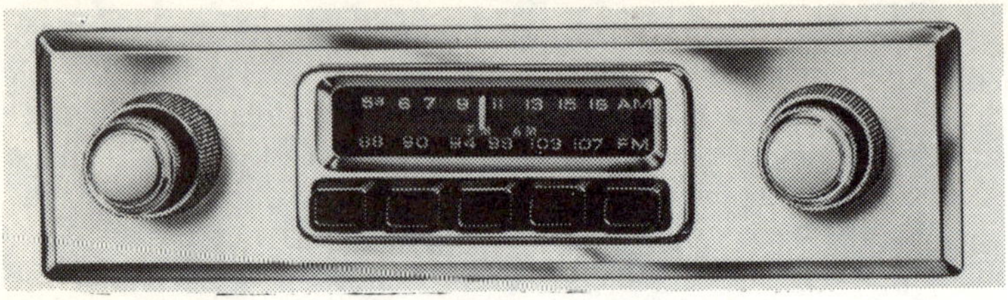

Fig. 20-1. Universal auto radio (courtesy of Motorola, Inc).

of the after-marker radios are as good as the OEM types, only a very few could be classified as better. Motorola, Bendix, Delco Electronics, and (in recent years) the Automobile Radio Division of Chrysler produce some of the best broadcast receivers in the world. The only real excuses for buying an after-market radio are price (OEM prices are usually higher) and features that fall into the "whoopee, wow-ding, whistles, and bells" category.

Figure 20-1 shows two typical car radios. The model shown in Fig. 20-1 is a Motorola "universal" car radio. It is specifically intended for the US aftermarket but also is found sold by many European car dealers as the custom "factory" radio. The front bezel of the radio, which fits through the decorative chrome panel shown installed, is a *European universal* size. Almost all of the European car makers (with the notable exception of Mercedes-Benz and Volkswagen) make a standard dashboard cutout for the radio. This radio is specifically designed to match that cutout. Automobiles made in the US tend to have custom cutouts, often being very different among cars of the same manufacturer made in the same year. This radio could be made to fit, often with very good success, by purchasing installation kits that were specific to the particular automobile in question.

The radio shown in Fig. 20-2 is a Delco Electronics model intended for one of the General Motors automobiles. Delco tends to produce similar radios for all models in a given year. It then custom fits them to the cars with specially molded black plastic bezels, knob packages, and shaft lengths (volume control and manual tuning). Delco, incidentally, has always been a leader in the design of auto radios and has pioneered such things as all solid-state design, AM/FM, windshield antennas, and the use of integrated circuits. In fact, Delco was the first to use the IC quadrature detector (their DM-11 was very similar to the Motorola MC1357P), and the nickname *ICQD* for such circuits is jargon direct from the early Delco service manuals on radios with the IC quadrature detector.

The environment in which a car radio must operate is poor. The temperature extremes vary from sub-zero to 180 degrees

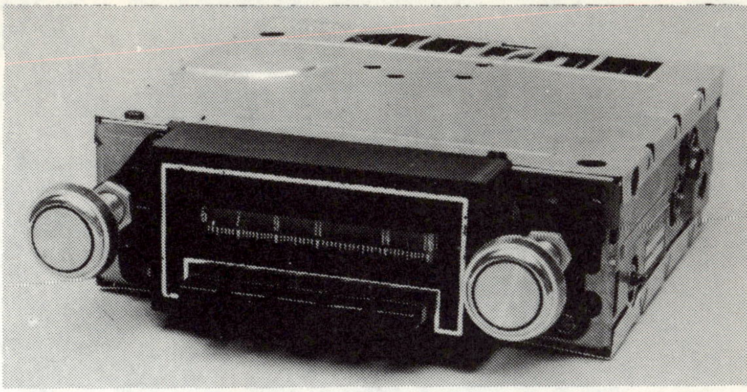

Fig. 20-2. OEM car radio (courtesy of Delco Electronics).

Fahrenheit. It is surprising just how hot it can get under the dashboard of an automobile left out in over 90 degree summer sunlight for even a few minutes. For this reason, the car radio must be made with higher quality levels than home radios. Auto radios must also operate in areas with a wide difference in signal strength. As a result, it is almost unheard of to make a car radio without an rf amplifier. Home radios are frequently made without rf amplifier stages.

THE POWER SYSTEM

The power system applied to a car radio is terrible. The voltage varies ±15 to 20 percent *normally*. Automobile voltage regulators are a bit crude in this respect, but the main problem in the auto power supply is noise. Where noise pulses on the AC mains used to power home radios are rare, occurring only a few times per day, they are normal on automobile power supplies. The line from the battery to the car radio must pass through the noisy engine

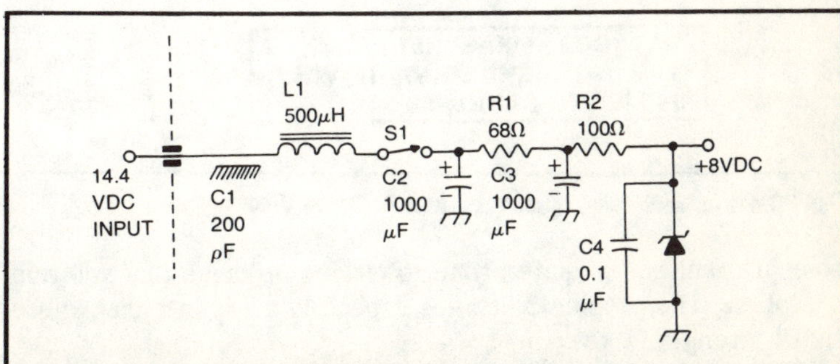

Fig. 20-3. Mobile power supply input circuit.

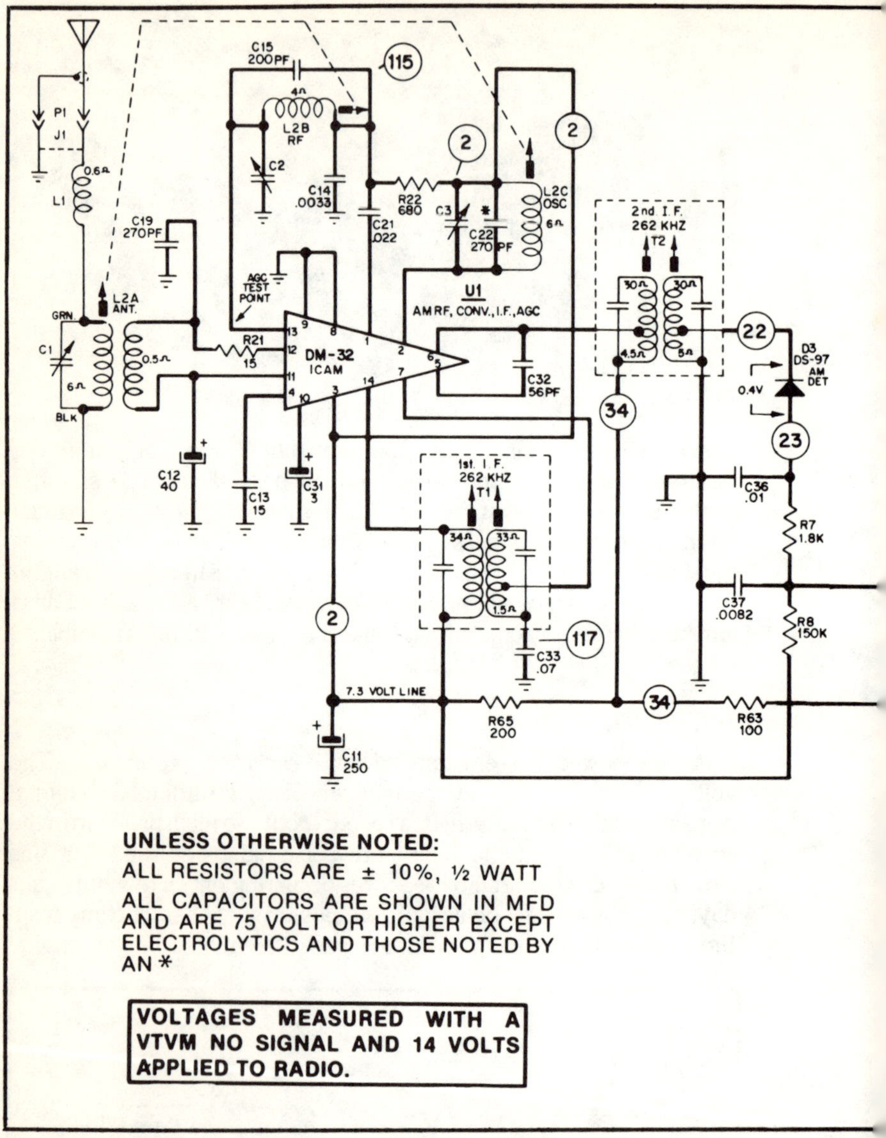

Fig. 20-4. Car radio schematic (courtesy of Delco Electronics).

compartment and is quite prone to picking up noise that will ruin reception. The problem becomes especially acute in areas where signal strength is low.

Figure 20-3 shows a partial circuit for a car radio DC power supply. Most of the components seen in this circuit are for noise

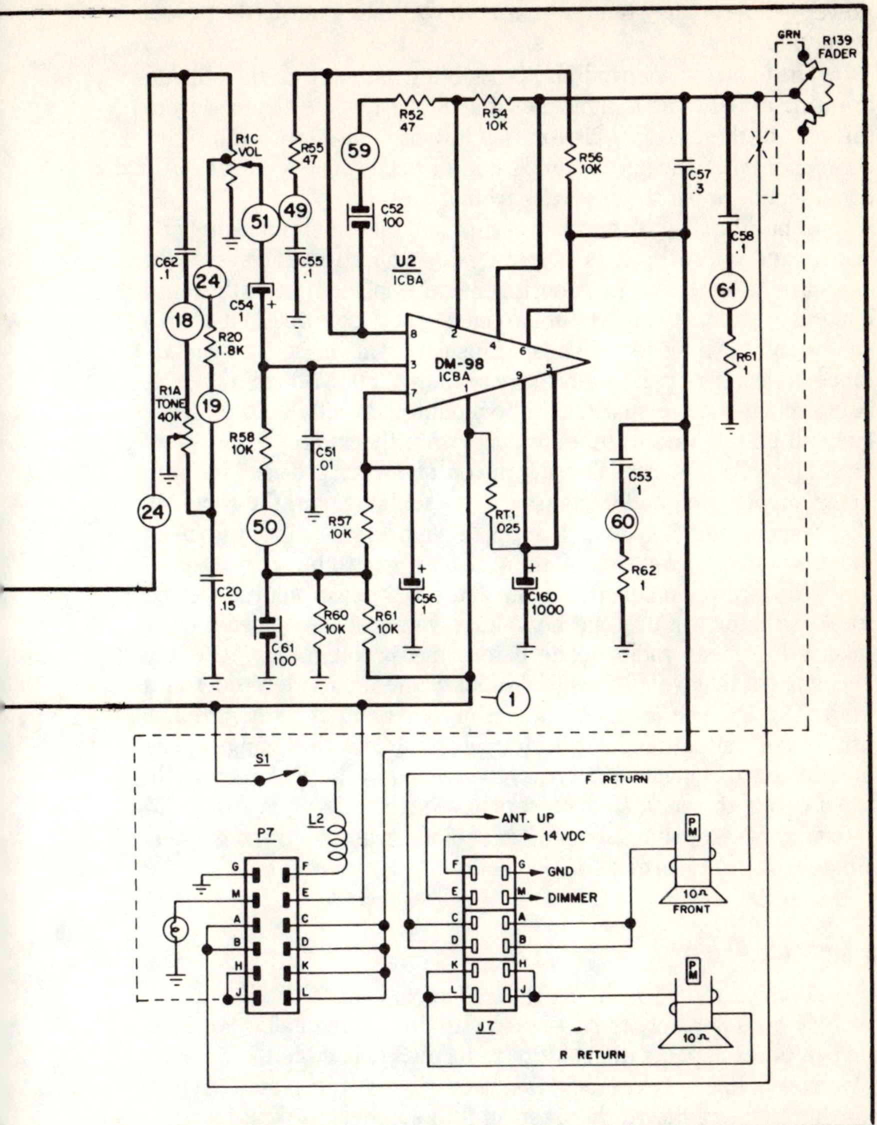

suppression. The normal battery voltage in a US automobile is nominally 12 volts, negative ground. In reality, however, the voltage when the car is running at a fast idel will be specified as 13.6 to 14.4 volts DC, depending upon manufacturer. This is why the maker of FCC type-accepted mobile radios specify the output power and frequency tolerance at 13.6 volts DC input to the power supply.

The inductor seen in Fig. 20-3 is often mounted on the outside or inside of the radio in a manner that allows it to be shielded from the rest of the circuitry. It forms a low-pass filter with the other components, reducing the amplitude of noise pulses and the AC component known as *alternator whine*.

Capacitor C1 is a special capacitor and may not immediately be recognized as such; it is a *spark plate* capacitor. This type of capacitor is effective in reducing noise impulses from the auto ignition system. It consists of two pieces of copper foil that form a sandwich with a piece of fishpaper insulating material. The capacitance of the spark plate will usually be in the 150 to 300 pF range. It is essential that the spark plate be mounted directly on the power line will be radiated to other wiring inside the radio.

The RC networks following the on-off switch are used for both noise suppression and decoupling between stages in the radio. In this respect, they look a lot like ripple-suppression capacitors used in home radios, even though there is no ripple to be suppressed.

The FM radio requires a regulated DC power source, or the local oscillator will drift as the voltage varies. I have seen a high-quality European radio, made before this problem was generally recognized, that would change stations as the car accelerated from a stop. The DC voltage from the battery charging system varied so much in that situation that the afc could not follow the changes! The solution is to regulate the DC voltage used in the FM section. In most cases, the car manufacturers have selected 8 volts as the FM circuit power supply. A simple Zener diode is usually the regulator. Some use three-terminal IC regulators today, however.

A MODERN RADIO

A typical modern AM car radio receiver is shown in Fig. 20-4. We will consider only the basic circuit here, incidentally, because certain other aspects in chapters to follow. This circuit is a Delco Electronics AM-only car radio that uses only two integrated circuits for the entire radio. In this respect it is somewhat exciting.

The entire AM radio section, less the audio, is contained within the DM-32 ICAM chip. This IC device contains the rf amplifier, converter, i-f amplifier and agc circuits needed to make the AM radio. A diode envelope detector is used at the output of the second i-f transformer to demodulate the signal to recover the audio. The recovered audio signal is passed through the volume and tone control circuits to a second IC which contains the entire audio section. This chip, the Delco DM-98, has the preamplifier driver and power amplifier stages all inside of one special IC package. The DM-98 is a so-called bridge audio circuit (see Chapter 10) in which the loudspeaker is bridged between two totem pole push-pull power amplifiers. It is important, incidentally, to not ground one side of the loudspeaker system in a radio that uses bridge audio. At the very least, distortion will result; at the worst, the bridge audio IC will be destroyed. Delco Electronics industrial distributors (non-auto radio semiconductor parts sales) sells a device that appears to be identical to the ICBA under the part number DA-101.

As is common in automobile radios, the AM intermediate frequency is 262.5 kHz rather than the 455-kHz signal used in home radios. Two i-f transformers are provided for this circuit and are external to the ICAM device. Note that the use of only two integrated circuits makes for a very small AM car radio; a blessing as radio complexity and available space becomes less and less.

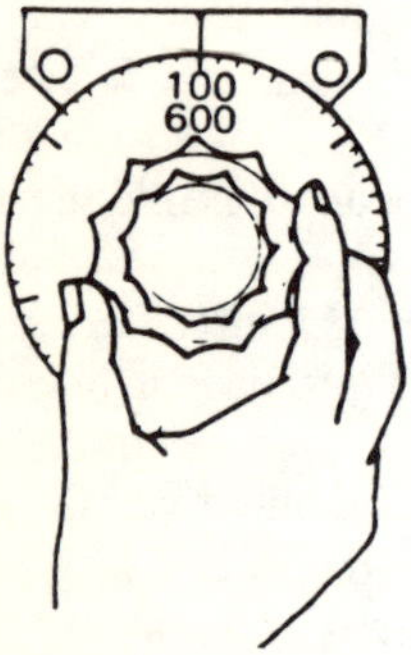

Chapter 21 Manual Auto Radio Tuners

The automobile radio is always a superheterodyne radio receiver in which several LC tank circuits are needed to tune in stations. There is a difference, however, between typical auto radio tuners and the tuners of home radios. In the home radio, typically a variable capacitor will tune the LC tank circuits. The rotary action of the capacitor shaft is easy to translate into a circular dial on the front panel. A little bit of dial cord, and a couple of pulleys will even translate the rotary action of the capacitor shaft to rectilinear motion for a slide rule dial. Somewhat more difficult is the problem of car radio tuners. If all that we wanted was a slide-rule dial, a variable capacitor would probably do nicely. But we also want pushbutton operation. A mechanical pushbutton system will select a station by moving the variable element of the LC tank circuits. The action of the pushbutton is rectilinear, so a lot of complex linkage is needed to make the rotational motion of the capacitor shaft work properly. This is one of the reasons why car radio manufacturers universally select a variable inductor to tune the car radio. All car radios currently being produced use a *permeability tuning mechansim* (PTM). The cores of the inductors are moved in and out of the coil forms to tune the receiver. This is also a rectilinear motion that is very easy to adapt to pushbutton operation.

There are other advantages to using a PTM. One is that inductors can operate in a dirty environment. A variable capacitor, on the other hand, would present service problems if the road dirt that frequently accumulates inside an auto radio were to get into the

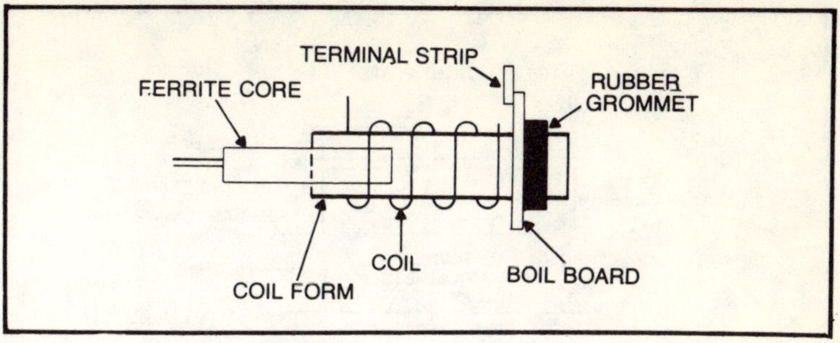

Fiq. 21-1. Slug tuning system.

bearings or underneath the rotor shaft grounding spring. Another advantage is that it is easy to *pitch wind* the coils so that the dial calibration is linear. This avoids the compression of the calibration points at one end of the range normally found in radios tuned with a variable capacitor.

Figure 21-1 shows a typical variable inductor from an auto radio PTM assembly. There will usually be at least three such coils in the PTM; one each for antenna, rf, and local oscillator tuning. There is usually some sort of Bakelite or other insulator forming a coil board to which the coils are mounted. A typical coil board will mount all three coils, even though only one is shown here. A rubber grommet and sometimes a dab of cement will hold the coil fast to the coil board. The winding of the coil is attached to terminal strips on one edge of the coil board.

Tuning of the PTM is performed by moving the position of the ferrite core (slug) in and out of the coil form. The ferrite core will be mechanically attached to a core bar that simultaneously drives all three cores in and out of their respective coil forms.

The simplest form of manual PTM is shown in Fig. 21-2. The individual coils are mounted to the coil board in the manner described above. The forms are seated in rubber retaining grommets and then usually cemented into place. The coil housing is designed so that the coils are parallel to one another, but shielded to eliminate mutual coupling between them. The three ferrite cores are attached to the core bar, which moves back and forth in order to tune the radio. When the core bar is at its maximum extension, the cores are almost all of the way out of their forms, so the inductance is minimum. This means that the radio is tuned to the high-frequency end of the dial. Similarly, when the core bar is closest to the housing, the cores are maximally inserted into the forms, and their inductance is maximum. In this case, the radio is tuned to the low end of the dial.

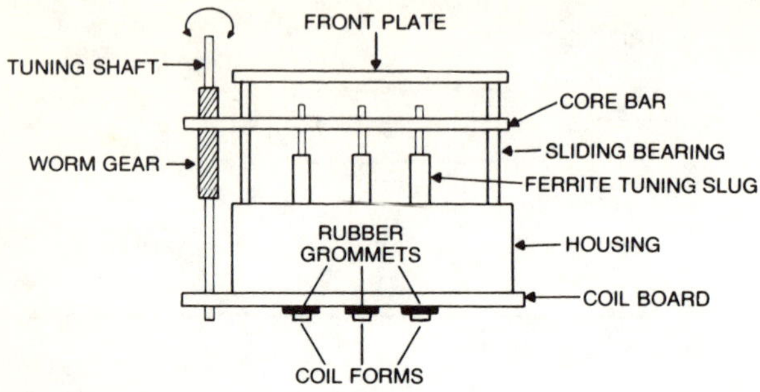

Fig. 21-2. PTM.

One end of the core bar contains a threaded hole that is fixed to a worm gear on the manual tuning shaft. When the operator rotates this shaft, the core bar begins to travel along the worm gear, varying the position of the cores inside the three forms. A standard slide-rule dial and dial cord arrangement provide the frequency readout.

PUSHBUTTON TUNERS

An example of a pushbutton tuner is shown in Fig. 21-3. This particular model is from Delco Electronics for use in General Motors auto radios. The principle components of this tuner are:

- ☐Manual tuning shaft
- ☐Worm gear
- ☐Antibacklash gear to the clutch
- ☐Clutch assembly
- ☐Treadle bar
- ☐Core bar
- ☐Core bar linkage (to clutch)
- ☐Cores housing

The purpose of the clutch assembly is to connect the manual tuning shaft to the core bar during manual tune operation, and to disconnect the manual tuning shaft (reducing drag and friction) during operation of the pushbuttons. When a pushbutton is depressed, the treadle bar is moved to a position determined by the presetting of the pushbutton. The pushbuttons are set as follows:

- ☐Pull out the pushbutton that you wish to set.
- ☐Tune in the station that you want on that pushbutton using the manual tuning shaft.
- ☐Push the button all the way back into the housing.

□Check the operation of the pushbutton by tuning off station a little ways and then depressing the pushbutton. The tuner should return to the same spot, with the station tuned in properly. If the station is a little bit off, repairs to the tuner or adjustment of the clutch assembly might be needed. A well made tuner in proper adjustment and good repair will show only a tiny amount of backlash.

The backlash of the pushbutton tuner is controlled by the antiblacklash gear that interfaces the worm gear and the clutch assembly. This gear is placed under tension when it is assembled, and it should remain under tension. Such a gear is made by having two identical gear faces placed side-by-side, with a spring between them. These gear faces are rotated a short distance opposite each other, making sure that the teeth align, just as it is being connected to the worm gear. This takes a little bit of mechanical cleverness, but is essential when you reassemble a tuner for any reason.

Figures 21-4 and 21-5 show two close-up views of the tuner mechanism, or at least parts thereof. Figure 21-4 shows the top of the tuner, including the pushbutton slides, the treadle bar, manual tuning shaft and worm gear. The view in Fig. 21-5 is the same mechanism in a partial state of disassembly.

The treadle bar is directly linked to the core bar, even though this is not shown in Fig. 21-4. The cam on the pushbutton slide is used to set the station. The position of the cam is set when the pushbutton is set. When the pushbutton is operated from then on, it

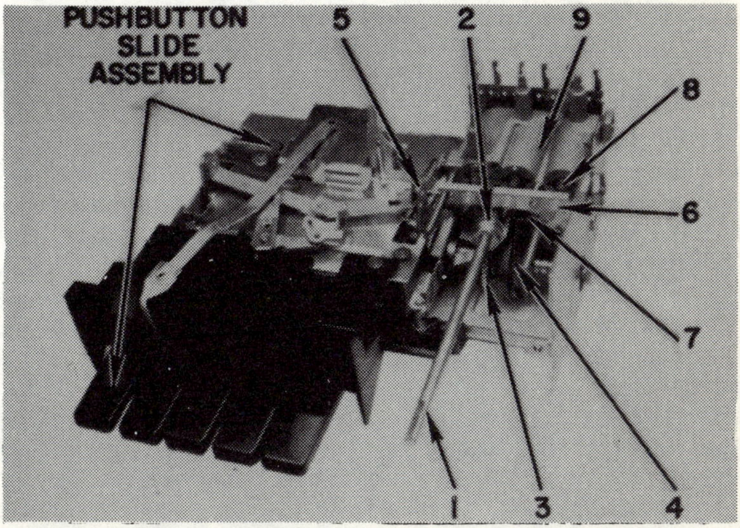

Fig. 21-3. Pushbutton tuner assembly (courtesy of Delco Electronics).

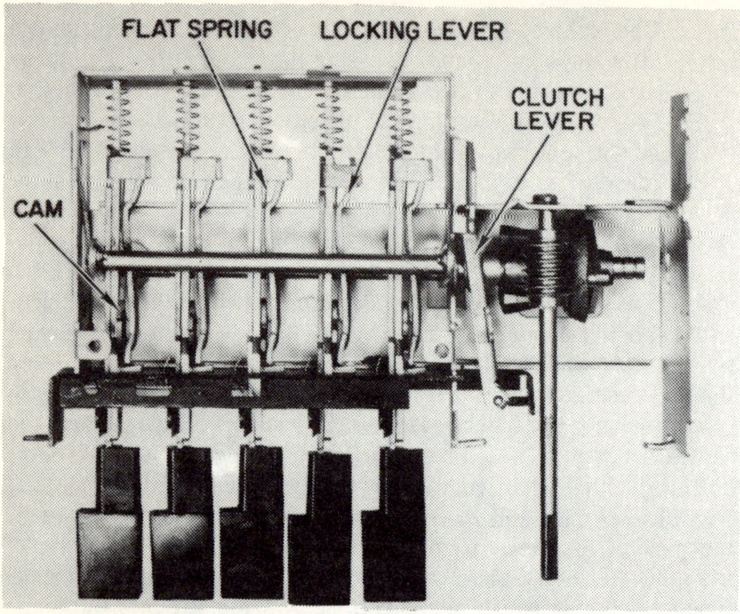

Fig. 21-4. Pushbutton assembly.

will push the treadle bar until it reaches the position dictated by the position of the cam. This should correspond to the tuning of the station originally set on that particular pushbutton. Several things must happen when the pushbutton is pressed. First, the manual tuning shaft must be disconnected so that the operation of the pushbutton is not too difficult. If this is not done the normal drag in the shaft and worm gear assembly makes it very difficult for even Tarzan to operate the button. The manual tuning shaft is coupled to the PTM through the clutch, so the clutch must be disengaged to disconnect the shaft. This is the job of the clutch lever (called by some manufacturers the clutch gate). When a pushbutton is pressed, the black bar on the front top side of the tuner slides to the left, bringing with it the clutch lever. This moves the clutch faces apart only a small fraction of an inch, but is enough to declutch the tuner. When the button is released, the black bar and the clutch level will return to their resting positions, with the manual tuner reconnected to the shaft.

The clutch assembly is the cause of much of the service required for the manual tuner. This device becomes worn and must be adjusted. Some clutches will have a rubber or cork clutch facing that must be periodically renewed. In the case of the Delco clutch assemblies, there is a screw on the side of the tuner (accessible form the outside of the radio in most models) that will adjust the clutch tension.

The principle symptom of clutch trouble is difficult in manually tuning the radio or excessive drag when the pushbuttons are operated. The former is caused by too little clutch tension; the latter is caused by too much.

Another frequently seen problem is a worn out of broken manual tuning shaft. In some models, other than Delco, the manual tuning shaft connects to the antibacklash gear through a light-duty spline gear on the end of the shaft. This gear will wear out, especially if the drive shaft is made of plastic or nylon. One major US auto marker experienced rather bad problems in this respect during the late 60s. The manual tuning shaft was a two-piece affair, with the piece connected to the antibacklash gear being called a *pinion shaft*. This shaft was made of nylon or plastic (depending upon vintage) and would wear out rapidly. The teeth of the spline gear would strip out, and the radio would not manually tune.

10-SLIDE TUNERS

The ordinary car radio one had but five pushbuttons, and these were divided between both bands in AM/FM models. But in the late 60s, several manufacturers developed tuners which had five buttons but 10 slides (the piece with the cam). A *shuttle bar* operated by the bandswitch assembly on each pushbutton connects the linkage from the various slides (five each for AM and FM) to their respective pushbuttons. This allows the user to select 10 stations, five on each of the two bands, from just five pushbuttons. An example of a pushbutton assembly from a 10-slide tuner is shown in Fig. 21-6.

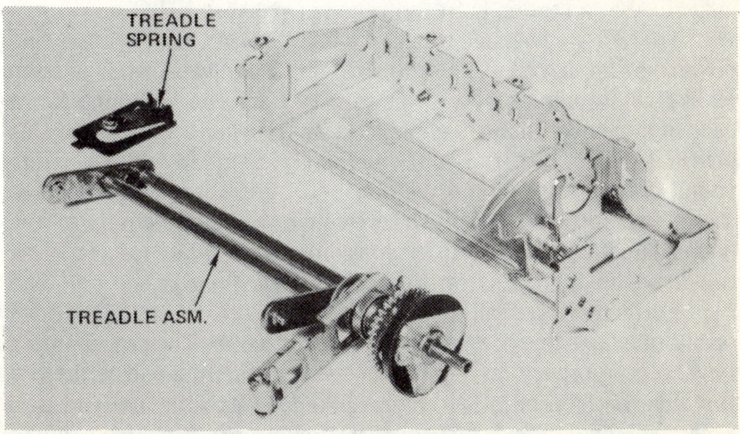

Fig. 21-5. Pushbutton mechanism disassembled.

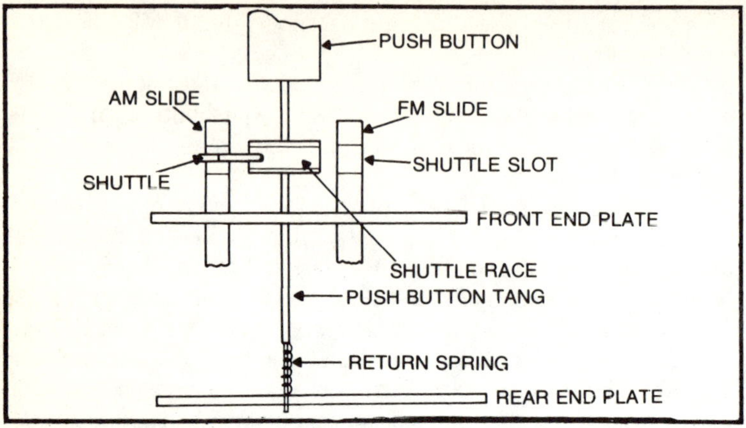

Fig. 21-6. Ten-slide tuner method.

The shuttle is a little movable linkage that mechanically connects the pushbutton to one of the two slides. This shuttle is moved back and forth between the two slides by a shuttle bar attached to the bandswitch assembly. Typical jamming problems with ten-slide tuners include poor alignment of the shuttles, broken shuttle races, and broken shuttle bars (the biased bars are mechanically warped so that only some of the shuttles are operated properly).

SIGNAL-SEEKING TUNERS

Signal-seeking tuners, which are covered in the next chapter in greater detail, are mechanical tuners which will automatically seek the next higher frequency station and then stop. The action is controlled by an electronic section that drives the PTM upband and then stops when it sees an agc signal that is strong enough to indicate that a station is in the receiver passband.

There are two basic types of signal seeker: motor driven and power spring driven. The motor driven types were once used in Ford radios made many years ago. They are currently popular in certain Japanese radios made for factory installation in automobiles sold in the US. These radios can be readily identified by the whirring sound of the motor as it drives the PTM assembly. Also, the DC motor can be reversed by interchanging its connections to the power supply. As a result, these radios use relay logic to reverse the direction of rotation for the motor at each end of the dial. The motor driven seeker will travel to one end of the dial and then travel back along the same path in the reverse direction. The power spring type, on the other hand, only travels from the low end of the dial towards the high end. It is then yanked back to the low end by a power solenoid, where it restarts its travel again from low to high.

Chapter 22

Signal-Seeking Radios

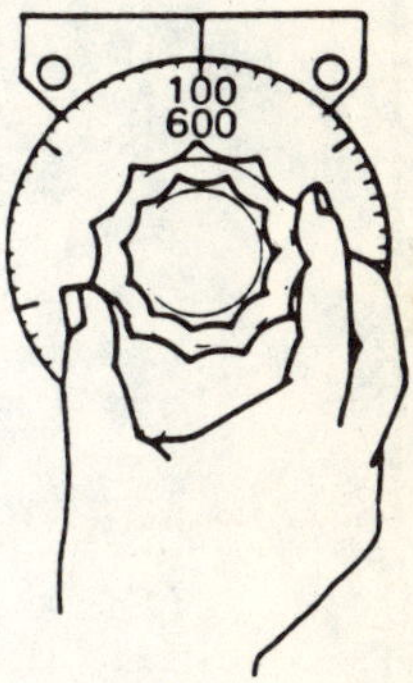

Signal-seeking car radios are designed to search out a station automatically merely by pressing a button, either on the radio front panel or the floorboard of the automobile. There are two general types of signal seeker tuners. The motor driven variety, which was introduced in the last chapter, uses a small DC motor to drive the manual tuning shaft of the radio in order to tune the radio. The other type is driven by a power spring and is used by Beckerautoradio of Germany and by Delco Electronics in their mechanically tuned seekers. (An electronically tuned seeker is in their digital radios covered in Chapter 23). The power spring type, pioneered by Delco in the early 50s) under the *Wonder Bar* trademark, is still essentially the same as it was in the early designs. True, it has been modernized using solid-state circuitry, and some of the mechanical components have been greatly improved and reduced in size over the years, but the essential operation of the seeker remains the same as it was in then.

Figure 22-1 shows a partially disassembled Delco signal seeker radio. The core housing has been raised in this photo to show the power spring and part of the seeker components. Visable are the power spring, solenoid (used to recock the power spring), control relay, and governor gear train assembly.

The power spring is used to propel the tredle bar in the PTM assembly (see Chapter 21) towards the high-frequency end of the dial. To limit the speed of travel of the tredle (the spring, under great tension, would cause the tredle to move instantly to the high

Fig. 22-1. Signal seeker mechanism.

end of the dial) a governor gear train is required. This is a special gear system that rotates at a constant speed in one direction and at a very high speed in the other. The forward travel of the spring and tredle in the seek mode is thereby limited, while the recock speed is rapid. The governor gear train contains a light weight paddle wheel that is used to stop the rotation of the gears. The gear ratio is such that this lightweight paddle wheel can be jammed, and stop the powerful forward thrust of the spring, using just a tang on the armature of an ordinary 12-volt DC relay. The relay is positioned such that the tang will drop through a hole in the housing of the gear train and jam the paddle wheel, stopping the rotation of the gears, when the relay is de-energized. The block diagram of the electronic control circuit for the seeker shown in Fig. 22-2. The relay is normally de-energized, so the tang is inside of the governor housing, jammimg the gears. When the start button is pressed, the start switch is closed. This action momentarily grounds one end of the relay coil. For the instant that the switch is closed, the pull-in current of the relay is supplied through the start switch (S1 in Fig. 22-3) to ground. One end of the relay coil is connected to the +14 volt DC power supply; the other end is grounded through S1. When the relays closes, one set of contacts (A-B) applies power to the relay control section. The voltage used in this section is regulated by an 8-volt zener diode (D1). When the 8-volt line becomes active, bias is applied to transistor (D1). When the 8-volt line becomes active, bias is applied to transistor Q1, the relay amplifier, which begins to conduct. As Q1 begins to conduct, its emitter current causes a voltage drop across resistor R2, which in turn is used to

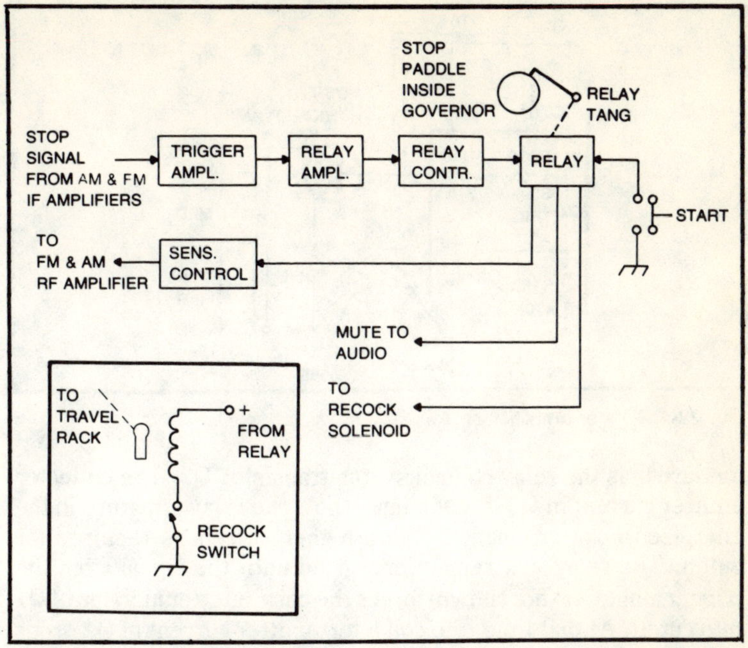

Fig. 22-2. Relay control system.

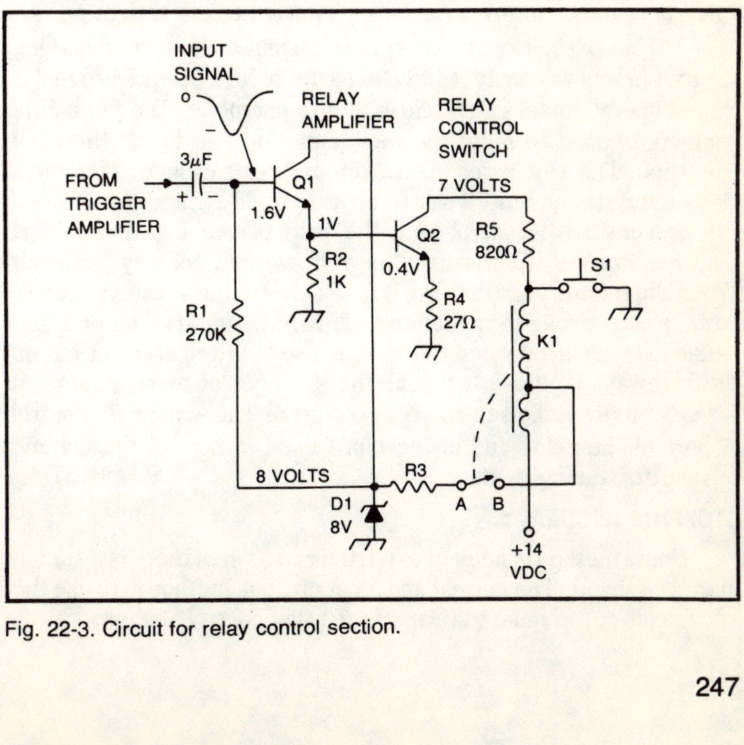

Fig. 22-3. Circuit for relay control section.

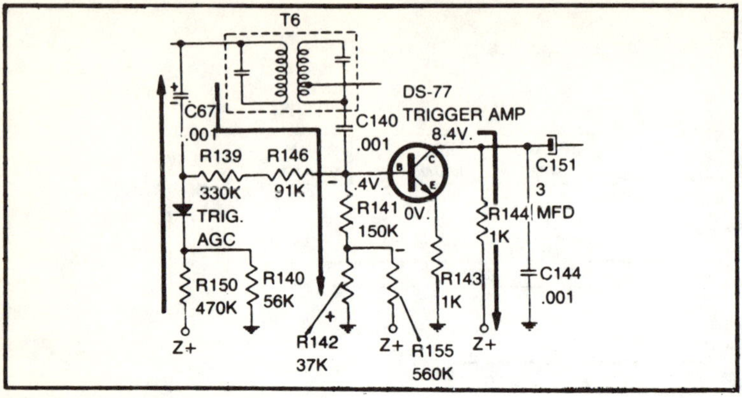

Fig. 22-4. Trigger amplifier on AM.

forward bias the relay control switch transistor Q2. The collector emitter current in Q2 is sufficient to hold the relay armature in the energized position even though a higher current is required for pullin. The relay will remain energized until the signal from the trigger amplifier (not shown) forces the base-emitter junction of Q1 into cutoff. At that time, the collector-emitter current of Q2 drops below the level needed to hold the relay energized, and the relay armature drops back to the resting position. This will reinsert the relay tang into the governor, stopping the seeker action.

The stop/seek relay also serves certain other functions. One set of contacts will apply +14 volts to the recock solenoid so that it cannot operate unless the radio is in the seek mode. Another pair of contacts is used to mute the radio audio output during the seek operation. This eliminates the background noise caused as the radio skips over stations too weak to provide good reception.

A sensitivity control sets the gain of the FM and AM rf amplifier stages. This is used as a "town and country" control. When the radio is used in town, it is wise to set the sensitivity lower so that only strong local stations will trip the tuner. The opposite situation is required when the radio is used in rural areas or out on the highway. In those locations the strength of most stations is lower so more radio sensitivity is required. The sensitivity control is part of the relay and is therefore used in normal manual and pushbutton tuning modes.

STOPPING ACCURACY

Some method is needed to tell the radio when there is a station in the passband. The easiest and most obvious method is to use the agc circuit of the radio to turn off the relay control section. This is not exactly a good method, however, unless some means is provided to make sure of the stopping accuracy. On AM, the radio must stop in the exact center of the passband of the radio station or mistuning will result. On FM, the tuner must stop close enough to the center of the passband for the afc to take over and snap the tuning into correct point. In this respect, at least, the FM stopping seems a little less critical than the AM stopping circuit. The trigger amplifier stage is used in both cases.

Figure 22-4 shows the AM stopping circuitry, while Fig. 22-5 shows the different signals applied to this circuit. Two signals are needed in Fig. 22-4 to insure proper stopping. One signal is taken from the primary winding of the 262.5-kHz output i-f transformer, T6. This signal is sampled by capacitor C67. The second signal is taken from the secondary of T6 and is supplied directly to the base of the trigger amplifier through capacitor C140. Agc action, coupled with these two signals, stops the seeker accurately on the AM band. Notice that the bias on the trigger amplifier transistor is 0.4 volts. Because this transistor is a Delco DS-77 silicon type, it wants to see approximately 0.6 to 0.7 volts for proper forward bias. With 0.4 volts on the base-emitter junction, however, the transistor is *almost* forward biased (another 200 to 300 millivolts will turn on this transistor).

The signal levels from the two signal sources are shown in Fig. 22-5, with respect to the trigger level needed to fire the trigger amplifier transistor. The signals from the i-f transformer primary and secondary are algebraically added at the base of the trigger amplifier transistor to form a resultant curve (C). This resultant rises above the trigger threshold level just an instant before the tuner reaches the center of the passband. The transistor suddenly begins to conduct, and this causes a sudden decrease in the voltage

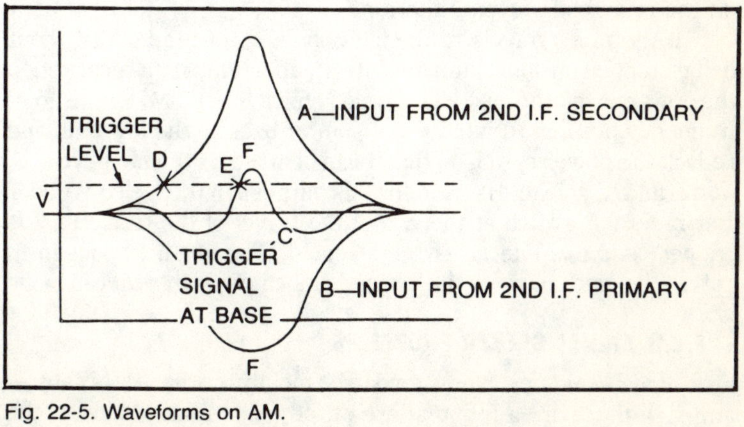

Fig. 22-5. Waveforms on AM.

at the collector of the trigger amplifier. This decrease looks like a pulse to the capacitor and is coupled through the capacitor to the base of the relay amplifier transistor (Fig. 22-3). The relay control switch transistor is then cut off, causing the armature of the relay to drop out. The time difference between the actual trigger point and the point where the radio is exactly center tuned is needed to allow for the drop time of the relay tank. This tank must drop from its energized position to the unenergized position, where it jams the paddle wheel inside of the governor gear train. The positive and negative signals from the two sides of the transformer in the i-f amplifier provide a sharply defined trigger point close to the center of the AM signal.

The FM stopping circuitry is shown in Fig. 22-6. Essentially the same method is used for FM as was used in AM, except that we have to "fake" the secondary signal. The primary signal is taken from the primary of the ratio detector (FM demodulator) transformer, which is supplied through capacitor C146 to the trigger agc rectifier. In the FM circuit, however, the signal from the secondary of the transformer cannot be used because it is part of the phase detection circuit needed to demodulate the FM signal. The problem is solved by using an extra tuned tank circuit, called the trigger transformer (T8). The signal from the primary of the ratio detector transformer is coupled to the trigger transformer through capacitor C147. This trigger transformer tank circuit is not tuned exactly to the resonant frequency until the DS-55 varactor (variable capacitance diode) is biased correctly, which is only during seek operations. With the DS-55 biased, a strong resonaant signal develops at the base of the trigger amplifier. When the signal at the base counteracts the bias, the voltage drop at the collector of the transistor is coupled through the capacitor into the relay amplifier, turning off the relay and stopping the tuner.

If the tuner travels to the high end stop of the PTM without being stopped on some intermediate station, it must be recocked to the low end of the dial. This will return the PTM to the low-frequency position, drive the dial pointer back to the low end, and recock the power spring so that it can be used on the next cycle. A powerful DC solenoid is used for this purpose in the Delco Wonder Bar radios. A switch at the end of the travel of the tredle rack is tripped, and this energizes the solenoid. The solenoid pulls in to return the mechanism to the low end of the range where it can begin anew.

TYPICAL SIGNAL SEEKER PROBLEMS

Signal seeker problems tend to be mostly mechanical in nature, although there are a few that are strictly electronic. Many of the

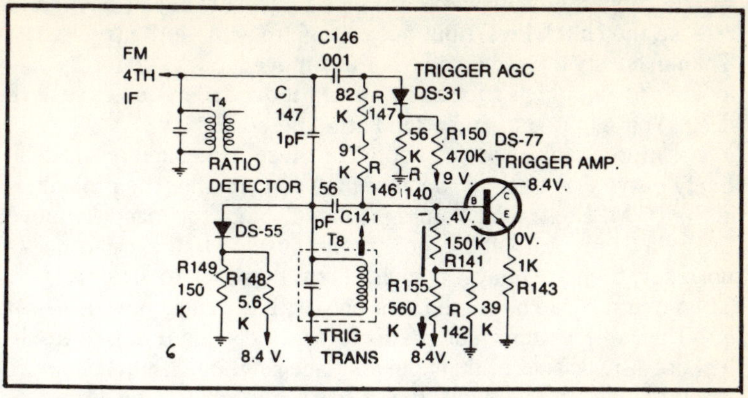

Fig. 22-6. Trigger amplifier on FM.

problems will involve merely an adjustment of some of the mechanical components. Before troubleshooting a signal seeker, however, you'll need a need for a heavy duty +14 volt DC power supply that is capable of delivering at least 18 to 20 amperes for a brief instant when the solenoid recocks. If the supply has poor regulation, the solenoid will hang up midway in its cycle. It often burns out at this point.

Failure To Seek

If the tuner fails to initiate a seek cycle when the wonder bar is depressed, look for either an electronic or mechanical problem. In most cases, it will be the mechanical problem that causes the trouble. Typical of the electronic problems will be an open transistor in the relay control section, such that the relay will not remain energized when the start button is released. If the relay falls back into the gear train governor when the start button is released, suspect a probable electronic problem. But there are several mechanical problems that will also mess up the mechanism. One is a misadjusted relay. If the angle of the tang is not correct, it will not pull out of the governor housing enough to release the gear train and initiate a seek cycle. A pair of screws on the radio side cover will allow adjustment of the relay position.

If the audio fails to mute when the start button is pressed, the problem could be the aforementioned electronic problem (the relay is not being pulled in), a defective relay, or the fact that the relay is misadjusted and is so tightly bound inside the governor that it cannot pull in.

Occasionally, there is a defective governor. If the gear train is jammed, the traveling rack that drives the tredle bar cannot be released so nothing happens. The PTM itself might be jammed. We can test this by manually depressing a pushbutton just enough to release the clutch lever, but not enough to actuate the tredle bar. Then manually take the tredle bar and move it back and forth while looking for any high friction (it should move freely) or excessive drag. You might also find a bind at some point.

Differentiating between some of these jam conditions is relatively easy. The method above will tell you whether the problem is in the PTM. Next, take a small insulated object and manually move the relay tang out of the governor housing. If the radio seeks normally, then the relay is at fault. If not, then look to the governor. The governor can be tested, but only with the radio power *turned off*. This is very important, so don't forget it when trying to troubleshoot a signal seeker that appears to be jammed. When the power is off, loosen the screws that secure and adjust the governor assembly. When the assembly is loose, the seeker mechanism should immediately advance rapidly to the high end of the dial, slamming into the high end stop.

If the radio neither mutes nor initiates a seek cycle, the problem might also be an open or misadjusted start switch. The nature of car radio construction makes it necessary to use leaf-spring switches for the start function. The open contacts of these switches become dirty and won't sustain the high current needed to initiate the relay pull-in. Sometimes, the leaves become misadjusted from being operated to often. A pair of long nose pliers and a little cleverness will often save the switch, while at other times the switch needs replacement.

Machine-Gunning

A broken or misadjusted governor gear train will cause the symptom known as machine-gunning. The tuner cycles back and forth rapidly between the low end and high end stops because the retarding action of the governor is no longer controlling the operation. This is why the power is removed from the radio when testing the governor in the previous section. Once the governor is loose, machine gunning would ensue! Machine gunning is corrected by either replacement of the defective governor gear train or be correctly adjusting it. But before going on to greater and more glorious things, carefully inspect the radio for additonal damage. Machine gunning is a violent malfunction that tends to damage and take out of adjustment other components. If allowed to continue too long, the duty cycle of the recock solenoid can be exceeded, and the solenoid will burn out. The PTM also jams, and the dial pointer breaks in radios that were allowed to machine gun too long. The term "too long" may be a little misleading because a remarkably short time is required to damage these parts.

Chapter 23
Digital Receivers

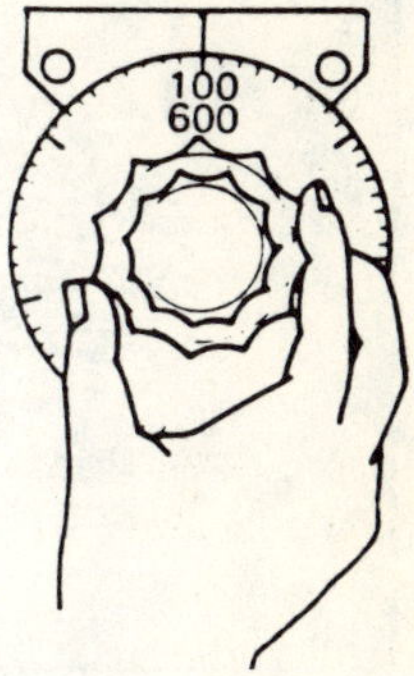

Digital technology has exploded in the last years of the 70s. The microprocessor is an integrated circuit that contains all of the elements needed to make a complete digital computer, or at least the CPU portion of the computer in some cases. Radio receivers now use the microprocessor and other digital circuitry to control the phase-locked loop tuning systems. Products from amateur radio transceivers, on most bands from DC to daylight, and consumer products alike have microprocessor-controlled front ends. Motorola, Delco Electronics, and Chysler's Huntsville Electronics Divison all have at least one microprocessor-controlled model in the line. Television receivers on the market also feature "computer tuning" of both VHF and UHF. Figure 23-1 shows the Chrysler computer-controlled radio receiver, an AM/FM broadcast model that fits into the dashboards of some of their automobile models. Note the absence of the ordinary control knobs associated with car radios—no manual tuning shaft! Tuning is by pushbutton selection or search tuning (which scans the entire band one station at a time). The frequency that the radio is tuned to is displayed on a digital readout.

PHASE-LOCKED LOOP TUNING SYSTEMS

The one key to making a digital radio tuner is to use a phase-locked loop (PLL) tuning system. Figure 23-2 shows the block diagram to a popular auto radio AM/FM PLL system. The local oscillator in each case is voltage tuned, using varactor diodes in their

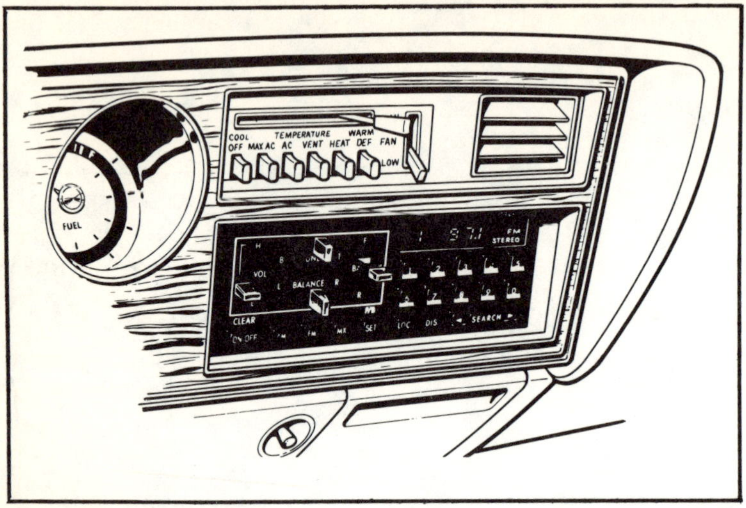

Fig. 23-1. Digital radio using microprocessor control (courtesy of The Chrysler Corporation).

respective tank circuits to set the operating frequency. The output of the local oscillator on each band is applied to their respective mixers to provide superheterodyne action. But the LO outputs are also applied to the input of a divide-by-N counter stage. The AM LO signal is applied directly to the counter, but the FM LO is first divided by 20 before being applied to the counter.

A divide-by-N counter is a frequency divider that will divide the input frequency by an integer factor *N*. In other words, the output frequency will be F_{in}/N, where N is an integer number. The division ratio N is set by a binary word of N-bits length applied to the counter. The value of N will determine the division ratio. The output of the divide-by-N counter is applied to one of two inputs of a digital phase detector circuit. A 10-kHz reference signal is applied to the remaining input of the phase detector. There are any number of ways to generate the reference frequency, but in almost every case it will be a frequency of 1, 2.5, 5, 10, 15 or 25 kHz. In the particular case shown, the reference frequency is generated by dividing the output of a 2.56-MHz crystal oscillator (XO) by a factor of 256.

When the radio is tuned to the frequency selected by the N-code, the output of the divide-by-N counter will be exactly 10 kHz. This matches the reference frequency, so the phase detector output will be zero.

The phase detector is more on the order of a coincidence detector, the normal situation for digital phase detector circuits. It is not like the analog phase detector discriminators in which the resultant output will be an analog voltage or current level. In this case, the output will be essentially anticoincidence pulses. Such a detector can be made using exclusive-OR gates that issue an output if either input is HIGH, but not if both inputs are HIGH. An example of the action of the anticoincidence circuit is shown in Fig. 23-3. The circuit will produce pulses that have a width proportional to the error between the two pulse frequencies. We can add a couple more gates to the basic circuit and produce a detector that will output pulses of one polarity if the unknown signal is above the reference frequency and quite another polarity if the unknown is below the reference frequency. This is the usual practice in PLL circuits, where we want to be able to drive the DC error control voltage in either direction to correct the LO frequency. We will output the coincidence (phase)detector pulses to an analog integrator, which will create a DC voltage that is proportional to the error in LO operating frequency. The polarity of this voltage is dependent upon whether the LO frequency is above or below the reference frequency. The amplitude of the DC control voltage is dependent upon the magnitude of the error.

Actually, two different methods are in operation. In one, which can be called a self-nulling method, we are always seeking a balance

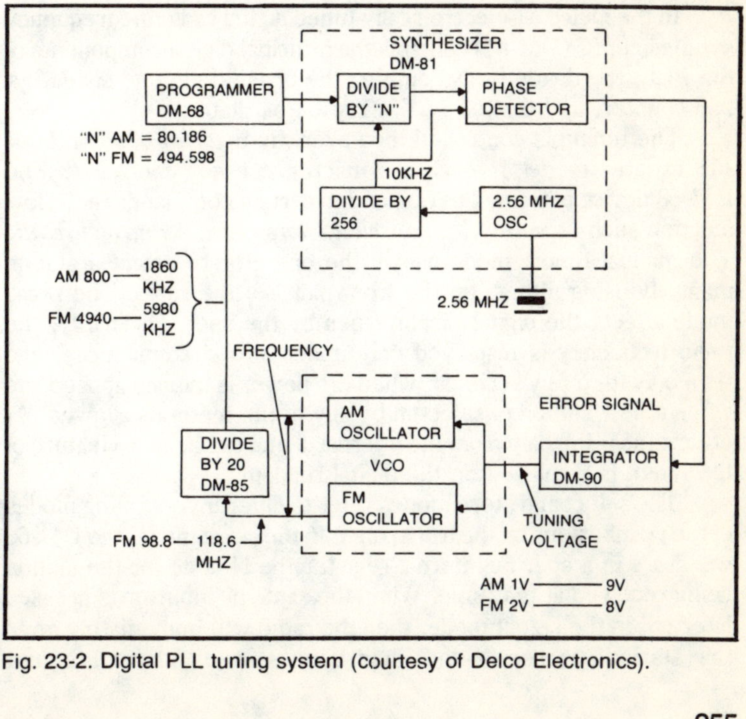

Fig. 23-2. Digital PLL tuning system (courtesy of Delco Electronics).

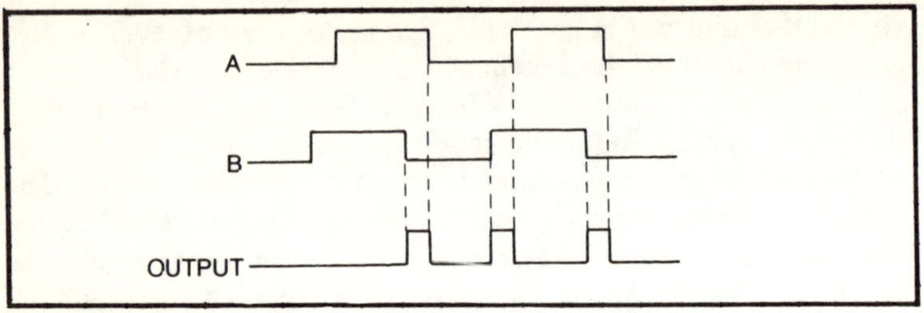

Fig. 23-3. Output of phase detector.

to cancel the error. In the other system, the error is used to generate the actual potential that is applied to the local oscillator circuits. We should be able to tell approximately what frequency the radio is tuned to by reading the voltage at the output of the DC level translator amplifier following the integrator stage.

Channels can be changed in this type of radio by changing the N-code applied to the divide-by-N counter. This code will be stored in a latching register and can be changed by the operator at will. In the description of the Delco system to follow, we will describe how their microprocessor tuner control chip will allow several different methods for determining the N-code that is actually stored in the main register.

In the Delco all-electronically tuned digital radio the frequency synthesizer IC (DM-81) contains the principal digital components of the PLL: the divide by-N counter, the phase detector, divide-by-256 counter, and the 2.56-MHz crystal oscillator.

The tuning is controlled, however, by the DM-68 chip (DM-118 in later models) which is a microprocessor controller. The DM-68 device is trained to respond to certain commands that allow function such as *manual tune, select, store, seek, scan,* and *recall*. The manual tuning mode allows the operator to select a station manually using a knob on the front panel of the radio. The recall mode affects the digital display seen by the user. Ordinarily, the radio frequency is displayed only under two circumstances, and then only for a few seconds: when the power is initially applied and when a new station is selected by any of the methods allowed. At other times, the data from the internal digital clock, also a feature of the radio, is displayed on the digital readout.

The select and store modes refer to different operating modes of the pushbuttons on the front panel. In the store mode, the DM-68 will store in a specific internal register the N-code for the station being received at that time. When the same pushbutton is pressed later on, in the select mode, then the radio will immediately go to the stations whose N-code had been stored. There are four select/store pushbuttons on the Delco radio. They are pushed in to select, and pulled out to store.

SEEK AND SCAN FUNCTIONS

Seek and scan are similar functions. Both are automatic tuning modes. In the seek mode, the tuner will advance to the next higher frequency station and stop. If the user wants to preview another station, then a new seek command must be initiated for the radio to go on. In the scan mode, however, the radio will increment to the next higher station and stop for only five seconds. It will then increment again, unless the scan command has been intentionally cancelled by the user. This mode allows the user to preview all of the stations in an area before selecting the one that is best suited.

The gating circuit that selects the different modes of operation for the DM-68 is shown in Fig. 23-4. There are two data buses associated with the DM-68: a three-bit input command bus and a four-bit timing or output bus. The output lines are active-HIGH, meaning that they normally sit LOW and snap HIGH only when that line is active. These four lines will go HIGH in sequence at an 80-Hz

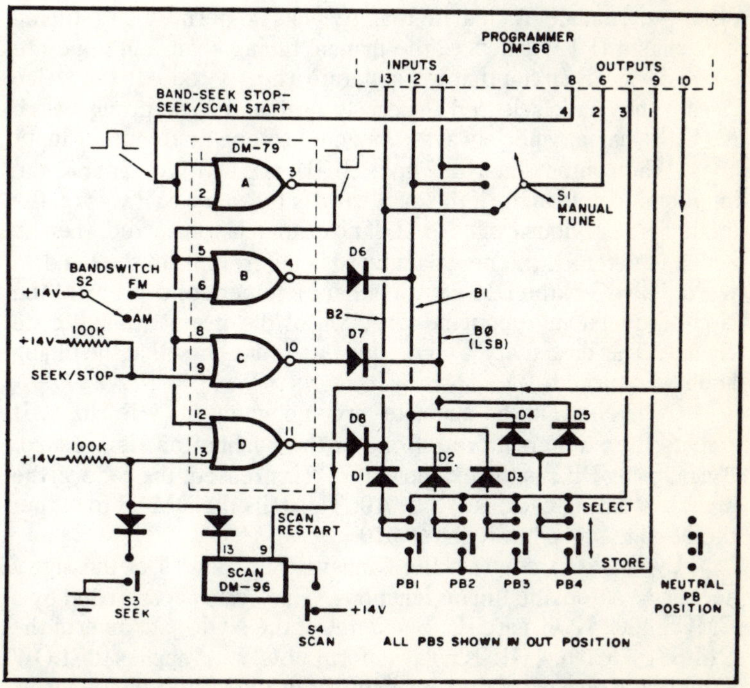

Fig. 23-4. Control circuit for the programmer chip (courtesy of Delco Electronics).

rate. Each output line causes the DM-68 to examine or poll a different facet of the tuning protocol. Each of the four cycles controls one of the four tuning modes. See Fig. 23-5.

The DM-68 knows what to do next by monitoring the data on the input buses during each of the four cycles. Timing is taken care of internally so no additional external synchronization circuits are needed.

Cycle No. 1 controls the operation of the front-panel pushbuttons in the store mode. If one of the front-panel pushbuttons is pulled out during cycle No. 1, the N-code applied to the synthesizer at that instant is stored in the internal DM-68 register corresponding to that button. The DM-68 examines the input bus during cycle No. 1. If a station is to be stored on, say, pushbutton *PB2*, one merely pulls out PB2 on the front panel. When output No. 1 goes HIGH (indicating cycle No. 1), a HIGH is passed through isolation diode D2 to bit No. 1 of the input bus. The DM-68 input bus, then, sees the binary code 010 during cycle No. 1, which it interprets to store the present N-code in register PB2. The store codes for the other pushbuttons are 100 for PB1, 001 for PB3, and 011 for PB4.

Cycle No. 2 controls manual tuning. This radio differs from other automobile radios in that it lacks a permeability tuning mechanism (PTM); hence, the manual tuning shaft must operate some device or circuit that will generate a binary code. In the Delco circuit, they have selected a specially-designed three-prong switch (S1) that has a wiper always connected to one or the other input lines. When output line No. 2 goes HIGH, the DM-68 examines the input bus to determine if the data word is the same as it was on the immediate previous cycle No. 2. If no change has occurred, then no action is taken. If the new data word is different from the old data word, however, the DM-68 will either increment or decrement the N-code depending upon the direction of the manual tuning code change. This directivity is controlled by the direction that the tuning knob was turned.

Cycle No. 3 is the opposite or complement of cycle No. 1. It selects the stations stored if one of the pushbuttons is pressed. Again, using PB2 as the example, if PB2 is pressed, the code on the input bus during cycle No. 3 is 010. This tells the DM-68 to output the N-code stored in register 010.

Cycle No. 4 controls the bandswitching and all of the signal seeker or automatic tuning functions. This cycle is controlled by a CMOS 4002 NOR gate IC. Section A of the NOR gate inverts the DM-68 cycle No. 4 HIGH pulse to form a LOW and applies the LOW to one input on each of the remaining three 4002 gates. This enables the gates only during cycle No. 4 and keeps it off during the other

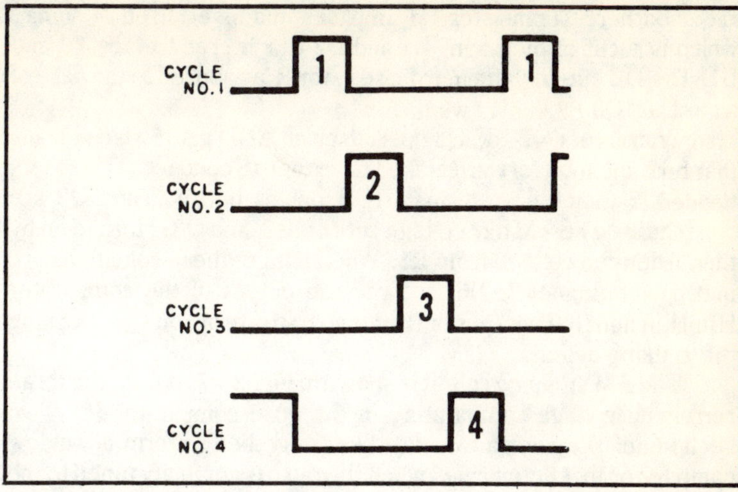

Fig. 23-5. Timing of the four cycles.

three cycles. When the inverted cycle No. 4 pulse goes LOW, it enables NOR gate sections B, C, and D.

Different N-codes are required for AM and FM tuning. Bandswitch S2 makes the remaining input of gate B HIGH for AM operation and LOW for FM operation. The DM-68 tells which has been selected by looking at input line B1 during cycle No. 4. If B1 is LOW, FM is selected. If B1 is HIGH, AM is selected. The N-codes are adjusted accordingly.

Seek and scan modes are controlled by gate D. The input (pin No. 13) of this section is normally HIGH. If it goes LOW, the DM-68 initiates seeking action. The scan circuit also causes pin No. 13 to drop LOW, but it contains a five-second RC timer to produce the scanning/stopping action described earlier. The DM-68 initiates a seek or scan operation if B2 is HIGH during cycle No. 4.

Seeker stopping is controlled by gate C. This input (pin No. 9) of the gate is normally held HIGH, but drops LOW if the stop circuits indicate that a station is tuned in. The DM-68 recognizes this command by a HIGH on B0 during cycle No. 4.

The AM and FM stop circuits are shown in Figs. 23-6 and 23-7, respectively. In both cases, transistor Q4 is the seek/stop switch that is connected to NOR gate C (Fig. 23-4). The collector terminal of transistor Q4 goes to ground when a stop command is issued. It remains HIGH at all other times.

The AM stop circuit of Fig. 23-6 uses an IC voltage comparator whose noninverting (+) input is biased by the +8-volt DC power supply. Two signals, the 262.5-kHz AM i-f and a DC agc voltage, are used here. Transistor Q1 amplifies and inverts the i-f signal, which is rectified by diodes D1 and D2 and filtered to C by R1 and C1. The DC output voltage from resistor R1 is applied to the base of transistor Q3.

Transistors Q2 and Q3 operate as an AND gate, which means that both must be forward biased for either to conduct. The collector of Q2 (point A) remains HIGH unless both Q2 and Q3 are conducting. The AM agc voltage cuts off Q2, and Q3 is turned on by the i-f-derived voltage from R1. When both of these conditions are met, point A goes LOW to force the output of the comparator HIGH. The HIGH output of the comparator turns on Q4, stopping the seeking action.

The FM stopping circuit is shown in Fig. 23-7. It uses the three remaining voltage comparators in the quad comparator IC. Two sections of the comparator are used together to form a window comparator that determines when the radio is correctly tuned to the center of an FM station by looking at the afc voltage. Recall that this voltage will hover about a fixed point when the tuning is correct. If the afc condition is not met, the radio is incorrectly tuned. The cathode end of diode D1 (point B) is grounded.

Point B is also controlled by the FM agc system through the last comparator. If the station has insufficient signal strength for proper reception, then pin No. 13 of this comparator remains LOW,

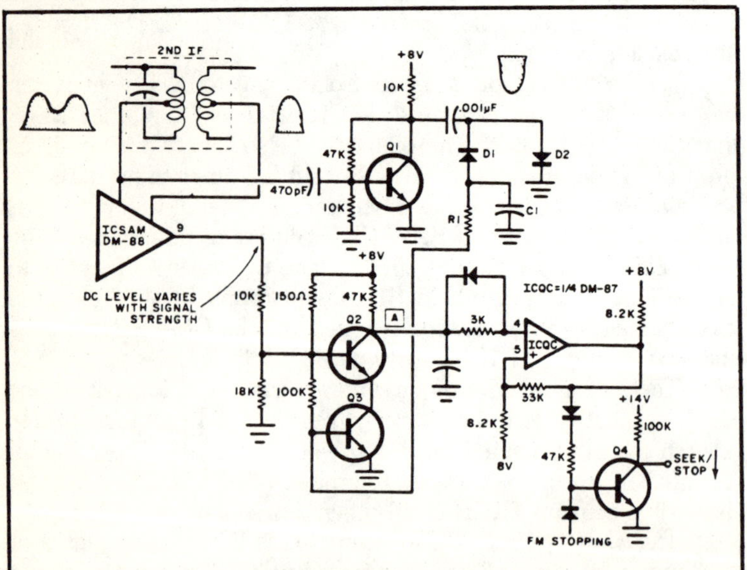

Fig. 23-6. AM stopping circuit (courtesy of Delco Electronics).

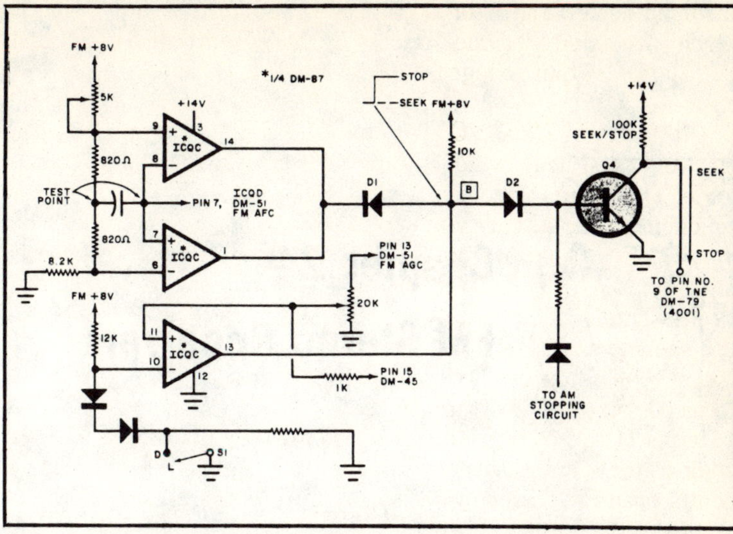

Fig. 23-7. FM stopping circuit (courtesy of Delco Electronics).

keeping point B LOW also. A local-distance switch changes the trip point of the comparator to allow for differences in average signal strength around town and out on the highway. If both afc and agc conditions are met, the outputs of all three FM stopping comparators will go HIGH, forcing point B HIGH also. This will forward bias transistor Q4, causing it to issue a stop command to the DM-68.

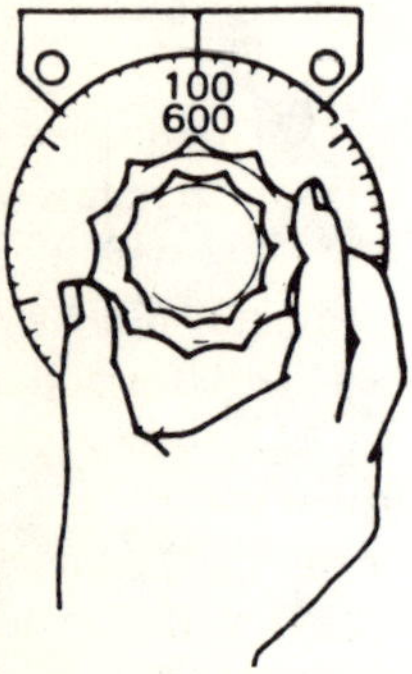

Chapter 24
FM Stereo Receivers

The advent of stereo broadcasting in the FM band made FM broadcasting a viable medium. Prior to the early 60s, the FM band was considered a wilderness for all but a few hi-fi fans. The technology did not exist to make a good radio in the FM band without high cost. But the advancing popularity of stereo, coupled with the introduction of stereo broadcasting and the development of lower cost solid-state VHF circuits, made FM broadcasting a somewhat more viable medium. Today, almost every radio sold will have an FM band, and all hi-fi receivers have an FM band. In fact, it is sometimes found that an AM band is lacking on some hi-fi receivers, but the FM band is present!

FM STEREO SIGNALS

The stereo information must be encoded onto the FM signal in such a way that monaural receivers will have no difficulty in receiving the program material in mono. This is the so-called compatibility that the FCC required. It was deemed bad practice to make a system in which the owners of existing mono radio receivers would be left out in the cold.

Figure 24-1 shows the FM stereo composite signal as adopted for use in the US. This system has been around in commercial use since about 1963. Two audio signals, L and R, represent the left and right stereo channels, respectively. The bandwidth for each channel is 15 kHz, which accounts for the high fidelity of the FM broadcasts. The L and R signals are added together in a linear combiner in the

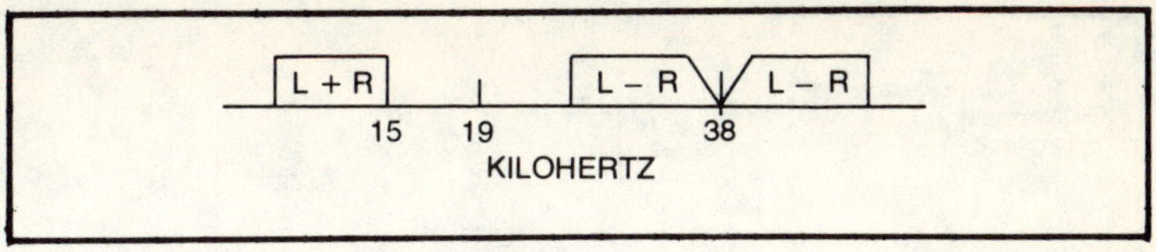

Fig. 24-1. Stereo composite signal.

transmitter to produce the L+R mono signal. This signal occupies a space in the audio spectrum from near DC to 15 kHz, and it contains the signals from both channels.

The encoded stereo information is formed from an L-R signal. This L-R signal is formed also in a linear combiner (a resistor adder network), but the right channel signal is first fed through a unity gain inverter stage that produces a – R signal. When the +L and – R are combined, then, the L-R signal will result. This signal is then used to amplitude modulate a 38-kHz subcarrier. In actuality, this is not a pure AM circuit, but a double sideband, suppressed carrier system. This is shown with the L-R sidebands extending ±15 kHz out from the suppressed 38-kHz subcarrier in Fig. 24-1.

One last signal is also added to the carrier. A precise 19-kHz pilot signal (one-half of the 38-kHz subcarrier) is used to synchronize the decoding process inside the receiver. The pilot signal has an amplitude that will cause approximately 10 percent modulation of the 38-kHz subcarrier.

The composite signal consists of the L+R mono signal, the 19-kHz pilot signal, and the 38-kHz L-R encoded signal. The composite is used to frequency modulate the FM transmitter at the broadcasting station.

When the composite is received in a monaural receiver, the roll off of the bandpass curve in the demodulator is very rapid, due to the need for de-emphasis to restore the normal audio balance. As a result, the 19-kHz and L-R components of the composite are attenuated substantially, leaving only the L+R mono signal. In a stereo receiver, though, these elements can be combined to decode the left and right channel stereo information.

STEREO DECODERS

Figure 24-2 shows the clock diagram of a typical "old style" stereo decoder section. The audio output from the detector is fed to a composite amplifier that has a bandwidth wide enough to accommodate signals up to 53 kHz, the upper end of the encoded L-R component. The output of the composite amplifier is split into two tracks. One is to an L+R amplifier, or delay line, which eventually supplies the L+R input to the decoder. The second branch is a

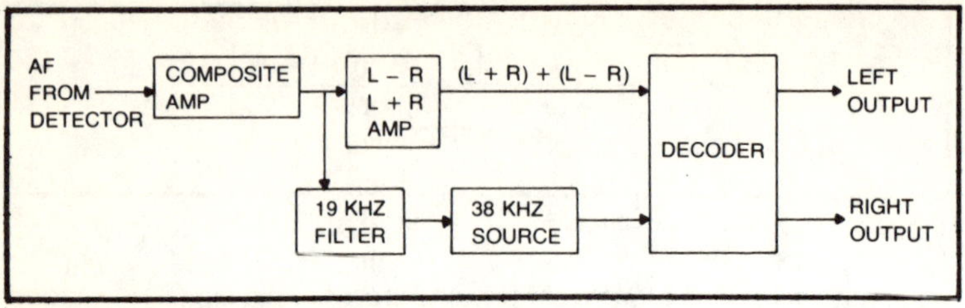

Fig. 24-2. Block diagram of stereo decoder.

19-kHz filter, or 19-kHz tuned amplifier. This picks off and amplifies the 19-kHz pilot signal, which is then used to reconstruct the 38-kHz subcarrier that had been suppressed at the transmitter. It is important for the proper operation of this decoder that the reconstructed 38-kHz signal have the exact frequency and phase relationship as did the original. Otherwise, a lot of the separation between left and right channels would be lost. Various methods of regenerating the 38-kHz signal have been used. One was an LC oscillator that took synchronization from the 19-kHz pilot. In most cases, the 38-kHz oscillator was a Hartley circuit. A second tap was added to the inductor, and the 19-kHz signal was used to excite the tank through this tap. Another method—probably the most popular—used a frequency doubler circuit. The output of the 19-kHz amplifier was fed through a transformer with a center-tapped secondary. The output of this transformer was fullwave rectified to form a standard double-humped full-wave rectification waveform of 38 kHz. The 38-kHz pulsating DC waveform was made into a sine wave by passing it through another amplifier in which the output was tuned to 38 kHz. The flywheel effect in the 38-kHz tank circuit took the 28-kHz pulses from the rectifier and converted them into sinewaves. The output transformer secondary would become the 38-kHz reference.

The decoder is basically a product detector circuit using a diode matrix. There are two input ports: one accepts the 38-kHz subcarrier reference signal and the other contains the (L+R) + (L−R) composite. When these signal are beat against the subcarrier the output will be the +L and +R (left and right) channel audio signals. These can then be output to the stereo audio amplifiers that follow. Note that a mono signal will be able to pass through this decoder without being changed, so the stereo receiver is also compatible with existing mono stations.

Most modern stereo receivers now have single chip stereo decoder sections. One of the first of these was the Motorola MC1304 device. This chip was introduced in the mid-60s and was considered very much an advance in the state of the art. One persistent rumor in the industry was that the old Scott company, a maker of high-fidelity equipment, contracted Motorola to make the chip, who then gave Scott a six-month or so exclusive on its use. Not long afterwards, however, a lot of companies were using the Motorola chip, including Delco Electronics who had pioneered FM stereo reception in automobiles. Today, the MC1304 is considered obsolete, and it has been replaced with other IC decoders in which a phase-locked loop regenerates the 38-kHz subcarrier. An RC voltage-controlled oscillator in the ship performs this function.

Figure 24-3 shows a typical "coiless" IC stereo decoder, this one by Delco Electronics using their DM-36 device. Notice the simplicity of the circuit. The composite input from the demodulator is applied through capacitor C7 to pin No. 1 of the DM-36. Output (L and R) is taken from pins 5 and 4, respectively. An RC network, which includes coarse and fine control potentiometers, is connected to pin No. 15 and is the frequency determining network for the PLL. In the first year that these were offered, an oscilloscope with Lissajous patterns or a frequency counter was needed to set the PLL frequency. But pretty soon some smart aleck notice that PLL circuits will pull into the correct operating frequency automatically if the technician just gets it close. In fact, this is so pronounced that later versions eliminated the fine potentiometer. All you have to do is adjust the frequency control until the stereo indicator light comes

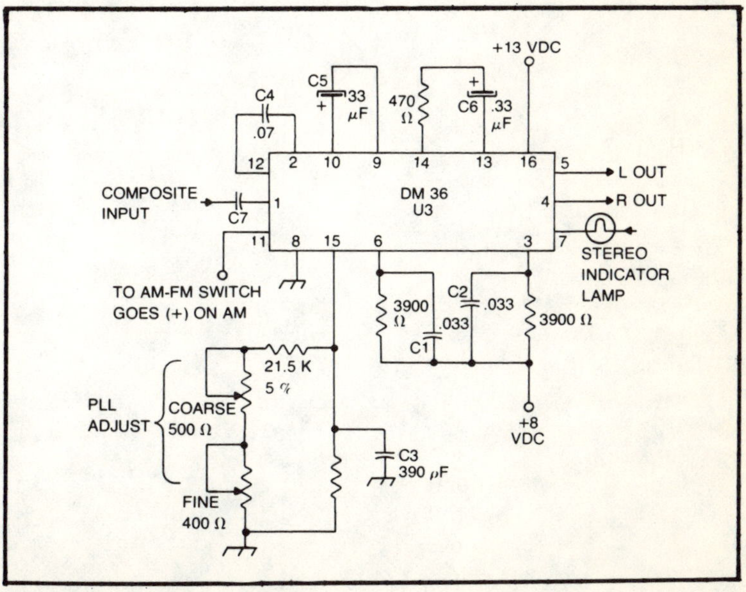

Fig. 24-3. PLL stereo decoder.

on. It was then advisable to tune the radio back and forth across several stereo stations to make sure that the stereo function would turn on under all circumstances. If the lockin was solid, the radio was properly adjusted.

Certain problems exist in the reception of a stereo signal. Stereo requires a much stronger signal than mono does to work properly. In fact, stereo signals become too noisy for proper listening once the received station becomes too weak. But the same station would come in properly if the stereo function were turned off and the receiver set for mono reception. The tongue in cheek rule of thumb was that 20 more miles (car radio) is gained on mono!

Chapter 25

Installing Mobile Receivers/Transceivers

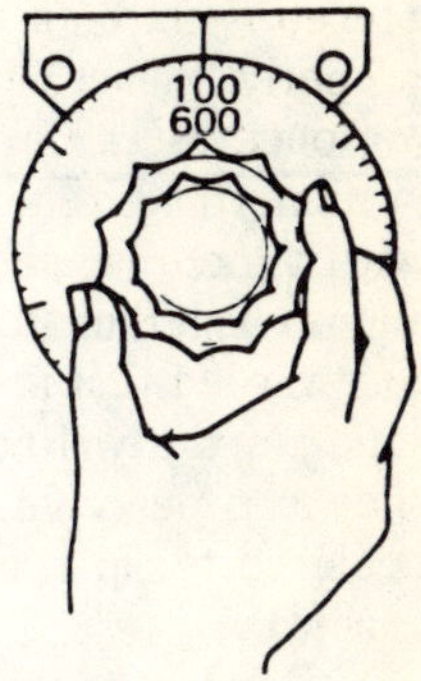

Americans have been accused of having a love affair with the automobile. This may or may not be true, but the fact is that many people do want to install radio equipment in their "go-machine." It is probably true that the vast majority of CB transceivers sold in this country are mobile models. Many—perhaps most—CB enthusiasts have only a mobile rig and do not bother with a base rig. More automobile broadcast radios are sold and installed in this country than anywhere else. Mobile installation is not at all tricky, although it helps to have some specific knowledge of the topic.

IN-THE-DASH INSTALLATION

The normal way to install a broadcast receiver in an automobile is in the dashboard. The makers of the automobiles create special places for the radio, even though some seem to be a little hard to get to. There are so many unique ways to get to the correct location, especially in American cars, that one is advised to seek out a car radio fixer, an auto mechanic from the dealer which sells that kind of car, or the factory RNR (remove and replacement) instructions for the car involved.

It is possible to buy a European-pattern universal car radio and a custom installation kit that makes it fit into the specific car that you own. These same kits can sometimes be modified by clever, mechanically inclined people to accept small CB rigs or 2-meter FM amateur radios. This will not theft-*proof* your installation, but may well discourage the less adventurous thief, especially when coupled

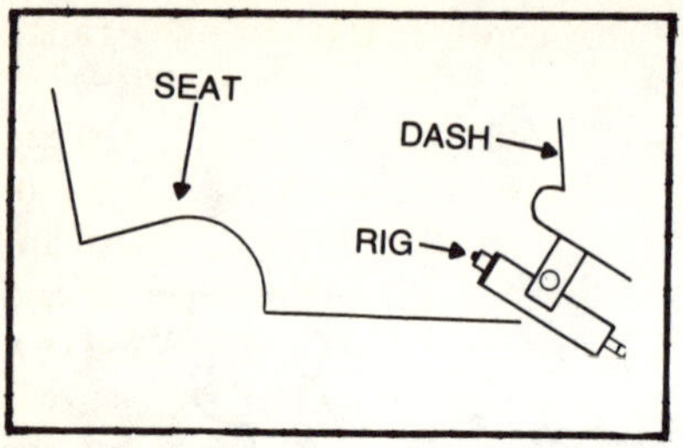

Fig. 25-1. Underdash mounting on gimbel bracket.

with a loud burglar alarm that is sure to attract attention. If the vehicle is left unattended in a dark area with no alarm, however, be forewarned that in-the-dash installation of two-way radios and/or tape players will not deter the more brazen thieves, especially if they can perceive "working" unmolested for very few minutes required to rip your dashboard apart with a crowbar (they do it, really!)

Don't expect "hidden" locations to protect the rig! You are simply not smart enough to find a hiding place that the thief cannot locate. The glove compartment is no-go, and under the seat is just plain stupid. That's how I lost a rig—it was under the seat hidden, but the antenna gave it away. Trunk-mounted rigs offer some additional protection, but it is minimal, despite what the ads claim. If you take the rig out of the car and place it in the trunk when you leave, or install it in the trunk in the first place, be well aware that it is a simple matter for anyone with a screwdriver to pry open most automobile trunk lids. Ask any police officer or insurance adjuster how long it takes even relatively inexperienced thieves to break into a locked trunk . . . and cry. The only solution seems to be *total* removal of the radio from the car when you leave it alone.

UNDERDASH MOUNTING

Two-way radios and tape players almost universally come with, or have available as an option, underdash mounting brackets (Fig. 25-1). These are U-shaped device that are designed to be attached with one thumbscrew on either side of the radio. Either holes or slots are in the top member of the bracket for fastening to the underside of the dashboard. In many cases, the mounting holes on the sides of the radio are positioned so that the radio will be balanced when the mounting screws are tightened, and no further bracing is needed.

The screws used to secure the radio must be substantial enough to hold the weight of the radio and not vibrate loose. This is sometimes a tall order. In most cases, use a No. 8, No. 10 or No. 12 machine screw if you can reach the top side of the dashboard lip. If this is not possible, use a sheet metal screw of the same size range to secure the bracket.

In many cars, the underdash lip will made of metal. There is little problem in applying sheet metal screws to this type of dashboard. But in the many cars which use a plastic dashboard lip, the screw will soon vibrate loose, causing the radio to fall off and possible damage the plastic lip. In these cases, a sheet metal screw might not be the correct choice.

Some amount of security is available in the form of the lockmount brackets popular with CB and tape player installers. These two piece brackets allow quick removal of the radio/tape machine if you use them the way they are intended (Fig. 25-20). If you are too lazy to take the radio with you, the lockmount is simply an open invitation to the thief, despite the "lock" used on the mount. But if you are religious about taking the radio with you when you go away from the vehicle, the lockmount is a blessing.

In CB installations, we can often get away with connecting the antenna through the terminals provided on the lockmount, but this is never good practice. It becomes almost impossible on high-power, solid-state amateur rigs, or on any VHF/UHF equipment. In those installations, one must attach the antenna directly to the antenna connector on the rig. It might help to either change the connector or a BNC (bayonet, snap-on action) or use an SO-239 coax to BNC adapter and fit the end of the antenna cable with a BNC connector.

The use of a floor mount is shown in Fig. 25-3. This type of mounting bracket is a little more costly that the other types but has its uses. It is bolted directly to the floor over the transmission hump. Take care in making the holes for this type of installation, however. If you attempt to drill directly through the floor carpet, then expect to wind many of the carpet threads around the drill bit,

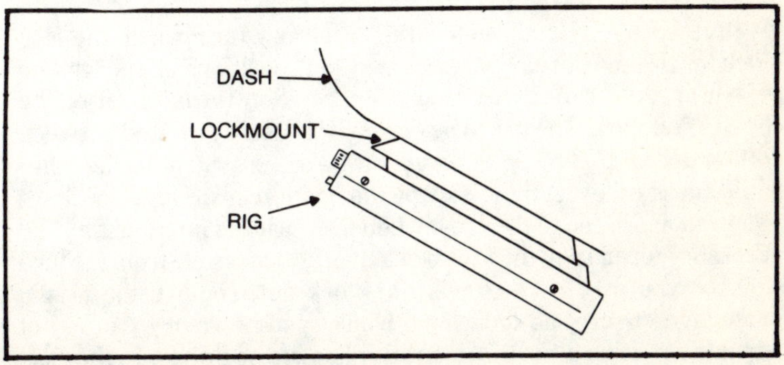

Fig. 25-2. Underdash mounting on lockmount.

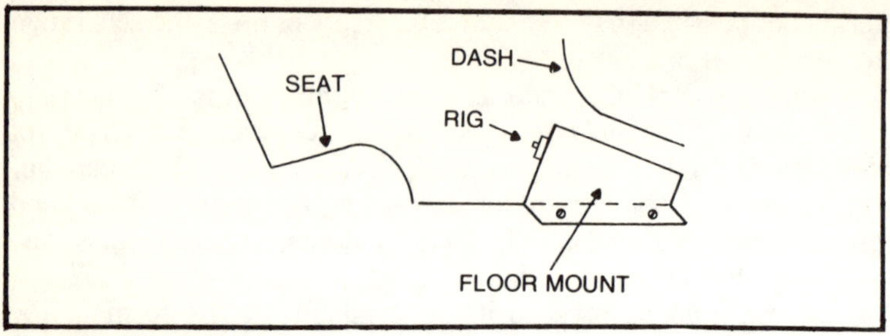

Fig. 25-3. Transmission hump mounting.

ruining the carpet. I generally either cut a hole in the carpet with a pen knife, and then drill through the underlying metal, avoiding the carpet threads, or use a short drift punch to beat a hole in the floor. This must be done with utmost care. If you let that punch go too far or hit it too hard, then the hole punched will be too large for any sheet metal screw that you are likely to be able to lay your hands on in short order. There is also a possibility of hitting the transmission bell and doing damage, although I have not ever seen that done. Floor-mounted brackets are available in both permanent and quick removal types, of which I prefer the latter. The problem with permanently installed radios is that they are *not* permanently installed—a thief can get them out faster than the best RNR mechanic.

POWER CONNECTIONS

The power connection to a small two-way radio is almost trivial. In the case of lower power rigs, such as some 2-meter FM and all legal CB rigs, simply connect the power line to the same wire the broadcast receiver is using. For years, I have had a 10-watt 2-meter FM rig connected to the AM radio A-lead. It is simply spliced into the radio power lead with solder and tape. No problems have resulted.

Alternatively, you can use the accessory terminal on the fuse block for this purpose. There are two "on" positions on the ignition switch in most cars. The regular *on* position turns on all of the circuits and allows the car to be driven. It is usually located between "off" and "start." The accessory position of the switch, on the other hand, turns on everything except the ignition system. This allows you to sit and listen to the radio when the engine is not running. The accessory terminal of the fuse block may be accessible from the fuse side, or you may have to pull the block out, remove the plastic protective cover, and look for a blank spade terminal that is hot when the ignition switch is in both *on* and *acc* positions. In the case where the terminal is available from the front panel, look for a spade terminal marked "AC," "ACC" or Accessory." The marking "A/C" usually means "air conditioner."

We can also sometimes use the fuse block when there is no accessory terminal. Sometimes, empty terminals operate in the accessory position. These are useful, even when only one of the electrodes is provided (the hot one, of course). First, before going to any extra trouble, look behind the dashboard to see if there are any insulated connectors that are hooked to one of the accessory fuse positions. If there are, use them and don't mess with the fuse block. These are intended, perhaps, for accessories which are not installed on your particular car. The wiring harness, however, must be reasonable universal for it to be economical for the car maker.

Figure 25-4 shows how to make connection to the fuse block using an impromptu homemade connector. Find which of the empty slots is hot and which connector of the slot is the hot side. Make sure that it is hot in the accessory position of the ignition switch. Next made the connector by sacrificing a fuse. Any fuse (not installed in the fuse block) intended for automotive use can be selected, because the rating is totally unimportant. Look for the glass body fuses with the metal end-caps that are common in autos. Smash the glass, being careful not to cut yourself. Now solder the exposed end of a piece of insulated A-lead wire to the inside of the fuse end-cap. This is then pressed into the slot intended for a fuse in the block. Since it is mandatory for a rig to be fused, use an in-line fuse holder for the A-lead wire. Dress the wire out of harm's way, considering the operation of the brake pedal and accelerator and secure it with tape so that it does not fall at all. While the ideal would have the in-line fuse holder accessible, the placement of the wire and holder must not interfere with the operation of the vehicle controls. This seems like a silly point but it has caused accidents when installations were made in an ill-advised manner.

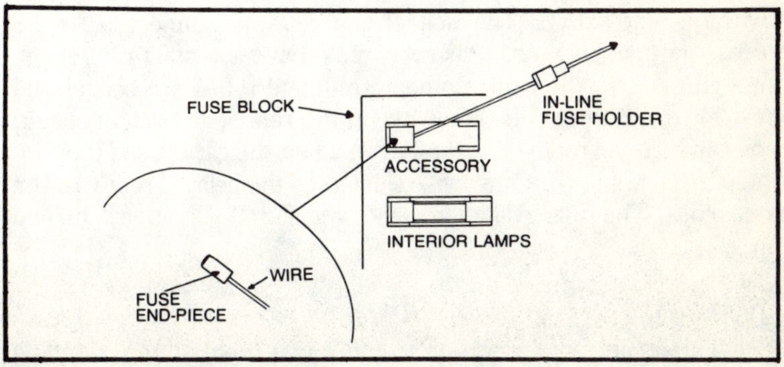

Fig. 25-4. Attaching to fuse block—the cheating way.

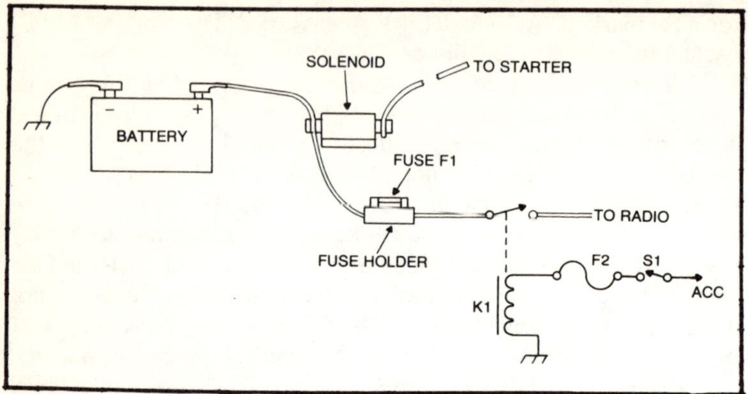

Fig. 25-5. DC power control for high-current rigs.

Higher power rigs used by amateurs and land mobile licensees require a different tactic. The accessory terminal is simply not equipped in most cars with enough current capacity to drive high-power radios. Remember that a typical 100-watt to 200-watt amateur rig will draw 15 to 22 amperes on transmit. If the main bus wire from the battery to the fuse block is hefty, you might want to try it. In most cases, however, you must go directly to the battery or to the starter solenoid to obtain power. Several connectors in the engine compartment contain the high current capacity needed for the rig. If you are willing to try remembering to turn off the rig every time you leave the car, connect it directly to the battery or solenoid terminals. It is very important to fuse this lead. Power is applied all the time so a short circuit is always a potential disaster, especially if it occurs when the car is left unattended. Figure 25-5 shows how a heavy-duty fust block is installed close to the point where power is obtained to protect against short circuits.

A relay (K1 in Fig. 25-5) can turn off the radio when the ignition switch is not active. The coil of the relay is connected to the accessory fuse position. This relay must have a contact rating high enough to carry the load. Some automotive types are considered best for this purpose. If the switch in the relay coil circuit is used, you can turn the radio on and off even when the car is used (but the radio is not wanted). Otherwise, just wire the relay directly to the fuse block. The radio will become active when the ignition is turned on.

ANTENNAS

The subject of antenna placement is a topic of never-ending debate in two-way radio circles. Even the experts seem frequently confused by bad information. In some tests run by engineers and antenna physicists at the NASA-Langley facility in Hampton, Virginia, however, some interesting things were discovered. The standard wisdom has always held that the best location for any two-way radio antenna was in the center of the roof. But—surprise of surprises—it was also learned that the gain and directional advantages of this location over the trunk or rear quarter panel locations was minimal. Actually, there was hardly enough difference to justify the extra trouble required to make a roof installation! There are too many hassles, too much danger of damaging the headliner of the vehicle interior and too much danger of water leak damage for the roof type. The trunk mount seems, then, the most *practical*.

The transmission line from the antenna in mobile installations is always coaxial cable. Because of the short runs involved, the smaller varieties of coax (RG-58, RG-59, etc.) are satisfactory. These have higher loss than the larger types, but in the 12-foot to 20-foot runs typical of mobile installations this difference does not amount to much. The larger cable is a lot harder to work with, so the smaller is somewhat more practical in mobile situations.

The cable is routed behind the rear seat and then under the carpet to the front of the vehicle where it connects to the mobile rig. The rear seat is easily removed. In most cases it is only necessary to remove the lower portion of the seat. You can then use a *snake* to draw the wire from the trunk compartment to the interior of the car. There is usually a cardboard barrier behind the seat, making it difficult to simply push the cable through. A cheap and quick snake is made by straightening a wire coat hanger. Tape the cable securely to the snake and pass it behind the top portion of the seat to the floor space over which the lower portion of the rear seat is normally place. You can get a good idea of where any available channels are located by noting where the cable harness for the rear lights passes. This is usually on the *driver's* side of the car in the corner of the trunk. It is not generally a good practice to pass the coaxial cable too close to the auto electrical cabel harness for any long run, because it might pick up too much noise. For the short run through the back seat, though, it is usually all right. The cable can then be routed under the front seat to the rear of the two-way radio. Many people simply pass the cable over the carpet and hope that no one tangles their feet in the cable or otherwise damages it.

A proper , professional installation, however, requires that the cable be routed underneath the carpet. The rear edge of the carpet will be exposed when the lower portion of the rear seat is removed. In some cases, it will then be possible to pass the cable through to the front edge merelv by manipulating the coat-hanger snake. But more often, it will be necessary to remove both the metal flashing that clamps the carpet at the door opening and the kick panel on the wall between the floor and the bottom of the dashboard. It is a lot easier to work from the passenger's side of the car in this case, because there will be no interference from the emergency brake and the control pedals. The kick panels are secured to the wall with Phillips screws and molded to fit. Some are a little difficult to remove and must be pried loose with a tool or screwdriver. Be careful not to damage these panels. Note well that it isn't always necessary to completely remove the panels for purposes of installing coaxial cable. You can often get away by merely pulling it away from the wall a half inch or so and pushing the cable in behind it. Be careful on both the kick panels and the floor molding when reassembling. Those Phillips screws can sever the coaxial cable and cause a short circuit. Be sure to dress the cable away from the screw holes.

The antenna mounting assembly usually requires one or more holes. Most of the permanently mounted 2-meter and CB antennas requires a 0.75-inch hole, while some of the trunk mounts for whip antennas require a hole from 1 to 2 inches in diameter. If you are working in the trunk, with access to both sides of the fender panel, the best bet is to use a Greenlee chassis punch to make the hole. Drill a pilot hole large enough to accept the power screw of the punch. This usually 5/16 to 3/8, depending upon the size of the punch. If you do not have access to either side or the cost of a punch seems too great for a single job, consider a power hole saw of the correct size. Remember that you are drilling into metal, so do not use one of the flimsy sheet metal or wood varieties. Use a real, heavy-duty, hole saw. Drill slowly, allowing the teeth of the saw to cool off every few seconds. Hole saws are a little dangerous, like any power tool, when not used properly. If you saw too fast, the teeth of the saw *will* (not *might*) wear out. There is a pilot drill bit in the center of the saw to guide the action. It is wise to drill a small hole at the center of the area where you want to make the larger hole, but make this hole slightly smaller than the pilot bit size. You don't want the saw oscillating around in a pilot hole that is too wide. Be light handed when drilling the final way through this hole using the pilot bit in the saw; the bit will suddenly break through and smash the saw teeth down onto the metal. This could crack the saw blade. If done when the drill is operating at high speed, there is a distinct danger that the saw blade will break, and a little piece will come flying off. This is dangerous to anyone around the vehicle. Proper procedure, however, virtually removes this particular danger from consideration.

Chapter 26

Mobile Noise: Sources and Suppression

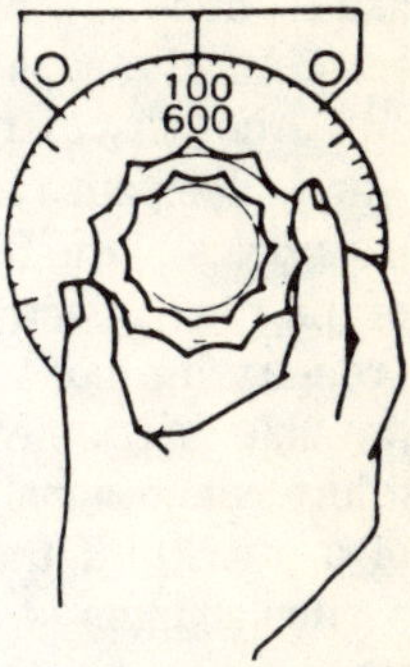

One of the first things that must be done to a new mobile installation is suppression of all of the noise sources in the vehicle. Techniques that are effective on AM and FM broadcast receivers are often less effective on high-frequency communications receivers. One of the main problems is that the frequency spectrum of the noise produced by various sources within the car tends to peak at different points in the high-frequency spectrum, and many of the noise suppression-components that proved to be successful on the AM band are less than effective in the high-frequency region. Capacitors, for example, should be coaxial types when suppressing the noise for a high-frequency mobile radio receiver, while ordinary noncoaxial types normally supplied with AM car radios work fine for the AM broadcast receiver. Fortunately, the problem of noise becomes less in the VHF region, because of the weakness of the harmonics in the noise spectrum at those frequencies.

WHAT CAUSES NOISE?

Any set of electrical contacts, any electrical spark, or almost any electronic or electrical appliance can cause noise in a radio. This especially true of radios using AM, and to some extent SSB and CW reception. FM radios can operate relatively free of noise if the signal strength is high enough to drive the limiter into hard clipping. Coupled with the VHF/UHF frequency on which FM transmitters normally operate, this usually solves much of the noise problems.

A large variety of noise generators are in a home. The furnace motor, vacuum cleaner motor, or a kitchen blender motor will produce noise. The ragged snow that these appliances cause in the TV picture is merely the video manifestation of the static noise that is heard on a communications receiver when such appliances are operated. The cure is usually to place a capacitor across the brushes in order to reduce the amplitude of the noise emission. This is usually done with a 0.5-capacitor, but should be undertaken with the advice of the appliance manufacturer's service department.

Another common source of noise in the home is the lightning. Three different types of noise are common, but they all produce harmonics of the 60-Hz power line frequency across a wide portion of the radio spectrum. One source of 60-Hz-harmonic noise is the ordinary incandescent lamp. If the fixture is in poor shape, it is possible that some minor arcing is occurring, and this can create the interference signal. Go around the tap lights (gently use the eraser of a pencil). If the noise stops or changes markedly when one particular lamp is tapped, then change the bulb. If this does not make the repair, change the socket (the problem is most often found in the socket). Another of the main causes of this noise is a fluorescent light fixture. These should be replaced with incandescent fixtures if the noise is great that it overcomes the reception. The last common source of 60-Hz interference to radio reception is the little SCR light dimmer that many homes use in place of the normal wall switch. Some progress in the design of these devices thas reduced the level of rf emission, but it still persists. Models with low rf emission use zero-crossing switching and brute force rf filter to achieve low-noise operation. If these are a "must" in your home, buy a model that has a switch detent position for full brilliance. This type usually has a switch that shorts out the SCR cathode-to-anode when full brightness is desired.

Interference also occurs from sources such as electric resistance heaters, especially older units that may have corrosion at the rivets that make the electrical connection to the heater element, and thermostats. These can be dealt with by repair in the first case and a capacitor in the second.

The power company also causes a certain amount of 60-Hz interference. If you live near high-tension transmission lines, don't even both calling the company about the problem. It is simply not feasible to transmit so high a voltage with some corona or arcing. If the noise tends to be intermittent, however, or seems to come from ordinary residential distribution systems, then by all means call the power company about the matter. Most local power companies have rf detectors that will help them locate the defective pole. The problem seems to be the tie wires that jump the electrical current over the cross bars on the poles. Tightening the tie wires and cleaning the junctions will cure the noise problem in most cases. Note that you can help locate the defective pole using an ordinary AM portable transistor radio. These radios have an internal loopstick antenna that is highly directional. A set of *nulls* will be noted as the radio is rotated through 180 degrees. You can predict where these nulls are going to be by taking the back cover off the radio and examining the loopstick. The nulls occur when the radio station or other rf source is in line with the ends of the loopstick. When the reception is off the ends, signal strength is minimal. (Note that a portable AM radio makes a handy direction finder for hikers and backpackers.) You can find the defective pole by noting the amplitude of the noise between stations. It should increase as you approach the defective pole, and decrease as you walk away from it. The directinal properties of the loopstick will help locate the pole almost 100 percent of the time. The expensive rf interference locator used by most power companies is nothing more than a radio receiver operating in the VLF-medium-wave range that has a direction finding loop antenna on top of the receiver case.

MOBILE INSTALLATIONS

The *first* step in noise-proofing a mobile radio installation is to secure the antenna properly. This goes for both broadcast and two-way radio receivers. The typical antenna used in mobile work is vertical fed with coaxial cable. If the coaxial cable is not *grounded well* right at the antenna base (see Fig. 26-1), there will probably be noise. In fact, this is the first point that experience troubleshooters examine when noise suddenly appears in a previously noise free installation.

Figure 26-1 shows a typical ground for a car radio broadcast antenna and for many of the small two-way radio antennas. By extension, we could also use this drawing to illustrate the slightly different system used to ground the mount of CB and amateur whip

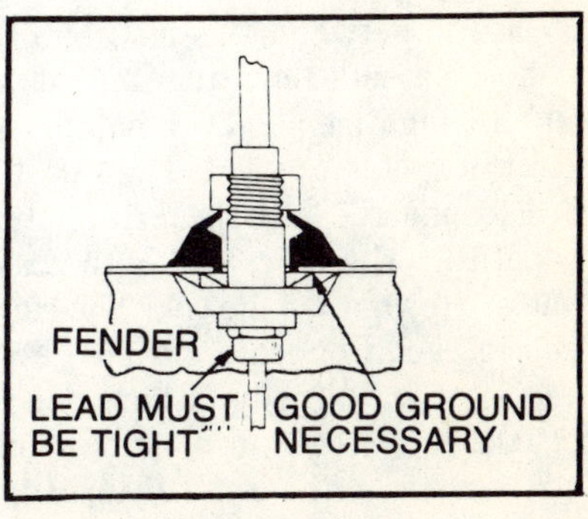

Fig. 26-1. Grounding the antenna properly for noise reduction.

antennas. The mechanical mounting of the antenna also provides the grounding. The ears that form what is essentially a toggle-bolt fastener are pointed, so that they will dig into the metal of the fender. A good *electrical* connection is necessary. Some installers find that it is necessary to scrap the underside of the fender clean in order to remove road crud, undercoating or just plain rust before a good ground is obtainable. I have seen some rather interesting homemade tools that allow the installer to clean the underside from the top while reaching through the hole in the fender. Be careful when mounting an antenna to the fender or quarter panel that serves as the outer wall of the trunk. It may appear to be unpainted, but that gray color is actually a primer that looks like metal. A shiny metallic surface should be apparent, even if scraping is needed. Don't even *start* any of the other troubleshooting methods if the antenna system is not squared away properly!

NOISE-SECURING THE VEHICLE

Noise problems in mobile installations are sometimes very difficult to suppress. Even *finding* the noise source is sometimes a bit of a problem. This is especially true when the noise if from the ignition system, and all of the "standard" remedies have failed. There are basically two paths into the radio: antenna and power lead. The antenna-born noise is usually suppressed by shielding and some bypassing of items like the low-voltage terminals of the ignition coil. But power line noise can be freaky. It has been characterized as being a lot like trying to nail hot butter to a wall.

The subject of mobile noise suppression is covered in some detail in TAB book No. 1194, *How to Troubleshoot & Repair Amateur Radio Equipment*. We will therefore concentrate mainly on the so-called standard remedies in this book. Figure 26-2 shows some of these remedies. At "A" we see the ignition coil capacitor. It is usually the practice to place a 0.5 μF capacitor between ground and the *battery* terminal of the ignition coil. Do not place the capacitor between the distributor terminal and ground. In AM radios, it is sufficient to use an ordinary noise suppression capacitor. But in high-frequency amateur and CB installations a coaxial capacitor must be used. The wire to the battery terminal of the coil is disconnected from the coil and then reconnected to one end of the capacitor. The other end of the capacitor is then connected to the battery terminal of the coil. The current must then pass through the coaxial capacitor center post to get to the coil. The other plate of the capacitor is its body, which must be properly grounded. Some coaxial capacitor brands use the same No. 10 thread as used on the coil terminals. Therefore, thread the capacitor directly onto the coil. The body is then grounded by soldering a piece of heavy braid to the mounting lug and then grounding it under the mounting screw that holds the coil to the engine block. The outer braid from a piece of large size coaxial cable (RG-8, RG-11) works well for this purpose, and it usually is much more available and less costly than a roll of real braid.

For really critical cases, it might be necessary to shield the entire ignition system. A firm called Estes Engineering advertises in the amateur radio magazine (usually classified sections) that they supply components for most makes of automobiles and—I believe—airplanes.

At point "B" in Fig. 26-2 one or more capacitors bypass the terminals of the voltage regulator in the battery charging system. The older relay-type regulators required two 0.1-μF capacitors, one each on the battery and armature terminals of the regulator. Modern solid-state regulators, on the other hand, often have a special terminal on one side especially for the noise suppression capacitor. In fact, solid-state regulators are somewhat less prone to

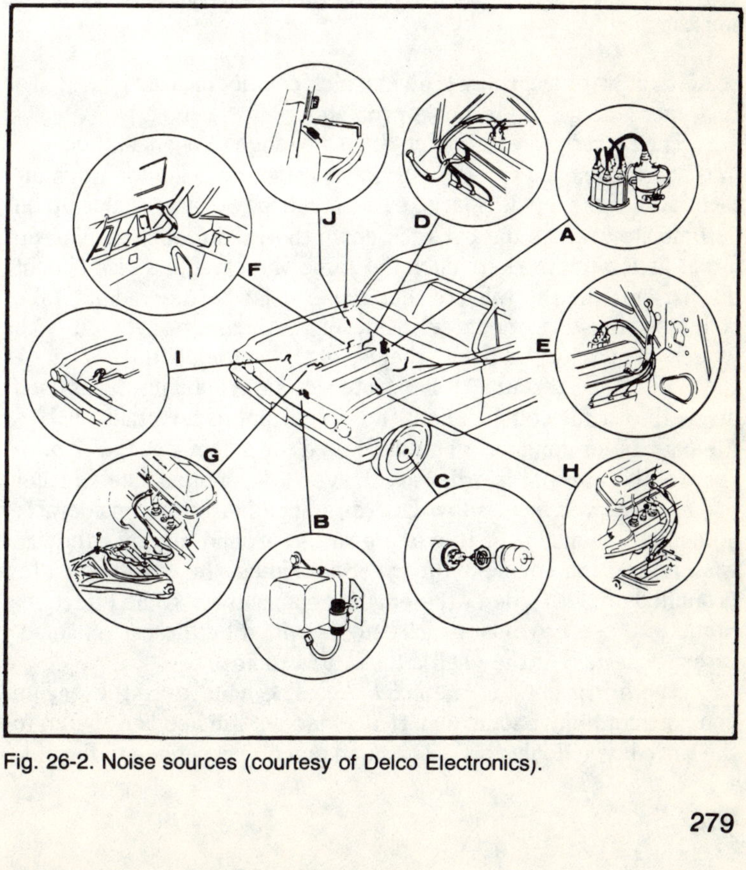

Fig. 26-2. Noise sources (courtesy of Delco Electronics).

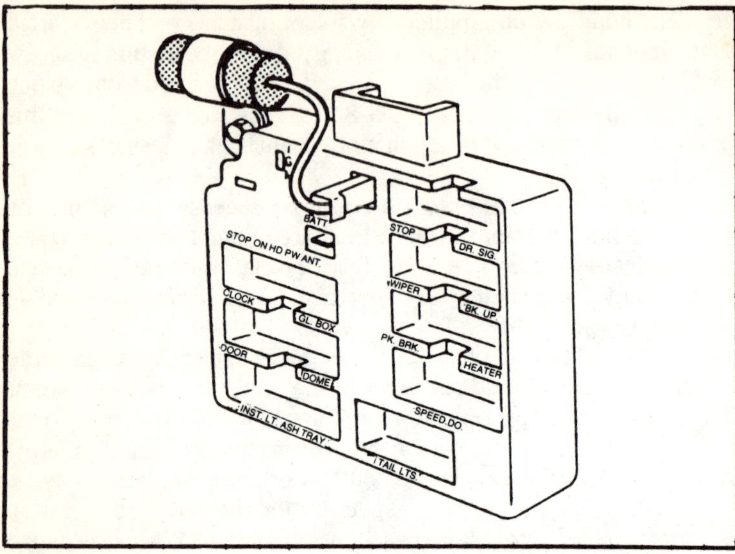

Fig. 26-3. Connecting capacitor to GMC fuse block (courtesy of Delco Electronics).

noise than were the relay type in which contact chatter provided a large percentage of the noise problems.

The remedy seen at point "C" is no longer a big deal, because most cars come equipped with these antistatic collector rings attached to each wheel. A large static electricity charge builds up on automobile tires as the car rolls down the road (doesn't a Van de Graaf generator work in much the same way?). If this static is not discharged, then it will arc and cause noise in the radio. Static collector rings are mounted inside of the bearing cover on each wheel, especially the front wheels, of the automobile.

Points "D" through "I" illustrate various grounding straps used inside the engine compartment. It is important to use braid (such as the coax outer conductor just described) or a heavy gauge (No. 8) wire for this purpose in order to achieve a low rf impedance. Smaller gauge wires will have a low DC resistance but offer considerable amounts of inductance. It is important to ground together the fire wall, frame, and engine of the car. Sometimes, the engine must be grounded on both side of the engine compartment to be effective. Some of these grounds are already provided by the car manufacturer, especially if they sell a lot of FM radios.

Another ground is the *hood clip*. The hood over the engine compartment must act like an rf shield (open the hood and listen to the radio if you doubt this). The hood must be grounded to the rest of the car body. Two methods are used. In some cars, little metal spring clips are mounted on the cowling where the rear of the hood meets the body. These clips are exposed when the hood is raised, but make an electrical contact between the hood and the cowling when the hood is down. The other method, used most often in fiberglass body cars, is to connect a piece of ground braid across the hood hinges. Fiberglass automobiles, incidentally, typically have a screen wire embedded in the hood to perform the shielding task.

Any small motors, such as the blower (point J in Fig. 26-2), should be bypassed to ground with 0.5-μF capacitor. The capacitor must be located as close to the point where the power line enters the motor as possible. Otherwise, noise radiation to other wires can nullify your effort.

General Motors automobiles often have a special terminal on the fuse block that will accept a capacitor for general noise suppression. If the car is equipped with a Delco FM radio, it is likely that this capacitor was installed at the factory. If not, then a 0.5-μF automobile noise suppression capacitor (available from many Delco or other auto electronics dealers) should be used. A typical example is shown in Fig. 26-3.

LOCATING TOUGH NOISE CAUSES

Some motor noise problems are really terrible to pin down. Almost any metallic component of the automobile will pick up and

Fig. 26-4. Using the noise locating Sleuth (courtesy of Delco Electronics).

reradiate noise pulses. The accelerator linkage, the emergency brake cable, and the tail pipe are common offenders, as are the dozens of cables coming through the fire wall to the interior of the car. There are several good ways to find them, but one of the best is to use a Channel Master *Sleuth* or a homebrew equivalent. This device (Fig. 26-4) is a directional rf detector that has a coaxial cable coupled directly to the antenna connector of the receiver. The receiver is then used as the rf indictor. Turn the volume up—off station—and probe with the detector until you locate the cause of the noise.

A similar device is made using an ordinary piece of coaxial cable. Install a connector on one end that will mate with the antenna terminal on the receiver. Strip back the outer shield and outer insulating jacket about 2 inches, leaving only the inner insulator and conductor exposed. You may then use this tip as the rf "sniffer" needed to ferret out the noise source.

Chapter 27
Troubleshooting Radio Receivers

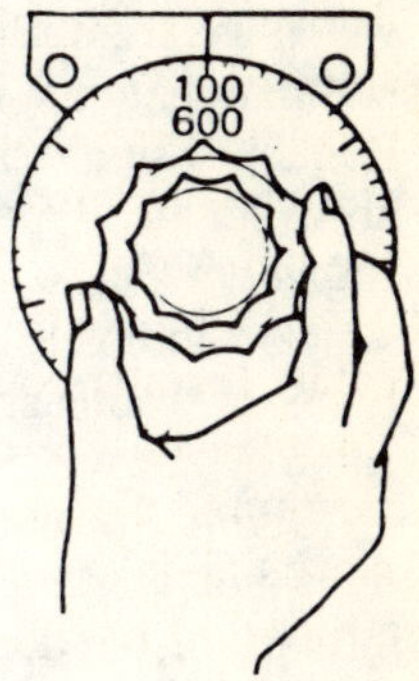

Troubleshooting is an activity that many, perhaps most, amateurs prefer to do without. Some view it as an arcane process, open only to the in-the-club professionals who inhabit the back rooms of the radio factory. It is popularly assumed that one must be a super technician and possess a large fortune in electronic test equipment. All of these things help, but a lot of troubleshooting is very straightforward. In fact, you would be surprised how many times the problem in the most complicated transceiver is little more than defective DC power supply. Many jobs, then, can be performed by any radio amateur who knows enough electronics to pass the General or Advanced class license examinations. To be sure, the person with 20 years of experience and a technical education will run rings around you as far as time goes and when the problem becomes really exotic. But there is little reason why you cannot be as successful given the time to do the job. The years of experience often serve to mature and wisen, there is something about being nailed to the workbench too many times for comfort that trains one better than any school book.

FIRST STEPS

It may seem like begging a point, but the first step in any troubleshooting activity is to observe the performance of the equipment. The test equipment needed for this all-important step is practically nil . . . most often your own God-given senses are sufficient. It is at this point that the seasoning of the professional becomes most apparent. An awfully large amount of data can be gleaned from this operation, and it has the chance of reducing significantly the time and trouble needed to make the job complete.

Let us simplify the topic of troubleshooting a little bit and pick on just one type of equipment. Let's assume that the equipment which we are dealing with is a single-conversion, superheterodyne communications receiver, or the receiver portion of a high-frequency single sideband transceiver. The principles which are laid down for this type of equipment can be extended to transmitters, amplifiers, and almost any other piece of radio equipment. It is simply too costly in space and time to cover all different possible types of troubleshooting, but the principles and practices are the same.

When you initially examine the receiver, determine just what is or is not working. The fact that the receiver is "dead" is very interesting but does not tell the whole story. Just what do you mean by "dead"? Is it dead after the fashion of seeming to be turned off; i.e., no pilot lights, filaments (vacuum tube models), and not even a squeak from the loudspeaker? Or are the filaments and pilot lamps normally operating? We can even further break it down by noting the level and type of noise coming from the loudspeaker. Is there a hiss? Do you hear static when the antenna terminals are touched with a screwdriver? How about when the bandswitch is rotated through its range? Do you hear static as the switch contacts make and break? Is there any scraping sound as you rotate the volume control knob through its range several times? All of these symptoms can give you a hint where to start. Volume control noise, for example, would practically exonerate everything from the volume control wiper to the loudspeaker because it appears that the audio amplifiers are amplifying the weak noise signal of the control wiper as it scrapes along the carbon element of the potentiometer. Subtle little things like this can go a long way toward pointing you in the right direction and will at least allow you a relatively fair chance of not wasting time on obviously good circuits and sections.

The next step might be to turn on the BFO, or set the mode switch to either CW or SSB positions. If the noise level increases or there is static when you turn the switch, it is a fair bet (not absolute that the problem is located in the stages prior to the detector. The next step after that indication is to rotate the bandswitch through its range. Look for either static crashes or signals as you set the switch to other bands. Static crashes will exonerate the circuits from the mixer to the loudspeaker, although one must be a little cautious here and take the indication with a grain of salt. If the receiver is not dead on all bands, then look first for components that are common only to the dead bands. Examples of such components might be some coils, some trimmer/padder capacitors, band-change crystals in multiple-conversion receivers, or even the switch contacts themselves. The matter of the switch contacts is often overlooked by the inexperienced troubleshooter. But corrosion on the switch contacts can easily interrupt the operation of the receiver. The corrosion is usually general, so it will normally affect all bands. But is it also all too common to see corrosion that affects only one or two bands. If the bandswitch operation seems intermittent, i.e., if the set seems to work temporarily when the bandswitch is first operated, or if you hold it in one position, then suspect corrosion *first*. These are sure signs. You can sometimes use an ordinary spray switch contact cleaner, or the type used by television repair technicians to clean tuner contacts, but this treatment sometimes isn't permanent. In really bad cases, it might be necessary to use a pencil eraser to clean the switch wipers. This must be done gently and with come care. It is all too possible to wreck the bandswitch if you are not careful. Lightly rub the exposed surfaces of the bandswitch wiper contacts with an ordinary pencil eraser. Be careful not to move any of the wires around the bandswitch (unless you hanker for an alignment job) and do not move any of the fixed contacts—they *will* break.

If the receiver is completely dead, you are actually ahead of the game and don't know it! To the uninitiated, the dead receiver seems like a nightmare rooted in some catastrophy. But the truth is that the deader the better. If the receiver is so dead that it looks like the AC power cord has been disconnected from the wall plug, the problem might be little more than a simple power supply repair. As the first step, you might want to check the AC power receptacle to see if it is hot. If you think that's absurd, you may be surprised at how often it happens. I have run many service calls over the years in which the problem was a disconnected line cord, a cold AC receptacle, or a wall switch that was in the off position. In some cases, the customer did not even know that the wall switch controlled the outlet where the TV was plugged in. You might live in a house for years and either forget—or never know—all of the details of your electrical system.

It is likely that when the receiver is totally cold, the problem is in the primary side of the DC power supply. Use an ohmmeter to check the fuse, AC line cord, power on-off switch, and the primary winding of the power transformer. If the rig is one of those transceivers with an external power supply, you might want to use a substitute power supply to make some checks. It might require 18 A for transmit, but the receiver can usually be powered from an ordinary 1 A DC bench supply. The operation might not even be very good, but it will at least tell you whether or not the external power supply is at fault. Another problem with the external type of supply is the unbilical cable that interconnects the transceiver and the supply chassis.

You might wonder about using an ohmmeter to check the fuse. After all, can't you *see* a blown fuse? Not all of the time, unfortunately. The fuse might not be blown but have either a defect in the end-caps where the fuse wire solders to the electrode or have a fatigue fracture that is difficult to see with the naked eye. Neither of these problems are visible, but they will show up with an ohmmeter. Of course, if the fuse is blown, please don't bother with the ohmmeter.

An open power on-off switch will also show up on an ohmmeter, but sometimes a little interpretation is needed. It must have a low resistance—less than 1 ohm. Many times corroded switch contacts will seem to pass an ohmmeter test only because the technician making the test used the highest ohmmeter range to make the test. A switch with a 100-ohm series resistance due to corrosion will *seem* to have a zero resistance if you use the X100,000 range on the ohmmeter to make the test. You can also test the switch dynamically by shorting across it and applying power if you want to risk burning up the rig. The only time that this is good practice is when the fuse has not been found to be blown. Even then it is best to short the two switch contacts with a piece of wire soldered in place (with the power off!). Then turn the set on.

As a safety reminder let me point out that it is dangerous to work the primary side of the power supply with the set plugged into an outlet. Always disconnect the AC power cord and keep the plug end in view so that you *know* the set is cold. After all, when you turn a set off, all you do is interrupt the AC line in the power supply primary circuit, but the 115 volts from the wall socket is still present on and about power supply components.

An old maxim regarding fuses must be understood: *Fuses don't cause trouble–they indicate and protect against trouble*. True, some fuses are defective. True, some fuses blow due to a transient on the line that will never be repeated. But in all cases where a blown fuse is found, you must suspect that something defective inside of the equipment caused it to blow. Under *no* circumstances is it proper to replace a fuse with one of a higher rating. I have seen amateurs who did this because of a recurrent fuse-blowing problem, only to regret it when they found out why the fuse kept blowing. In one case, a mobile rig was connected into an automobile battery system in which the boltage regulator failed, placing +17.5 volts DC on the 13.6-volt line of the 2-meter mobile rig. This was not enough to pop the internal +18 volt "corwbar" protection circuit, but was quite sufficient to destroy some of the components in the final amplifier and power supply. The only time when I see it as warranted to replace a fuse with one of a higher value is when the manufacturer of the equipment tells you to do it. The onus then falls on the manufacturer, who, after all, is the *de facto* expert on the equipment design. It does happen incidentally, that fuses are *sometimes* (very rarely) underspeced, so a 1 fuse is used in a 999 mA circuit, causing fatigue failure or blowouts when reasonable transient level occur But only undertake this dangerous step with the advice and consent of the engineers who designed the equipment.

When making your preliminary inspection, look for and note any unusual odors that might indicate a burned out component. The odor of burning transformer insulation is unmistakable. If the set looks totally dead and you find a blown fuse and smell transformer, then look immediately to the power transformer as the likely problem.

Also note any unusual sounds from the loudspeaker or any sounds from inside of the equipment cabinet that might indicate arcing, etc. This information could lead you directly to the source of the problem.

A visual check of the insides is also in order, all prior to bringing out the first piece of electronic test equipment! Look for charring, burned components (resistors are big on looking burned when overloaded), cracked capacitors, broken connections, or any of a dozen different types of mayhem that seems to affect radio equipment innards. Note that, while the description of this process takes all morning to write and takes you several dozen moments to read the actual process on the bench is merely a few moments of careful reflection and observation . . . perhaps saving dozens of minutes, or even hours and days, or hard labor later on (well. . . . . maybe?).

Before proceeding with the discussion of troubleshooting practices, let me drop a couple of words of wisdom on you (guys with 18 years on the bench get to utter words of wisdom to people just starting out . . . even when the receiver of such gems are bored stiff),: It is true that all troubleshooting of electronic circuit problems in equipment which has once worked properly (in newly constructed equipment, all bets are off) boils down to finding one of two situations: we have lost a desired path for AC or DC current (an *open* circuit somewhere), or we have gained an undesired path for current (a *short* circutt). The problem might be in the AC current path or the DC current path, but it will fall into one of the two above catagories. In the vast majority of cases, a simple step-by-step procedure will locate the defective component. When you see the

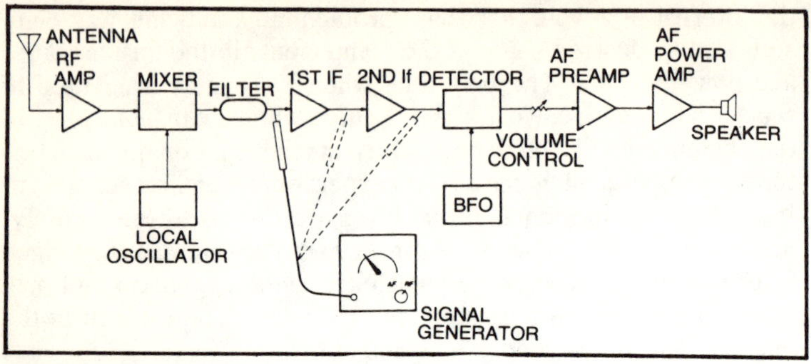

Fig. 27-1. Signal injection.

experienced old pro make a quick, lightning-like diagnosis of some obscure problem, it will impress you. But rest assured that it is not some super intellect that caused that to happen. It was probably possible because that oldster managed to get nailed to the wall many years before on a similar problem. But when the seasoned pro comes up against a new situation, where old experience serves only to embolden, note well that he almost universally reverts to some logical step-by-step procedure. If he didn't, then he would never have become an old pro in the first place and would have been fired 10 years ago!

BITE-SIZED CHUNKS

Almost any technical problem can be solved eventually if we divide it into bit-sized chunks and solve the little pieces one at a time. We might also call this philosophy *divide and conquer*. We will examine each stage of the equipment in its turn, going on to the next stage if the present stage being tested seems to work properly. In radio receivers, audio amplifiers,and other equipment that uses cascaded analog stages, two different (but related) methods usually yield results: signal injection and signal tracing. You will find that pros will generally prefer one method over the other, but the truth is that they are both used *best* under certain circumstances. It is, then, wise to keep both in mind until you have the experience to arrogantly dismiss one on favor of the other (the preference is usually based on available favorite test equipment rather than any real consideration.

The signal injection process is shown in Fig. 27-1. We must first have some means of detecting an output signal. In the case of a radio receiver, the "indicator" might be your ears and a loudspeaker. But it could just as easily be an AC voltmeter, oscilloscope, or anything else that will tell you when a signal is present.

A signal is injected from a signal generator, or other source (noise generator) into each stage in succession, starting with the output. In some cases, it is found that the audio power amplifier lacks sufficient gain to be driven with the output of the signal generator. In those instances, the output will seem very weak (making you suspect—horror of horrors—*trouble* in the output amplifier). We typically start at the input of the audio preamplifier, probably because the wiper of the volume control potentiometer is so available. If the output level seems normal, we can be sure that the problem lies elsewhere in the rig. We then switch to the intermediate frequency and begin injecting signal into the inputs of the i-f amplifiers, again in succession from the output toward the input. Keep making this test until you find a point at which the signal either fails to go through at all or is highly attenuated. It is a reasonably sure (if not absolute) bet that the problem is in the circuitry between the last point where signal was normal and the first point where the signal level at the output was not normal.

An example is shown in Fig. 27-2. This is the i-f amplifier of a radio receiver. Ceramic band-pass filters control the intermediate frequency and bandwidth. Let us say that an i-f signal injected to point A, the input side of i-f amplifier IC U2, produces a normal output level at the loudspeaker. But when the signal generator output probe is connected to point B, the output level drops to barely audible. We are reasonably sure that the problem will be found somewhere between points A and point B. In all likelihood, the problem is in the IC itself, with the ceramic filters and capacitors running a close second in the probability sweepstakes.

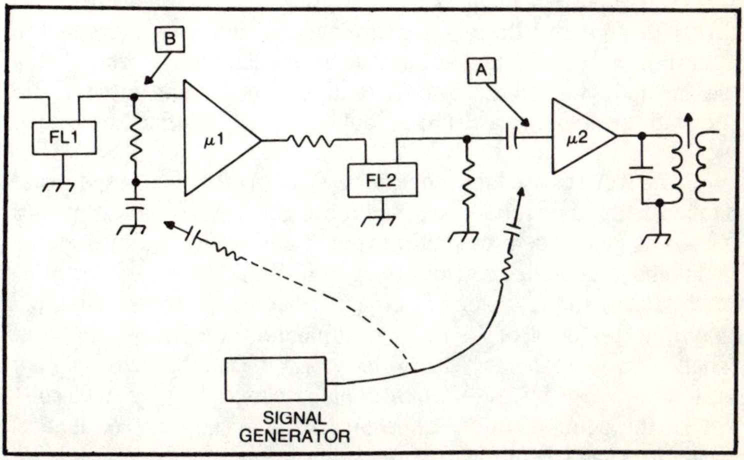

Fig. 27-2. Signal loss indicates problem area.

The other principle troubleshooting method is signal tracing, shown in Fig. 27-3. This technique is exactly the opposite of signal injection. Here we connect a signal source to the input of the device and then take an output indicator and look for the signal at each stage in succession beginning with the input stage.

Different kinds of signal tracing equipment are available. The simplest is an amplifier connected to a loudspeaker, which is what you get when you buy a signal tracer from a test equipment manufacturer, such as Heath or Eico. If the signal tracer is to be used in an rf or i-f amplifier, it will be equipped with a demodulator rf probe. This is still nothing more than a high-gain audio amplifier with a high input impedance. The instrument which I prefer is the oscilloscope. You can see a lot more on the 'scope than you really need to know, but it is a lot handier to use. If you plan to go into troubleshooting professionally, you will find why many pros use the 'scope even in relatively minor jobs. By seeing the waveform, rather than hearing it, you save a lot of wear and tear on the old nerves. That day-in-and-day-out blaring from the radio output can drive one up the wall quickly.

MORE WORDS OF WISDOM

Permit me to take advantage of my senior position. In this section I want to discuss some matters that will make things a little easier for you in the long run. Some troubleshooting problems are a lot like trying to skin an amoeba, so some insight is occasionally useful.

Vacuum Tubes

These ancient devices are not gone, merely obsolete in receiver design. But there are many vacuum tube receivers still in operation and still floating around the used equipment market. They will be there for years to come, and it is likely that anyone contemplating becoming a troubleshooting wizard will find some tube work.

The word of wisdom concerning vacuum tubes is to never trust a tube tester when it tells you the tube is good. A *bad* indication on a mutual conductance tube tester is generally absolute; they are long on finding gross defects. But a *good* indication must be taken with a grain of salt. Many dozens of times, I found customers telling me that they "tested all of the tubes, so it must be something else!" But when we went in the set, the problem was a bad tube that could not be found except by *substitution*. This is especially true in circuits where the actual gain is important, or in circuits like oscillators which may fail to start with a tube that passes the tube checker's criteria for "good."

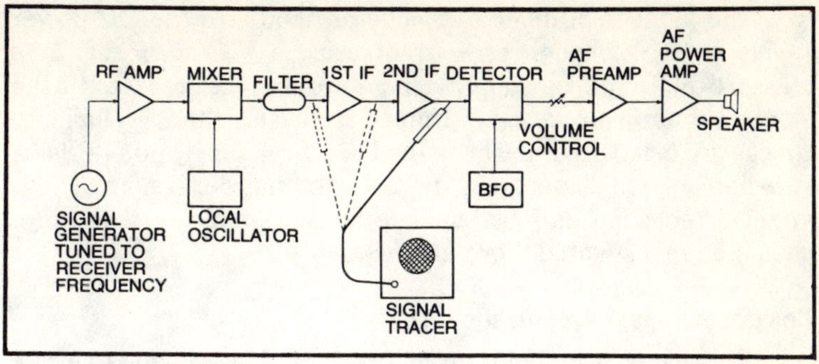

Fig. 27-3. Signal tracing.

Don't believe much of anything that a simple emission-type tube tester tells you, unless it tells you the tube is *bad*. These devices are useful as screeening devices, if you lack a mutual conductance checker. Professional tube testers are universally mutual conductance types, though.

But even with substitution, one must be wary. Although I am a hamfest shopper myself, I never beleive that a hamfest component bought used out of the trunk of someone's car is good until it is proven so. When you place a tube of unknown lineage into a set in an attempt at troubleshooting by substitution, you might be led astray by a bad tube. Use new tubes only. In fact, there are so few tubes in most amateur radio stations that you are well advised to keep a supply of at least one of each type used in your equipment in stock all of the time. Many times, the replacement of a single tube will solve the problem, although probably not at the 80 to 95 percent rate claimed by the tube tester ads. If you test all of the tubes in your rig, and find a few weak or with minor shorts, then go ahead and replace them with new, or known-good surplus, tubes.

Power Supply

For several years I was responsible for keeping the service records of a repair and maintenance shop that worked on all manner of electronic equipment. Once, an older hand told me that *most* problems in electronic equipment would be either in the power supply or would affect the DC voltages on the stage.

It turned out, after keeping a thumbnail survey for six months, that he was quite correct. Almost 50 percent of all problems were *in* the DC power supply, and a considerable number of the remaining problems would affect the DC voltages in the defective stage. A bad transistor, for example, is often first noted because the DC potentials on the base, emitter, and collector are not normal (see Chapter 28).

The number of defects directly attributed to the DC power supply is large enough to warrant using a VTVM or VOM to measure the DC power supply voltage as one of the first steps in the troubleshooting procedure. Immediately after the preliminary checks are completed, examine the DC power supply potentials. A variation of 15 to 20 percent in unregulated supplies is normal (less in regulated supplies), but voltage levels outside of that range should be investigated more carefully.

Component-Level Troubleshooting

Once you have used signal injection or signal tracing to locate the defective stage, isolate the defective component. Although one can often work without a schematic, it is usually better to try and locate this important diagram. Most amateur radio equipment manufacturers are pretty good about supplying a service manual with the equipment, or making them available for a modest fee. But, of course, there are times when no manual is available. If the set uses vacuum tubes, there are some aids: *RCA Receiving Tube Manual*, General Electric's *Essential Characteristics,* or the rear of an ARRL *Radio Amateur's Handbook*. These publications will give you the base diagrams for the tubes and some idea of the normal operating voltage levels.

Another tactic when troubleshooting without a schematic is to obtain one from a similar model of the same period, or from either previous or succeeding versions of the same model. For instance, my SX-28 was a lot like the later SX-28A. It turns out that engineers and production people tend to go with what worked in the past, so their products bear a sameness over any given period of time, barring company upheavals.

When there is no schematic available, try comparing the voltages measured in the circuit with what you know should be approximately correct. In a vacuum tube rig, for example, when you find only a couple of volts on the screen grid or a high positive voltage on the control grid, you *know* something is wrong in that circuit. A shorted screen bypass capacitor, and/or an open (or greatly increased value) screen dropping resistor would cause the first situation, while a leaky coupling capacitor from the plate of the previous stage would cause the second.

Recognize that there are always two paths for current in electronic circuits: AC and DC. The AC path includes any component through which the *signal* passes, and will include bypass capacitors, coupling capacitors, transformers, etc. The DC path includes components in which the DC bias currents or output currents flow. In some instances, the same component is in both paths, such as the primary of a coupling transformer in a transistor amplifier. The collector current of the transistor (DC path) and the output signal (AC path) all flow in this component. It is generally wise to check the DC path first, as most defects tend to reflect as incorrect DC voltages and currents in the stage. With the possible exception of open bypass capacitors, the first indication might be an error in a DC voltage level on an IC pin or transistor terminal.

Mixer/Oscillator Problems

When using signal injection at the mixer try the intermediate frequency first. This will tell you whether or not the mixer is amplifying. If an intermediate frequency signal is injected into the input of the mixer that normally receivers the rf signal, the mixer will act like a tuned amplifier to the i-f signal. But this does not tell us the whole story. The local oscillator could be inoperative or be operating on the incorrect frequency. In a later section we will deal with the possibility of troubleshooting a dead oscillator. But you can tell if there is reason to suspect the LO by injection an rf signal into the mixer and tuning the radio to that frequency. If the LO is operating properly then the radio will tune in the signal.

You can also tell whether or not the LO is running by substituting the signal generator output signal for the LO signal. Tune the signal generator to a frequency approximately equal to the radio frequency (to which the receiver is tuned) *plus* the intermediate frequency. If the stations are then heard in the output, and the radio can be "tuned" using the frequency control knob on the signal generator, you have a good case for suspecting the local oscillator.

Alignment

Alignment of the rf, LO and i-f stage tank circuits is not a troubleshooting tool and cannot be used as such. Even in the few cases where it is possible to glean information from alignment adjustments, it is futile. We can get the same information from other techniques that do not cause us to screw up the operation of the radio. We seldom, if ever, find a problem in which a receiver suddenly went awry due to an alignment problem. Even over the long run, the alignment of a quality communications receiver does not deteriorate sufficiently to warrant using alignment as a troubleshooting technique. Be aware that the minor improvement in older receivers gained by realignment is overshadowed by the fact that the adjustment screws tend to become loose. There is a general rule that tells us that the more times alignment is done, the more frequently it will be *needed* in the future. Leave general, overall alignment alone unless there is some very positive indication that it might be needed.

Bizarre Repairs And Offbeat Components

Avoid them like the plague! Every now and then I hear of somebody who wants to do (or did!) something like bending the plates of the main tuning capacitor to improve the tracking of the local oscillator. This is a fool's method that should not be done. It might be a better idea to find out why the receiver no longer tracks properly. It is a reasonably sure bet that it did work properly at one time. In one case involving a friend of mine, one of his other buddies wanted to bend up the capacitor plates to make the tracking better. I told him to bring the receiver to me before that pliers-happy nut ruined it. The problem turned out to be a small capacitor in one of the tank circuits.

The other gremlin to avoid is oddball, off-the-wall, or cheap components. Another friend of mine used a junk-box, paper-dielectric capacitor to replace a bypass capacitor in the i-f amplifier of a radio transceiver. The inductance of that capacitor made it as effective as a block of wood at the 9-MHz intermediate frequency. It might have been OK to use it to test the circuit, but when one makes the repair permanent, a proper component is a better choice. Some hobbyists and amateurs seem allergic to spending money for new components. I have seen people buy hamfest junkers of the worst order and then try to use these components in the repair or modification of a "kilobuck" sideband rig. They get only what they deserve. It is true that many components brought at auctions and hamfests are first grade. This may even be especially true of components bought from dealers. But in general it is wise to go to the wholesaler and buy a proper part.

Do not be swayed too much by the claims made of the parts sold in some retail hobby outlets. Some of them are blister-packed versions of top-grade stuff, but a lot of the material is factory seconds that didn't get past the inspection given by the manufacturer. This is especially true of unbranded components. At the same time, do not be totally afraid of using the so-called replacement grade semiconductors sold by Motorola (HEP), GE, Sylvania (ECG), RCA (SK) and others. It is true that many of these lines are totally inept. But the big companies have been in the game for almost three decades now and have a pretty good handle on what substitutes for what. Yes, there are some problems. If the specs of the replacement number called out in the crossover guide seem a little out of line with what you know the transistor must do (frequency, power, operating voltages, etc), use it with a grain of salt.

The transistors sold in this form are usually as good a quality as any other source. The only thing is that they are costly compared with quantity prices. I have seen a 2N3904, which sold for $22 per 100 in bulk, sell for $1.69 in a blister pack. But what you are paying for is the expensive distribution of the "onesy-twosy" replacement market and the cost of preparing that big crossover manual (which they give away free many times). I have used replacement semiconductors for almost two full decades, and I have been burned only a few times. The "hit rate" of good crossover data is now better than ever, so there is little to fear.

TROUBLESHOOTING DEAD OSCILLATORS

The oscillator circuits of a receiver, of which a superhet will contain at least one, are sometimes a little difficult to troubleshoot. For this reason they deserve treatment of their own.

We can sometimes use the conduction voltage drops in vacuum tube and transistor oscillators to tell if the circuit is oscillating. Consider a vacuum tube circuit such as shown in Fig. 27-4. The grid voltage will be slightly negative when the oscillator is running (in most circuits, at least). If the circuit is a crystal oscillator, remove the crystal from its socket, while measuring this voltage. The voltage should change suddenly when the rock is removed. Similarly the cathode resistor voltage drop will change suddenly when the rock is removed or disabled. If the oscillator is a VFO, on the other hand, look for a changing grid voltage as the VFO tank circuit is tuned from one end of the band to the other. It is seldom that this voltage drop is constant across the whole band, so the change is an indication of oscillation.

The voltmeter used to measure these potentials must have high impedance to not load the circuit. Also, use a low-capacitance probe, such as found on an oscilloscope, so the oscillator is not detuned so far that the act of making the measurement causes the oscillation to cease.

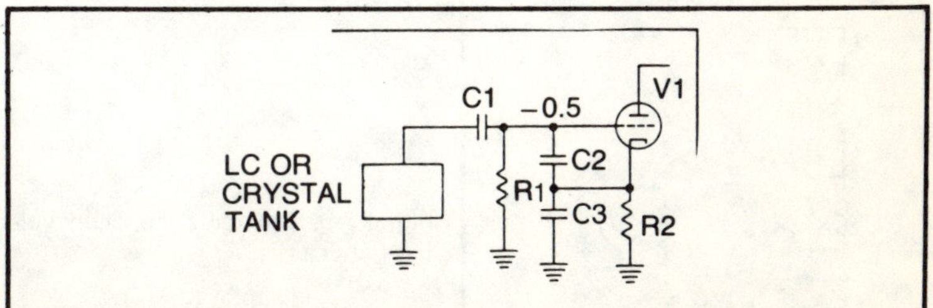

Fig. 27-4. Typical oscillator using tube will show a negative grid voltage when oscillating.

We can also use an rf demodulator probe for the DC voltmeter to tell if the oscillator is running. The output signal is an rf sine wave, so we should be able to measure it on an appropriately equipped voltmeter. The existence of rf output voltage is *de facto* evidence that the oscillator is running.

In a transistor oscillator circuit, the emitter conduction voltage drop (Fig. 27-5) tells if the oscillator is running. Again, this voltage should vary as a VFO is tuned from one end of the band to another, or change abruptly when a crystal is removed or disabled.

If the oscillator element is an IC, you might have to use one of the signal detection methods, of which the rf voltmeter probe is but one example. You could also use an oscilloscope to check the output, or another receiver. If you loosely couple a wire from the antenna terminals of the other receiver to the vicinity of the oscillator, and then tune the second receiver to the anticipated frequency, you should be able to hear the oscillator signal as an unmodulated signal on the second receiver. I have seen LO signals propagate *across* a room with little difficulty, so this technique works well.

The simple checks for oscillation tell us only that the oscillator is running. It does not tell us that it is running on the correct frequency. A superheterodyne radio will appear to be totally dead if the LO frequency is so far off that it is trying to tune in signals from outside of the normal rf passband of the radio. The oscilloscope method allows us to make rough checks of operating frequency, and the receiver method allows us to make even more precise checks. We might also want to use a frequency counter for this purpose. We must couple the input of the counter to the output of the LO (or LO port of the mixer) through a capacitor small enough in value so that it does not change the LO frequency significantly. In some cases, the radiation of the signal from the circuitry to a gimmick wire will suffice (Fig. 27-6). Sufficient signal can be developed to drive the

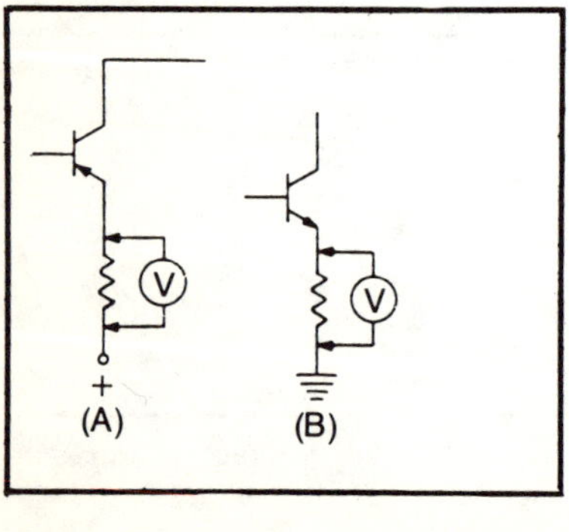

Fig. 27-5. Emitter conduction voltage checks.

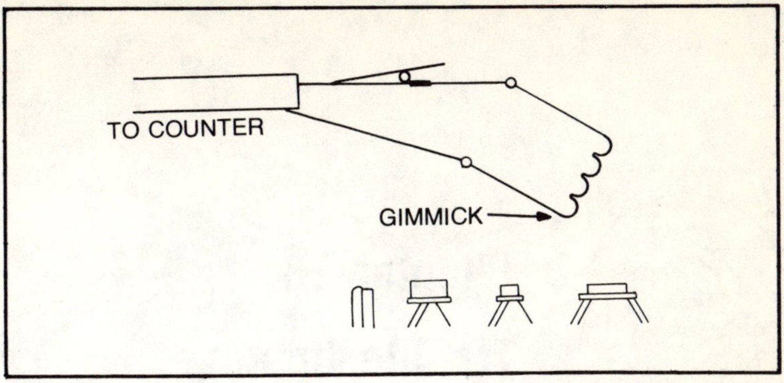

Fig. 27-6. Gimmick.

inputs of some counters. Remember that the proper frequency is the radio frequency to which the dial is tuned *plus* or *minus* the intermediate frequency. Do not expect down-to-the-hertz precision of the LO frequency in most receivers. The dials are not calibrated *that* accurately. But suspect the circuit when big changes are evident (tens of kilohertz, for example). Few oscillator defects that would leave the LO running, but at a frequency so far removed that the radio is dead, result in a small frequency shift.

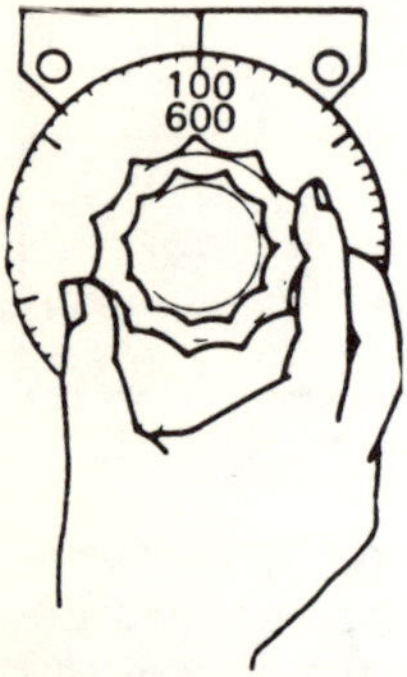

Chapter 28 Troubleshooting Transistor Circuits

All modern amateur radio and commercial communications equipment uses transistors or integrated circuits instead of vacuum tubes. While it may be argued that a tube circuit is easier to troubleshoot, the truth is that transistor circuits present little in the way of real difficulty. In some respects, IC circuits are even easier.

TRANSISTOR BASICS

We must understand a little of transistor operation before we can successfully deal with troubleshooting transistor circuits. I can recall—when transistors first came into the consumer electronics market—the reaction of some older technicians who couldn't understand how 0.2 volts could make much difference. But 0.2 volts is the forward bias junction potential of a germanium transistor. We have covered transistor theory to a little greater depth in the companion book to this volume, TAB book No. 1224, *The Complete Handbook of Radio Transmitters*, so only need a quick listing of some basic facts here. One of these concerns the junction voltages. In the germanium transistor, the junction potential will be 200 to 300 millivolts (0.2 to 0.3 volts) when the transistor is properly forward biased. In the silicon transistor, on the other hand, it will be 600 to 700 millivolts (0.6 to 0.7 volts).

The polarity of the voltages in PNP and NPN transistors will be opposite each other. The PNP device requires that the base be more negative, or less positive, than the emitter. Just the opposite is found in NPN devices: the base must be more positive, or less negative. than the emitter.

Table 28-1. Relationships to Expect from Normally Biased Transistors.

Junction	NPN	PNP
1. Base-to-emitter	+0.6 to +0.7 +0.2 to +0.3	−0.6 to −0.7 (Si) 0.2 to −0.3 (Ge)
2. Collector-to-base	+++	−−−
3. Collector-to-emitter	++++	−−−−

Note that we have used terms such as *more negative* or *less positive*. These are equivalent ways of saying the same thing. But in any given circuit, we might find that the power supply is negative-grounded, yet uses PNP devices. We want to make the base voltage negative with respect to the emitter voltage. But with a positive ground power supply the only way to do this is to make the positive voltage on the base less than the positive voltage on the emitter. As long as the junction sees what appears to be a negative potential, the device will work properly. Table 28-1 shows the approximate relationships to expect from normally biased transistors.

DC VOLTAGE CHECKS

Most of the problems encountered with a transistor circuit will upset the DC voltages on the stage. Only those faults that are strictly AC-path defects will maintain the DC voltages, while causing the stage to fail to operate. It is, therefore, extremely important to make DC voltage checks when troubleshooting transistor equipment. But we must guard against certain problems. In a vacuum tube circuit, we often find it reasonable to use a low-sensitivity VOM to make the measurements. But in a transistor circuit, the VOM frequently "falls down." Here we need to use a voltmeter with a high input impedance, such as a VTVM, FETVM, EVM, DVM, DMM, etc. Almost any of these instruments will work fine in transistor circuits, and not give us the erroneous readings that are sometimes obtained with a VOM.

TRANSISTOR TESTERS

Once we have decided that a particular transistor is a likely suspect, then it behooves us to test the transistor to see if it is actually defective. There are many different forms of transistor testers on the market—some good and some useless. Some of the really low cost "go-no-go" testers are about as effective as the emission-type tube tester. They make good screening testers to find the really gross errors. You can believe them when they tell you that a device is bad, but they have a high "false positive" result that could lead you astray when the transistor tests "good." The best type of transistor tester is one that allows you to vary the base current while measuring gain. If the collector current varies when the base current is varied, and there is no appreciable leakage, then it is a fair bet that the transistor is good.

The absolute best transistor tester is a curve tracer. This is an instrument which traces the family of operating curves for the transistor onto the screen of an oscilloscope. We can tell if the transistor is good or bad by examining the curves against what we know they should be.

An ohmmeter can be used as an impromptu transistor tester to find the kinds of gross faults that normally are found in receiver troubleshooting. This method can give you quite a bit of information about the transistor, although it is necessary to observe certain rules and precautions. One of the first and most important is to know the battery voltage used in the ohmmeter section of your meter. Many older types and a few newer imported types use high voltage batteries (7.5 to 22.5 volts) in the ohmmeter section. These can destroy the transistor being tested. Use an instrument that uses a 1.5-volt battery in the ohmmeter. Some FETVM and digital instruments will not allow us to test transistors. They have low-voltage ohmmeter reference supplies. The purpose of this type of supply is so that PN junctions are not forward biased. This makes it easier to make measurements of DC resistance in-circuit without removing semiconductor devices. But for the same reason—it will not forward bias PN junctions—it renders the instrument useless for testing transistors, except for clearly shorted devices. Some modern instruments use two ohmmeter power supplies. One is the low-power type that permits in-circuit DC measurements, while the other is a higher power type that permits the testing of PN junctions. A few models output a constant current in the "diode" or "high power" ohmmeter mode, so we can measure the various PN junctions by noting the number of millivolts on the meter. A proper silicon PN junction, for example, will read 600 to 700 if good.

Let us assume that we are using an ordinary ohmmeter to test PN junctions. A sample situation is shown in Fig. 28-1 where the collector-base junction of an NPN transistor is being measured. Recall that a PN junction will pass current when it is forward biased, and it will not pass current when it is reverse biased. Use the DC supply inside of the ohmmeter to bias the junction being tested. When the probes are connected such that the junction is forward biased, the resistance reading will be low. When the probes are connected such that the PN junction is reverse biased, however,

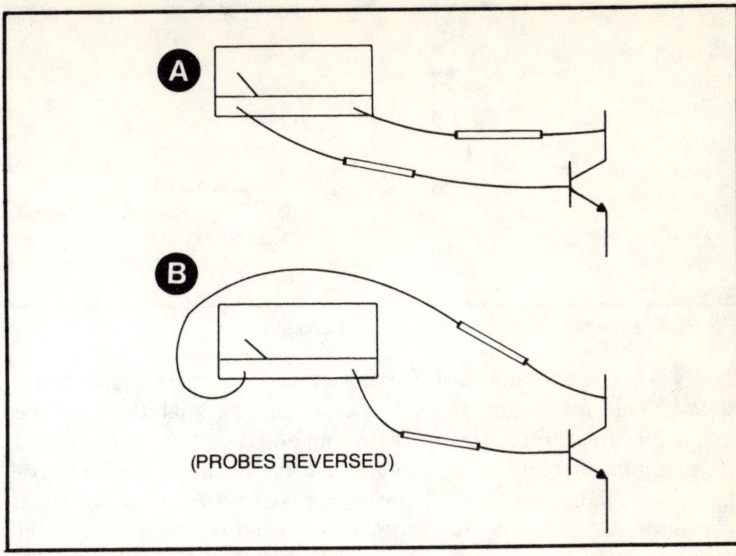

Fig. 28-1. Checking a transistor with an ohmmeter.

the resistance reading is high. A 10:1 ratio between high and low readings is ordinarily expected, but the higher the better. Test each junction of the transistor in the same manner and make a measurement of the junction resistance. Then reverse the probes and make another measurement. Compare the results, looking for a high forward/reverse ratio.

It is important that you know the scale of the ohmmeter being used to make this test. If you use too high a scale, the reading might not produce results (too many false negatives), and if it is too low you might burn out the device under test. As a general rule of thumb, use the X1 or X10 scale for power transistors, and either the X100 or X1000 scale for small-signal transistors.

This above is relatively crude, but it serves as a quick screening method. It does not, however, tell if the base is capable of controlling the collector current, which is what the current-amplifying transistor is all about. A little better test uses the method shown in Fig. 28-2. Here the ohmmeter gives a deflection proportional to the collector current—the collector resistance when the voltage of the ohmmeter is constant. When the base-emitter junction of the transistor is not biased, then the collector-to-emitter resistance of the transistor will be high. But when the b-e junction is forward biased, the collector-emitter resistance drops very low. Use the ohmmeter scales indicated above for this test. Connect the ohmmeter probes such that the polarity of the ohmmeter power

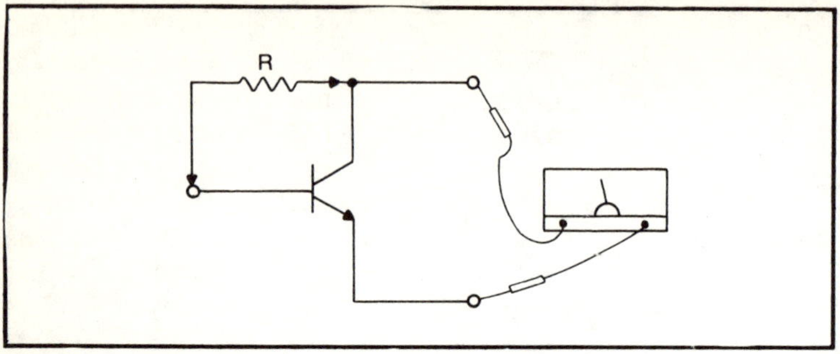

Fig. 28-2. A better method of checking a transistor.

supply is correct for normal operation of the type of transistor being tested. This means, in the NPN case shown, that the positive terminal of the ohmmeter must be connected to the collector and the negative terminal to the emitter. Exactly the opposite is required in PNP cases. Take a high-value resistor (over 10K for small transistors and under 1K for power transistors) and connect it between the base and collector, while watching the ohmmeter. When the connection is made, the resistance should drop substantially.

You can screen unknown transistors for PNP or NPN by knowing the ohmmeter polarity and using either of the two tests given above. You only need to apply the normal rules for transistor polarity of bias to determine which type of device is at hand.

WHICH STAGE IS BAD?

We can modify the signal tracing scheme of the last chapter in transistor circuits so that a DC voltmeter is used. This, in fact, may well be desired because injecting signals sometimes shocks the set back into operation temporarily. According to most of the corrolaries of Murphy's Law, a set so excited will remain operating until you tire of the game and try putting it back in the cabinet. Only after the last screw out of 34 is in place will the set malfunction.

Fortunately, the DC conduction voltage drops in transistor stages are often useful as an indicator of DC path problems. The DC conduction voltage is the voltage drop across the emitter resistor caused by the collector-emitter current. In most circuits, the emitter resistor is bypassed to AC, so the voltmeter probe will not upset any part of the circuit operation. Something similar was in Chapter 27 when we dealt with troubleshooting solid-state local oscillator circuits.

Figures 28-3 and 28-4 show how to isolate the defective stage in a cascade chain, using only a voltmeter. This method depends upon the fact that most problems show up in the form of modified DC potentials, so it will not work if only the AC signal is affected. But it is good enough for most applications. In the example of Fig. 28-3, the receiver uses a cascade chain of PNP stages. We connect one probe of the voltmeter to the DC power distribution line and then use the other to probe the emitters of the respective transistors. If the circuit is a negative ground system, as shown, then connect the positive terminal of the meter to the distribution bus and use the negative probe as your tester. If the set uses a different polarity, reverse this order. Start with the output stage and measure the voltage drop across the emitter resistor, and then work backward to the input stage. You will be able to tell from the schematic most of the time when the reading is correct or approximately so. Any wide deviation from the correct value, remembering the ±20 percent rule, should be cause for concern.

The situation in a negative-ground radio using NPN transistors is shown in Fig. 28-4. In this case, the negative terminal of the voltmeter is grounded, and the positive terminal becomes the troubleshooting probe. But the procedure is the same: measure the emitter voltage drop starting at the output stage, working back toward the input. The local oscillator emitter voltages will vary when the VFO is tuned or the crystal is removed from the circuit, as the case may be.

In the rf amplifier circuit, the emitter voltage drop is sometimes too small to be of use. We can, however, measure the fluctuation of the voltage as the radio is tuned across signals, taking

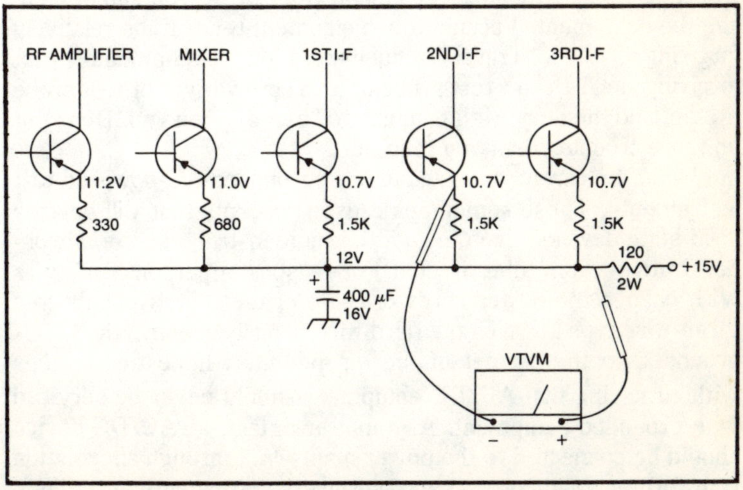

Fig. 28-3. Conduction checks in radio using PNP devices.

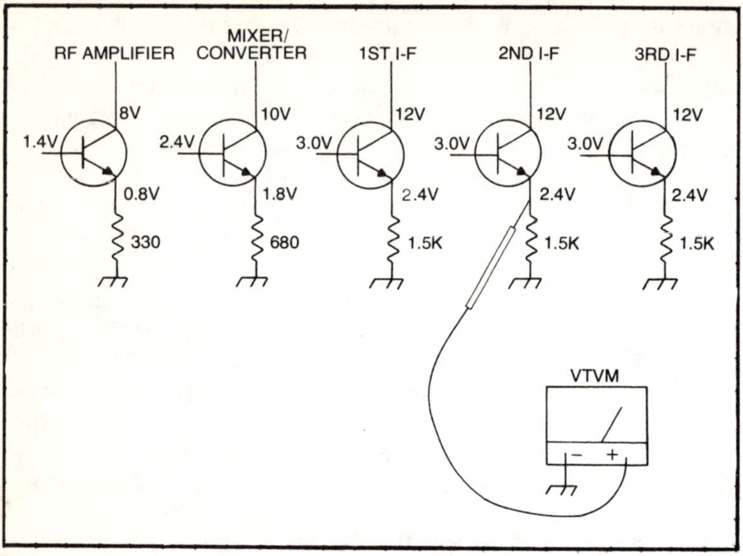

Fig. 28-4. Conduction checks in radio using NPN devices.

advantage of the action of the agc circuit. In the case of negative-ground radios using PNP transistors, incidentally, the collector conduction drop is usually greater in the rf amplifier stage and may be substituted for the emitter drop.

CAUTION

Always be on guard for matters of safety when dealing with troubleshooting situations. You might electrocute yourself or damage the equipment. Vacuum tube equipment tends to be relatively forgiving of "slips." Transistor equipment, on the other hand, is not forgiving at all! Drop a test probe or let a screwdriver or test probe tip slip, and the device it hits might go "bye-bye" on you. Don't put on more troubles than you take off.

Also, be careful of using old or ungrounded AC-powered test instruments. These sometimes cause transients that will destroy solid-state devices, even though the same instrument works wonders with vacuum tube circuits. For reasons of personal safety as well, change the power cords of some of these instruments to a three-wire type in which the third wire (usually green in the US) is grounded to the cabinet of the equipment. I hope that it goes without saying that AC/DC equipment should never be serviced with grounded equipment. As a matter of fact, an AC/DC device should be connected to the power mains *only* through an isolation transformer. Many professional servicers permanently install isolation transformers on their workbenches as a safety precaution.

Chapter 29

Receiver Alignment

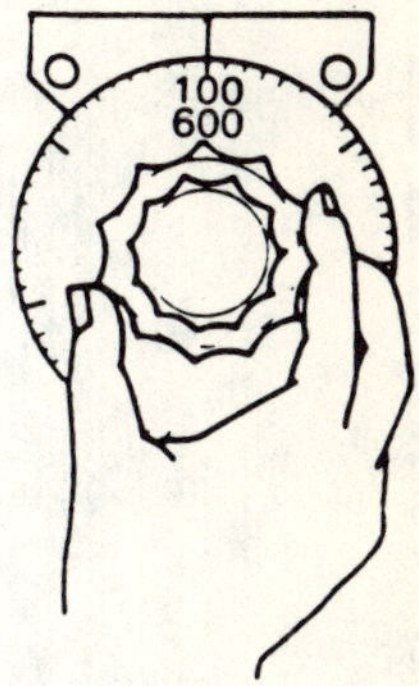

The alignment of radio receivers is one of those topics which seems to provoke fear from some amateur radio people and is dismissed as crudely simple by others. Of the two, the former creates less damage. Do not engage in alignment tasks lightly. Contrary to the advice one published by a respected amateur radio organization, receivers are not materially improved by periodic alignment. It was once recommended that amateurs have their station receivers aligned once each year. But after having performed literally hundreds of alignment jobs over an 18-year period, I find *rarely* any measurable improvement, much less any improvement that one can notice. There is a real hazard to frequency alignment. One danger is the possibility that it is done incorrectly. This happens all too often especially where inexperienced personnel are employed or when insufficient equipment is used. The second hazard is independent of the quality of the test equipment or the skill of the technician. It is the fact that the tuning adjustments become worn out rapidly. They are not designed to be adjusted frequently, so will become loose after only a few alignments are performed. The result of this is that the performance will deteriorate somewhat over the years, but in a manner much faster than if the receiver had been left alone. A general rule of thumb is that the more often alignment is done, the more often it will be *needed*.

There should be a conservative approach to deciding when to realign a radio receiver. One time that alignment is needed is when the receiver as a kit has been first constructed. Another time is

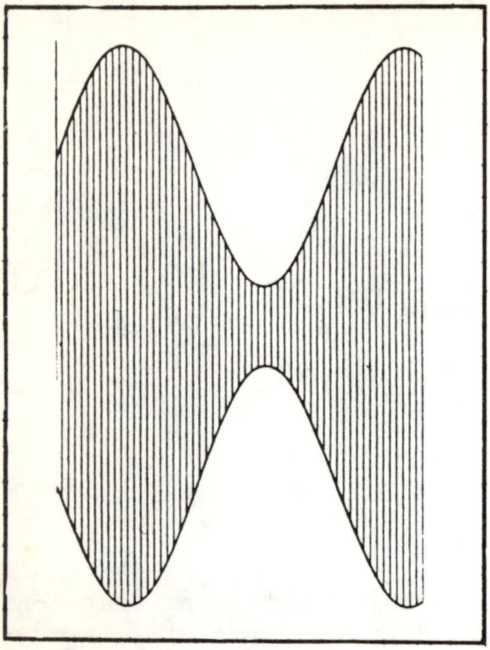

Fig. 29-1. AM waveform

when any major component changes, repairs or modifications are made to the rf, LO or i-f circuits. Of course, keep in mind that replacement of a bad i-f or rf tank circuit component is not an excuse to realign the whole radio. Realign *only* the component that was replaced. If you must replace or repair one of six different cascade i-f transformers, for example, don't both with a realignment of the remaining five. Align only the one that you replaced.

Another indication for an overall alignment is, after years of operation, the performance has deteriorated quite a bit. But before performing the alignment make sure that the loss of sensitivity is not due to some other problem such as weak tubes or bad transistors. You may have heard that transistors don't go bad, but this is not true. There might be sufficient gain in a multistate i-f amplifier to overcome the effects of a single dead stage, so it might not be noticed until you try to measure the sensitivity, or the individual stage gains using techniques specified by the original manufacturer. It is also possible that leaky transistors will cause a loss of sensitivity. This problem is especially true of older solid-state receivers in which germanium transistors are used. We might also find that the loss of sensitivity is due to a loss of power supply voltage. These problems must be addressed prior to making an attempt at the alignment. If the repairs indicated bring the reciever well within the manufacturer's specifications, then do not bother with the alignment; the improvement is not worth the hazards.

Never use alignment as a troubleshooting method. The kinds of problems that are caused by alignment difficulties are never *sudden*, and they take place over the course of *years*. If the receiver failed suddenly, the problem is not misalignment. If you use alignment as a troubleshooting technique, you will find yourself in a heap of trouble and get what you deserve. The "galloping diddle stick" (alignment tool) is the prime identifying characteristic of the novice troubleshooter. The only thing that diddling the alignment will tell you is if *that* tank is working, and that can be gotten from other measurements.

ALIGNING AM RADIOS

An amplitude-modulated signal (Fig. 29-1) contains an rf carrier whose amplitude is varied by the modulating signal. We can either a straight unmodulated signal, or a modulated signal, to align most AM receivers. Of course, different techniques would be used in the two cases. When a modulated signal is used, we rarely want a 100 percent modulated signal. We might also want to use a given audio frequency to modulate the carrier. It is common to select either 400 or 1000 Hz, modulating the carrier 30 percent, for AM alignment tasks. In a few applications, the manufacturer will specify 100 percent modulation.

Figure 29-2 shows the simplified instrument setup needed to make the most elementary alignment on an AM radio receiver. We need a signal source, which will be a signal generator of good quality. The signal generator is very important to the success of the alignment, and no attempt should be made to align any really good radio receiver with a junk generator. Unfortunately, most of the low-cost "service grade" signal generators are not suitable for use in the alignment of a really good radio receiver. The frequency

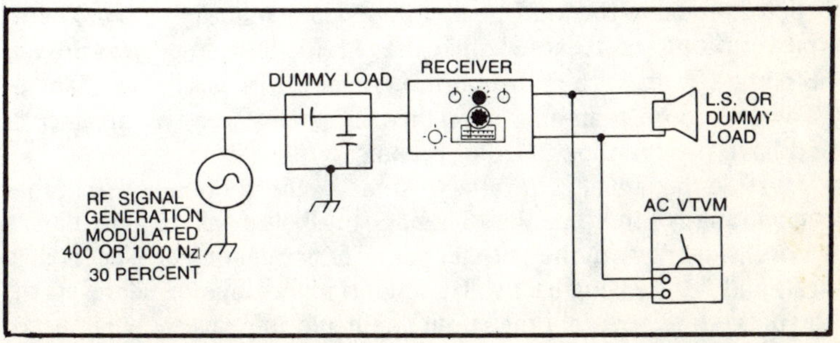

Fig. 29-2. AM alignment set-up.

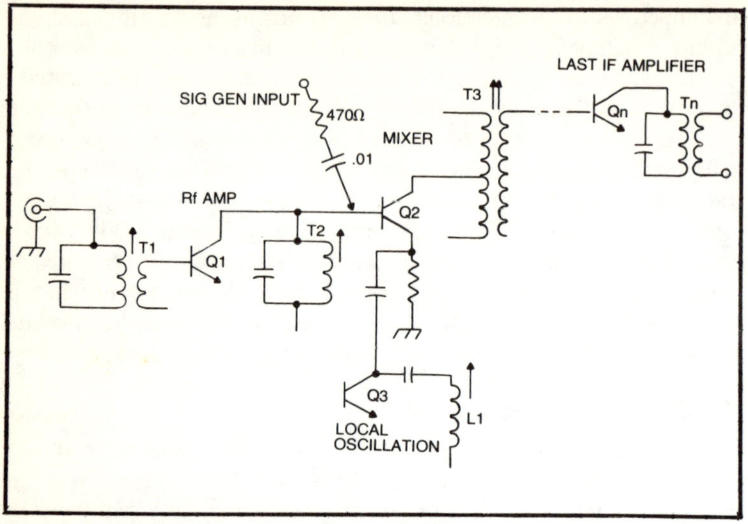

Fig. 29-3. Injecting the i-f signal into the mixer.

calibration is not sufficient, and there is so much leakage of rf around the cabinet flanges that the undesired signal will often be higher in amplitude than the signal passed to the receiver through the rf output cable. On the other hand, it is fortunate for amatuers that some good quality signal generators by Hewlett-Packard, Measurements, Inc., and Marconi are now on the surplus market. It is not the "in" thing in commercial laboratories to have these old clunkers but they still work fine and are a lot better than the service grade generators that are within the amateur's budget.

We also need an output indicator. Use an AC VTVM across the loudspeaker for this purpose. An oscilloscope will also work for this function.

You will only align one radio receiver with the loudspeaker connected before you figure out what a dummy load is for. Obtain a resistor, equal to the loudspeaker impedance (usually 8 ohms) with a power rating sufficient to handle the receiver output power. Seven 56-ohm, 2-watt, carbon composition resistors in parallel will handle 14 watts (more than any communications receiver will produce), and have a total impedance of 8 ohms.

Tune the signal generator to some exact frequency within the tuning range of the receiver. Then, peak all of the alignment adjustments, starting with the output i-f amplifier transformer (secondary first) and progressing backward to the rf input transformer. Repeat the procedure several times until no further improvement is possible.

It is possible that we will see some reaction from the agc circuit. The agc must either be disabled, or we must keep the output level of the signal generator low enough that the agc does not drive the rf amplifier into saturation. This means that we must keep turning down the output level of the signal generator as the alignment progresses. There are two ways to disable the receiver agc (other than turning the *on-off* switch to *off*; if the receiver has such a switch, then use it!). One is to ground the agc line or clamp it will an outside power supply. Do not try this unless you have the service manual and then follow the manufacturer's recommendations.

Note that there is a "dummy load" network between the signal source and the receiver input terminals. This dummy load might be a resistor network, capacitor network or an RC network.

We often find it advisable to apply the signal to the i-f amplifier first, and then tune these stages to a precise frequency before aligning the rf amplifier and the local oscillator. If the receiver lacks a test point intended for the injection of signals during alignment, apply the signal generator to the base of the mixer transistor. Use an RC network such as shown in Fig. 29-3 between the generator and the transistor base.

If it is impossible to find the correct point of if it is buried beneath some imponderable shielding or other circuitry, then you might want to rough the alignment by using a gimmick inside the top side of the first i-f transformer. The gimmick is a short sliver of insulated wire (see Fig. 29-4) dropped inside the transformer. Don't

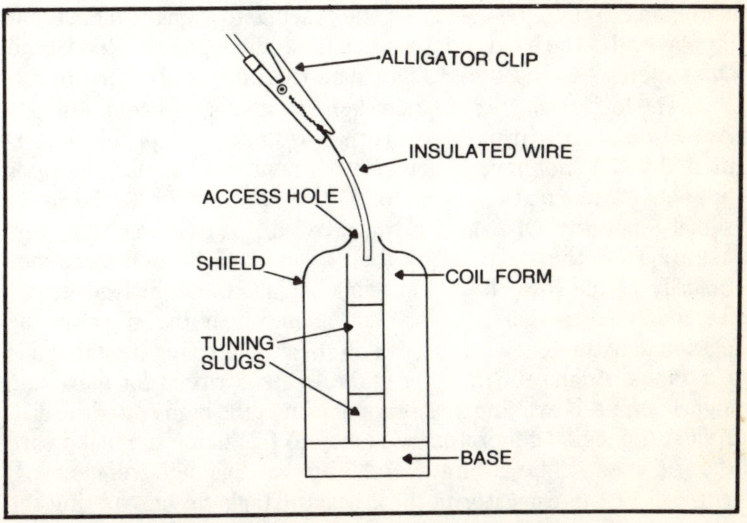

Fig. 29-4. Injecting the i-f signal gimmicked into the first i-f.

let the wire fall too far—only deep enough to spray a little signal into the windings. The top winding is usually the primary, so this will work nicely for both windings. We can tune both windings of the input transformer if the slugs have hexagonal adjustments holes and a bottom access hole. We can then slip the alignment tool up through the bottom.

When the signal generator is coupled to the i-f amplifier, we can then make the alignment more precise. Adjust the signal generator to the exact intermediate frequency (a crystal-controlled generator, or a marker oscillator, can be useful here, as can a digital frequency counter). If, as is true in most communications recievers, there is a band-pass filter in the i-f amplifier, we must vary the procedure a little bit. It is rare that the center of the filter bandpass is precisely on the nominal frequency marked on the label. The actual frequency will be close to the rated frequency, but not exactly on the mark. In this case, couple the signal generator as in Figs. 29-3 and 29-4 and then rock the signal generator frequency dial back and forth to find the maximum output level. Leave the frequency at this point, once found. Tune the i-f amplifier LC tank circuits up on the actual center frequency of the i-f band-pass filter.

Once the i-f alignment is completed, using the instructions given for Fig. 29-2, proceed to the rf amplifier and local oscillator. Peak the rf amplifier coils for maximum. There are usually at least two adjustments for each band: one at the input and the other at the output of the rf amplifier.

Sometimes, we must also align both the coil and the capacitor in the tank circuits. It is generally the practice to align the inductor at the low end of the band and the capacitor at the high end. Repeat the adjustments several time to eliminate the effects of interaction.

The local oscillator must be aligned with a signal source with a known frequency; it must be precise. Do not attempt this adjustment if you do not have either a crystal-controlled signal generator or some form of marker oscillator with which to calibrate the actual signal generator. Of course, if you have a digital frequency counter, use it. Tune the radio dial to the oscillator calibration frequency (usually at the low end of the dial). Adjust the signal generator frequency to this same point. You should hear the signal in the passband of the receiver, unless it is so far out of alignment that it is ridiculous. Adjust the inductor in the LC tank circuit for maximum signal output. Next, find a calibration point at the high end of the dial. Adjust the signal generator frequency to this point and make sure the receiver will be accurately calibrated at this high frequency. If not, adjust the capacitor in the oscillator tank circuit to bring the calibration into alignment. This adjustment and the previous one of the coil are highly interactive, so make them several times until no further improvement is possible.

We can modify these two procedures to use a DC VTVM as the output indicator. Connect a DC voltmeter to the agc line. It is important to make sure that the rf signal strength is high enough to make a small DC voltage on the agc line, but not so high that it will saturate when the alignment is made. If the receiver is far out of alignment, it will then be necessary to keep readjusting the signal generator output lower and lower as the alignment progresses.

In more complex receivers, especially double-conversion or triple-conversion single sideband models, be very care to not attempt an alignment job unless you have the manufacturer's service manual, *and* the test equipment specified by the manual. Having nerve is no substitute for proper equipment.

ALIGNING FM RADIOS

The best way to perform the alignment of the FM receiver is to use a sweep generator and an oscilloscope. A nonswept signal generator can also be used in the alignement of an FM receiver, especially communications receivers.

Sweep Alignment

A typical sweep set-up for an FM receiver is shown in Fig. 29-5. The sweep generator most nearly simulates the signal from the transmitter. The marker generator provides small pips which help determine the frequency at a particular point on the response curve traced out by the oscilloscope. The marker and the sweep

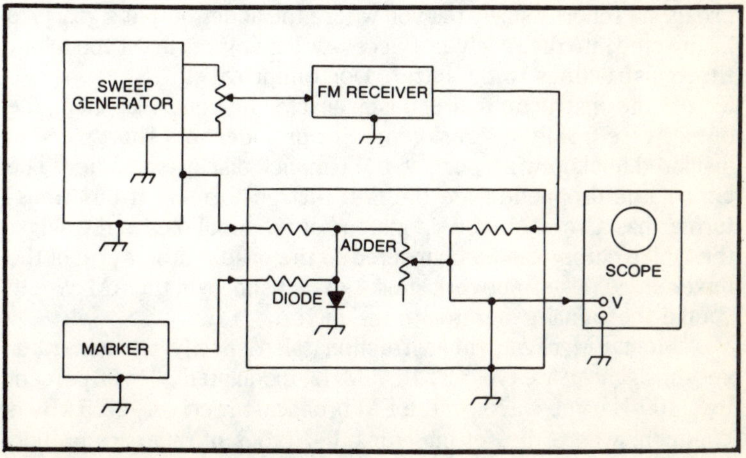

Fig. 29-5. Sweep alignment equipment setup.

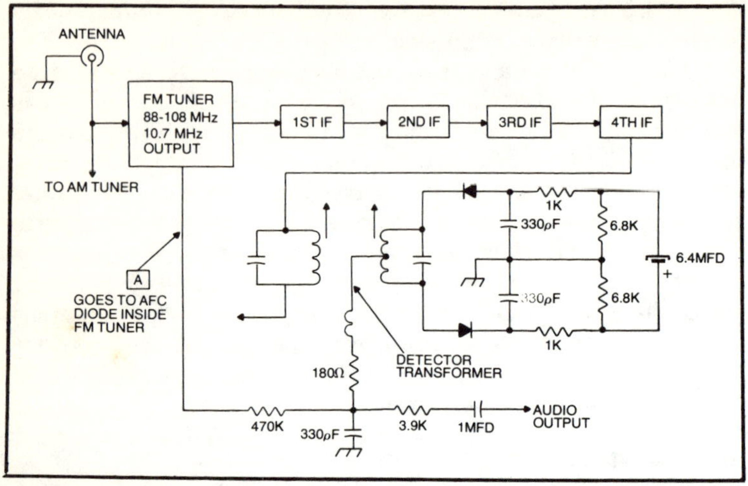

Fig. 29-6. FM alignment points.

generator outputs are fed into an adder circuit. This device combines all of the required signals into one composite that can be fed into the FM receiver. Direct connection, without the adder, will cause interaction between the sweep and marker generators which will distort the response curve. Many test equipment manufacturers now offer sweep, marker and adder functions inside one cabinet. Many of these instruments also include bias supplies (for agc) and other functions needed by television servicers in the alignment of color TV receivers.

The alignment instructions for most FM receivers and/or broadcast tuners usually tell you where the adder output signal is to be injected into the receiver. Success will be more sure if you follow these instructions to the letter. Don't improvise!

If the instructions are not available, however, connect the output wire from the signal generator or adder to a short piece of insulated hookup wire, such as the gimmick discussed earlier. The gimmick is dropped inside the first i-f transformer. If this transformer has a screwdriver slit instead of the hex holes in the slugs, the adder output can be connected to the base, gate or grid of the mixer stage. Use a network similar to that shown in the AM case to couple the signal generator to the mixer.

Most FM alignment instructions also specify the generator settings. These are typically 10.7-MHz, modulated 30 to 50 percent by a 400-Hz sine wave. In the FM broadcast receiver, 22.5 kHz is common, where the settings for other types of radio are proportionally less.

A simplified block diagram of an FM receiver is shown in Fig. 29-6. Connect a high-impedance, zero-center voltmeter to point A. Note that most ordinary service-grade VTVMs or FETVMs can be adjusted to zero center with a hearty twist of the *zero* knob. With a 10.7-MHz signal (or whatever the intermediate frequency is in the particular receiver) applied, adjust the secondary of the detector transformer to produce zero volts at point A. The meter will display a positive voltage on one side of the correct point and a negative voltage on the other side.

Next, connect the meter across the loudspeaker or dummy load of the receiver. Peak all other i-f tuned circuits to produce a response curve similar to that in Fig. 29-7.

Front end alignment requires that an appropriate signal be applied to the antenna terminals of the receiver. If the receiver is channelized, obtain an FM signal source on the channel frequency. Use a signal generator and digital frequency counter, or a crystalized signal generator. Apply the on-channel rf signal to the antenna terminals and then adjust the local oscillator frequency adjustment trimmer until the voltage at point A returns to zero. It will be off zero at this point, unless the receiver is perfectly netted to the correct frequency.

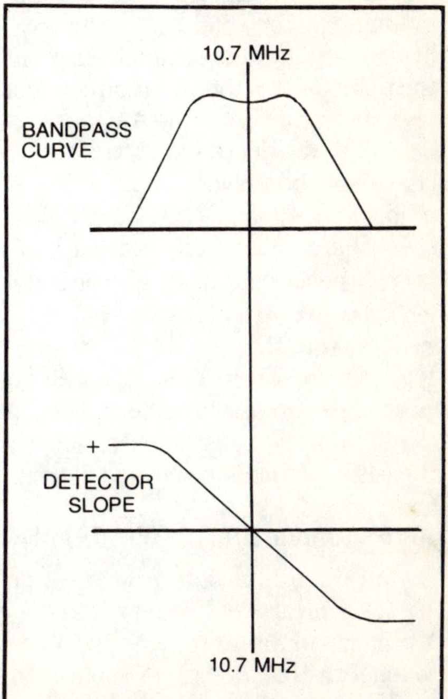

Fig. 29-7. I-f and discriminator response curves.

Next peak the rf amplifier adjustments while inspecting the bandpass response on the oscilloscope. You want a double-humped response curve, as shown in Fig. 29-7. Note that the ideal curve is flat, but this is not obtainable in actual practice (at least not often). Adjust the rf and i-f alignment adjustments for a symmetrical curve, with a center frequency dip not exceeding ten percent of the overall amplitude. In addition, the two peaks should have amplitude within ± 10 percent of each other. Ideally, they have the same amplitude. Where a trade-off is necessary, it is usually preferable to sacrifice a little bit of overall gain in favor of a symmetrical response curve.

Nonswept Alignment

Nonswept alignment requires an unmodulated signal generator. This job can be performed with a homebrew crystal oscillator, provided that the rf leakage is low, and that the frequency is known precisely. The same type of zero center voltmeter is needed to align the detector, but a level indicator is also required. If the receiver has a limiter stage at the output end of the i-f amplifier chain, a resistor in this stage often produces a DC voltage proportional to the input signal amplitude. Similarly, you can sometimes use the agc voltage. Both of these are contingent upon being able to adjust the input signal level so that the receiver is not into saturation or limiting. If any of these are not available, then use an rf voltmeter at the output of the i-f amplifier chain. If a ratio detector is used as the demodulator then you can connect the DC voltmeter used as the level indicator across the AM suppression capacitor (see Fig. 29-8).

Zero the detector first and then peak the i-f transformers. The signal applied should be an accurate intermediate frequency, unmodulated signal. Next, adjust the local oscillator with a known on-channel signal. Sometimes a subharmonic oscillator is useful for this purpose, especially if its output is rich in higher harmonics. To align 146.91 MHz, for example, a 12.2425 MHz crystal might work nicely. Note that the local oscillator trimmer capacitor is usually a low-value ceramic type that will rotate the full 360 degrees to produce a potentially confusing double peak. So watch what you are doing. Finally, align the rf tank circuits using one of the level indicator methods discussed above.

ICQD (QUADRATURE DETECTOR) ALIGNMENT

If the manufacturer does not provide an alignment procedure to the contrary and you don't know exactly where they intend for the indicator meter to be placed, it is still possible to align the ICQD. Connect a 10.7-MHz signal source to the input of the i-f amplifier by using one of the methods specified earlier. Peak the i-f amplifier

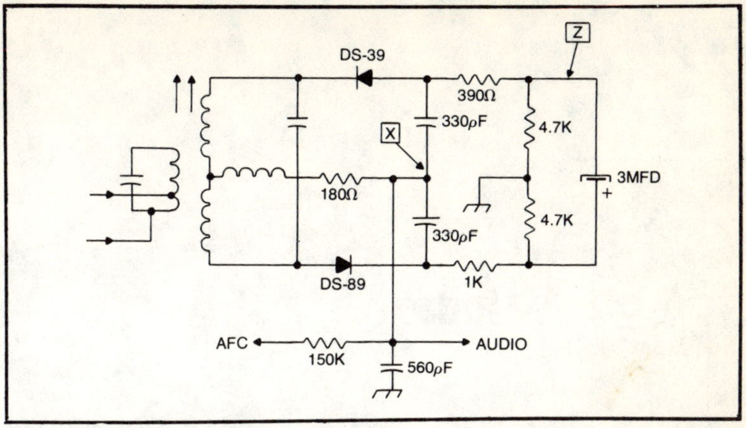

Fig. 29-8. Ratio detector alignment points.

coils as normal, or use an rf voltmeter, or a detector probe on a DC voltmeter, to monitor the level of the i-f signal. When adjusting the phasing coil, however, a little skill is required. Place an AC voltmeter across the speaker output and inject the 10.7-MHz signal. Reduce the rf output of the signal generator until the background noise just begins to overcome the quieting of the receiver. Adjust the phase coil through its entire range. You should note two noise peaks. The correct alignment point is the null approximately midway between the two noise peaks.

If you have a sweep generator available, then apply a modulated signal and measure the total harmonic distortion (THD) of the audio output. The phasing coil is correctly aligned when the THD is minimum.

Index

THE COMPLETE HANDBOOK OF
RADIO TRANSMITTERS

Other TAB books by the author:

No. 787 *OP AMP Circuit Design & Applications*
No. 901 *CET License Handbook—2nd Edition*
No. 930 *Servicing Medical & Bioelectronic Equipment*
No. 1012 *How to Design and Build Electronic Instrumentation*
No. 1070 *Digital Interfacing with an Analog World*
No. 1152 *Antenna Data Reference Manual—including dimension tables*
No. 1182 *The Complete Handbook of Radio Receivers*
No. 1194 *How to Troubleshoot & Repair Amateur Radio Equipment*
No. 1250 *Digital Electronics Troubleshooting*
No. 1271 *Microcomputer Interfacing Handbook A/D and D/A*

THE COMPLETE HANDBOOK OF RADIO TRANSMITTERS

BY JOSEPH J. CARR

TAB BOOKS Inc.
BLUE RIDGE SUMMIT, PA. 17214

FIRST EDITION

FIRST PRINTING—JUNE 1980

Printed in the United States of America

Library of Congress Cataloging in Publication Data

Carr, Joseph J
The complete handbook of radio transmitters.

Includes index.
1. Radio—Transmitters and transmission—Amateurs' manuals. I. Title.
TK9956.C352 621.3841'31 80-14458
ISBN 0-8306-1224-6 (pbk.)

Contents

Introduction

During the dawn hours of the radio era, transmitting stations smelled heavily of battery acid and ozone. Ozone, that smelly oxygen product of the pyrotechnics is created by high-energy electrical sparks. Radio transmitters of that time were spark gaps. In fact, many early amateur radio operators went on the air for the first time using the spark coil salvaged from the family Model-T Ford. They transmitted by keying the low-voltage primary of the coil, and coupling the secondary to the antenna system. A little later, high-power radio signals were sent on their way from an Alexanderson alternator, a mechanical AC generator of high frequency.

It was not until after World War I that vacuum tube transmitters came into their own, and even then they did not completely displace the spark transmitter until 1939 when sparks were outlawed by the FCC. Those early tube transmitters were puny affairs by today's standards. By the outbreak of World War II, however, multikilowatt transmitters were on the air, and the vacuum tube generation of rf energy was firmly established.

For the first two decades of the transistor era, the solid-state power amplifier could not keep pace with the vacuum tube at radio frequencies. Only a few watts could be generated from transistors. But today, a whole new look is upon low-power to medium-power transmitters. It is almost unheard of to see a tube-type transmitter mobile radios operating from CB through UHF. All of these low-power transmitters are now solid state. Similarly, transistors have made serious inroads into the medium power (100 watt to 300 watt) high-frequency transmitter market. Although a few rigs are solid state up to the final amplifiers, where tubes are used, most of them now use transistors. Solid-state HF linear amplifiers with 100 to 300 watts of power are common, and 1000-watt amplifiers are not unknown.

In this book we are going to consider the basic circuitry and design practices used in radio transmitters in the high-frequency through low UHF region. Although we would like to cover from DC to daylight, space considerations limit us to the most popular ranges. The main thrust of this book will be practical, with just enough theory to lead you into a better understanding of the device called transmitter.

Joseph J. Carr

Chapter 1

Radio Transmitters—An Overview

What is a radio transmitter? A radio transmitter is an instrument that will produce an rf signal with sufficient power to be transmitted to a distant location via an antenna system. Radio transmitters abound in dozens of formats. A radio transmitting station might consist of a multikilowatt transmitter, such as the Collins 40,000-watt FM broadcast model shown in Fig. 1-1. This transmitter requires seven large racks to hold the circuitry and consumes huge amounts of electrical power. A large multikilowatt rig can actually heat a small building. One of the standard practices of radio station owners in this energy-conscious era is to divert the transmitter heat into the forced air ducts of the station buildings to supplement—in some cases supplant—the regular heating equipment.

SHORTWAVE TRANSMITTERS

A radio station might also be a multimegawatt installation such as the Voice of America transmitting location in Greenville, NC (Fig. 1-2). This station produces radio frequency power that is measured in the *millions* of watts. The antenna farm for this installation is, to say the least, extensive.

TWO-WAY RADIOS

Alternatively, a radio station might be a small hand-held portable (Fig. 1-3) producing 1 to 5 watts, or a mobile installation (Fig. 1-4), also producing only a few watts. These land mobile, amateur

Fig. 1-1. 40-kW Collins FM broadcasting transmitter (courtesy of Collins/Rockwell).

Fig. 1-2. Megawatt antenna farm (courtesy of The Voice of America).

radio and marine devices are usually frequency modulated (FM) and operate in the VHF and UHF bands. Regardless of the power level and physical configuration, however, they are still radio transmitters. In these cases, though, they are *transceivers*—combined *trans*mitters and re*ceivers*. The transceiver is a self-contained two-way radio station, which sometimes even includes the antenna.

Fig. 1-3. Portable hand-held tranceiver (courtesy of Yaesu).

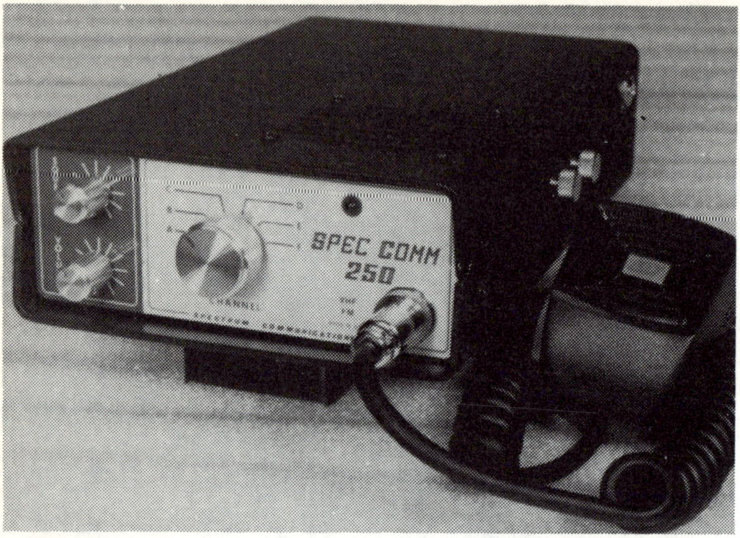

Fig. 1-4. Mobile unit (courtesy of Spectrum Communications).

It is difficult to place any one type of transmitter in the category of a typical amateur radio transmitter. Any transmitter class, save only the multikilowatt broadcast models, are used by radio amateurs in one activity or another. Amateurs use radiotelegraph (CW), several forms of radiotelephone (AM, SSB, and FM) radioteletype (RTTY), and even television. But if one were to

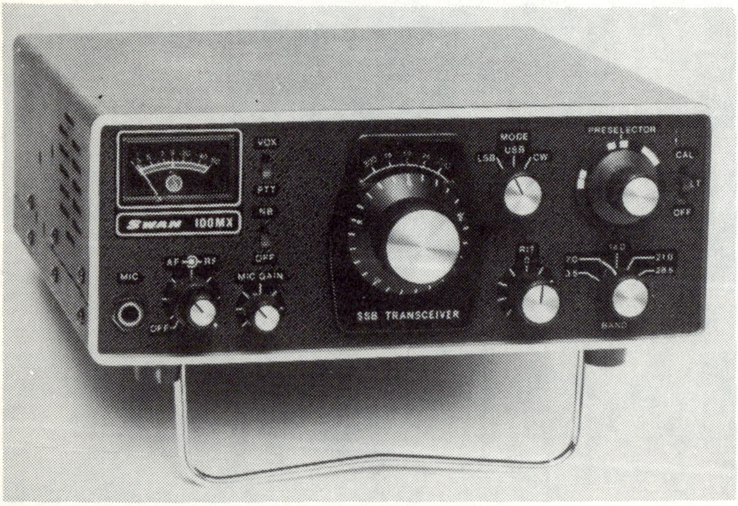

Fig. 1-5. Solid-state HF SSB rig (courtesy of Swan).

Fig. 1-6. Solid-state HF rig with digital readout (courtesy of Drake).

venture out and proclaim any one transmitter as the *standard* or *typical* type, it would have to be the single-sideband 100-watt to 300-watt transceiver. Figures 1-5 and 1-6 show modern SSB transceivers that use analog and digital readout, respectively. At one time, amateur transmitters were relatively simple devices using low technology. But the modern SSB transceiver will use such high technology circuits as digital readouts, phase-locked loop frequency control and solid-state circuits (including the final amplifiers).

The simplest amateur radio transmitters have only two stages: an oscillator and a final power amplifier. The oscillator generates the radio frequency, while the power amplifier boosts it to the desired power level. The output of the final amplifier is passed through an impedance matching circuit to the antenna, where it was radiated in the form of an electromagnetic radiation. This type of simple transmitter is called a *master oscillator power amplifier* (MOPA) transmitter. If the carrier, as the rf output signal is called, is turned on and off to form a code the transmitter is a radiotelegraph transmitter. But if some form of modulation, such as speech, is applied to the carrier, then a radiotelephone (or one of the other classes) is realized.

In this book we will consider radio transmitter theory and practice. Both practical circuits and radio transmitter theory are given. We hope that a fair balance is realized. The principal thrust will be toward the radio amateur, but radio broadcast engineers and two-way radio service technicians should find the book equally useful.

Chapter 2

Transistor Fundamentals

Solid-state devices made their commercial appearance in the form of diodes during World War II, spurred on by radar and electronic research in general. Natural semiconductor diodes were known prior to World War I, though. The *Galena* crystal is a mineral that will act like a rectifier diode; in fact, many were used as crystal set detectors. Only certain points on the Galena crystal were able to detect radio signals. This meant that a "cat's whisker" electrode was needed to find the correct point, which was a tedious matter to say the least.

Transistors grew out of the semiconductor and metallurgy research of World War II. The bipolar transistor, the most common type, was postulated in the late 30s, but was not feasible until the late 40s when a trio of physicists at Bell Laboratories successfully built the first transistor.

Today, transistor devices of one description or another have all but eclipsed the vacuum tube. In fact, only a few products designed many years ago are still produced using low-power vacuum tubes. Only in high-power devices, such as broadcast radio transmitters, are vacuum tubes still used extensively.

SEMICONDUCTOR BASICS

The study of solid-state electronics must necessarily begin with a description of the popular model of the atom. We will not dwell on the physics of semiconductor devices, but leave that to other textbooks (TAB book No. 717, *Transistor Theory for Technicians & Engineers* by Andy Veronis.) We will, however, have to consider some of the basics.

The atom is believed to have a nucleus, made of positively charged particles called protons, negatively charged particles called neutrons, and a cloud of negatively charged particles surrounding the nucleus called electrons. The electrons orbit the nucleus in *shells* at fixed distances from the nucleus. The physical principle of electroneutrality requires that the number of negatively charged particles equal the number of positively charged particles, the basic charge levels being equal. This means that the electrically positive charges will exactly cancel the negative charges and produce a neutral atom.

The electrons in the outer shell control most elementary chemical reactions and the flow of electrical current. These electrons may be tightly bound or loosely bound. There is a critical number of electrons in the outer shell that makes it very tightly bound. In the case of a one-shell atom, for example, two electrons will make the outer shell stable. Only two elements have just one shell: hydrogen (H) and helium (He). The hydrogen electron has one electron in the shell, so it is very unstable. Touch a match to some hydrogen gas, and you will find out just how unstable it is! Helium, on the other hand has two electrons in the single shell. This is a stable configuration, and helium is a stable gas. This is why balloons have helium instead of hydrogen. In more complicated atoms, the magic number of electrons for a stable electron shell is eight. A filled shell—a stable shell—is not easily persuaded to give up electrons. In fact, it will tend to lose or accept electrons to maintain stability. Examine the carbon atom, for example. The carbon atom contains six electrons. Two are in the inner shell, completely filling it and therefore making it stable. The remaining four electrons are in the second shell. Since the second shell is completely filled with tight electrons therefore, this element is unstable. It conducts electrical currents and readily participates in chemical reactions. Those electrons available for electrical current flow and chemical reactions are known as *valence electrons*.

There are three general classes of electrical ability among materials: *insulators, conductors,* and *semiconductors*. An insulator is a stable element which has a filled outer shell. It takes a large potential, or a high voltage, to electrically strip off these electrons, so little current will flow through an insulator. There is only a small leakage current passing in insulators. A conductor, on the other hand, has loosely bound electrons in the outer shell of each atom. These electrons are easily given up to become free electrons, available for an electrical current flow. Semiconductors, on the other hand, are usually neither good conductors nor good insulators. Examples of semiconductor materials used to make diodes and transistors are germanium (Ge) and silicon (Si). Germanium has an atomic number of 32. It has two electrons in the innermost shell, eight in the second shell, 18 in the third shell, and four electrons in the fourth shell. Silicon has an atomic number of 14, so will have two and eight electrons in the first and second shells, respectively, and four electrons in the outer shell. Because both of these elements have four valence electrons, they are *tetravalent*.

The semiconductor elements such as Si and Ge form covalent bonds with nearby atoms in an attempt to fake a stable octet (eight-electron outer shell). Figure 2-1 shows this arrangement. The electrons from the various atoms form electron pairs with electrons of nearby atoms. One electron comes from each atom. Because these materials are tetravalent, each will form a bond with four adjacent electrons; each bond formed of two electrons, one

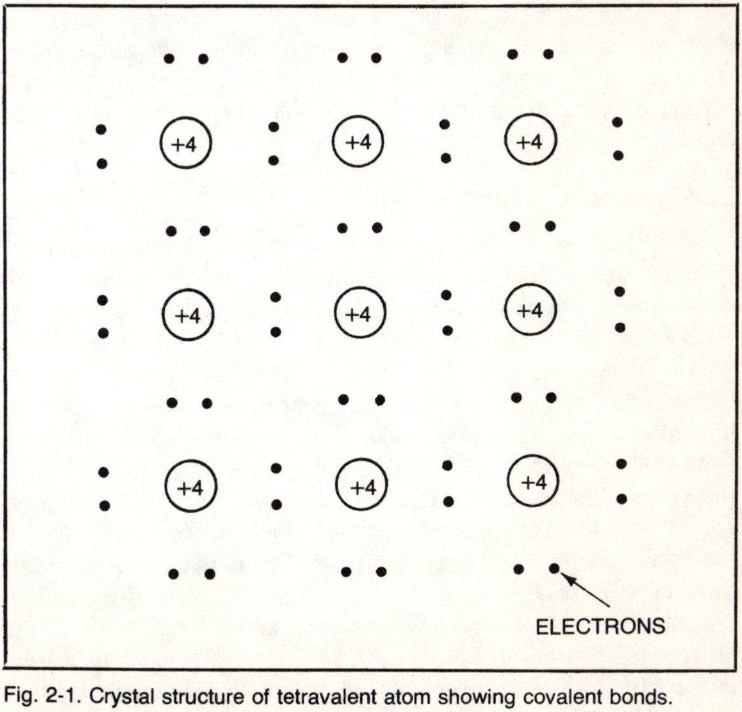

Fig. 2-1. Crystal structure of tetravalent atom showing covalent bonds.

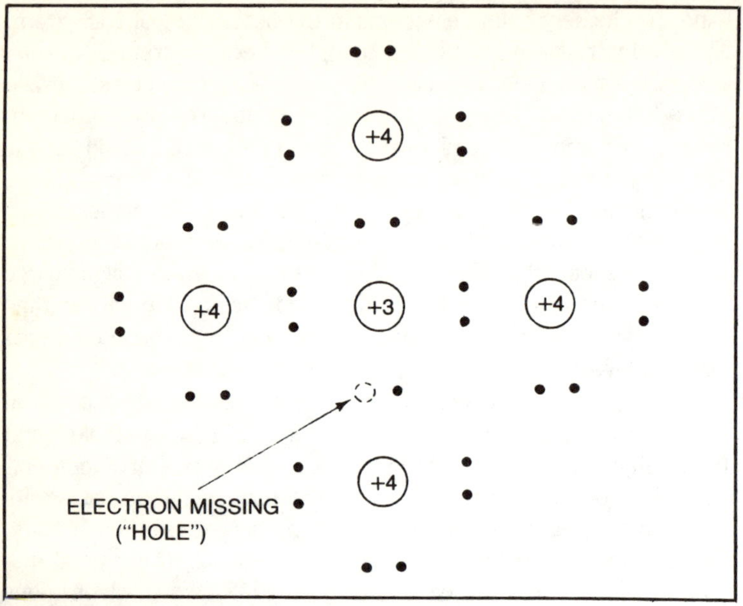

Fig. 2-2. Creation of a hole in P-type material.

from each atom. Our example shows a simplified two dimensional picture, but in the actual atom there will be a three-dimensional crystalline structure in which literally billions of atoms form such bonds.

CONDUCTION IN SEMICONDUCTORS

The pure semiconductor is a pretty poor conductor; it is, after all, a *semiconductor*. But we can make a semiconductor into a good conductor of either positive or negative current carriers by introducing some impurity element into the crystal structure of the semiconductor material. Consider the structure in Fig. 2-2. The tetravalent semiconductor material has been with a doped trivalent impurity (aluminum, boron, indium, etc). This creates a P-type semiconductor. The trivalent material forms covalent bonds with the tetravalent material. Because there are points where there should be but isn't a bonded electron, a *hole* is created.

This concept of a hole often gives newcomers to radio a difficult time. It is really quite simply a place where an electron should be, but isn't. One possible cause of the confusion is the fact that many textbooks tell the student that the holes flow. This isn't true; it only *appears* to be true. In reality, it is electrons that do the flowing. A hole may be filled with an electron, but this is at the expense of creating a hole in another atom nearby. The hole can be treated as if it were a positively charged particle, so it appears that the positive hole flows in the opposite direction of the electron flow. This is shown in Fig. 2-3. The original hole existed on atom B. This atom attracted an electron form atom A to fill the hole. This in turn created a hole at atom A. A little while later, an electron from atom F is attracted to fill the hole of atom A, and so forth. It appears that the hole drifted from A to B to F to D to C. Actually, there were a series of *electron* movements: A to B, F to A, E to F, D to E, and C to D. This appears to be a flow of positively charged particles and is considered a positive current flow.

We create N-type semiconductors by adding pentavalent impurities to the totravalent atoms (Fig. 2-4). This results in covalent bonds in which some atoms will have one extra electron. The fifth electron of the pentabalent atom has no electron from a tetravalent atom to pair with, resulting in a loosely bound free electron. This will cause an electron flow when an electrical field is placed across the crystal.

Common practice is to speak of *charge carriers* regarding semiconductor materials. The negative charge carrier is the electron, while the positive charge carrier is the hole. There is no such thing as a perfectly doped semiconductor; i.e. all P-type or all N-type. In actual devices, some negative carriers are in the P-type material, and some holes are in the N-type material. A material designated N type, however, has four more free electrons than

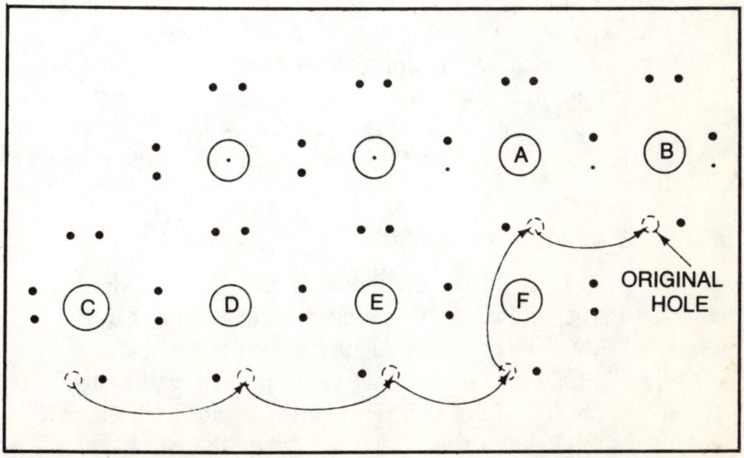

Fig. 2-3. Hole migration.

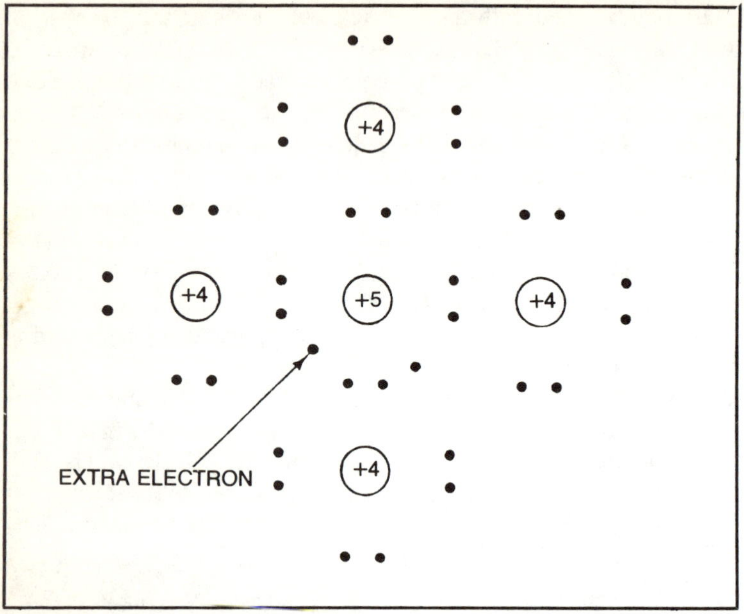

Fig. 2-4. Creation of N-type material.

holes. A material designated P type has far more holes than free electrons.

Keep in mind that the holes are really just places. We can assign a polarity to them by noting that the combination of a tetravalent material (4) with a trivalent material does not create an electrically neutral pair. There will be one more proton in the bonded pair than electrons, so the net effect is a single positve charge.

Keep in mind a convention followed in this and most other texts: Electrons flow from negative-to-positive, while holes flow from positive-to-negative. Since the hole flow is just the opposite of the electron flow, we are further justified in calling them *positive* charges.

DIODES—THE BASIC PN JUNTION

The most basic solid-state device is the simple diode. It consists of a PN junction, which is a junction of P-type material and N-type material. The original structure was a physical mating of the two types of material, while in all modern devices the mating is caused by diffusion of the different materials into each other.

Figure 2-5A shows the PN junction when there is no electrical potential applied across the junction. The distribution of holes and

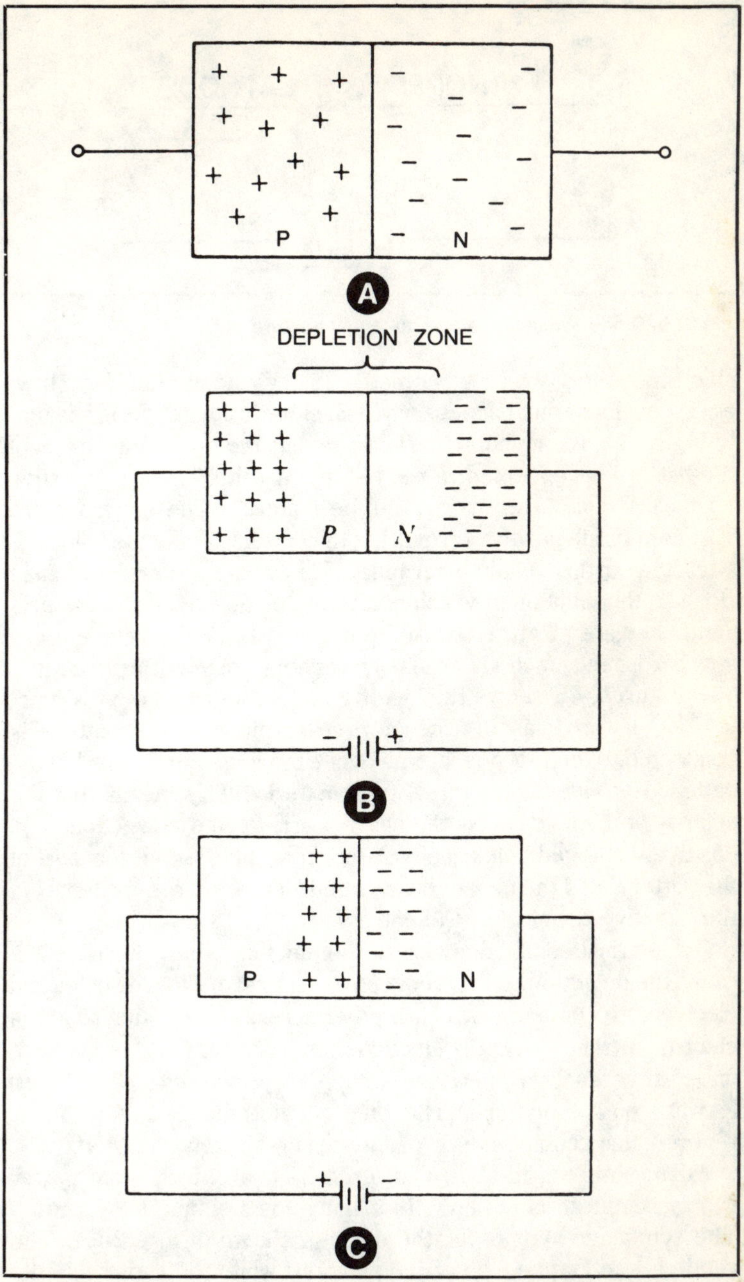

Fig. 2-5. PN junction unbiased at A, reverse biased junction at B, and forward biased junction at C.

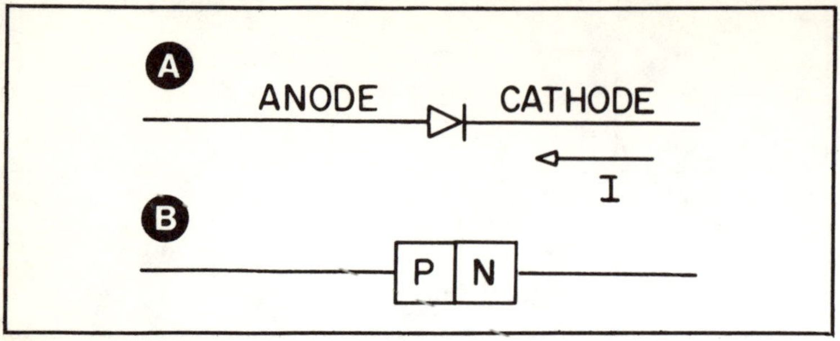

Fig. 2-6. Diode symbol at A and material structure at B.

electrons in the two types of material is essentially random. However, consider what happens when an electrical potential is applied. In Fig. 2-5A, the junction is *reverse biased*; the positive terminal of the battery is connected to the N-type material, and the negative terminal is connected to the P-type material. The electrons are attracted to the positive terminal of the battery, while the holes are attracted to the negative terminal. This creates a *depletion zone* close to the junction in which neither holes nor electrons are present; they are all attracted to the battery. Hence, no current can flow across the junction because no carriers are *near* the junction.

Figure 2-5C shows the electrical situation when the PN junction is forward biased. The positive terminal of the battery is connected to the P-type material, while the negative terminal of the battery is connected to the N-type material. This means that the polarity of the field repels the charge carriers in each section. The positively charged holes are repelled from the positive terminal of the battery, and the negatively charged electrons are repelled from the negative terminal of the battery.

Electrical conduction across the junction is now possible because the depletion zone is reduced to near zero, allowing holes and electrons to combine with each other across the junction to attain electroneutrality. The PN junction then has a very high resistance under reverse bias conditions and a very low resistance under forward bias conditions. The only current that will flow under reverse bias conditions is a minute reverse *leakage current*.

The simple solid-state diode used in signal detection and power supply rectification (among other things) is a simple PN junction. The schematic symbol for the diode is shown in Fig. 2-6A. The anode is the P-type side of the junction, while the cathode is the N-type side of the junction (see Fig. 2-6B). Electron flow is *against* the arrow while *hole flow* is in the same direction as the arrow.

Like its vacuum tube counterpart, the PN junction diode will conduct current, or is forward-biased, when the anode is positive with respect to the cathode. It will not conduct, or, is reverse-biased, when the anode is negative with respect to the cathode.

The current-vs-voltage characteristic of the PN junction diode is shown in Fig. 2-7. In the forward bias direction, the current will increase as the voltage increases. In the reverse direction, only a small leakage current flows, and the current is almost zero in quality device.

The ideal diode characteristic shown in this figure is not realizable. Real diodes will not begin to conduct current exactly as the voltage begins to increase. The current in the region below some certain potential E_b is a nonlinear function of the applied voltage. But at potentials higher than E_b, the curve becomes linear—a straight line—and the diode approximates the ideal.

Figure 2-8 shows resistor models of the diode. In the forward bias region, the diode appears to be a small value resistor, R, in series with a small, variable voltage. The maximum value of this voltage is 0.2 to 0.3 volts in Germanium diodes, and 0.6 to 0.7 volts in silicon diodes. The reverse bias model consists of a single very high value resistor that permits only a very tiny leakage current to pass. An ideal diode would have an open switch as the reverse bias model, but real diodes contain some reverse leakage current, and hence the resistance.

The current rating of a diode tells us the maximum current that can be safely passed in the forward direction. There are two current

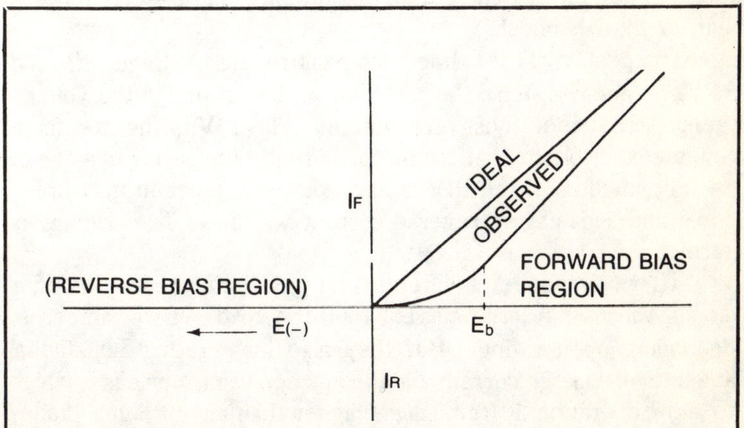

Fig. 2-7. I-vs.-E characteristic of a PN junction diode.

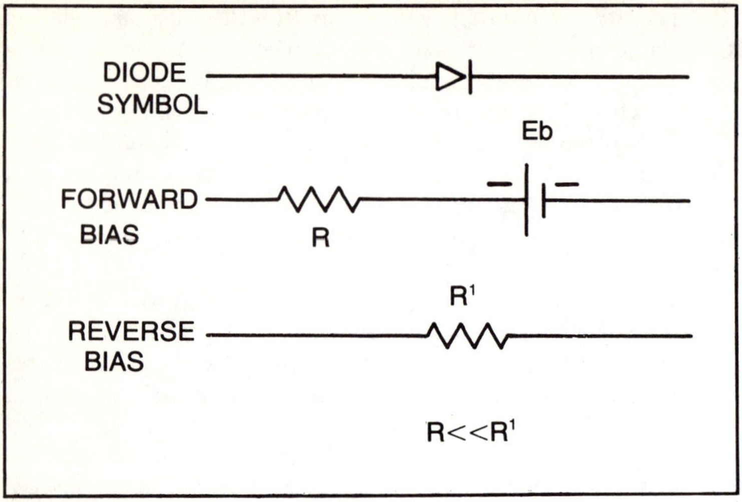

Fig. 2-8. Diode equivalent circuit models.

ratings seen: average and peak (or surge). The average current is the rms value that can be safely handled by the diode. It is the most often encountered rating. A 1-ampere diode, for example, will pass a DC or rms AC current of 1 ampere without damage. The surge current is the maximum peak current that can be passed as a nonrepetitive transient. In most cases, this current is specified to exist over no more than one input cycle, or 1/60 second (16.6 ms). The surge current can be quite high. Some 1 ampere diodes will withstand a surge of 25 amperes, but this bears no relation to its ability to pass such a current on a continuing basis—only for less than 17 milliseconds!

The peak inverse voltage and peak reverse voltage (PIV and PRV, respectively) are the same thing. This rating is the voltage rating of the diode most often quoted. The PIV is the maximum reverse bias voltage that can be applied without causing damage to the PN junction. If the PIV is exceeded, the junction may break down and a damaging reverse current will flow. The damage is permanent.

There are several different catagories of diode. Rectifier diodes will pass a large current (500 mA to dozens of amperes, depending upon rating). But they also have rather substantial amounts of leakage current and the junction capacitance is too high to allow operation at frequencies higher than audio. Signal diodes usually have lower PIV ratings than rectifiers and always pass less current. The maximum forward current of the small-signal diode is

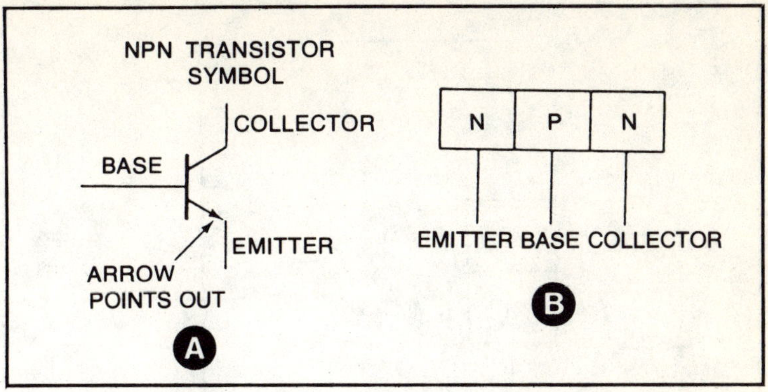

Fig. 2-9. NPN transistor with schematic symbol at A and structure at B.

usually a few milliameres. Some signal diodes are good only into the VHF or low-UHF region, while others will operate into the microwave region.

BIPOLAR TRANSISTORS

The bipolar transistor was the first practical transistor manufactured. This device is likened to two PN junctions connected back to back. This is actually too much simplified, and you cannot obtain transistor action by physically connecting real diodes. But transistors are made with two PN junctions, as shown in Fig. 2-8. Two types are recognized NPN (Fig. 2-9) and PNP (Fig. 2-10). The symbols for these two different transistors are also shown. Note that the difference is that the arrow representing the emitter points in different directions for the two types of transistor. A little mem-

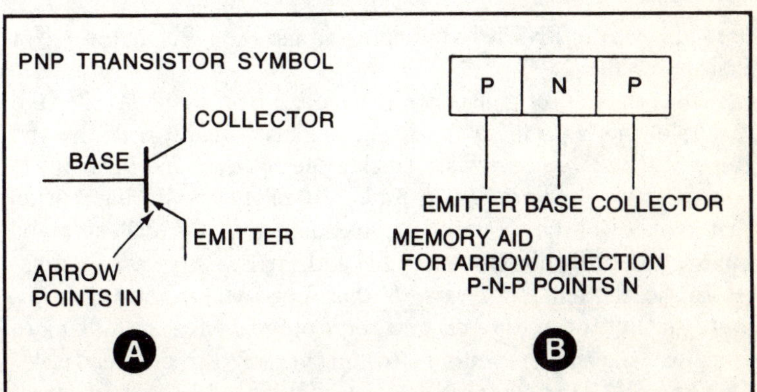

Fig. 2-10. PNP transistor with schematic symbol at A and structure at B.

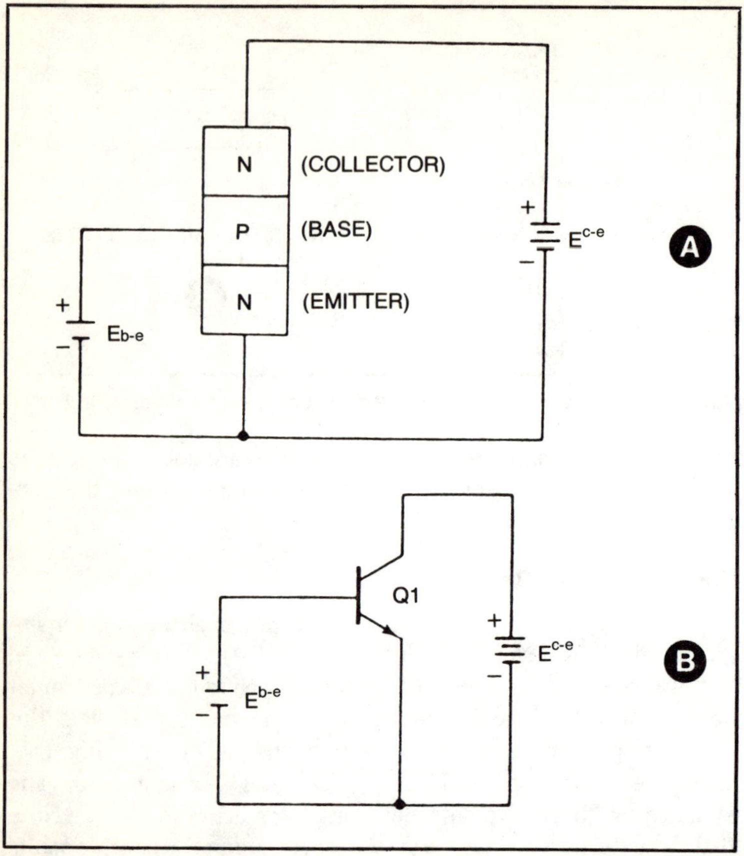

Fig. 2-11. Forward-biased NPN transistor with structure at A and schematic at B.

ory aid is that in PNP arrow Points iN, and the NPN arrow is Not Pointing iN.

In both types of transistor, however, a central region called the *base* is of the opposite type of semiconductor material as the end regions. The two regions are labeled the *collector* and the *emitter*. The diagrams shown in Figs. 2-9 and 2-10 might give the impression that the device is symmetrical, meaning that the collector and emitter terminals could be interchanged. In the early days of transistors, the elementary devices of that time were somewhat symmetrical. But today, the real geometry of the device is a little more complex, and this prevents us from reversing the end terminals.

Figure 2-11 shows an NPN transistor biased for normal operation. The battery polarities, incidentally, will be *reversed* in the base of the PNP device. The protocols that are followed for transistor biasing are shown in Table 2-1. Note that the collector-base junction is reverse biased and the base-emitter junction is forward biased. Although this description is a bit oversimplified, the current carriers from the emitter are attracted into the base region and will flow across the base region into the collector. The base region is very thin relative to the collector and emitter regions. A better picture is gained from the book referenced earlier, and you are referred to it if you desire a deeper understanding.

In most cases, when the transistor is biased for normal operation, the base current will be 0.5 to 6.0 percent of the total emitter current. The remainder of the current will pass through the base region to the collector. This action is the basic mechanism of amplification: A large current is controlled by a low current. The transistor is said to be a *current amplifier*.

But let's fix in our minds that the transistor is *not* just a current amplifier. It will amplify voltages as well; Ohm's law has not been repealed. If we apply a voltage to the base of the transistor, a current will flow. The transistor voltage amplifier is shown in Fig. 2-12. The signal voltage (a sine wave source) is connected in series with the bias voltage V_b. The collector is not connected directly to the collector-emitter supply V_{cc}. It is connected with a series resistor to the V_{cc} (+) supply terminal. The bias is adjusted such that collector voltage V_c is approximately $\frac{1}{2}V_{cc}$. The current flowing in the collector (I_c) also flows in the collector load resistor R_c. This current creates a voltage drop equal to V_{cc}-V_c. When the signal input voltage is zero, the output voltage (E_o is the collector voltage V_c) is $\frac{1}{2}V_{cc}$.

When the signal increases in the positive direction, more current flows in the base of the transistor. This is because base current I_b is proportional to $V_b + V_s$. This increased bias current will in turn increase the collector current, I_c. Increasing the collector current will increase the voltage drop across R_c, causing V_c to drop

Table 2-1. Transistor Biasing Protocols.

	NPN	PNP
V_{CE}	COLLECTOR POSITIVE WITH RESPECT TO EMITTER	COLLECTOR NEGATIVE WITH RESPECT TO EMITTER
V_{BE}	BASE MORE POSITIVE THAN THE EMITTER	BASE MORE NEGATIVE THAN THE EMITTER

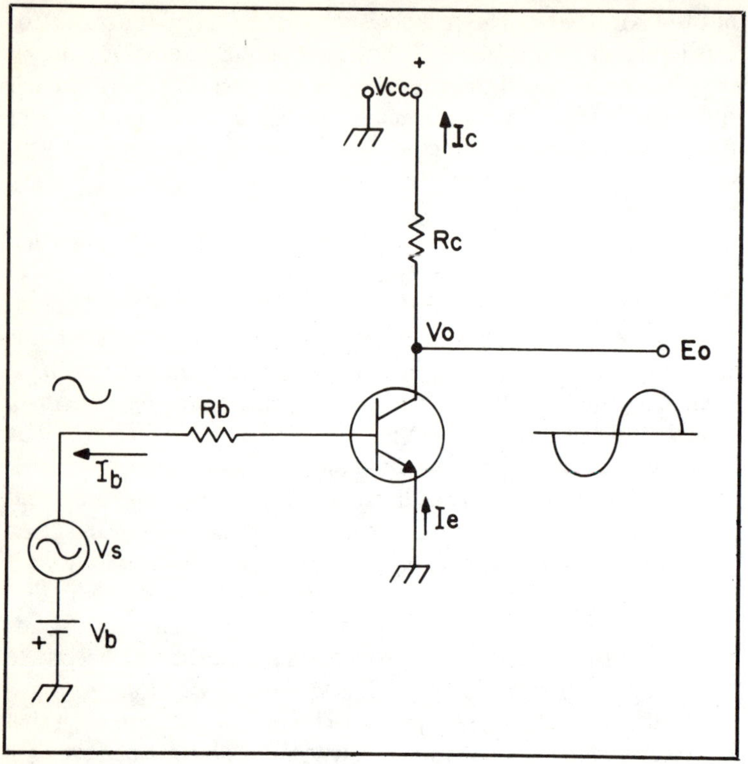

Fig. 2-12. NPN common emitter circuit.

(see Fig. 2-13). The voltage at the collector of the transistor can drop as low as zero without distorting the output sine wave. Similarly, when the signal voltage increases in the negative direction, the total base bias voltage is less ($V_b - V_s$). This will reduce the collector current, thereby reducing the voltage drop across R_c. When this occurs, the output voltage at the collector will increase. The output, then, is a voltage under the control of the input signal voltage. In other words, this is a true voltage amplifier.

TRANSISTOR GAIN

There are two basic methods for specifying the gain of a transistor, and these are designated *alpha* and *beta*. The alpha specification is the ratio of the collector and emitter currents:

$$\alpha = \frac{I_c}{I_e}$$

where α is the gain, I_c is the collector current, I_e is the emitter current (both emitter and collector currents specified in the same units). Typical values of alpha gain are always less than unity, with most falling in the range of 0.65 to 0.99.

The beta gain is perhaps most commonly employed. This gain is defined as the ratio of the collector and base currents:

$$\beta = \frac{I_c}{I_b}$$

where β is the beta gain of the transistor, I_c is the collector current, and I_b is the base current in the same units as I_c.

The alpha and beta gains are related through the expressions:

$$\alpha = \frac{\beta}{1 + \beta} \qquad \beta = \frac{\alpha}{1 - \alpha}$$

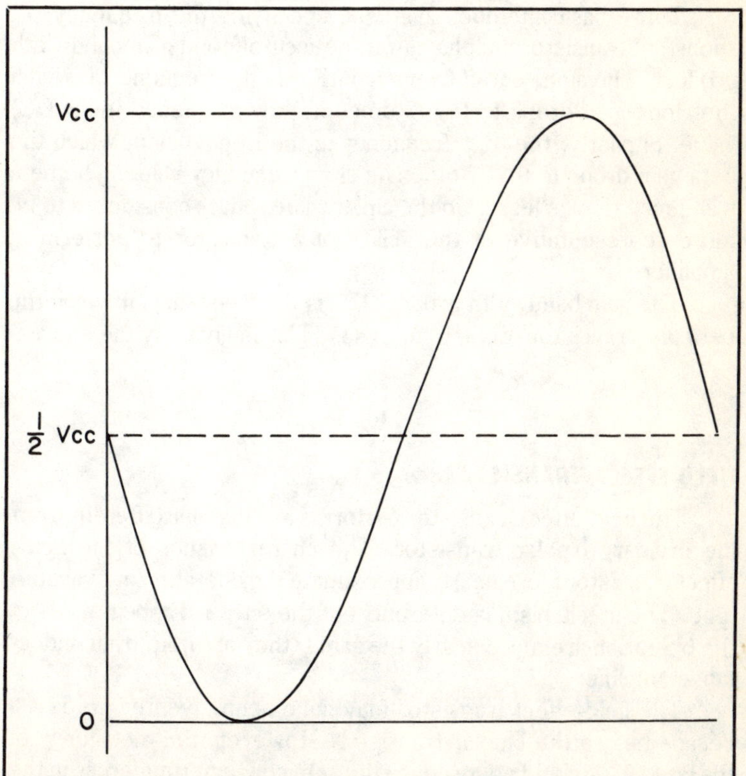

Fig. 2-13. Collector voltage swing.

The beta is sometimes represented by the letters H_{FE}, which refers to the DC (static) current gain. Similarly, the DC alpha is sometimes referred to as H_{FB}. The AC gains might be a little different, however, and are expressed by the equations:

$$h_{fb} = \Delta I_c / \Delta I_e$$

and

$$h_{fe} = \Delta I_c / \Delta I_b$$

Δ (Greek letter *delta*) means a small change in the associated parameter. In other words, ΔI_c means a small change in the collector current. This is interpreted in the last equation to mean a "small change in collector current caused by a small change in base current I_b."

FREQUENCY RESPONSE

Three basic methods are used specifying the frequency response of transistors: alpha cutoff, beta cutoff, and gain bandwidth product. The alpha cutoff frequency, F_{ab}, is the frequency at which the alpha gain drops to 0.707 times its low frequency (near DC) value. Similarly, the beta frequency is the frequency at which the beta gain drops to 0.707 times its slow frequency value. The beta frequency (F_{ae}) is less than the alpha value, but is considered to be more representative of the ability of a transistor to perform in amplifiers.

The gain bandwidth product, F_t, is the frequency at which the beta of a transistor drops to unity (1). This is given by the expression:

$$F_t = F_{ae} \times h_{fe}$$

FIELD-EFFECT TRANSISTORS

The field-effect transistor performs in a manner different from the ordinary bipolar transistor. The characteristics of the field-effect transistor more nearly approximate the operation of a vacuum tube. The mechanism is different, but the outward appearances of the operation are more nearly the same (they are both tranconductance amplifiers).

The field-effect transistor may have been invented at least a decade before the bipolar transistor. However, the metallurgy of the late 1930s could not produce the semiconductor materials in the purities needed for proper operation of the field-effect device.

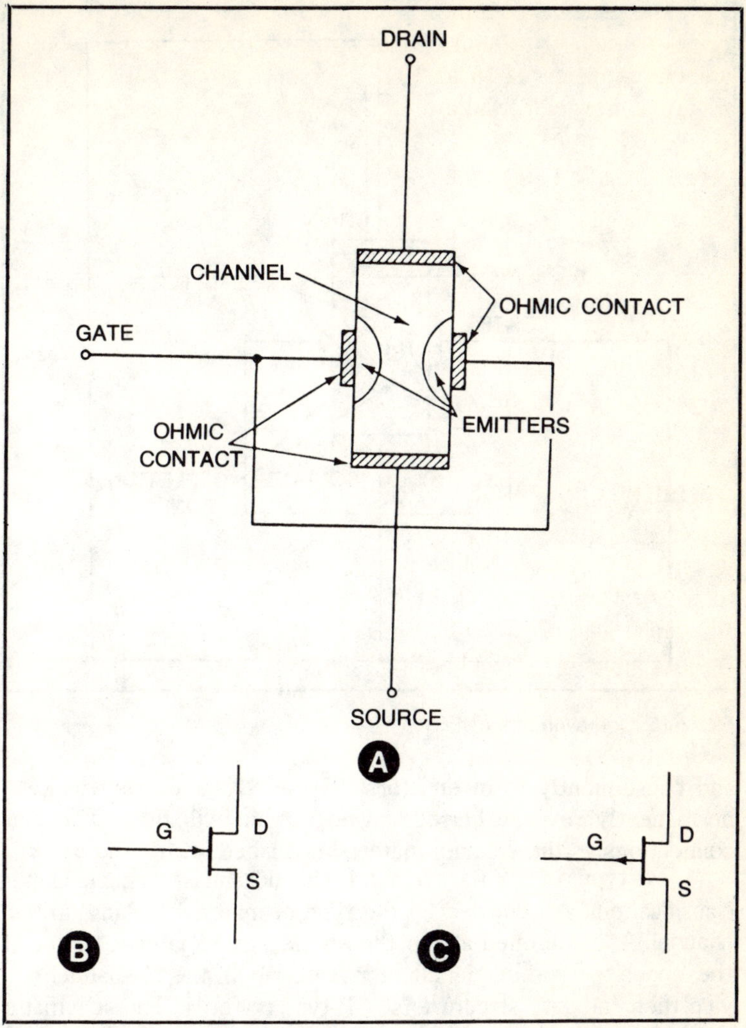

Fig. 2-14. JFET model at A, N-channel type at B, and P-channel at C.

There are two basic types of field-effect transistor: junction field-effect transistors (JFET) and metal oxide semiconductor field-effect transistors (MOSFET). The MOSFET is also sometimes called the insulated gate field-effect transistor, or IGFET.

A model of the junction device (JFET) is shown in Fig. 2-14A. It consists of two semiconductor regions, the channel and the gate. The gate regions are made of the opposite semiconductor material from the channel. These structures are diffused into the channel

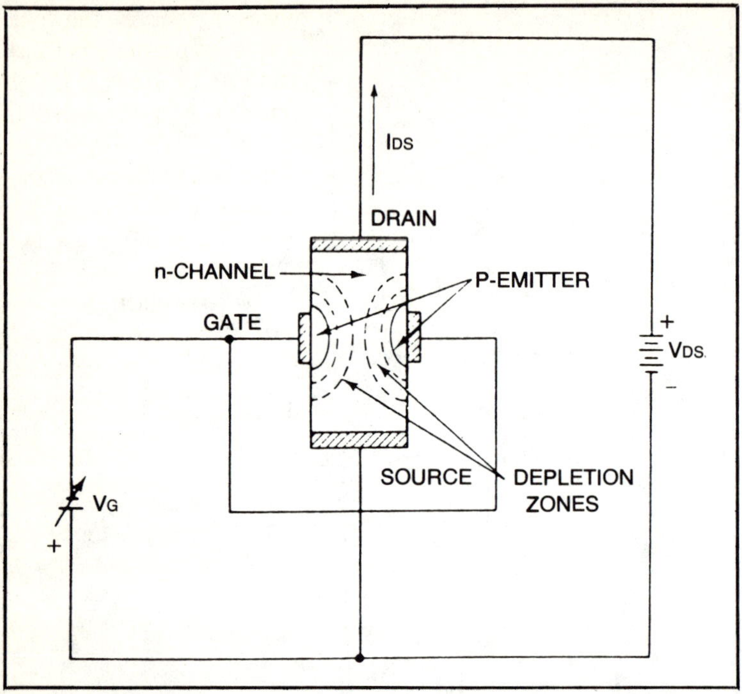

Fig. 2-15. Operation of the JFET.

and consequently form emitters. These PN junctions are kept permanently reverse biased, so no current will flow. The end connections to the channel material are labed *source* and *drain*.

Two types of JFET are found: P-channel and N-channel. These classifications are decided by the type of material making up the channel. As mentioned above, the emitter/gate material will be of the opposite type semiconductor material. In the N-channel device, then, the gate structure is of P-type material. The schematic symbols for the N-type and P-type JFETs are shown in Fig. 2-14B and 2-14C, respectively.

The operation of the JFET (in this case an N-channel device) is shown in Fig. 2-15. When the emitter-channel junction is reverse biased, no current will flow across the junction. The normal operation of the JFET requires that this junction be kept reverse biased.

When the bias is applied to the gate, however, a depletion zone is created in the channel due to the electrostatic field caused by the bias. When the gate voltage is zero, there is no depletion zone, so the channel resistance is minimum. As the voltage becomes more negative (more positive in P-channel devices), though, the field creates a wider depletion zone, thereby increasing the electrical resistance of the channel. There is a certain voltage in any JFET at which the gate potential will cut off all current flow through the channel: the *pinch off voltage*.

When a voltage is connected across the end terminals of the channel (called the source and drain), a current will flow in the channel. By Ohm's law, this current is proportional to the applied voltage and inversely proportional to the channel resistance ($I_d = V_{ds}/R_{ch}$). Since this resistance is under the control of the gate voltage, V_g, the drain current will be controlled by V_g also. This is the prescription for a transconductance amplifier. The gain, or transconductance, is given by:

$$g_m = \frac{\Delta I_c}{\Delta V_g}$$

The basic units of the transconductance is the *mho*, although it is usual to find the sub unit micromho (μmho) used instead because it results in more reasonable numbers.

Chapter 3

Vacuum Tube Fundamentals

The invention of the vacuum tube was the great leap forward that really launched the electronic age. Electrical devices had been known for several decades prior to the introduction of the vacuum tube, but it took the electronic age to usher in the advanced use of electricity in communications. Today, the vacuum tube has been all but eclipsed in radio transmitter design at power levels up to several hundred watts. But the tube is still alive, both in older designs and in high-power radio transmitters. Most kilowatt and higher power radio transmitters use vacuum tube final amplifiers. This includes almost all broadcast transmitters.

EDISON EFFECT

In the late 19th century workshop of electric incandescent light inventor Thomas Alva Edison, a peculiar effect was noted. Edison had placed an electrode in the evacuted glass envelope containing one of his filament electric lights. When a positive electrical potential was applied, a brief electric current occurred. Edison duly jotted down this phenomenon in his notebook and went on to other things. It was not the vacuum tube that interested Edison, but a method for improving the performance of the electrical light. Because the anode placed in the lamp provided no such improvement, Edison tried other investigations. What Edison had noted was the flow of thermionic electrons produced by the lamp filament.

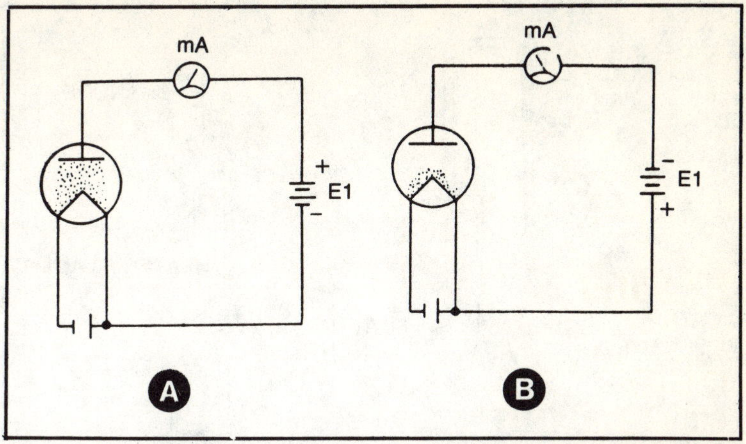

Fig. 3-1. Thermionic emission with the plate positive at A and the plate negative at B.

Thermionic emission of electrons occurs when the filament is heated to incandescence. The increased temperature of the filament causes the electrons to increase their energy level to a point where some of them can "boil off" the surface of the filament. This action forms a *space charge* in the vicinity of the filament wire.

The space charge would remain in the vicinity of the filiment unless some electrical force causes them to move. An Englishman named Fleming produced the first practical vacuum tube, called a diode. It had two-electrodes. Fleming's diode consisted of a thermionic filament called the *cathode* and a positive plate electrode called the *anode*. Figure 3-1 shows the basic construction of the diode. When the anode is made positive with respect to the cathode (the filament in this case), electrons from the space charge will flow toward the anode and out of the tube in the anode circuit. This is shown in Fig. 3-1A. But when the external battery polarity is reversed, the negative side is applied to the anode. This will repel the space charge back toward the cathode, thereby causing the flow of current to cease. The diode, then, is a one-way valve for electrical current. In fact, the British still refer to vacuum tubes as *valves*.

The one-way valve phenomenon is useful for several purposes. These include but are not limited to rectification, detection, mixing, waveshaping (clipping), etc.

The filament cathode of Fig. 3-1 is called a *directly heated cathode*. Many high-power transmitting tubes use the directly heated cathode, but many also use the indirectly heated cathode, but many also use the indirectly heated cathode of Fig. 3-2. In this

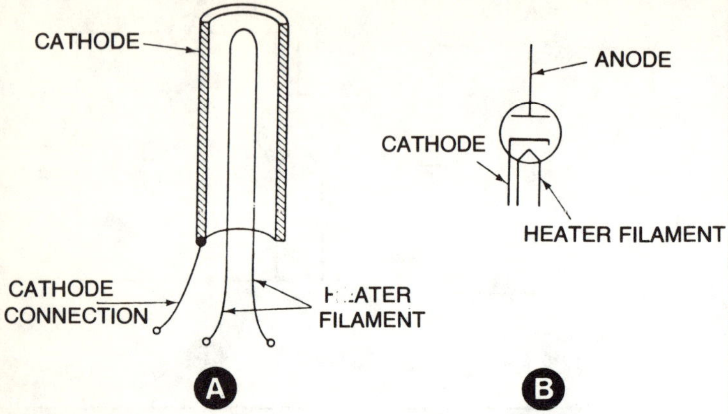

Fig. 3-2. Indirectly heated cathode at A with the circuit symbol at B.

design, the actual cathode is a metal cylinder that surrounds a heater filament. In theory, it doesn't really matter where the heat comes from—a blow torch would do—but in practice, this heater is used to electrically provide the thermionic heating required of the cathode. When the cylinder surface becomes hot enough, the electrons on the surface will leave to create a space charge. In many cases, the cathode cylinder will be made from thorium-coated tungsten. The circuit symbol for an indirectly heated cathode diode is shown in Fig. 3-2B.

DIODE OPERATION

Basic diode operation can be demonstrated by connecting a circuit such as shown in Fig. 3-3A. A variable DC voltage source is connected across the anode-cathode circuit, with the positive terminal of the power supply connected to the anode. A plate milliammeter will measure the diode current.

The characteristic plate-current-vs.-anode-voltage curve is shown in Fig. 3-3B. The plate current will be zero when the plate voltage is either zero or negative. Current flows, as indicated on meter M1, when the anode voltage becomes positive. But the plate current increase is not linear in the range where the plate voltage is low, because the electrons boiled off the cathode will have different levels of kinetic energy. At low potentials, only those electrons with the greatest kinetic energy will flow towards the plate. As the voltage increases, however, all of the electrons will be drawn to the plate. At some high potential, the diode becomes once again nonlinear. In this case, however, the nonlinearity is due to the fact that all of the electrons emitted by the cathode emitted by the cathode are drawn to the anode; increasing the anode potential cannot attract any additional electrons, so the curve flattens out.

We must revert now to a little matter of terminology. It became common in the years before World War II to designate the

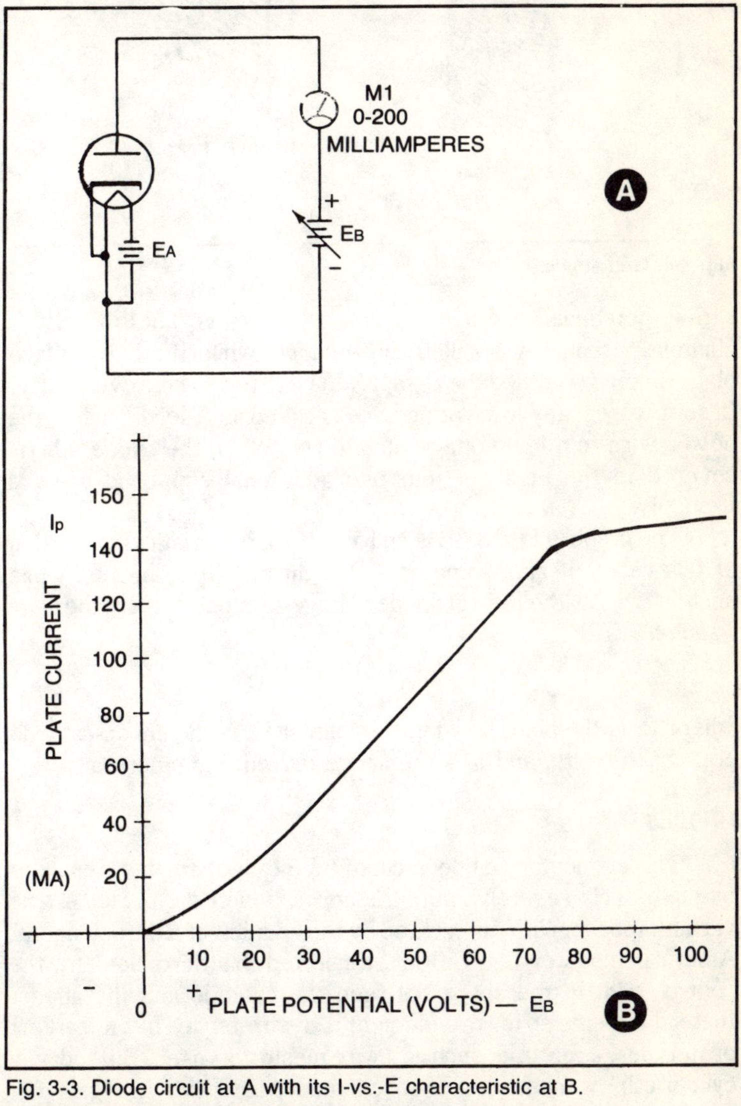

Fig. 3-3. Diode circuit at A with its I-vs.-E characteristic at B.

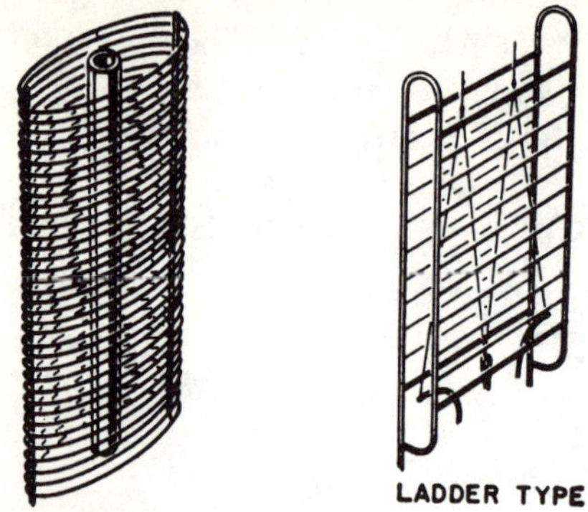

Fig. 3-4. Grid structures.

various potentials used in the vacuum tube circuits. The low-voltage filament potential was called the *A-supply* while the high-voltage plate supply was called the *B-supply*. These terms hold even today. In some cases, any low-voltage line is called an A-lead, such as the power wire to a low-voltage mobile rig. When the triode tube is covered in the next section, one additional supply is used: a C-supply.

The graph in Fig. 3-3B is an I-vs.-E characteristic. In a portion of this curve, it approximates a straight line; it is linear. In this section, we call the ratio of anode voltage to anode current the *plate resistance* (R_p):

$$R_p = E_B/I_p$$

where R_p is the plate resistance in ohms, E_B is the anode-cathode potential in volts, and I_p is the anode current in amperes.

TRIODES

The elementary diode vacuum tube cannot amplify signals or oscillate. This severely limits the application of the diode. A true vacuum tube amplifier is created by the addition of a third element. American inventer Lee DeForest inserted this third element, the control grid, in the space between the cathode and the anode. Instead of being solid, this element is a wire fence. In the earliest prototypes, a real grid, such as a wire mesh, was used. But today, a cylindrically wound spiral of wire is used, as shown in Fig. 3-4.

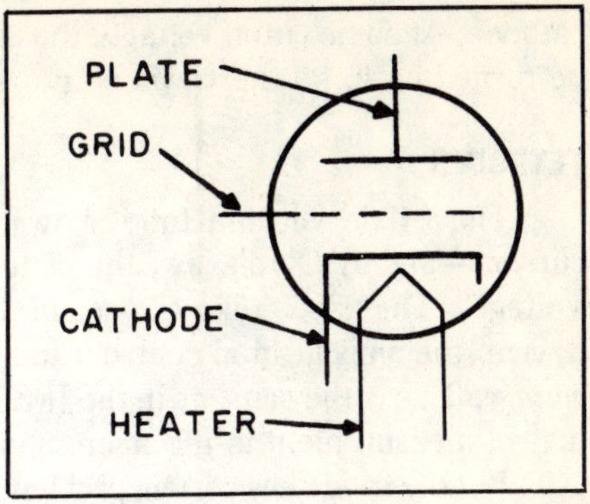

Fig. 3-5. Triode symbol.

The circuit cymbol for the triode tube is shown in Fig. 3-5. An indirectly heated cathode is used in this case. The circuit in Fig. 3-6 is a typical triode amplifier. Like the diode, the B-supply is connected across the anode and cathode connections. A negative voltage, the C-supply, is connected between the cathode and the grid.

The idea of the grid is to place a negative potential between it and the cathode in order to repel electrons. The flow of current between the cathode and the anode is controlled by controlling the grid voltage. When the E_{cc} supply voltage is zero, the current from the cathode will be attracted unimpeded to the anode. When a negative potential is applied, however, some of the less energetic electrons will be repelled from the grid and back toward the

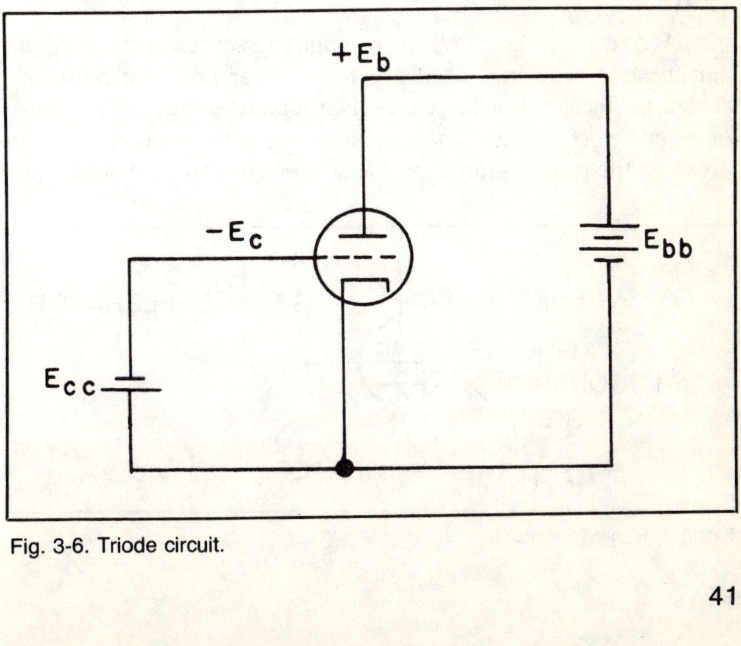

Fig. 3-6. Triode circuit.

cathode. At some cutoff voltage, the entire stream of electrons will be turned back, so there will be no anode current flow.

TETRODES

The triode vacuum tube allows us to control a large plate current—and by Ohm's law, the plate voltage—using a small grid voltage. The triode, like all vacuum tubes, is a transconductance device; the amplification controls and output *current* with an input *voltage*. This is the same as in the JFET of the last chapter, and the unit of measurement is the micromho.

But the triode has certain problems. The requirements for high amplification include placement of the control grid close to the cathode, and this severely limits the voltage level that can be applied between these elements. It will also increase the input capacitance of the device. We can reduce this problem by placing a second grid between the control grid and the anode. The purpose of this *screen grid* is to accelerate the electrons once they are past the control grid. For this reason, it is common to find the screen grid called the accelerator grid (sensibly enough). With four elements, this tube is called a *tetrode*.

The screen grid will be biased with a positive potential. This voltage might be as high as the anode potential in some applications. In most cases, though, it is lower than the anode potential.

PENTODES

Accelerating the electrons through the screen grid solves the amplification problem, but it creates another problem: *secondary emission*. Acceleration increases their kinetic energy. When these energetic electrons strike the anode, some of them will collide with atoms in the plate material and cause electrons to be shaken loose

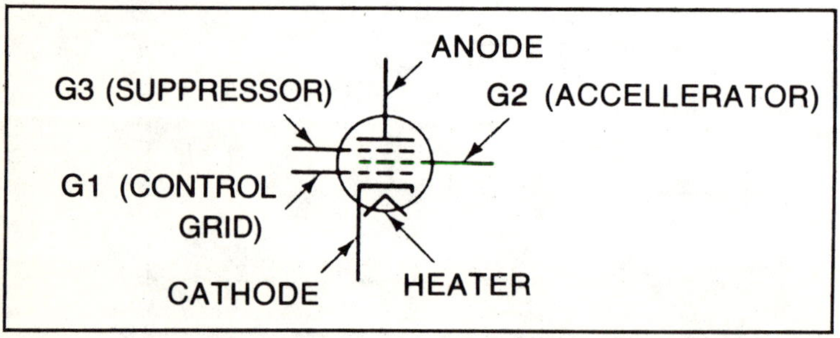

Fig. 3-7. Pentode symbol.

into the space between the anode and the accelerator grid. The accelerator is positively charged, so some of these secondary electrons are attracted to the grid.

The solution to this problem is to insert one additional grid in the space between the screen grid and the anode. This is called the *suppressor grid*, and it is given either a zero bias or is placed at the same potential as the cathode. Figure 3-7 shows the circuit symbol for a pentode vacuum tube, showing the five elements.

Chapter 4

Fundamentals of Amplifier Circuits

Most of the transmitter stages which we will discuss are merely amplifier circuits of one form or another. Even oscillators are merely amplifiers with positive, instead of negative, feedback. The subject of amplifiers is broad enough to encompass several large textbooks, so only the fundamental concepts can be given here.

The examples given in this chapter are transistor amplifiers, and *bipolar* transistors devices at that. The same general concepts also apply to both vacuum tube and field-effect transistors. Just remember the analogies between the device elements, given in Table 4-1. When we talk about a grounded-base, or common-base amplifier, the vacuum tube equivalent is the *grounded-grid* circuit.

THE TRANSISTOR AS AN AMPLIFIER

Figure 4-1 shows one of the basic circuit configurations required to make a transistor amplifier circuit. In this case, the circuit is a voltage amplifier that is in the common-emitter configuration. The input signal is applied between the base and emitter terminals of transistor Q1, while the output signal is taken across the emitter-collector terminals. The input signal is in series with the base bias voltage, E_b.

The collector load resistor, R1, is in series with the collector voltage supply, such that collector current I_c must flow through resistor R1. The output signal, E_c, is the voltage that actually appears across the collector-emitter terminals of Q1.

Table 4-1. Transistor and Tube Analogies.

BIPOLAR TRANSISTOR	VACUUM TUBE	FIELD-EFFECT TRANSISTOR
COLLECTOR	ANODE (PLATE)	DRAIN
BASE	GRID (CONTROL)	GATE
EMITTER	CATHODE	SOURCE

The base bias of the transistor is set to make voltage E_c exactly ½-E_{cc}. This will place the operating quiescent point of the transistor in the middle of its linear range. In this condition, half of the collector supply voltage is dropped across the resistor, and half is dropped across the transistor, assuming zero signal conditions ($E_{sig} = 0$).

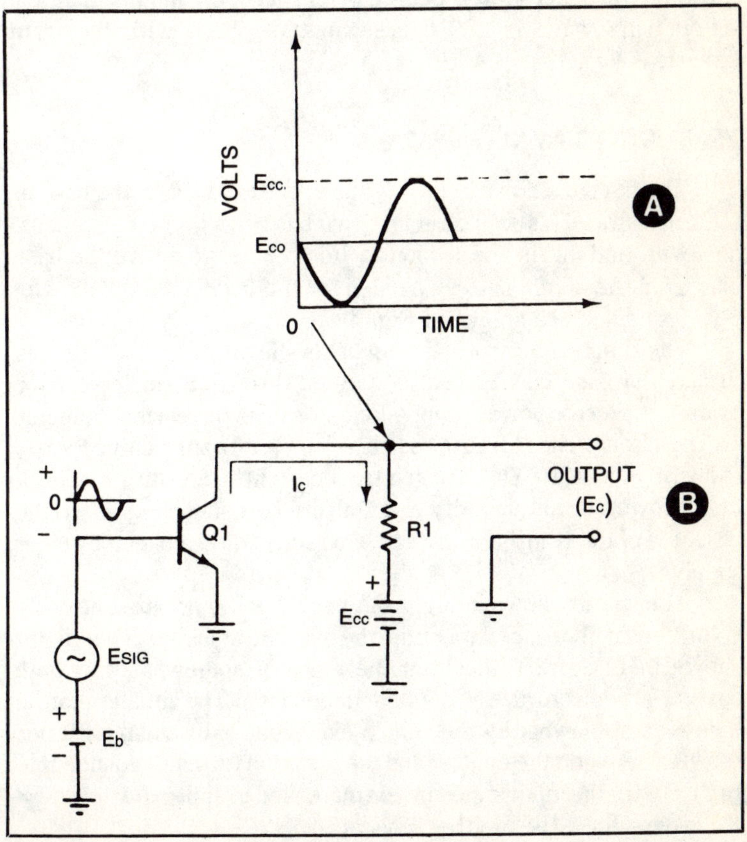

Fig. 4-1. Common emitter circuit as a voltage amplifier. Sine wave output is shown at A.

When the input sine wave increases from zero to a positive value, the transistor is turned on harder, and this causes collector current I_c to increase. When I_c increases, the voltage drop across resistor R1 increases also. Because collector voltage E_c is the difference between supply voltage E_{cc} and the drop across R1, E_c decreases as increasing E_{sig} causes I_c to increase. When the input signal reaches its maximum allowed value, E_c will drop to zero. If the input signal attempts to drive the transistor any harder than this level, distortion (in the form of negative peak clipping) of the output waveform occurs.

Similarly, when the input signal E_{sig}, does negative, the transistor is turned on less hard, causing I_c to decrease. This will cause voltage E_c to increase. The common emitter amplifier of Fig. 4-1, therefore, produces a 180-degree inversion of the input signal; the output sine wave is 180 degrees out of phase with the input sinewave.

TRANSISTOR BIAS NETWORKS

The circuits shown in Fig. 4-1 uses battery bias on the base of the transistor. This would require two batteries in the equipment, however, and this is not economic. It is cleaner to derive the base bias from the single battery needed for the collector-emitter supply. Examples are shown in Fig. 4-2.

The simplest form of bias network is shown in Fig. 4-2A. In this circuit, the base bias current is derived through a single resistor from the collector power supply. Resistor R1 is the normal collector load resistor, while R2 is the base bias supply. An alternative form is shown in Fig. 4-2B. This circuit introduces a little negative feedback to improving circuit linearity and stability by connecting R2 to the collector of the transistor instead of directly to the collector power supply.

The circuit shown in Fig. 4-2A provides the greatest possible voltage gain, but lacks much in the way of stability. The circuit shown in Fig. 4-2B improves the stability somewhat, but both circuits suffer from a very low input impedance. The situation can be improved somewhat by the simple expedient of providing a series resistor between the emitter and the common (ground) connection. In that case, the input impedance is increased to approximately $h_{fe} \times R_e$, where R_e is the emitter resistance.

The current gain of this stage is approximately h_{fe}, the AC beta. The voltage gain is approximately h_{fe} (R_L/R_e).

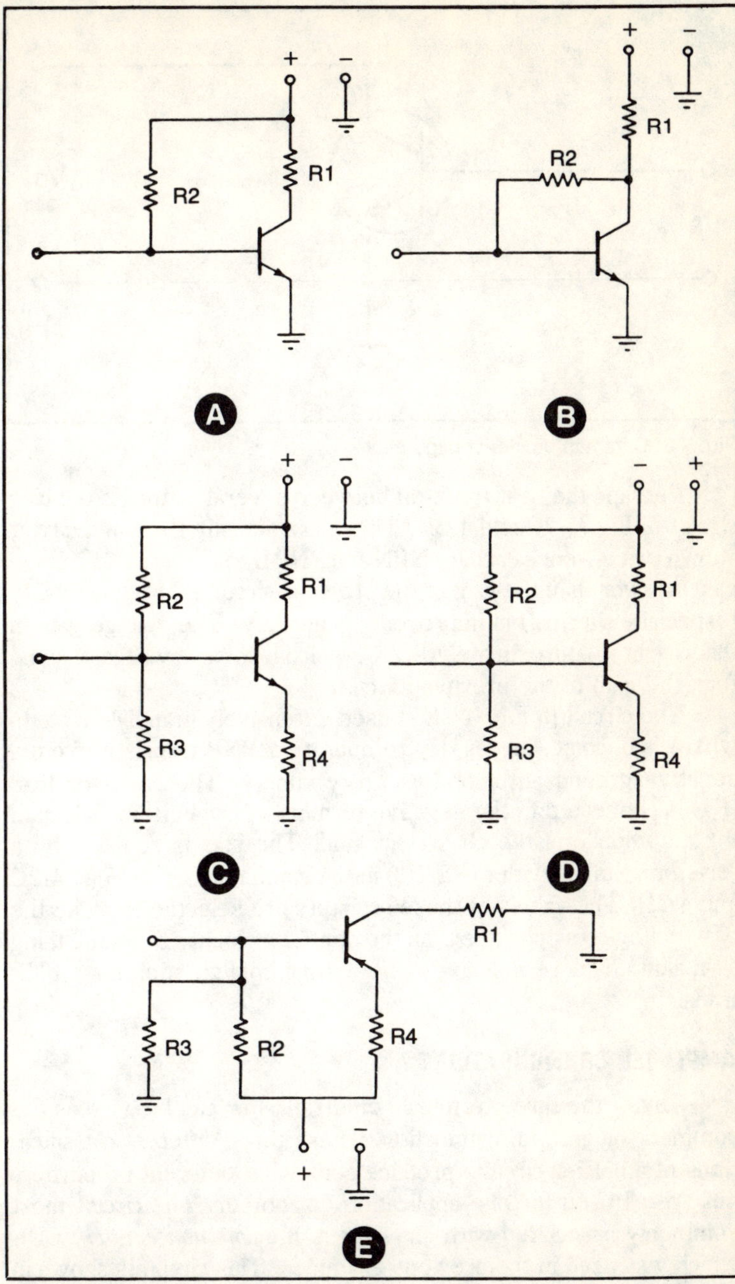

Fig. 4-2. Bias circuits for bipolar transistors. Simplest circuit at A, negative feedback at B, NPN and PNP trade-offs at C and D, and a negative ground circuit at E.

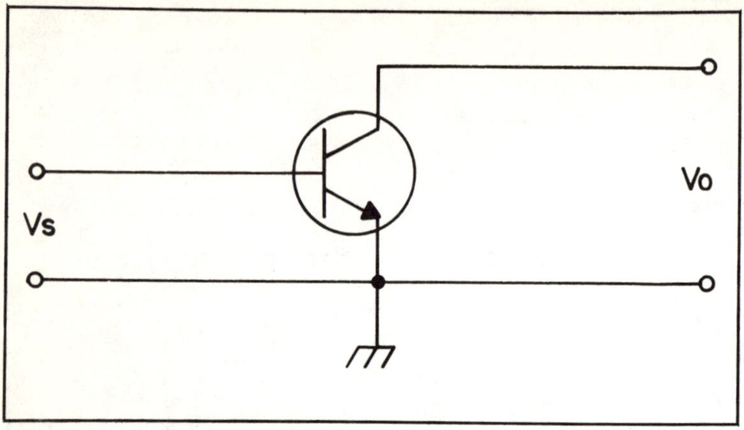

Fig. 4-3. Common emitter configuration.

Perhaps the best trade-off between several factors is the bias circuit of Fig. 4-2C and 4-2D. These are actually the same circuit shown twice—once each for NPN and PNP transistors. The voltage gain is lower than the gain of the previous circuits, but the stability (especially thermal) is improved significantly. The voltage gain in this circuit is approximately R_L/R_e, which is lower by a factor of h_{fe} than the gain of the previous circuit.

The circuit in Fig. 4-2E is used extensively in mobile circuits where it becomes necessary to operate a PNP transistor from a negative ground automobile battery supply. The collector load (R1) is connected to the negative terminal of the battery, which in the automobile is the chassis ground. The bias is supplied by a resistor voltage divider (R2/R3) in the same manner as Figs. 4-2C and 4-2D. The values of these resistors are selected to make the base voltage less positive than the emitter, which is the same thing as making it more negative—the normal configuration for a PNP transistor.

AMPLIFIER CONFIGURATIONS

Any of the three terminals of a transistor can be used as the common, or ground, connection. These three different classifications of amplifier circuits produce somewhat different properties, but are all used for one application or another. The circuit most commonly associated with amplifier is the *common emitter* (CE), which was used in the previous examples. This circuit is shown in simplified form in Fig. 4-3. The input signal V_s is applied to the base-emitter terminals, while the output signal is taken from across the collector-emitter. The emitter, then, is common to both input and output circuits of the common emitter amplifier.

The *common collector* circuit is shown in Fig. 4-4. In this type of amplifier, the input signal is impressed across the base-collector terminals, and the output signal is taken across the emitter-collector terminals of the transistor. The collector terminal, then, is common to both input and output terminals. This circuit is usually called an *emitter follower*. It is analogous to the cathode follower and source follower circuits (tubes and FETs).

The voltage gain of the emitter follower is less than unity, although in most cases it will approximate unity (0.98, etc.). The current gain, however, can be considerable, being on the order of h_{fe} +1.

Note that the input and output signals in the CC amplifier are in phase with each other. There is no phase inversion between input and output, as there was in the common emitter case.

There are two main applications in which the CC amplifier is used extensively: buffering and impedance transformation. A buffer amplifier is one that is used to isolate two circuits from each other. Typical transmitter application might include the output amplifier from the oscillator. Changing load conditions at the base of the first power amplifier could easily pull the frequency of the oscillator. These changes are eliminated by using the emitter follower as a buffer between the oscillator and the first power amplification stage.

The emitter follower can also provide impedance transformer action. If we have a low-impedance input and a high-impedance source, the input tends to load the source. Trying to drive a low-impedance circuit from such a source can be a real bear, so some form of impedance transformation is needed. The input impedance of the emitter follower is on the order of 100K ohms, or more

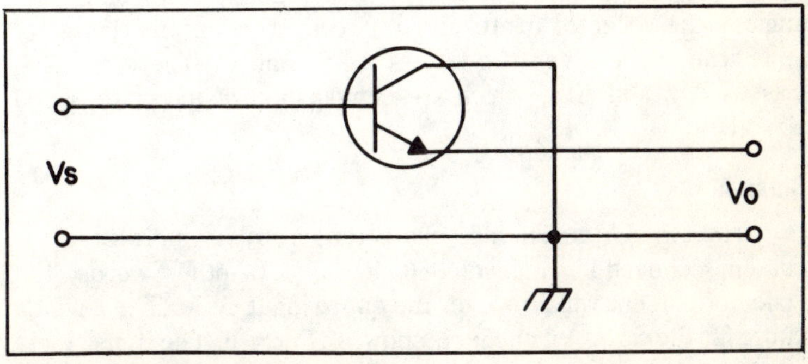

Fig. 4-4. Common collector circuit.

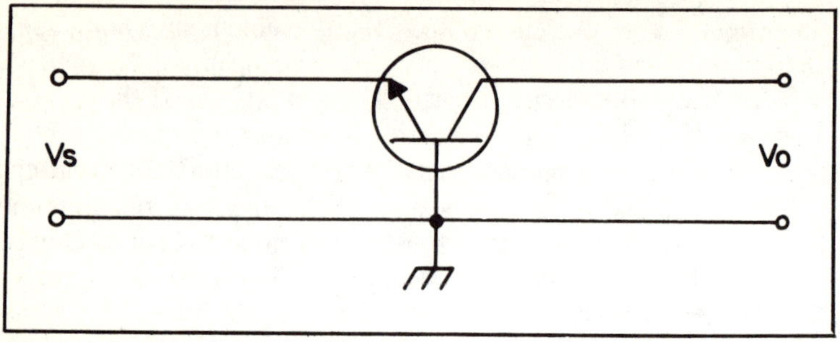

Fig. 4-5. Common base circuit.

(in most cases), while the output impedance is typically very low (being only the emitter resistance R_e). This circuit will provide impedance transformation for voltage source inputs, with little loss in amplitude to the driving signal.

The common base amplifier is shown in Fig. 4-5. In this circuit, the input signal is applied between the emitter and the base, while the output signal is taken across the base and collector circuits. The base terminal, then, is common to both input and output circuits.

The voltage gain of the CB circuit is typically high, on the order of 100 or more, compared with the current gain of less than unity. The main use of the CB circuit is at high frequencies and VHF/UHF, because these amplifiers are stable without neutralization. The CB amplifier provides good AC isolation between output and input, but it does not produce any phase inversion. Of these three, in fact, only the CE amplifier produces phase inversion. Table 4-2 synopsizes the characteristics of the three different classes of circuit configuration for bipolar transistors.

CLASSES OF AMPLIFICATION

Amplifiers can also be classified according to the conduction angle in the collector (plate, drain) circuit. The various classes of amplification are given the letters A, B, and C. There are also classes AB_1 and AB_2, which are combination of the class-A and class-B.

Class-A

The class-A amplifier is characterised by having a collector current conduction angle of 360 degrees. In other words, the collector current will flow over the entire input cycle. The I_c-vs.-I_b curve for the class-A amplifier is shown in Fig. 4-6. The transistor is biased to some midpoint in its operating range, so that input signal swings above and below the quiescent and point will not cut off or saturate the collector current. The result is an output signal that is an exact representation of the input signal. The amplifier is therefore *linear*.

The class-A amplifier suffers from a couple of problems, however. One is the fact that collector current flows all of the time, so the transistor must be capable of dissipating a considerable amount of power. The nomral, zero-signal, power in the circuit is ($\frac{1}{2}$-V_{cc}) (I_{cq}), where the current is the quiescent value. This means that the class-A amplifier will give off a surprising amount of heat. In fact, the power dissipation is constant under a wide range of conditions.

Class-A amplifiers are used mostly as current amplifiers or voltage-amplifiers in which the collector dissipation is normally low. But a number of class-A amplifiers are used in power amplification also. The efficiency of the class-A amplifier cannot exceed 25 percent (some authorities claim 50 percent for vacuum tube amplifiers). The linearity of the class-A amplifier, however, makes it the amplifier of choice for voltage amplification, as in audio preamplifiers in transmitter modulator stages.

Class-B

The class-B amplifier operates only over one-half of the input signal cycle; it has a 180-degree conduction angle. This means that

Table 4-2. Characteristics of the Basic Amplifier Types.

	Common Emitter	Common Collector	Common Base
Voltage Gain	High	Low ($\approx$1 or less)	High (>100)
Current Gain	High	High (β + 1)	Low (<1)
Z_{in}	Medium ($\approx$1 K)	High (>100 K)	Low
Z_{out}	Medium to High (50 K)	Low (<100Ω)	High (>500K)
Phase Inversion ?	Yes	No	No

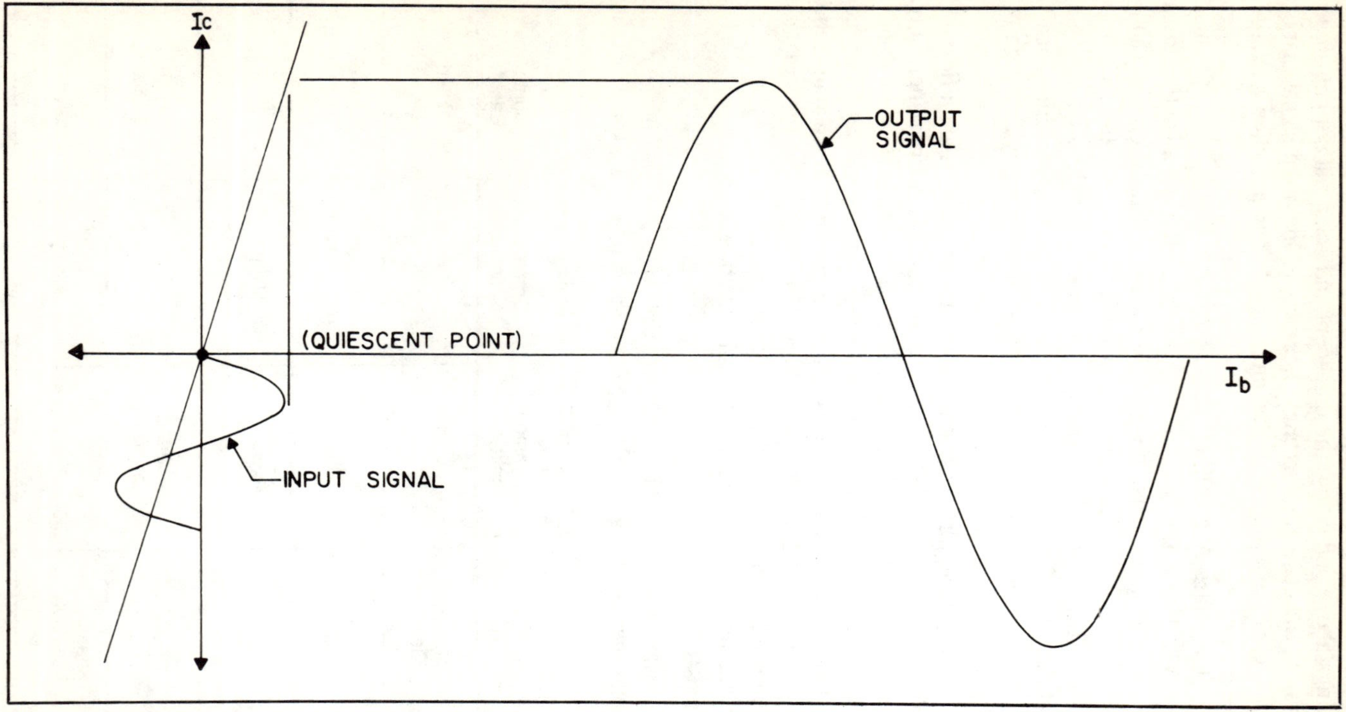

Fig. 4-6. Class-A amplifier transfer curve.

the collector current will flow over half of the input signal cycle. The output transfer property, and a typical class-B amplifier circuit is shown in Fig. 4-7. The transistor in a single-ended class-B amplifier is given no bias, so the signal is the sole source of bias for the transistor. When the signal voltage is zero, the transistor is cut off, so no current will flow. As the signal level increases in the direction that forward biases the transistor, however, the collector current will increase also. When the signal level increases in the direction that reverse biases the transistor, however, no current will flow. In

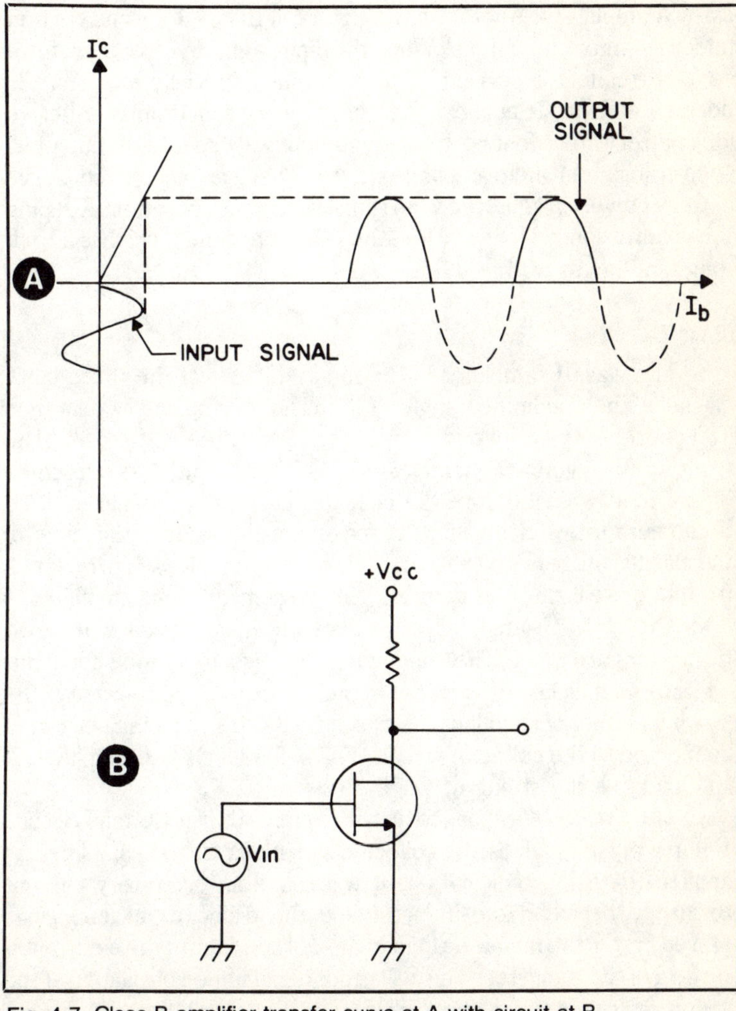

Fig. 4-7. Class-B amplifier transfer curve at A with circuit at B.

the NPN case shown, therefore, the output signal exists only when the input signal is positive. The reverse is true in a class-B PNP stage.

The class-B amplifier produces a significant amount of output distortion, so it cannot be used as a single-ended (one-transistor) linear amplifier. By combining two transistors, however, we can make a linear class-B amplifier. These will be discussed shortly.

The one saving grace of the class-B amplifier is that it is much more efficient than the class-A amplifier. The maximum theoretical efficiency is 78.5 percent, as opposed to less than 50 percent for class-A amplifiers. Another saving grace is that is draws no current under no-signal conditions. When the input signal is zero, therefore, so is the collector current. This becomes especially important in portable and mobile applications where the current drain is critical to long battery life. Most portable equipment will use class-B push-pull (two-transistor) audio amplifier stages. These allow us to conserve battery power on squelched (no signal received) operation. Keeping the volume control low also lengthens battery time , which is a trick long familiar to walkie-talkie owners.

Class-C

The class-C amplifier is the most efficient of the three basic classes. The conduction angle of the class-C amplifier is less than 180 degrees, typically on the order of 120 degrees (Fig. 4-8). The transistor is given a slight *reverse bias* ($-V_b$) that must be overcome by the input signal before the collector current will begin to flow. Some authorities claim that the maximum theoretical efficiency of the class-C stage is 100 percent, but other circuit losses prevent us from ever realizing the perfect, 100 percent efficient amplifier.

The class-C amplifier can never be linear, even when two transistors are used. The conduction angle is simply too short. But we can use the class-C amplifier in radio transmitters because of the flywheel effect of resonant LC tank circuits. If such a tank circuit is connected to the collector circuit (Fig. 4-9) of a class-C amplifier, it will linearize the rf output waves.

The *flywheel effect* has to do with a *pulse-excited* LC tank circuit. When a pulse, such as the collector signal of a class-C amplifier, is applied to an LC tank circuit, it will cause an oscillatory current exchange between the inductor and the capacitor as energy is stored first in the magnetic field of the coil and then in the electrostatic field of the capacitor. This will produce a damped oscillation at the resonant frequency of the tank circuit. The oscillation will die out

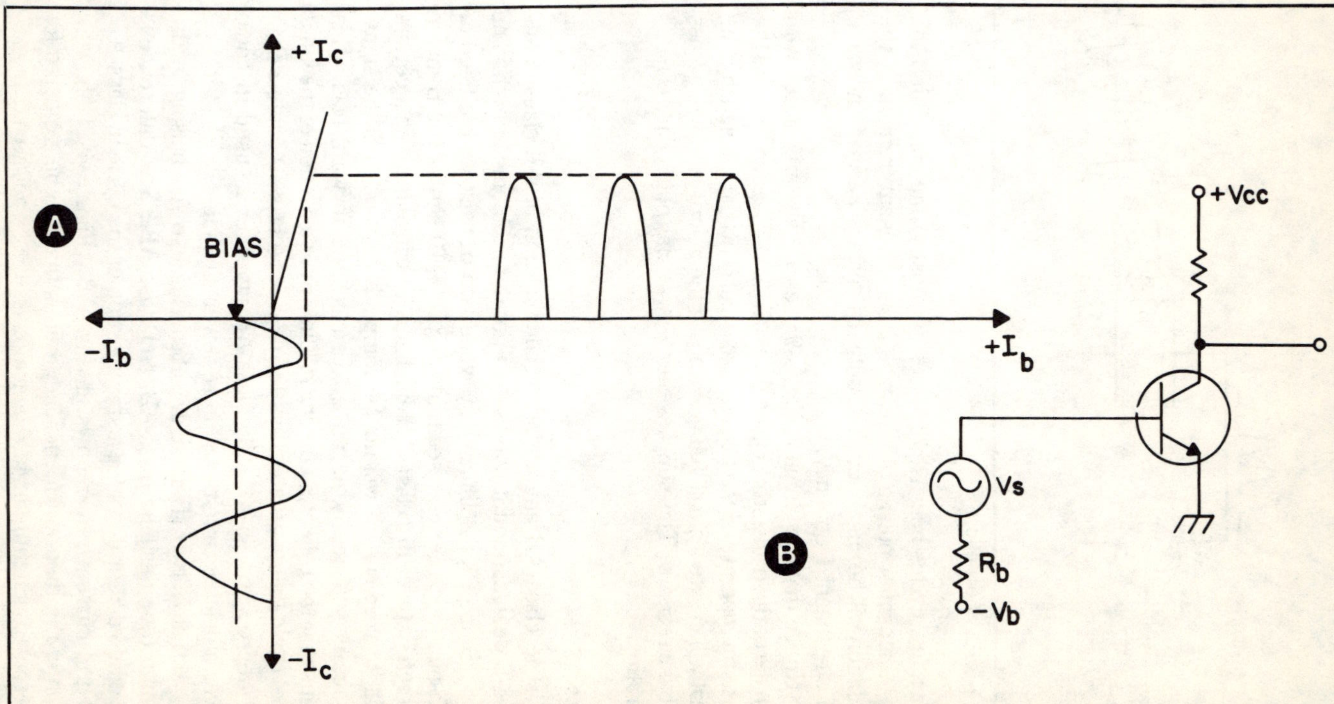

Fig. 4-8. Class-C amplifier transfer curve at A with circuit at B.

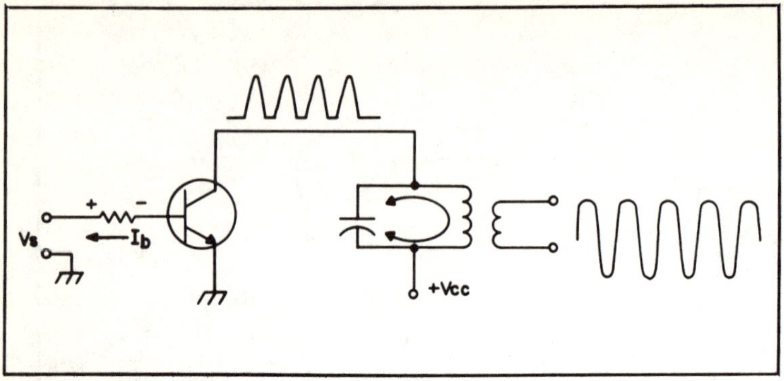

Fig. 4-9. Flywheel effect.

after a few cycles because of losses in (mostly) the coil. But if the class-C amplifier pulses the LC tank at its resonant frequency, the oscillation will be sustained. The output signal will be an rf sine wave at the resonant frequency.

Notice that the circuit in Fig. 4-9 uses a resistor in series with the base of the transistor. Often this is all that is needed to make the transistor operate class-C. The base current (I_b) caused by the signal will create a small voltage drop, of the polarity shown, across the base resistor. This supplies the reverse bias needed for class-C operation.

Class-AB

The class-AB amplifier is a combination of the class-A and class-B designs, and it solves some of the problems of both. The class-AB amplifier is linear only when two transistors are used in push-pull (see the next section). Figure 4-10 shows the transfer characteristic of the class-AB amplifier. Because is given a certain small forward bias, it will not be completely cut off until the signal voltage is sufficiently negative to cancel the forward bias. This means that the collector current will flow over a portion of the input signal cycle that is greater than 180 degrees (class-B) but less than 360 degrees (class-A). The class-AB amplifier is used in audio power amplifier circuits and in rf power amplifiers.

Actually, two rf amplifier subclasses use AB biasing in the vacuum tube designs: class-AB_1 and class-AB_2. The difference is that grid current flows for a small portion of the cycle in class-AB_2 but grid current never flows in class-AB_1. Figure 4-11 shows a graph of the theoretical efficiency values that can be expected for different conducting angles; hence, classes of operation.

PUSH-PULL AMPLIFIERS

The only power amplifier that is linear using only one transistor is the class-A circuit. But this circuit is also terribly wasteful of power, a decided deficiency! By using the class-B or class-AB, however, greater efficiency and linear operation is possible with two transistors. Recall that the class-B amplifier operated over only one-half, or 180 degrees, of the input cycle. If we configure two class-B amplifiers such that they supply output power to the load on different half-cycles, we will obtain linear operation and class-B efficiency. An amplifier that operates in this manner is called a *push-pull amplifier* (one transistor is pushing when the other is pulling).

Figure 4-12 shows the classical push-pull configuration. This same circuit has been used for years in vacuum tube designs, even though others have more or less eclipsed it in solid-state equipment. In this circuit, two identical transistors or tubes are used. These transistors are driven from a center-tapped interstage transformer (T1), which in turn drive a center-tapped output transformer.

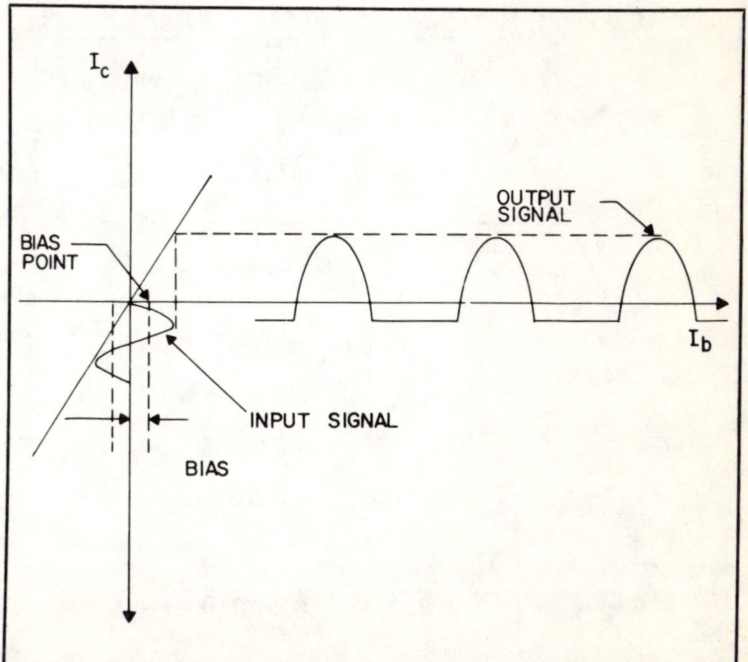

Fig. 4-10. Class-AB transfer characteristic curve.

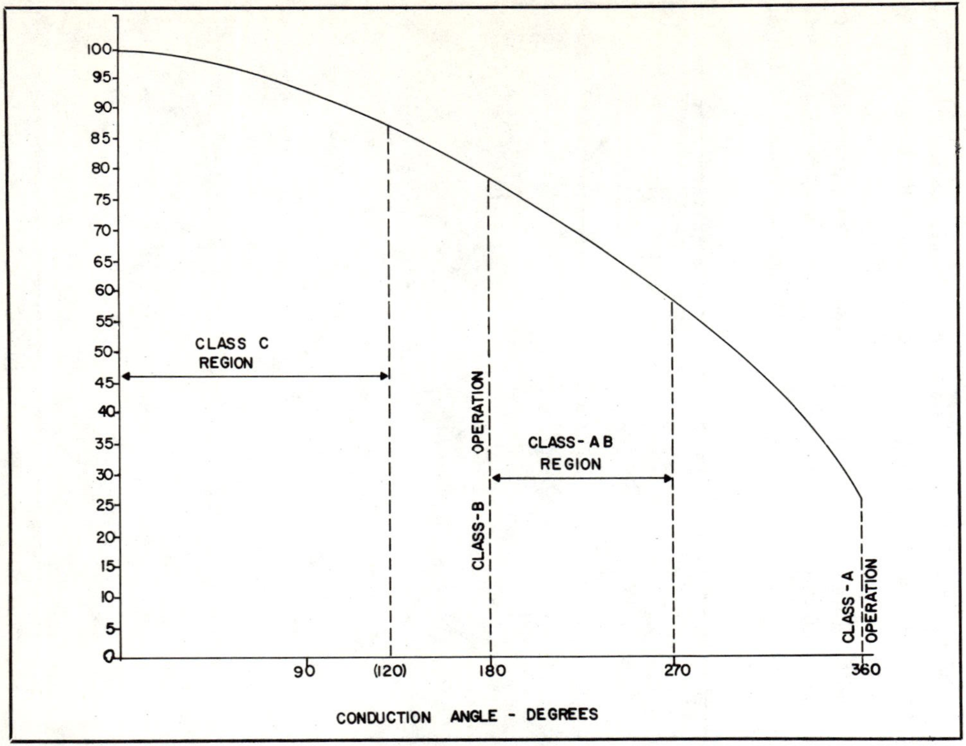

Fig. 4-11. Conduction angle-efficiency curve.

The transformer connections are very similar to those of a center-tapped transformer used in a full-wave power supply. In both T1 and T2, the zero reference point is the center tap, so one end will be positive with respect to the center tap, and the other negative, on each half cycle.

Consider the situation when a sine wave is applied to the input side of T1. The sine waves applied to Q1 and Q2 will be 180 degrees out of phase with each other, so one transistor is turning on while the other is turning off. These same relationships occur in the primary side of output transformer T2. On the secondary side of T2, however, the half cycles recombine to form a pure sine wave output. The output sine wave is made from two half-cycle contributions, one each from Q1 and Q2.

The circuit of Fig. 4-12 is not used much in audio amplifiers anymore. In the early days of transistors, however, it was quite common. The vacuum tube version of this circuit is used in rf power amplifiers. In those designs, the two transformers are tuned tank circuits. It is the usual practice to tune the primary of T2 with a split-stator variable capacitor, and then use a tuned primary on T1 (the T1 secondary is often untuned).

A popular modification of the basic push-pull amplifier is shown in Fig. 4-13. In this circuit, the output transistors, again identical to each other, are connected in series with each other. The voltage at the junction (point A) is $\frac{1}{2}V_{cc}$, so a DC blocking capacitor is needed between the junction and the load. In some cases, a negative voltage power supply is applied to the emitter of Q2, instead of ground. The net potential at point A would be zero, so the capacitor may be left out of the circuit. This type of push-pull amplifier is very popular in medium-power automotive audio amplifiers that might be

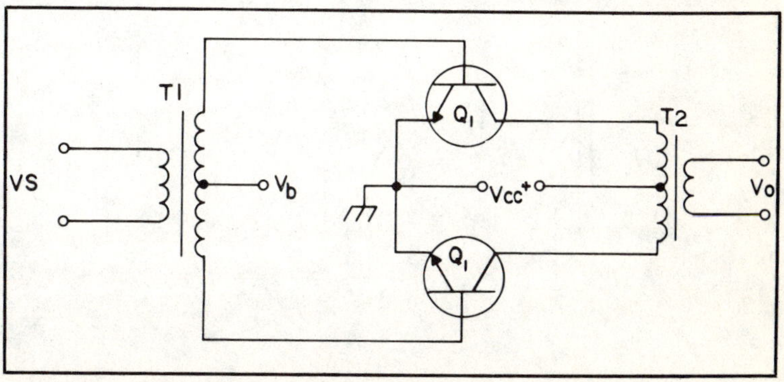

Fig. 4-12. Classical push-pull amplifier (transformer coupled).

found in car radio sound systems. It is sometimes called the *totem pole* power amplifier, possibly because the output transistors are stacked, or series connected.

One final push-pull circuit is shown in Fig. 4-14. This circuit uses two transistors—one is PNP and the other is NPN. Their electrical characteristics, however, must be matched. Semiconductor manufacturers call such matched, opposite-polarity, transistors *complementary pairs*. This type of push-pull amplifier is called a *complementary-symmetry push-pull amplifier*.

One necessary condition for push-pull operation in the previous circuits was a method for driving the two transistors out of phase with each other. This was done with transformers in Figs. 4-12 and 4-13. It could also be done with a preamplifier with complementary (push-pull) outputs. Such a circuit is called a phase inverter. One signal is taken from the inverted collector, and the other is taken from the noninverted emitter of the transistor. But in the complementary-symmetry amplifier, the transistors are driven in phase with each other, and it is the *transistors* themselves that provide the phase inversion. Recall the operation of the two classes

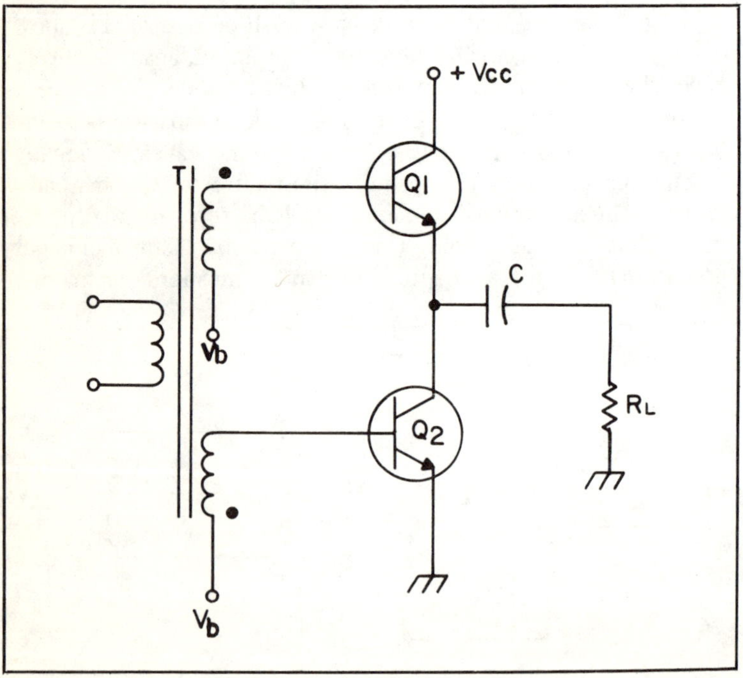

Fig. 4-13. Totem pole push-pull amplifier.

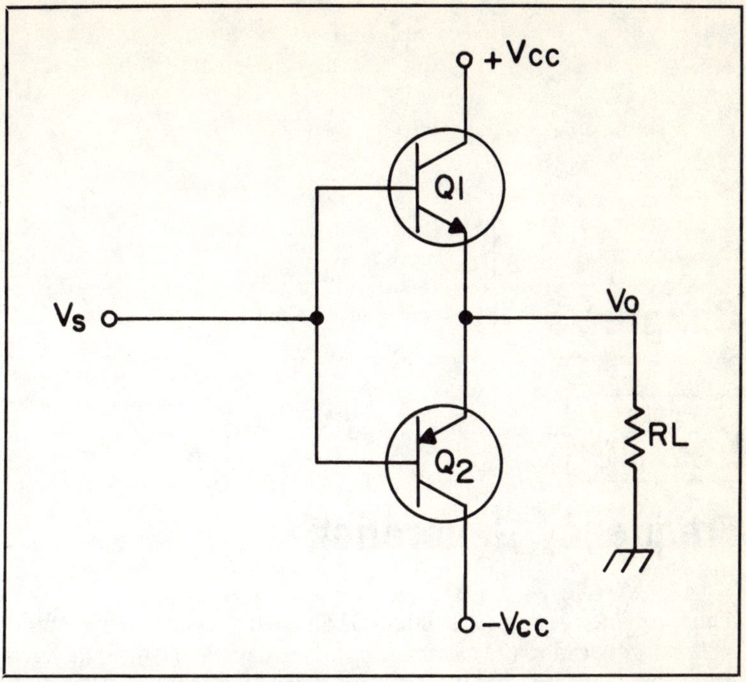

Fig. 4-14. Complementary symmetry.

of transistors. The PNP turns on when the base becomes more *negative* than the emitter. The NPN turns on when the base becomes more *positive* than the emitter. This is a perfect prescription for push-pull operation.

The complementary-symmetry amplifier is basically two opposite-phase (NPN and PNP) emitter followers driving the same load (R_L). When the input signal is positive, NPN transistor Q1 turns on, and PNP transistor Q2 turns off. On the negative excursions of the input signal, the opposite situation occurs: NPN transistor Q1 turns off, and PNP transistor Q2 turns on. The signal across the load is the combined contributions of Q1 and Q2. Because each operates over one-half of the input cycle, the output waveform is the full sine wave.

There is a subclass of complementary-symmetry called *quasi-complementary symmetry*. At one time it was difficult to obtain high-power complementary pairs, but medium-power pairs were readily available. In those days, many designers would make a complementary driver amplifier, and use it to feed (direct-coupled) a totem pole circuit without the transformer.

Chapter 5

Frequency Generation

The frequency generator section(s) of a radio transmitter is responsible for generating the carrier signal. A decade ago we would have called this chapter "oscillators," and even today much of the frequency generation is done solely in an oscillator circuit. But modern radio transmitters make extensive use of techniques such as the phase-locked loop (PLL), heterodyning, and frequency muliplication/division. For the modern user of transmitters, the simple rubric "oscillator" tells only part of the overall story.

OSCILLATOR BASICS

The block diagram to an elementary oscillator circuit is shown in Fig. 5-1. The circuit consists of two principal sections: an active device that acts as an amplifier with gain A, and a feedback network with a gain B. The gain of the feedback is usually a loss, so may well be described as a negative gain. This feedback network usually consists of R, L, and C elements. The amplifier in most circuits is an inverting amplifier, so the gain is negative. In other words, the output signal is 180 degrees out of phase with the input signal.

Two conditions must be fulfilled for the circuit to actually oscillate, and these conditions are known as the Barkhausen criteria:

- □The feedback must be in-phase with the input signal.
- □The loop gain AB must be greater than, or equal to, unity (1).

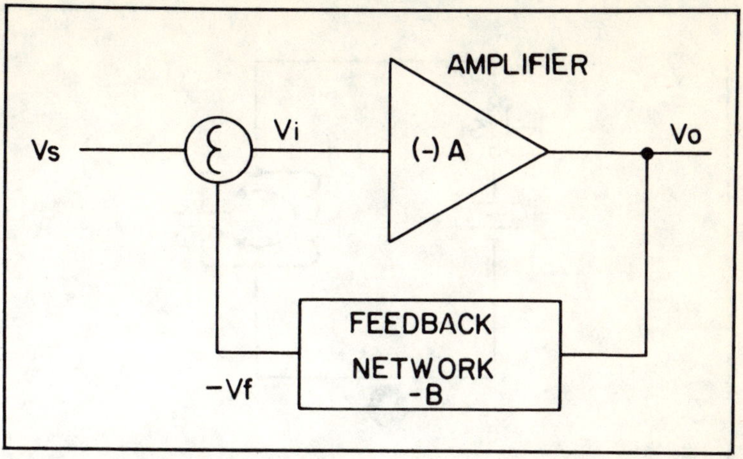

Fig. 5-1. Block diagram of an oscillator.

The first of these criteria is met by producing a 180-degree phase shift in the feedback network at the desired frequency of oscillation. The remaining 180 degrees of the required 360 is provided by the inverting gain of the amplifier.

The feedback network, being a passive circuit, produces an attenuation of the feed back signal. If V_o is the output signal and V_f is the feedback signal, V_f will be less than V_o. We note that V_f is inverted due to the 180-degree phase shift of the feedback network, but only at the desired oscillating frequency. The amplifier gain is set to just overcome the loss of the feedback network, and the second criterion is then satisfied.

We insure single-frequency operation of the oscillator because most RLC networks produce the needed 180-degree phase shift at only one frequency. A LC tank circuit, for example, produces such a phase shift only at its resonant frequency. The capacitive and inductive reactance are equal, but of opposite sign, so they cancel. All of the oscillator circuits presented in this chapter will be rf types, so the frequency control elements will be either LC tank circuits or piezoelectric crystals.

SHOCK-EXCITED OSCILLATIONS IN LC TANK CIRCUITS

Let's consider a circuit such as the LC tank shown in Fig. 5-2. If switch S1 is momentarily depressed, current will flow in the tank. This current immediately charges capacitor C, but as soon as the switch is opened again, the charge in electrostatic field of the capacitor is dumped back into the circuit and flows through the

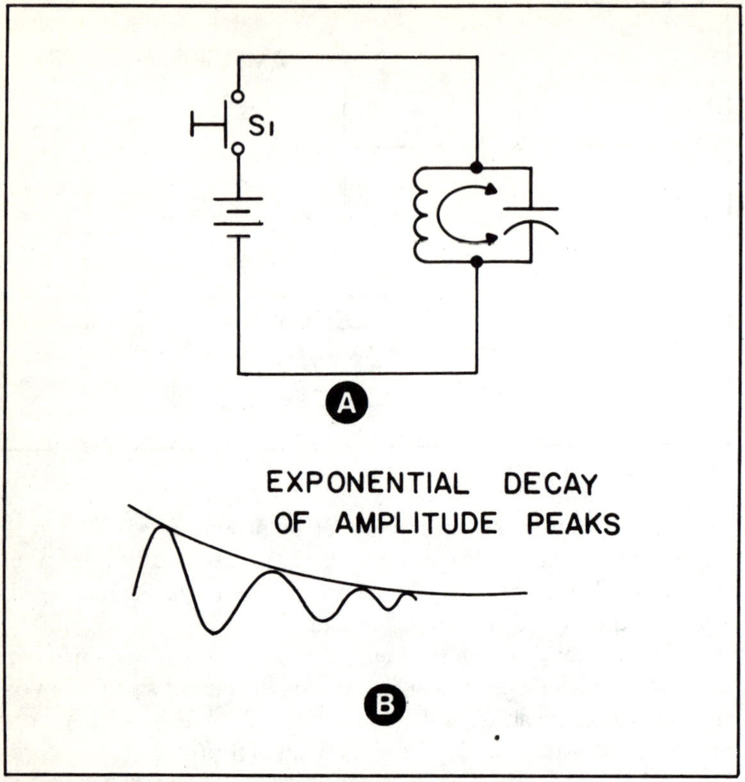

Fig. 5-2. Shock exciting an LC tank. Circuit at A with effect at B.

inductor. This current then sets up a magnetic field about the inductor. When this field collapses an instant later, it generates a countercurrent that once again charges the capacitor. This exchange of energy takes place back and forth between the electrostatic field of the capacitor and the magnetic field of the inductor, producing a sinusiodal oscillation at the resonant frequency of the tank circuit. If the tank were perfect, then this oscillator current, once induced, would flow forever. Several experiments have been performed with supercooled tank circuits in which the temperature was kept to –250 °F. The current was sustained for a long period of time. But in real room-temperature tank circuits there are substantial losses, such as the series resistance of the coil, the connecting wires, and the parallel resistance of the capacitor. These cause the current to die out in an exponential manner. Figure 5-2B shows this damped oscillation. In order for the oscillation to be sustained, the impulse current that shock-excited the tank circuit must be repeated at intervals. The function of the oscillator circuit that uses the LC tank is to maintain this shock excitation. Shock-excited oscillation is also called the *flywheel effect*.

LC OSCILLATORS

LC shock excitation is used to build oscillator circuits. If the capacitance or inductance of the LC tank is variable, it is a *variable frequency oscillator*, or VFO. Several different circuits are commonly used. Those covered in this text include the Armstrong, TGTP, Hartley, Colpitts, and a version of the Colpitts oscillator called the Clapp oscillator. Incidentally, if you are studying for a higher class amateur radio license than Novice, be able to identify these oscillator circuits. The examination might have such an identification question. As this chapter progresses, the salient features that identify the different oscillator types will be detailed.

Almost any of the oscillator circuits can be drawn in several different manners. The two classes for each type are usually classified as *series feed* and *shunt feed*. These same characterizations also apply to rf amplifiers, and are shown in Fig. 5-3. A series feed circuit is shown in Fig. 5-3A. The main feature of this circuit is the fact that the collector (plate, drain) current must flow through the rf

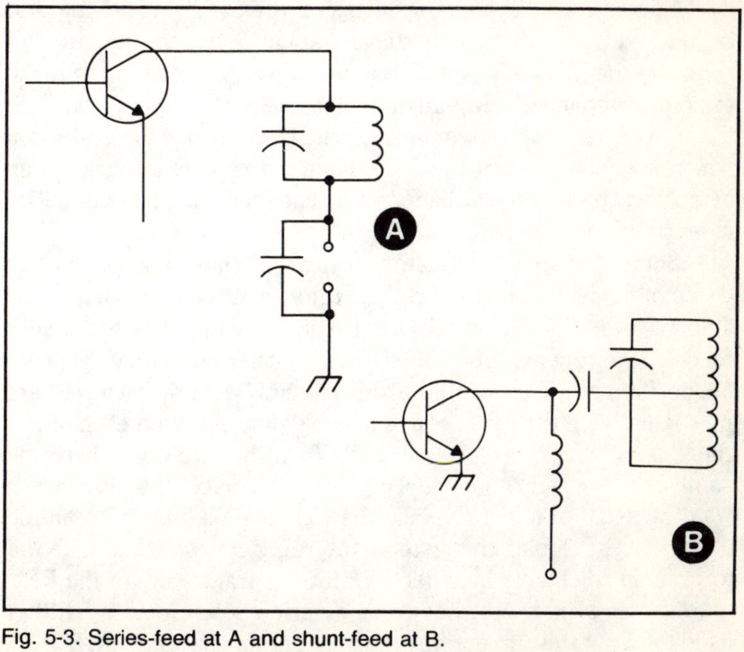

Fig. 5-3. Series-feed at A and shunt-feed at B.

tank circuit. In the parallel feed system of Fig. 5-3B, the collector current does not flow through the rf tank circuit.

Armstrong Oscillator

The Armstrong oscillator is one of the earliest oscillators. It is named after its inventor, Major Edwin Armstrong. In most classical versions of this circuit, the active element was a vacuum tube. Here a junction field-effect transistor is used. Any active device can be used, including bipolar transistors and integrated circuits.

An LC tank circuit L1/C1 is connected between the gate of the JFET and ground. Capacitor C2 serves as a DC blocking capacitor and provides some grid leak bias (actually, *gate leak bias*) for the JFET. The parallel resonant tank circuit shunts to ground all frequencies except its resonant frequency.

Feedback from the drain of the JFET is provided by a link-coupled *tickler* coil (L2). When power is first applied to the JFET, the drain current is rising from zero to its normal quiescent value. This changing current creates a changing magnetic field around L2. Because L2 is normally closely coupled with L1, this magnetic field will shock-excite the L1/C1 tank circuit. These oscillations are developed across the high impedance of the parallel resonant tank circuit and become an input signal voltage between the gate and source of the JFET. The oscillatory voltage is amplified by the JFET to create a drain current variation identical to the gate signal. This signal in turn sets up variations in the magnetic field of the tickler coil which are induced into the L1/C1 tank circuit. Once this action begins in earnest, the oscillations are sustained, and the tank will be re-excited once every cycle.

Some Armstrong oscillators, especially those used as a radio detector circuit, contain a regeneration control that adjusts the amount of feedback needed to sustain the oscillations. In some older models, the regeneration control was a tickler coil with an adjustable position viz. The coil in the tank circuit. When the two coils are positioned with their respective axes 90 degrees from each other, there is practically zero feedback. When the two coils share the same axis, however, the coupling is close to unity (100 percent). It was the usual practice to place the tickler on a gimbel mounting inside L1. In bipolar transistors, the regeneration could take the form of an adjustable bias network for the transistor. In the FET case shown in Fig. 5-4, however, a series potentiometer (R3) is used to adjust the current flowing in the drain circuit of the JFET.

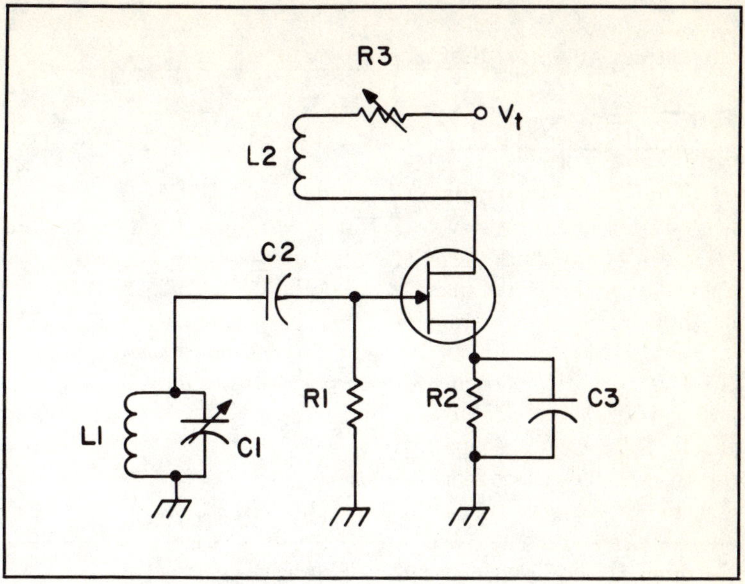

Fig. 5-4. Armstrong oscillator.

The Armstrong oscillator is not extremely popular today. But it is used in receivers/transceivers and certain other cases in the form of fixed frequency oscillators such as the BFO, etc.

TGTP Oscillator

This class is actually misnamed for modern circuits. It is the *tuned-grid—tuned-plate* oscillator circuit (Fig. 5-5). The name was appropriate in vacuum tube circuits, but is inappropriate for solid-state circuits. The name none the less persists, however. So you please forgive us if we keep calling the circuit a TGTP oscillator, even though you know it is a tuned-base—tuned-collector (TBTC) or tuned-gate—tuned-drain (TGTD) circuit in reality.

The oscillator works because of the interelement, or interelectrode, capacitances. All tube elements, or transistor junctions possess some amount of capacitance. In an rf amplifier circuit, these must be cancelled in the process called neutralization. But in TGTP oscillator we take advantage of these capacitances. Capacitances C_{cb} (collector-base) and C_{be} (base-emitter) form a capacitive voltage divider inside the transistor. The collector-emitter current, then, in effect creates an in-phase voltage drop across the base-emitter capacitance. This forms a valid input signal for the transistor and causes it to be amplified. After a few cycles, this oscillation is

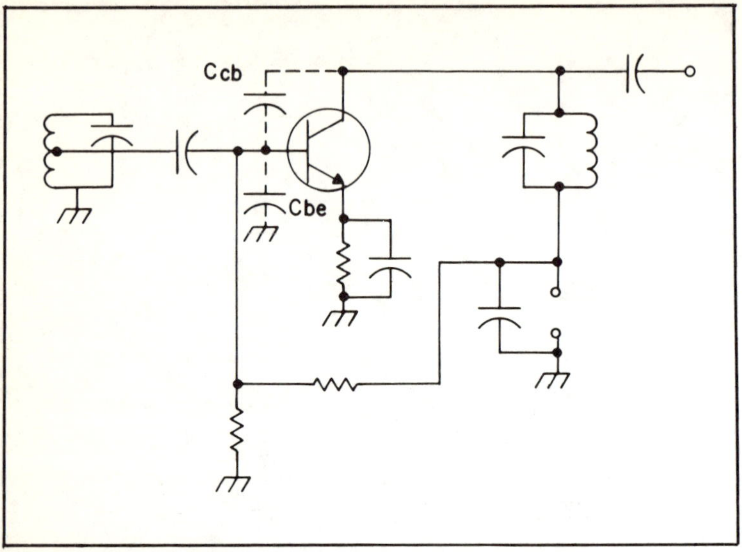

Fig. 5-5. TGTP oscillator, although this really uses a transistor.

sustained. The signals are selected by the collector and base LC tank circuits. Only those signals at the resonant frequency of the tanks (which are shock-excited by the transistor currents) will oscillate.

The collector and base tank circuits are tuned to approximately the same frequency. They are at substantially the same frequency, but are usually detuned from each other by a small amount in order to make the oscillation more stable. When they are identically tuned to the same frequency, the oscillation will tend toward instability.

In vacuum tube circuits, the interelectrode capacitances are usually quite sufficient to maintain oscillation. In fact, it is usually *so* sufficient, that rf amplifier designers who tune both grid and plate have a fit trying to neutralize the circuit! But in transistor circuits, the interelement capacitances are usually somewhat smaller. In those cases, there might be an external capacitor connected either from collector to base or collector to emitter.

Hartley Oscillator

Figure 5-6A shows the Hartley oscillator circuit. The principal feature of this type of oscillator is the *tapped coil* in the LC tank circuit as part of the feedback network. Don't confuse the tapped coil used in some rf amplifers as a means for impedance matching with the tapped coil used in some rf amplifiers as a means for

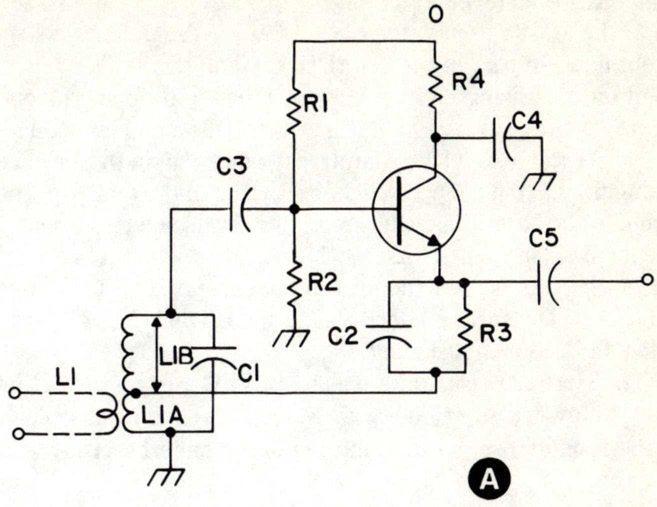

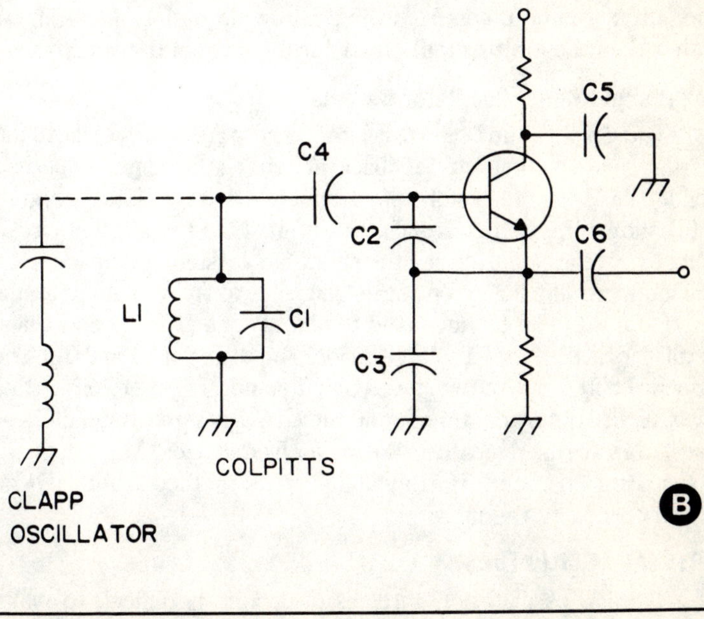

Fig. 5-6. Hartley LC oscillator at A and Colpitts LC oscillator at B.

impedance matching with the tapped coil of the Hartley oscillator. If the coil is part of a feedback network, then it is an oscillator, not an impedance matched amplifier.

In the Hartley oscillator circuit shown, the emitter-collector current flows in part of the coil (L1B). When the oscillator is first turned on, the rising emitter current creates a varying magnetic field around the portion of L1 in which it flows. This induces a current into the rest of L1 by autotransformer action, thereby shock exciting the LC tank. The oscillatory current in the tank presents a voltage to the base of the transistor, and this is seen as a valid sine wave input signal. When this signal is amplified, it creates a varying emitter-collector current identical in waveshape to the base voltage (a sinusoid). This further shock excites the LC tank via L1A, and the oscillation becomes sustained.

The Hartley oscillator is popular in VFO designs, and is used mostly in low-frequency circuits. Many AM-band radio receivers use the Hartley oscillator as the LO in the superheterodyne process.

Output can be taken from any of several points in the Hartley oscillator. In the example shown, the output is taken from the emitter of the transistor through DC blocking capacitor C5., We could also take an output signal from a link coupling to L1. Similarly, the output can be taken from the emitter tap on L1, a special tap (other than the emitter tap) on L1, or the base of the transistor.

Colpitts and Clapp Oscillator Circuits

The Colpitts and Clapp oscillator circuits are both similar to the Hartley, except that the feedback network is a tapped capacitor voltage divider instead of a tapped inductive voltage divider. Figure 5-6B shows the Colpitts oscillator circuit. Like the Hartley circuit before it, we are considering the shunt fed version, so capacitor C5 is used to set the collector of the transistor at AC ground potential for rf signals. The feature used to identify the Colpitts and Clapp oscillators is the tapped voltage divider consisting of C2 and C3. The principal difference between the Colpitts and Clapp versions is that the *Colpitts* is *parallel* tuned, and the *Clapp* is *series* tuned. These oscillators are used mostly in the higher frequencies. Many amateur transmitters use both Hartley and Colpitts methods, but the Colpitts is probably the more popular.

CRYSTAL OSCILLATORS

The LC oscillator, regardless of design, is difficult to make sufficiently stable for most high-frequency (and up) oscillator circuits. But a certain form of oscillator frequency control provides the needed stability: the *crystal oscillator*.

Certain naturally occuring crystals—namely quartz—exhibit a phenomenon called *piezoelectricity*. When such a crystal is deformed mechanically, an electrical charge appears across the crystal faces. If we shock-excite the crystal by hitting it with a mechanical impulse, the slab of piezoelectric crystal will *vibrate* at a resonant frequency that is determined in part by its physical dimensions. The charge appearing across the crystal faces, then, will be an AC potential at the resonant mechanical frequency of the crystal slab.

Similarly, when an electrical charge is placed across the surfaces of the crystal slab, mechanical deformation takes place. If an AC signal is applied across these plane surfaces, the crystal will attemp to vibrate at the frequency of the AC signal. These oscillations don't become large enough to become really useful unless the AC frequency is at, or near, the mechanical resonance of the crystal.

Figure 5-7A shows how a resonant slab of crystal material is physically mounted to a holder for use in radio transmitter fre-

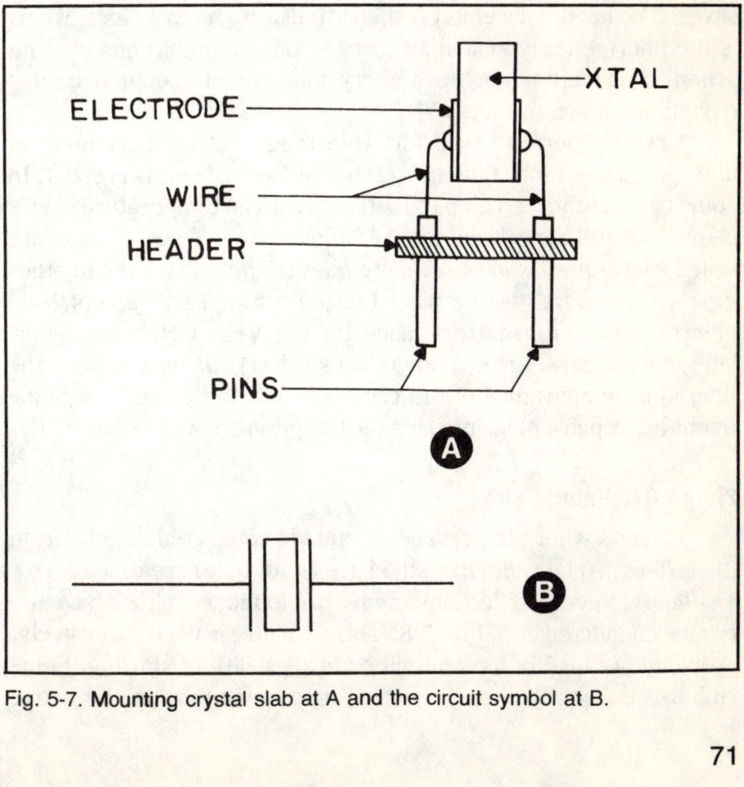

Fig. 5-7. Mounting crystal slab at A and the circuit symbol at B.

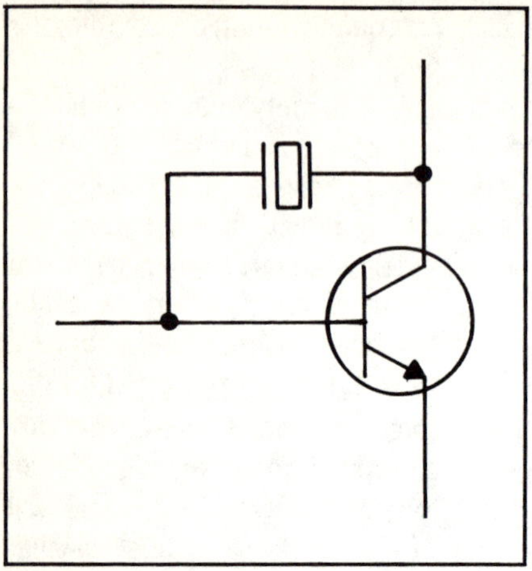

Fig. 5-8. Pierce oscillator.

quency control. The slab is placed between two electrodes. In older crystals, such as the FT-243 holder, the electrodes were held to the surfaces with spring tension. In all modern crystals, however, the electrodes are made of silver that is vacuum-deposited onto a portion of the crystal surface. The wires are then soldered to the silver. The header assembly is then fitted into a metal holder, which is then hermetically sealed against the outside environment. The schematic circuit symbol for the crystal element used in oscillator circuits is shown in Fig. 5-7B.

Crystals perform well for wide ranges of temperature and mechanical vibration. But still a certain amount of *drift* is present. In some cases, other circuit parameters are given temperature coefficients that will cancel the drift of the oscillator. These circuits are called *temperature compensated crystal oscillators* (TCXO). In other cases, the crystal will be placed inside a temperature controlled environment. Transmitters made for the VHF/UHF commercial land-mobile market frequently use a small *crystal oven* to keep the operating temperature of the crystal at 75° to 85° celsius. Some ovens are capable of maintaining the temperature within ±0.01° C.

Pierce Oscillator

Circuits where the crystal element is connected directly from the collector (plate, drain) to the base (grid, gate) are called *Pierce oscillators*. Several different varieties are found, but all will have the crystal connected as in Fig. 5-8. This oscillator is used extensively, probably because of its simplicity and reliability of starting. Some circuits, especially those in which a vacuum tube is the active element, will use a DC blocking capacitor between the plate and the crystal in order to prevent the high voltage DC from stressing the crystal. In low-voltage transistor circuits, however, it is quite common to find the crystal directly connected between the two terminals.

Miller Oscillator

The Miller oscillator circuit of Fig. 5-9 is one of the most commonly found in VHF-FM transmitters. The crystal is found connected between the gate (base, grid) and ground. The output is tuned to a frequency approximately equal to the crystal frequency by parallel resonant tank circuit L1/C1. This oscillator is essentially a crystal version of the TGTP oscillator circuit.

The Miller oscillator circuit will not oscillate if the output tank circuit is mistuned. The stability of the circuit is dependent also upon the tuning of the tank. The tuning will change the drain current of the oscillator, and the slope of the current change as the tank is tuned will be different on different sides of resonance. If we ap-

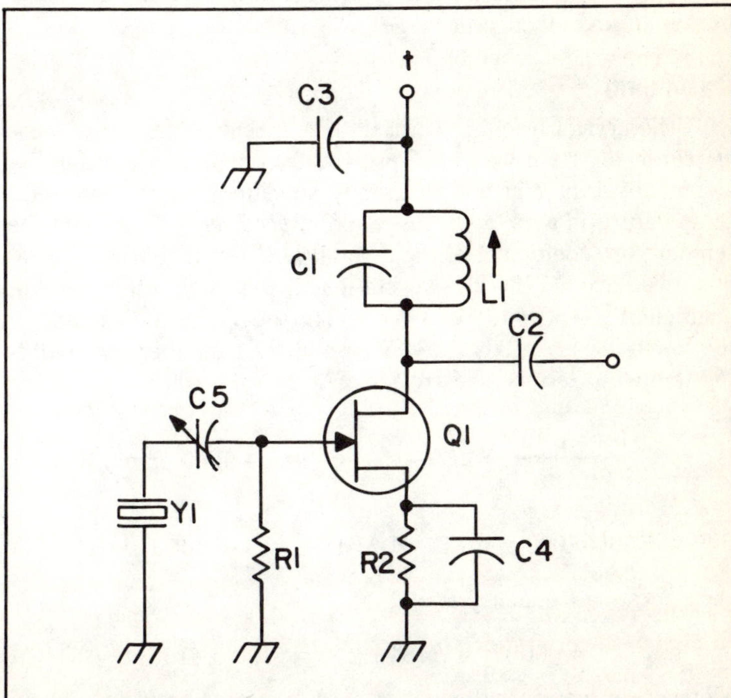

Fig. 5-9. Miller oscillator.

proach the resonant point from below, for example, the slope is less than if we approach the resonant frequency from above. The correct point is approximately 50 to 75 percent up the shallow side of the tuning slope. This is the point of maximum stability.

The variable capacitor in series with the crystal is used to vary the actual operating frequency of the oscillator. One of the factors that determines the oscillating frequency is the capacitance seen by the crystal when it looks into the circuit. By placing a series, or shunt, capacitance in the crystal circuit, the frequency can be pulled a small amount. This feature is used to net the transmitter onto the correct frequency. This same technique could be used in any of the oscillator circuits, but is shown only in one for simplicity's sake.

Colpitts Crystal Oscillator

An example of a Colpitts crystal oscillator is shown in Fig. 5-10. This circuit is also very popular in amateur equipment, quite possibly more popular than the Miller circuit. Again, we note that the Colpitts oscillator is identified by the capacitive voltage divider C1/C2. The crystal forms a tank circuit in the base circuit, and is excited by the feedback voltage.

Crystal Drift

The crystal frequency will drift with temperature. the manufacturer of the crystal will provide information regarding the amount of drift, and this is usually specified as the change for temperature in ***parts per million*** (ppm) ***per degree C***. This is the temperature coefficient of the crystal. Let's try an example. Suppose we have a 7184-kHz crystal oscillator, with a temperature coefficient of –50 ppm/°C. What will be the operating frequency if the frequency is exactly 7.184 MHz at 60° C, and then the operating temperature increases to 80° C?

$$\frac{-50\ \text{ppm}}{^\circ\text{C}} \times (80 - 60)^\circ\text{C} = -1000\ \text{ppm}$$

$$(-1000\ \text{ppm})(7.184\ \text{MHz}) = -7184\ \text{Hz}$$

The operating frequency is

$$7{,}184{,}000\ \text{Hz} - 7184\ \text{Hz} = 7{,}176{,}816\ \text{Hz}$$

We must also consider the matter of temperature coefficient when trying to keep a transmitter on the correct frequency. We need to know the temperature range of the operating environment, and the frequency tolerance of the transmitter frequency. An AM broadcaster, for example, must operate within ±20 Hz of the assigned operating frequency. If a crystal has a temperature coefficient of −15 ppm/°C, the operating temperature must remain within 2° of the calibration temperature. In that case, it would be necessary to operate the crystal in a 0.1° (or better) oven.

Ultraudion Oscillators

Many VHF and UHF LC oscillators are *ultraudion* oscillators. This is the standard oscillator used in TV tuners. The oscillator is actually a Colpitts type, but the circuit schematic seems to contain only an inductor. The capacitances are the interelectrode capacitances of the tube or solid-state device. The capacitances are present, but are hidden in the form of the electrode capacitances of the active device.

Overtone Oscillators

An *overtone* is a frequency that is almost a harmonic of the crystals fundamental frequency. The overtone is always approxi-

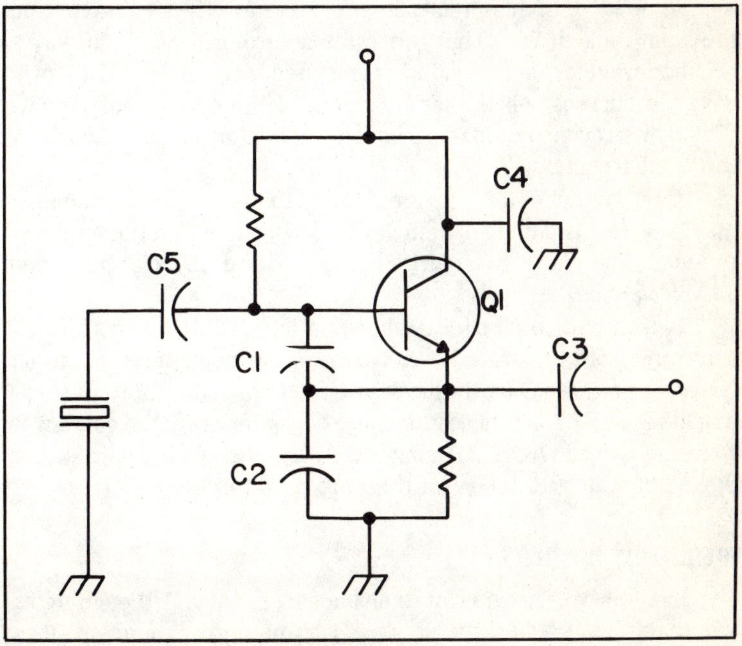

Fig. 5-10. Crystallized Colpitts oscillator.

mately an odd integer multiple of the fundemental. But if the frequency is an exact multiple it is a harmonic. Some crystals can be made to oscillate in the overtone mode, which means that they will produce a signal at the overtone frequency. Overtones of 5, 7, and 9 are commonly available, and this type of crystal is usually used at frequencies above 20 or 25 MHz. Fundamental mode crystals become too thin at these frequencies, and this makes them too fragile for common use.

Overtone oscillators are built with LC tank circuits that are tuned to the overtone frequency, not the fundamental frequency. Note that the frequency marked on the case of an overtone crystal is calibrated at the overtone, not at the fundamental frequency.

FREQUENCY MULTIPLIERS

It is common practice in VHF transmitters and receivers to generate the carrier at a low frequency (5 to 20 MHz and then multiply it in a circuit that is designed to emphasize the harmonic content. Figure 5-11 shows two common frequency multiplier stages. The tank circuit at the input of Fig. 5-11A is tuned to the lower frequency F1. To generate harmonics, some nonlinear element is needed. In this circuit, diode D1 rectifies the sine wave from the input tank circuit. This generates a harmonic-rich, half-wave-rectified waveform. This signal is amplified in transistor Q1 to form collector current pulses that shock excite the output tank circuit. This tank circuit is tuned to a frequency that is an integer multiple of the input frequency.

If the output of a multiplier is tuned to the second harmonic, it may be listed as a doubler. Similarly, circuits in which the output is tuned to the third and fourth harmonics are called *triplers* and *quadrupler*, respectively.

A transistor multiplier is shown in Fig. 5-11B. In this circuit, the transistor is operated either class-B (without the base circuit resistor) or class-C (with the base circuit resistor). Both of these amplifier classes are highly nonlinear, so the collector current is extremely rich in harmonic content. Again, the desired harmonic is picked off with the tuning of the tank circuit in the collector.

FREQUENCY DIVIDERS

In a few cases, we want to divide a frequency. Although there are several ways to do this, almost all modern equipment uses digital logic ICs called flip-flops (FFs) to perform the frequency

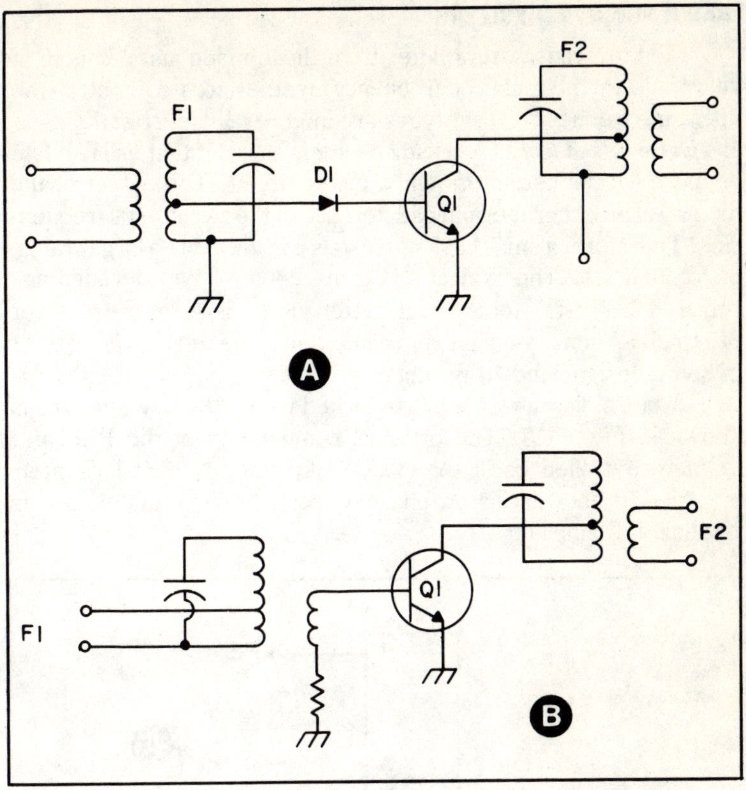

Fig. 5-11. Diode frequency multiplier at A and transistor multiplier at B.

division. A clocked J-K flip-flop (Fig. 5-12A) is a device that will change output state only on negative-going transitions of the input waveform (which must always be a square wave). This means that the waveform at the Q output will be exactly one-half of the input waveform frequency (see the timing diagram in Fig. 5-12B). By cascading several J-K flip-flops, we can obtain division ratios of 2,4, 8 or any power of two. If we provide a means for resetting all flip-flops to zero, we can obtain non-power-of-two ratios by resetting the FFs after the required number of pulses have passed. Some TTL and CMOS ICs have been designated as *counters* in base-8, base-10, base-12 or base-16. The bases, which are powers of two, are merely cascaded flip-flops. It takes three FFs to make a base-8 counter $2^3 = 8$), and four to make a base-16 ($2^4 = 16$) counter. The output of these counters is F-8 and F/16, repectively. Similarly, devices such as the 7490 base-10 counter will output a frequency of F/10.

FREQUENCY SYNTHESIZERS

Modern transmitters differ from those made just a couple of years ago in their use of frequency synthesizers to control the operating frequency. Two types are in general use: *crystal plexers* and *phase locked loops*. A crystal plexer is a circuit that will use two or more crystal oscillators and a mixer circuit. The output of the mixer will be either the sum of difference of the two crystal frequencies. Therefore, a small bank of crystals can generate a larger range of frequencies. The typical CB transceiver of two decades ago required 23 crystals for the transmitter and 23 more for the receiver (46 total). When crystal plexers came along, the total dropped to 11 in some designs and 14 in others.

A block diagram to a phase-locked loop (PLL) synthesizer is shown in Fig. 5-13. The principal components of the PLL are: voltage-controlled oscillator (VCO), divide-by-N digital counter, crystal reference oscillator, phase detector, low-pass filter, add an optional DC amplifier.

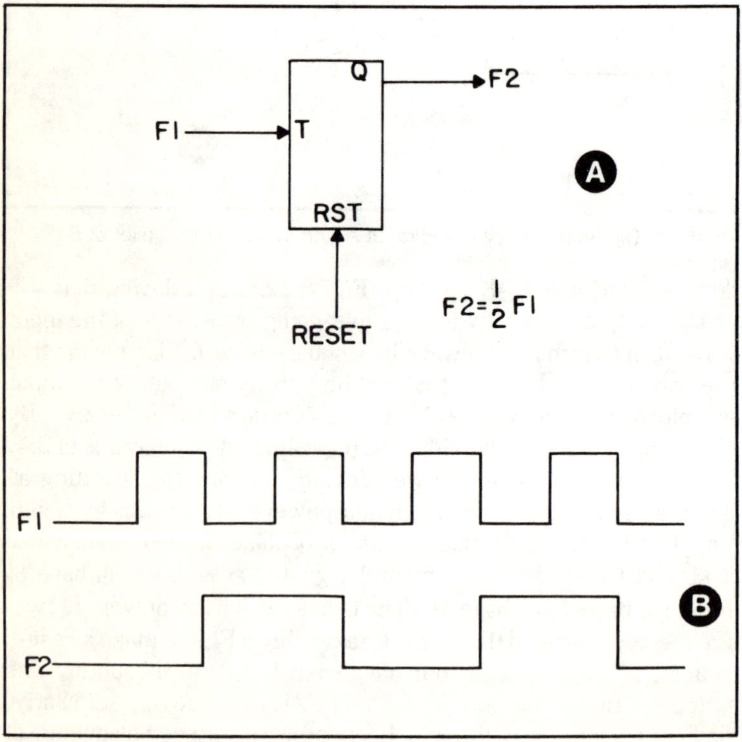

Fig. 5-12. J-K flip-flop divider.

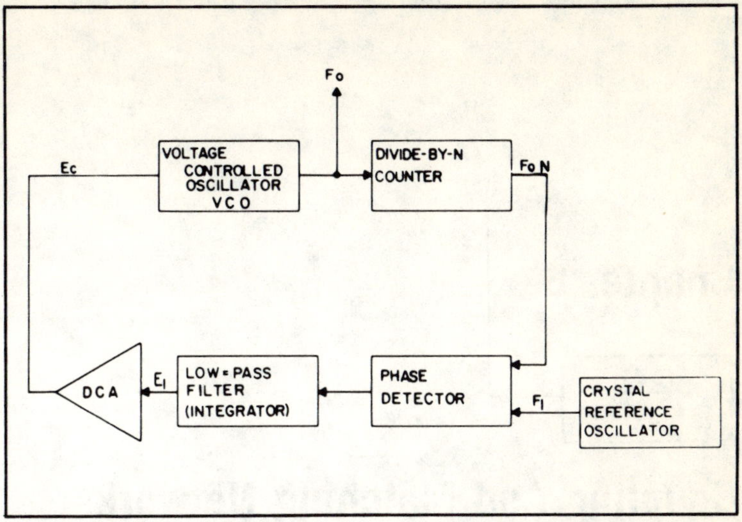

Fig. 5-13. PLL frequency synthesizer block diagram.

The VCO oscillates at a frequency that is determined by control voltage E_c, and its output is the output frequency F_0. This frequency will also be applied to the input of the divide-by-N counter. The factor N is an integer and is the division ratio of the counter. The binary word representing N is applied to a set of binary inputs. The output of this counter is F_0/N, and this frequency is applied to a phase detector circuit. The other input of the phase detector receives F1, the output of the crystal reference oscillator. This oscillator typically has a frequency in the 1-to-25-kHz range.

The output of the phase detector is a series of thin pulses of width that is proportional to the difference in frequency between F1 and F_0/N. When these pulses are integrated in the low-pass filter, the resulting DC voltage E1 will be proportional to the error between the two frequencies. This may be applied directly to the input of the VCO or be amplified in a precision DC amplifier. If there is an error, the DC error voltage will increase, thereby pulling the oscillator to the correct frequency. The accuracy and stability of the output frequency is essentially the accuracy and stability of the low-frequency crystal reference oscillator. Frequencies are changed *digitally* by changing the N-code applied to the divide-by-N counter stage.

Chapter 6

Coupling And Matching Networks

The input and output impedances of amplifier stages are not the same. If we want to couple two stages together, or an amplifier and any other load, and obtain the maximum power transfer, some matching must be provided. Recall from elementary electronic theory that maximum power transfer between source and a load occurs when the load impedance is the same as the source impedance.

In other cases, the DC voltage levels present at the output of one stage are not compatible with the DC requirements of the next stage. Consider the situation when two transistor CE voltage amplifiers are connected in cascade. In many cases, the collector potential on the first stage will be considerably higher than the base voltage of the second stage.

In both cases, some form of coupling network is needed. In the first situation we are looking for impedance matching; in the second, isolation of the two stages.

SIMPLE COUPLING CIRCUITS

Consider Fig. 6-1. This is a three-stage audio amplifier circuit specifically selected such that three different coupling methods are illustrated (we cleverly drew it that way!). Transistor Q1 is the preamplifier stage, and it is direct-coupled to the drive stage Q2. This means that the collector of Q1 is connected directly to the base of Q2. The DC voltage levels must be selected such that the normal

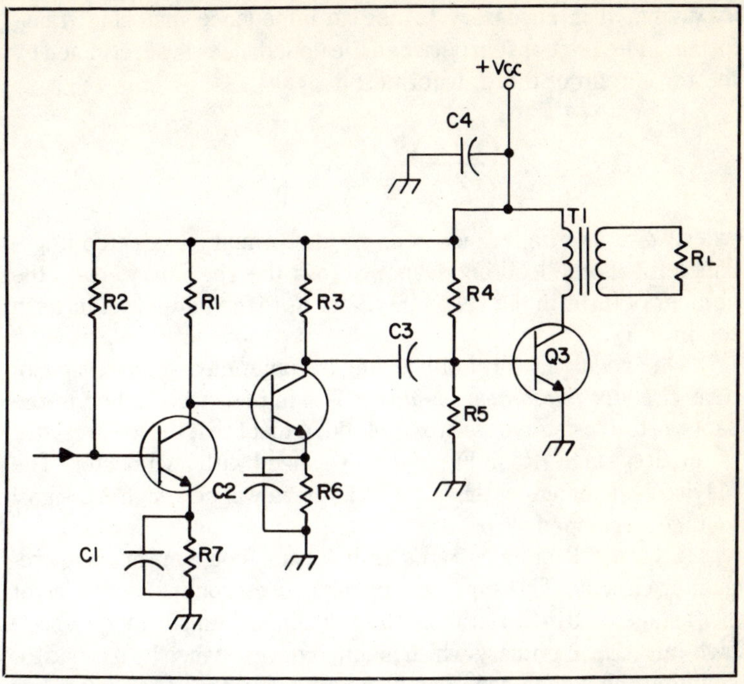

Fig. 6-1. Three different types of interstage coupling.

DC potential on the collector of Q1 will be the voltage selected for Q2. Careful design is needed. It is typical to find 1.5 to 3.5 volts used at this point in a direct-coupled circuit. Incidentally, it is common practice to use direct coupling in audio hi-fi amplifiers and in car radios.

The second form of coupling is *resistor-capacitor coupling*, which is shown between Q2 and Q3 in Fig. 6-1. The elements of the coupling network are the load resistor of Q2 (R3), the base resistor of Q3 (R2), and the coupling capacitor (C3) between them. The reactance of C3 at the lowest frequency in the amplifier passband must be low compared with the resistance of R2. This form of coupling is used in AC voltage amplifiers. Here, it isolates the base of Q3 from the collector of Q2.

Transformer coupling is shown in the collector circuit of power amplifier Q3. The purpose of the transformer is to provide impedance matching between the power amplifier and the load. The output impedance of the amplifier is typically hundreds or even thousands of ohms, where the load impedance is usually 4 to 16 ohms (in audio amplifiers). In order to keep the power transfer maximum, it is necessary to use an impedance matching transformer. The relationship between the impedances is determined by the turns ratio of the transformer:

$$\left(\frac{Z_p}{Z_s}\right)^{1/2} = \frac{N_p}{N_s}$$

where Z_p is the impedance seen by the primary (reflected impedance), Z_s is the load impedance across the secondary, N_s is the number of turns in the secondary, and N_p is the number of turns in the primary.

One additional form of coupling is sometimes seen in vacuum tube circuits: *impedance coupling*. The impedance method is the same as the resistor-capacitor method, except that the load resistor of the first stage (R3 in Fig. 6-1) is replaced with an inductor. The only real advantage of this method is there will be a high AC signal voltage developed across the choke.

Figure 6-2 demonstrates transformer coupling in radio frequency circuits. This type of coupling is used routinely in receiver rf/i-f stages and in SSB transmitter i-f stages. Transformer T1 has a high-impedance primary which is tuned to resonance by a capacitor shunting the inductance. The secondary is untuned and has a low impedance for proper matching to the base of amplifier transistor Q1.

Similarly, the collector output circuit for this stage is tuned. But in this case, both the primary and the secondary are tuned. The tuned tank circuits on the input and output sides of this amplifier make it possible to pass only a narrow band of frequencies around the resonant frequency of the two tank circuits.

AC AND DC PATHS

With signal amplifiers, such as those in Figs. 6-1 and 6-2, it becomes necessary to break the circuit into two portions: AC and DC paths for current flow. Keeping these two paths in mind, incidentally, is critical to troubleshooting a defective stage. All problems with equipment that once upon a time worked properly involves one of two possibilities: a desired path for current has been lost (an open) or an undesired path for current has been gained (a short).

Consider the DC path, using Fig. 6-2 as the example. The DC path for current in this stage includes resistor R1, the secondary of T1, the Q1 b-e junction, and resistor R2 on the input side. On the

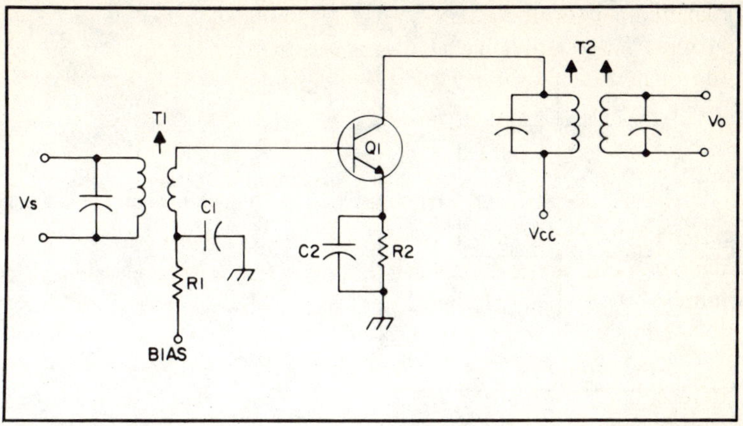

Fig. 6-2. Rf/i-f amplifier.

output side, the DC path includes resistor R2, the collector-emitter path of Q1, and the primary of transformer T2. The entire DC path is necessary for the proper operation of the circuit, and defects in this path are *usually* reflected as changes in the DC potentials on the three elements of the transistor.

The AC path contains some elements of the DC path, most notably the secondary of transformer T1 and the primary of transformer T2. But the AC path also contains other elements. Two of these elements are the primary of T1 and the secondary of T2. Both of these circuits are tuned tank circuits. The AC path also contains capacitors C1 and C2. These are bypass capacitors designed to provide a low-impedance path to ground for the AC signal, while keeping the DC path at high resistance. It is the general practice to make the capacitive reactance of a bypass capacitor one-tenth of the resistance it bypasses. For example, X_{C2} should be not more than R2/10 at the frequency of operation. In the case of the wideband stage in Fig. 6-1, the reactance of the bypass capacitors should be low, such as 0.1R, compared with the resistance at the *lowest* frequency of operation. Similarly, the reactance of the coupling capacitor in Fig. 6-1 (C3) should be one tenth of the resistance of R5.

TUNED TANK CIRCUITS

A tuned tank circuit consists of an inductor and capacitor connected either in parallel (Fig. 6-3) or series (Fig. 6-4). The tank circuit exhibits a phenomenon called *resonance*. In this condition, the inductive reactance has the same magnitude as the capacitive

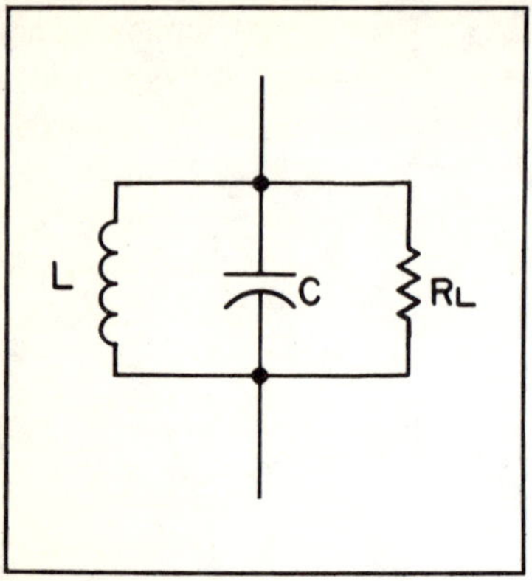

Fig. 6-3. Parallel resonant tank circuit.

reactance: $X_C = X_L$. Recall that an inductor will store energy in the magnetic field, while a capacitor stores energy in an electrostatic field. One simplified view of the circuit action at resonance is to view the circuit as two buckets for energy. Assume that a brief pulse is applied to the tank circuit. It will create a magnetic field about the coil, which will collapse as soon as the pulse passes. When the field collapses, it creates a tank circuit current that will charge the capacitor. But as soon as the current ceases, the capacitor dumps it's charge back into the circuit, creating a magnetic field around the coil. In a perfect, ideal tank circuit, this seesaw action between the coil and the capacitor would continue indefinitely. In real tank circuits, however, losses cause the amplitude of each successive oscillation to be lower than the last. After a few cycles, the oscillation dies out. Such an exponentially decaying oscillation is said to be *damped*.

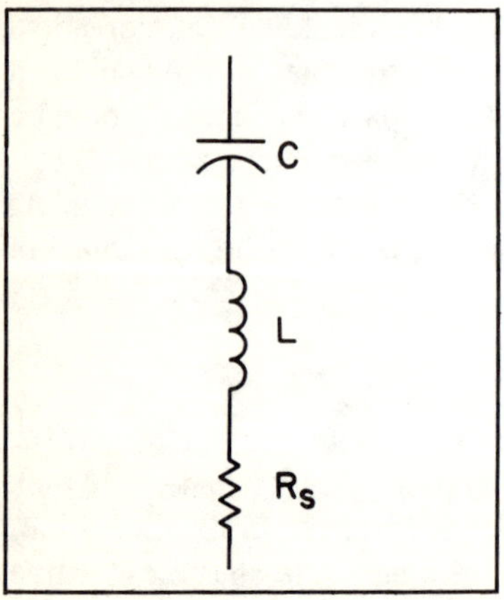

Fig. 6-4. Series resonant tank circuit.

The parallel resonant tank circuit exhibits a high impedance at the frequency of resonance, and a low impedance at frequencies removed from resonance. If a parallel resonant tank circuit is placed in shunt with a signal line, it will bypass to ground all frequencies passing down the line, except the resonant frequency. Similarly, when a parallel resonant tank circuit is placed in series with a signal line, it will pass all frequencies except the resonant frequency, which it blocks. The relationship between impedance and the applied frequency in a parallel resonant tank circuit shown in Fig. 6-5 as curve A.

At resonance, a tank circuit will exhibit such a high impedance in most cases that the resonant impedance of the tank is essentially the load resistance (R_L in Fig. 6-3). The impedance at frequencies away from resonance is given by:

$$Z_p = \frac{R_I}{\sqrt{1 + (R_L)^2(X_L - X_C)^2}}$$

where Z_p is the impedance of the parallel resonant tank at frequencies other than F_r, R_L is the load resistance, X_L is the reactance of

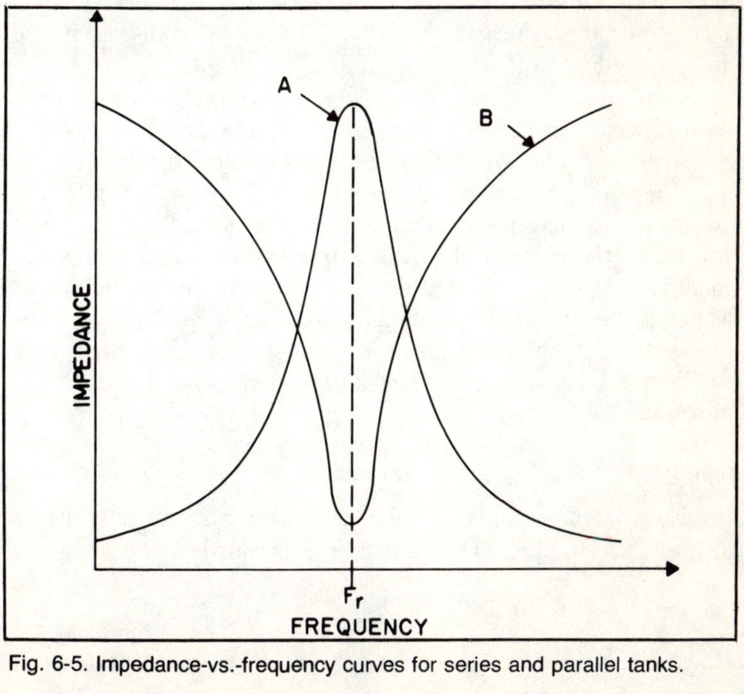

Fig. 6-5. Impedance-vs.-frequency curves for series and parallel tanks.

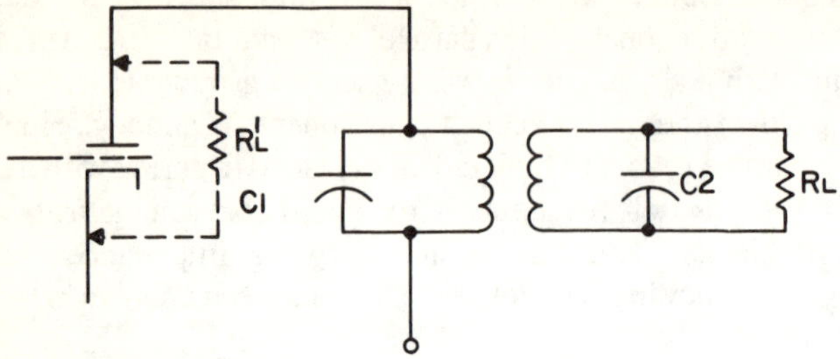

Fig. 6-6. Tank used to tune plate circuit.

the inductor at the frequency of operation (other than F_r), and X_c is the reactance of the capacitor at the frequency of operation (other than F_r).

When the tank circuit is in the plate or collector circuit of an amplifier, the load resistance R_L is the transformed resistance seen by the tube or transistor (see Fig. 6-6). In this circuit, the actual load resistance such as a resonant antenna, causes a reflected impedance across the transformer. To the tube that drives the tank circuit, therefore, the load resistance is the transformed resistance (R_L'), according to the first equation of the chapter.

The series-tuned tank circuit (Fig. 6-4) consists of an inductor and a capacitor in series with each other. The properties of the series resonant tank are exactly the opposite of the parallel properties. The impedance across the series tank (curve B in Fig. 6-5) is low at the resonant frequency and high at frequencies away from resonance. If a series tank circuit is placed such that it shunts the signal line, therefore, it will bypass signals at the resonant frequency and allow off-resonant signals to pass along the line. Similarly, when a series resonant tank is placed in series with a signal line, it will pass only the resonant frequency and attenuate all others.

Tank Q

Tank circuits can be rated according to a figure of merit or quality figure (*Q*). The *Q* of a tank is expressed by:

$$Q = \frac{R_L}{X}$$

In most cases the Q of the capacitor is so high compared with the Q of the inductor that the inductor predominates. In those cases, we can rewrite this equation in the form $Q = R_L/X_L$.

The Q determines the bandwidth of the tank circuit. All tank circuits will pass some frequencies close to resonance; it is not a sharp response only at the resonant frequency. Figure 6-7 shows the frequency response curves of tank circuits with high, medium, and low values of Q. Note that the high-Q tank circuit passes a much narrower band of frequencies than do the other two. We can control the Q of a tank circuit by choosing coils of appropriate Q. For example, with a high-Q tank, we would choose a coil with a very low resistance winding. We can also vary the Q, which is desirable in some cases, by placing different value resistances in parallel with the coil. This is sometimes done in receiver circuits.

The Q is defined by the relationship of the bandwidth of the tank circuit to the resonant frequency. The bandwidth is measured between the points where the response falls off to 0.707 times its value at the resonant frequency. The bandwidth (see Fig. 6-8) is F_h-F_L, and the Q is defined as:

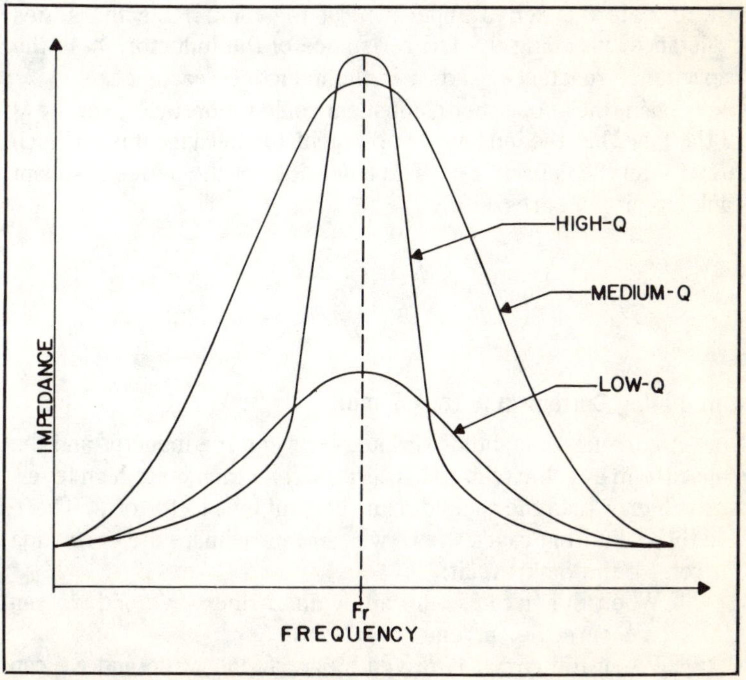

Fig. 6-7. Effect of Q on bandwidth.

$$Q = \frac{F_r}{F_h - F_L}$$

We can also specify bandwidth as:

$$BW = \frac{1}{2\pi R_L C}$$

It is considered good engineering practice to make the *Q* high enough to attenuate harmonics and other responses removed from resonance. But there are also some losses in very high-*Q* tank circuits. They are also a bit difficult to tune manually. These situations call for a trade-off between high and low *Q*, so most transmitter designers pick a *Q* between 10 and 20. A *Q* of 12 is common.

The impedance of a series-tuned tank circuit is a little different from the parallel resonant tank and is given by:

$$Z_s = \sqrt{(R_s)^2 + (X_L - X_c)^2}$$

where Z_s is the series impedance of the tank, R_s is the series resistance, including the DC resistance of the inductor, X_c is the capacitance reactance, and X_L is the inductive reactance.

Again, the *Q* of either component could theoretically set the *Q* of the tank, but the inductor *Q* predominates because it is so much lower than the capacitor *Q*. The bandwidth of the series resonant tank circuit is expressed by:

$$BW = \frac{R_s}{2\pi L}$$

Circulating Current In A Tank Circuit

A current constantly circulates between the inductor and the capacitor in a tank circuit. This current can reach quite high levels and is higher than the applied rf line current by a factor of *Q*. There are three different cases where we can approximate the circulating current of the tank circuit:

- ☐When the circuit is driven by an rf sine wave and we can measure the current
- ☐When the circuit is driven by an rf sine wave and we can measure the voltage across the inductor

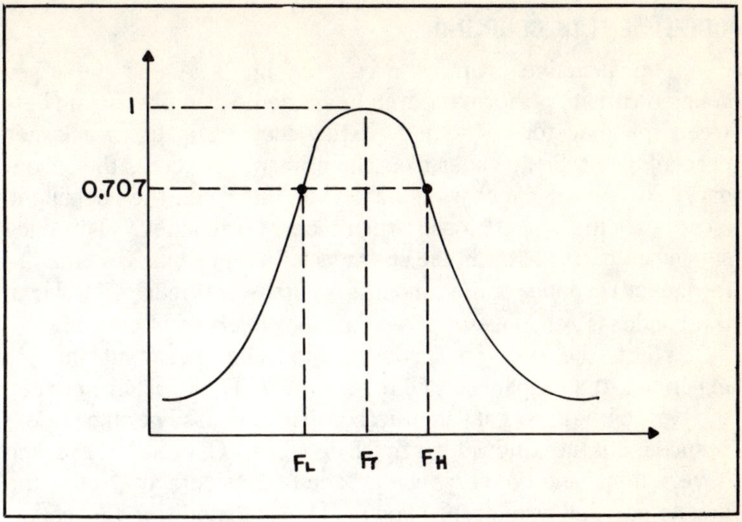

Fig. 6-8. Half-power points.

□When the tank is connected across a tube anode circuit and we know the DC current level.

These are expressed respectively, by:

$$I_c = I_{rf} \times Q$$
$$I_c = E_{rf}/X_L$$
$$I_c = 2 \times I_{DC} \times Q$$

Figure 6-9 shows several different tuned tank circuits used as either interstage or output coupling in transmitters and receivers. The circuit in Fig. 6-9A is a double-tuned tank circuit. This design will produce a wider frequency response than will the single-tuned parallel resonant tank circuit. The bandwidth of this circuit is:

$$BW = \frac{0.707C_2}{2\pi R_L' C_1}$$

where BW is the bandwidth in hertz, C_1 is the capacitance of C1 in farads, C_2 is the capacitance of C2 in farads, and R_L' is the transformed load resistance (see Fig. 6-6). The value of R_L seen by the tube is $(2R_LC_2/C_1)$, where R_L is the load impedance connected across the secondary.

COEFFICIENT OF COUPLING

The inductive coupling circuit of Fig. 6-9A and the other coupled circuits perform differently as the degree of coupling between the inductors is varied. If the coupling is tight, a larger percentage of the flux lines from the primary cut across the secondary, so a wider bandpass is achieved. But when the coupling is loose, only the signals close to the resonant frequency will produce significant current flow in the secondary. In those tank circuits, the frequency response of the bandpass is narrower. Figure 6-10 shows the bandpass response of three different degrees of coupling.

When the voltage across the parallel resonant tank is maximum, the coupling is said to be *critical*. The coupling between the two coils is looser at subcritical coupling levels. The amplitude is reduced, but the bandwidth is much narrower. Overcritical coupling is very tight, and the response broadens considerably. Notice the double-humped waveform shown. The amplitude at the center frequency is reduced from that of the critical case, but the amplitudes of the two humps are approximately equal to the amplitude in the critical case. We often adjust the degree of coupling in order to vary the selectivity of the tuned circuit.

We express the degree of coupling using the *coefficient of coupling*, which can be found from:

$$k = \frac{L_m}{\sqrt{L_s L_p}}$$

or

$$k = \frac{X_m}{\sqrt{X_s X_p}}$$

where L_m is the mutual inductance between primary and secondary, L_p is the inductance of the primary coil, L_s is the inductance of the secondary coil, X_m is the reactance of the mutual inductance, X_s is the reactance of the secondary coil, and X_p is the reactance of the primary coil.

The circuit in Fig. 6-9B is a link-coupled tank circuit. The secondary is untuned and usually consists of a few turns of wire wound over the primary coil and coupled to it. This circuit was once used extensively in radio transmitter and is still found as the interstage coupling in many rigs. In this particular example, the primary is tapped so that the low impedance of transistor collector circuits can be matched while maintaining the higher *Q* properties of

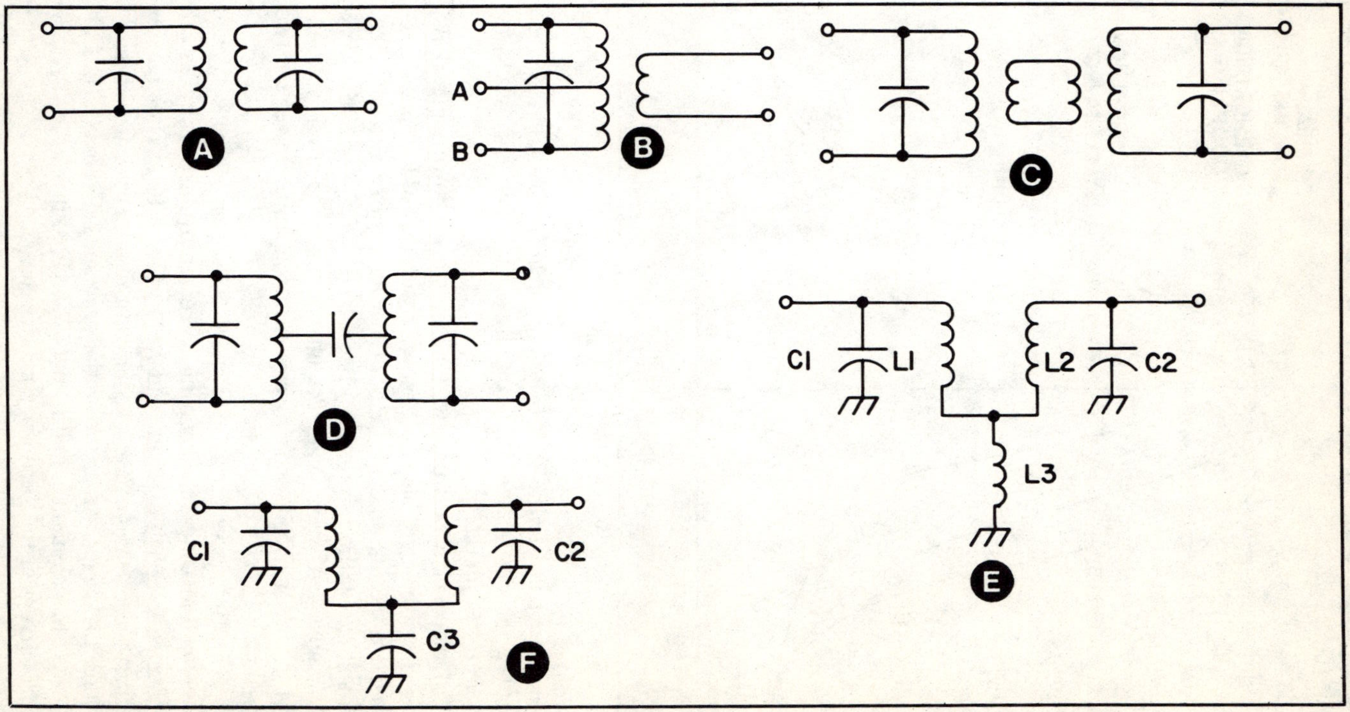

Fig. 6-9. Assorted tuned coupling networks.

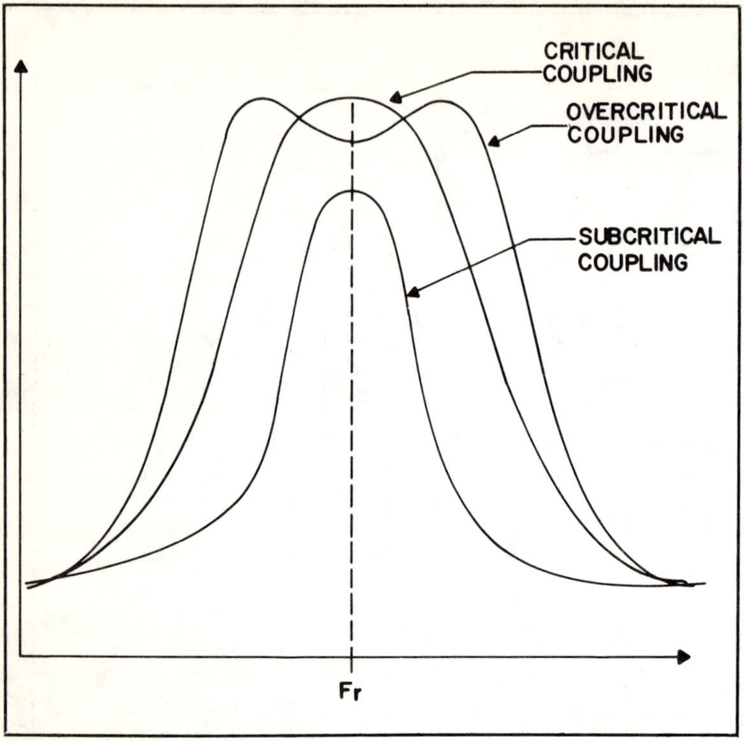

Fig. 6-10. Effects of coupling on bandwidth.

the entire tank. If the low transistor collector resistance were shunted directly across this coil, then the overall Q would be lowered considerably.

Link coupling between two tank circuits is shown in Fig. 6-9C. This circuit behaves almost identically with that of Fig. 6-9A, provided that the link windings for both coils have the same inductance. In most cases, the two tuned coils will be identical—have the same inductance—so we can make the link windings also physically identical. A similar circuit involving capacitor coupling between the two tuned tank circuits is shown in Fig 6-9D.

The circuits of Figs. 6-9E and 6-9F are also similar to each other. In both cases, the coupling is created by introducing a small reactance (X_{L3} in Fig. 6-9E and X_{C3} in Fig. 6-9F). The transfer occurs because the current flowing in the primary causes a small voltage to develop across the mutual reactance element (L3 or C3). Because this element is mutual to two circuits, it will then set up a small current in the secondary tank circuit.

L-SECTION NETWORKS

The L-section coupling network is shown in Fig. 6-11. Two forms are given here. The circuit in Fig. 6-11A is used to transform a lower impedance to a higher load impedance. This form is sometimes seen in solid-state transmitter buffer amplifiers, where the collector impedance of the first stage is less than the input impedance of the second stage. It was also commonly employed in vacuum tube transmitter final output stages, where the plate resistance of the tube is greater than the 50-ohm to 75-ohm antenna impedance. This circuit was common 2-MHz marine transmitters at one time.

You can calculate the required values for the two circuit elements from the expressions below:

$$X_L = \sqrt{(R_L R_i) - (R_L)^2}$$

$$X_c = (R_L R_i) / X_L$$

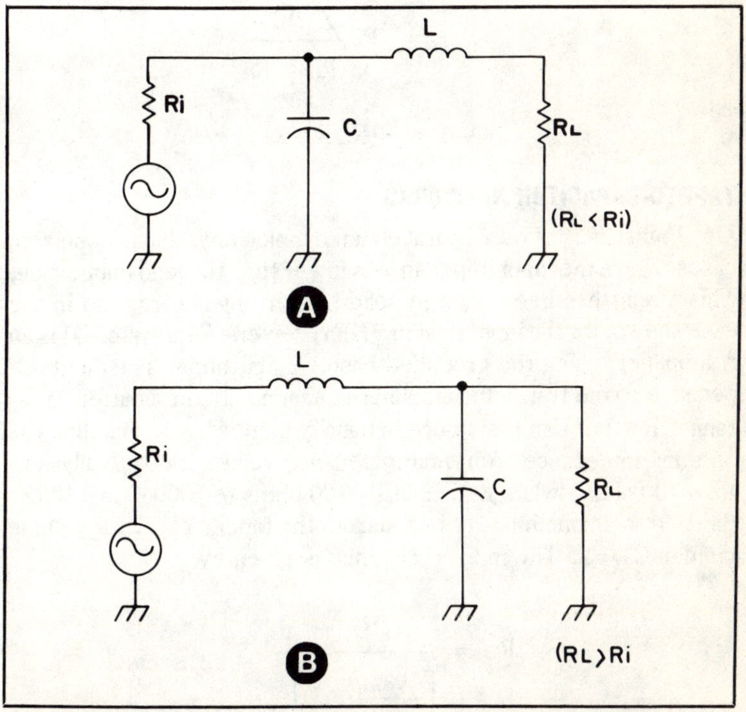

Fig. 6-11. L-section impedance matching networks for a load less than the input impedance at A and a load greater than the input impedance at B.

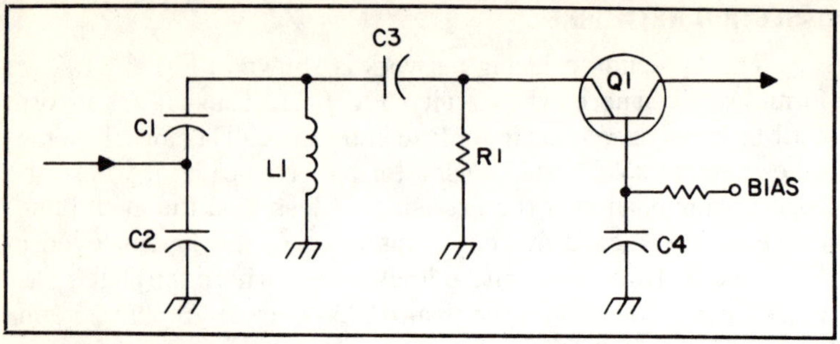

Fig. 6-12. Tapped capacitor tuned network for impedance matching.

The circuit in Fig. 6-11B is exactly the opposite configuration from Fig. 6-11A and is used to transform a lower input source resistance, R_i, to a higher load resistance, R_L. This network form is sometimes used to couple the output of a low-impedance generator, such as a transistor power amplifier, to an antenna load. The component values are:

$$X_c = R_L \sqrt{\frac{R_i}{R_L - R_{in}}}$$

$$X_L = (R_L\ R_i)/X_c$$

TAPPED-CAPACITOR NETWORKS

Figure 6-12 shows a parallel tuned tank circuit that is popular in cases where the input impedance is lower than the load impedance. This circuit has been used in solid-state transmitters, and in the case shown, for the input tank in VHF receivers. Transistor Q1 is an rf amplifier, using the grounded-base configuration. The input impedance of the transistor amplifier is essentially the emitter resistance, R1, but this resistance is usually somewhat larger than the antenna impedance. Antenna impedance values are typically less than 100 ohms, while R1 is usually 100 ohms to 1000 ohms. Impedance transformation is obtained from the tapped capacitor voltage divider C1/C2. The input resistance is given by:

$$R_i = \frac{Q_L\ X_L}{\left[\frac{C_2}{C_1} + 1\right]^2}$$

THE PI NETWORK

One of the most popular output coupling networks used in amateur transmitters is the pi network of Fig. 6-13. This circuit obtains its name from the fact that its circuit resembles the Greek letter π (pi). This circuit is used to transform the low impedance of the antenna load (R_L) to the higher impedance of the plate resistance of the vacuum tube, R_p. The values of the components are given by:

$$X_{C1} = R_p/Q$$

$$X_{C2} = \sqrt{\frac{(R_p/R_L}{Q^2 + 1 - (R_p/R_L)}}$$

$$X_L = \frac{QR_p + (R_pR_L/X_{C2})}{Q^2 + 1}$$

Select the *Q* of the circuit to be between 10 and 20, with 12 being a common value. In addition, X_L be less than or equal to $(R_pR_L)^{1/2}$.

OUTPUT COUPLING FOR SOLID-STATE TRANSMITTERS

In older vacuum tube designs, it was always true that the plate resistance of the tube was matched to a much lower antenna resistance. This is assumed in the previous equations for the pi network. In solid-state amplifiers, however, especially as the power levels increase, the situation is just the opposite. It is very

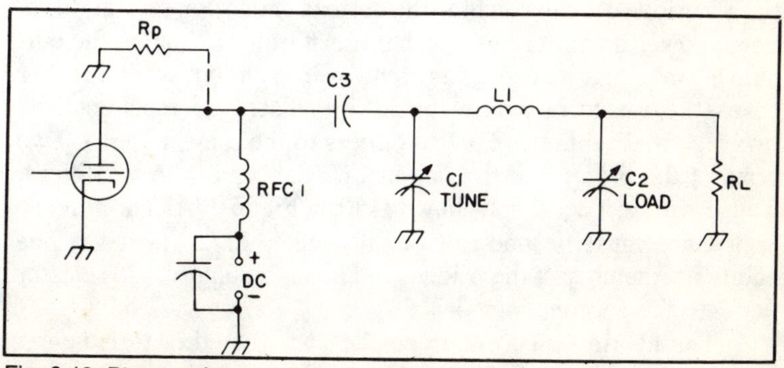

Fig. 6-13. Pi network.

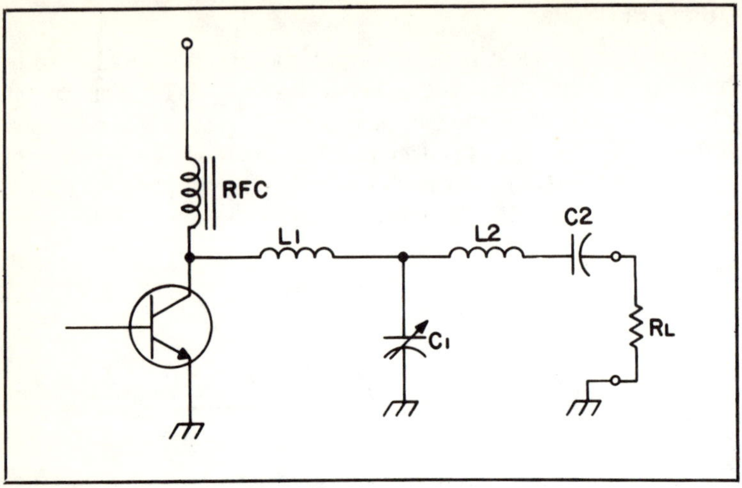

Fig. 6-14. Solid-state output matching network.

often true that the collector resistance of the transistor is less than the nominal 50-ohm to 100-ohm antenna impedance. This resistance is V_{cc}/I_c. Assuming a 13.6-volt collector potential (mobile rigs) and a collector current of 4 amperes, the collector resistance is only 3.4 ohms. Figures 6-14 and 6-15 show several popular output coupling networks for solid-state power amplifiers. Note that circuits which have several variable capacitors often have a specified adjustment sequence that must be followed when loading the amplifier into the antenna.

Figures 6-16 through 6-18 show the use of hybrid transmission line transformers for output coupling. The circuit in Fig. 6-16 is tuned, while that in Fig. 6-17 is untuned. The hybrid wideband coupler has been generalized in Fig. 6-17A. This circuit is essentially a bridge used to combine the outputs of two power amplifiers. Some power amplifiers use such a transformer to couple the outputs of two single-ended sides of the overall circuit, while in other cases, the transformer can be made into a discrete component in its own right to couple two final amplifiers to the same antenna. Two arms of the bridge are the output impedances of the amplifiers. These arms are equal and shown as R_L in Fig. 6-17A. The antenna impedance must be one-half of this value, $R_L/2$. There is one additional member of the bridge, and this is a dummy load resistor equal to the antenna impedance.

The hybrid transformers can be constructed of toroid cores and wire, as shown in Fig. 6-17B, when operating in the high-

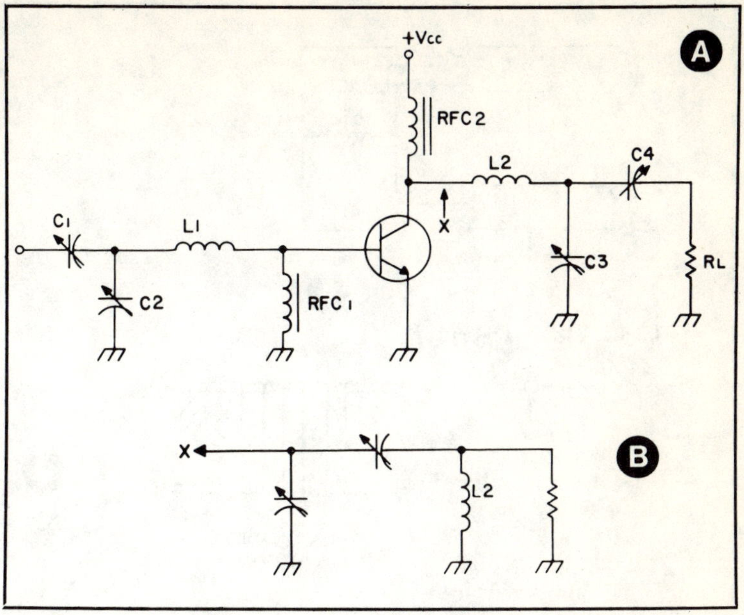

Fig. 6-15. Another solid-state output matching network.

frequency range. As the frequency increases, though, stripline construction or a coaxial combiner such as shown in Fig. 6-18 must be used. This combiner is made from ordinary coaxial cable cut to the operating frequency. Three arms of the coaxial bridge are

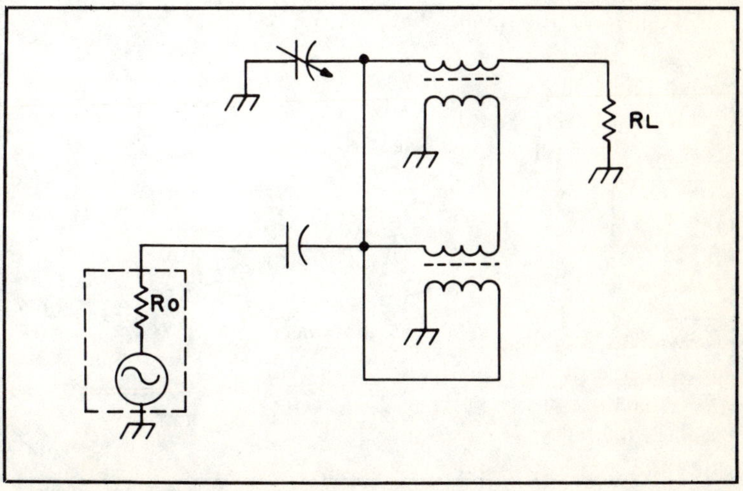

Fig. 6-16. Broadband transformer.

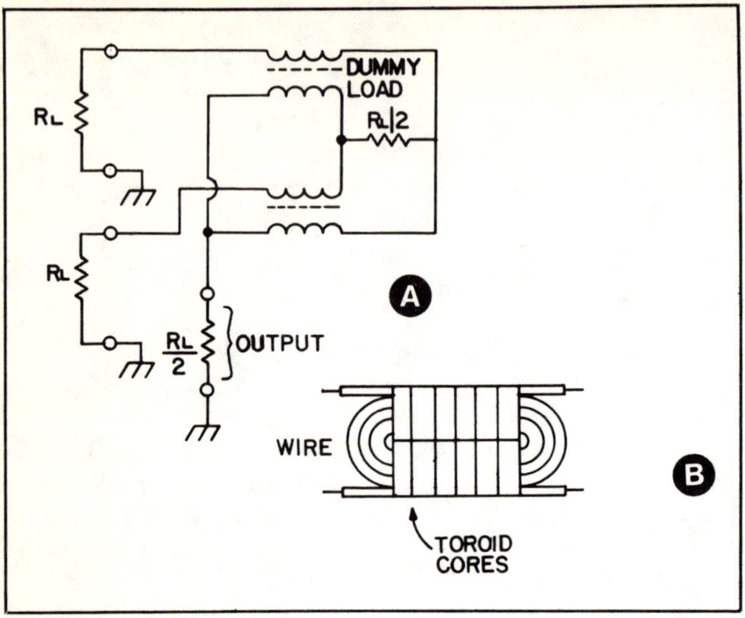

Fig. 6-17. Hybrid transformer coupling at A and typical transformer construction at B.

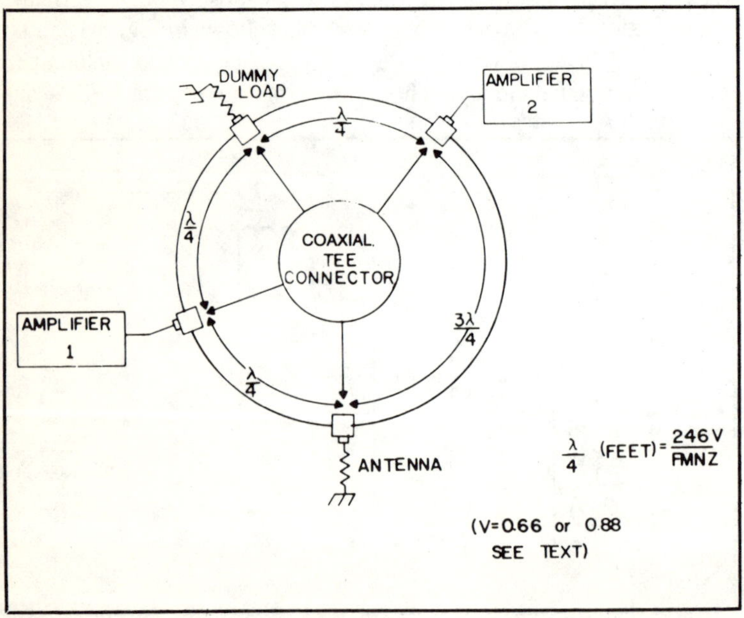

Fig. 6-18. Coaxial hybrid ring method.

quarter wavelength, while one section is three-quarter wavelength. The purpose of this is to combine the signals at the dummy load out of phase so that they cancel. This will allow the use of a low-value dummy load resistor. The only problem with this is that the circuit operates only at those frequencies close to the quarter-wavelength frequency for which the coaxial cable is cut. These lengths, incidentally, must reflect the velocity factor of the coaxial cable:

$$\frac{1}{4} \text{ wavelength} = (246 \times V)/F_{MHz}$$

where the length is in feet, V is the velocity factor of the coaxial cable, and F_{MHz} is the frequency in megahertz. Velocity factor is different for different types of transmission line. In general:

- □Foam dielectric is V = 0.8
- □Regular polyethylene is V = 0.66

Chapter 7

Vacuum Tube RF Power Amplifiers

One of the first accessories acquired by amateur radio operators, after the initial low-to-medium power transceiver or transmitter/receiver combination is obtained will be a power amplifier. The power amplifier will boost the signal output considerably, although not as much as some people think. But more of that in a moment. Some years ago, it was necessary to have a lot of space to support a power amplifier. These instruments took a 19-inch wide rack, at least four-feet high (most were seven footers, it seems!). But modern designs with advances in vacuum tubes and power supplies require only a small amount of desk top space. Figures 7-1 through 7-3 show three examples of commercial power amplifiers commonly used by amateurs.

Two different types or classes of power amplifiers are commonly used. Many commercial and homebrew amplifiers will provide both kinds: linear and class-C. The linear (typically class-B or class-AB) is used with single sideband systems because the modulation is added to the carrier in a low-level (prefinal) amplifier stage. All stages following the balanced modulator in an SSB or DSB rig must be linear. But with a radiotelegraph (CW), frequency-modulated, or high-level amplitude-modulated system, the higher efficiency of the class-C amplifier can be taken advantage of. In the case of the AM transmitter, the modulation to the carrier must be added in the final amplifier, or the same rules apply as for the SSB.

Fig. 7-1. Heath linear amplifier.

HOW MUCH POWER?

How much output power is needed? For the US amateur radio operator, it is permissible to use a power amplifier that has a DC input to the final amplifier stage of 1000 watts or less. This is 2000 watts peak-envelope-power (PEP), assuming a 50 percent duty cycle for speech.

What we mean with our rhetorical question is more a matter of economics, rather than regulations. Most people seem to think that the more you can crank out of a transmitter, the better the received signal at the other end. 'Tain't so! Some years ago, many CB operators risked breaking the law by modifying their tube-type transceivers to output more power by changing the value of screen-dropping resistors or some power-supply resistors in most cases. A couple of less scrupulous manufacturers even designed the rigs to be 10-watt to 15-watt rigs, and then cut them back with lowered screen voltages in order to comply with the law for CB operation. But they let the dealers know just how to make the changes!

Did this do any good? Is 15 watts any better than 5 watts? How much better is 1000 watts than the 250 watts put out by a heavy-duty amateur transceiver? Let's see. The expression for power gain in decibels is:

$$dB = 10 \log_{10} (P_1 / P_2)$$

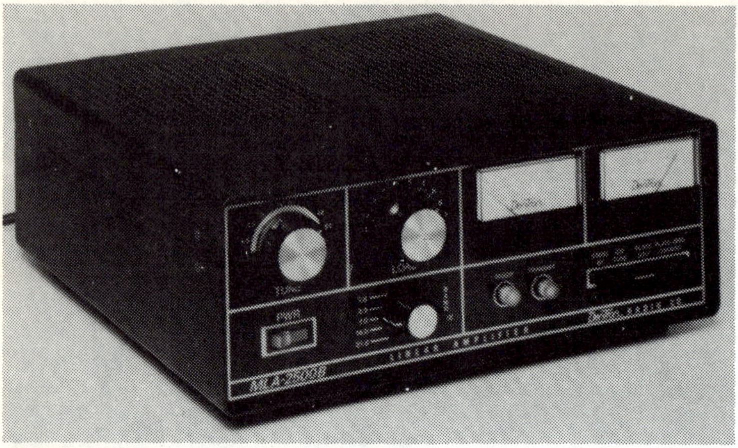

Fig. 7-2. Dentron linear amplifier.

If we solve this expression for a 2:1 power increase, the gain is 3 dB. Examining the service manual for most amateur grade receivers tells us that each S-unit on the S-meter of the receiver represents a 6 dB change in the input power to the antenna terminals, with S-9 represented by 100 μV across 50 ohms, or 2×10^{-10} watts. This means that doubling the power from your transmitter results in a mere half S-unit difference at the receiving end—not much when you consider the claim by the receiver manufacturer that one S-unit represents the smallest increase that the operator can hear. A 4:1 increase in transmitter power results a 6 dB change, which is the minimum that will do you any good. Some authorities require a power change of 5:1 as the minimum that is economically feasible. If you have a 100-watt exciter or transceiver for your transmitter, the gain is substantial (1000/10 = 10:1, or 10 dB). But if you already have a 300-watt to 750-watt rig (some transceivers go 750 watts), then examine whether the added cost of a linear amplifier will do you any good.

TUBE SELECTION

The tube used in a power amplifier greatly determines its maximum operating capability. The principal specification to watch is the anode (plate) heat dissipation, expressed in watts. The efficiency of any power amplifier is given by:

$$n = \frac{P_o(100\%)}{P_{DC}}$$

where n is the efficiency of the amplifier in percent, P_o is the rf output power delivered to the load (antenna), and P_{DC} is the DC power applied to the anode of the final amplifier tube(s).

Let's see what this means to the final amplifier tube. Assume that we have a linear amplifier with an efficiency of 45 percent, which is not an unreasonable value. The DC power applied to the anode is 1000 watts. First, calculate the output power by rearranging the last equation:

$$P_o = nP_{DC}/(100\%)$$
$$P_o = (45\%)(1000 \text{ watts})/(100\%)$$
$$P_o = (0.45)(1000 \text{ watts}) = 450 \text{ watts}$$

The output power is 450 watts. What happened to the other 550 watts? It is dissipated in the anode of the tube in the form of heat. The tube manufacturer will specify for each tube type the total amount of heat that can be dissipated and under what conditions this value is considered valid—convection or forced air cooling.

The tube or combination of tubes used in the final amplifier must have a total heat dissipation that is equal to the lost power or is *higher* than the lost power. In this case, we need to select a tube or series of tubes that will dissipate at least 550 watts. But it must be recognized that it is not really good practice to select a tube that barely meets the requirement. The tube will then run to its maximum ratings all of the time, and that is not good. A tube that is

Fig. 7-3. Heath linear amplifier.

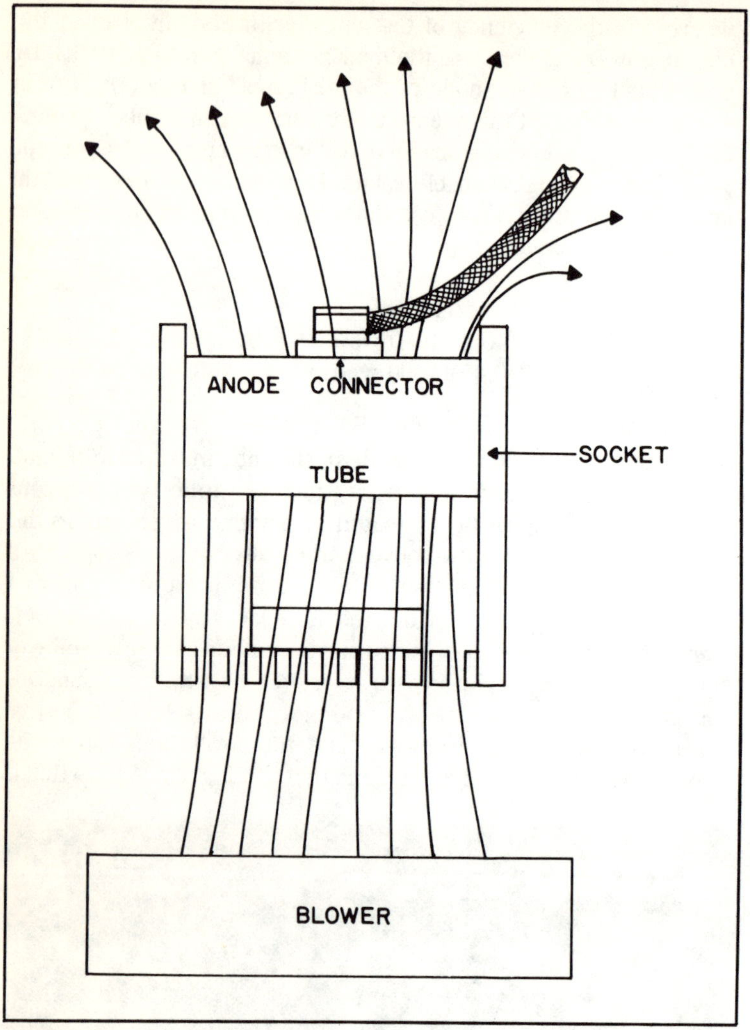

Fig. 7-4. Use of air-system socket with external anode tubes.

run "hard" all the time can be expected to have a shorter life expectancy than one which "loafs" along.

COOLING

Most transmitting tubes will dissipate the full amount of anode heat only when the envelope is sufficiently cooled. The glass seals around the anode, grid and filament terminals will break if the glass is heated too much. Part of the problem is the difference in the thermal coefficient of expansion between the glass used in the envelope and the metal electrode connections. Another part of the problem is that the glass can melt. I have seen a linear amplifier in which the blower fan failed, and the operator ignored the failure. The tube was a 4-400 (capable of 400 watts of plate dissipation), operated in a "pair of 4-400s" linear amplifier. The envelope melted and was blown inward from the action of atmospheric pressure on the vacuum inside of the tube. The explosion sounded like a handgun report! Incidentally, *I* wasn't the operator who ignored the failed blower.

All high-power amplifiers should have a blower to force air over the final tubes. A blower is cheap insurance against expensive tube failure, but it must be applied properly. If the Eimach-series power tubes are used, then the same company will also sell you an air-system socket that allows proper air flow around the tubes. Figure 7-4 shows the type of air system socket frequently used for power tubes with an external anode (4X145, 4X250, 4CX250B, etc). These tubes have a large metal finned heat sink for the anode. The air system socket usually has a cylindrical wall that snugly fits

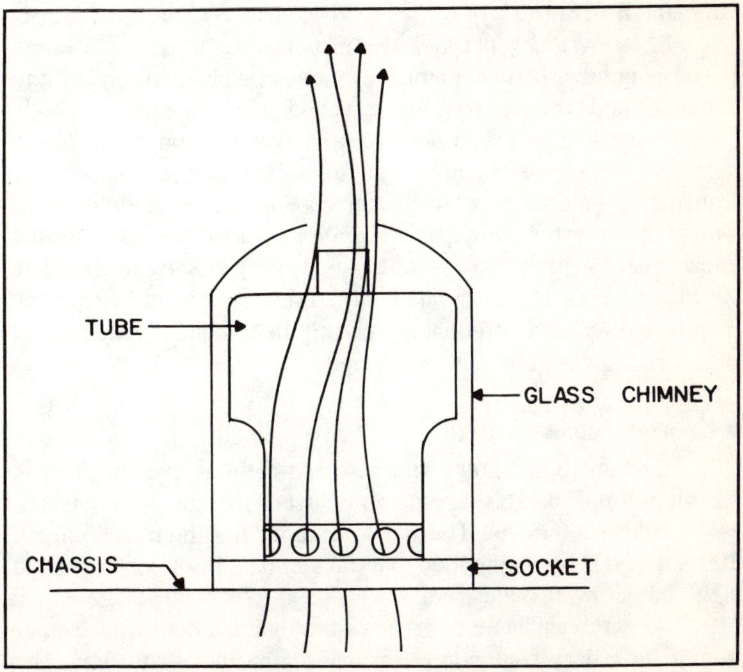

Fig. 7-5. Use of air-system socket with glass envelope tubes.

around the anode and has holes in the base to admit forced air. The blower is positioned (or plumbing provided) to force cool air from the bottom of the tube through the air system socket and out through the metal heat fins of the anode. The tube envelope seals on the bottom portion of the tube are kept cool enough to prevent their breaking (which would destroy the tube by admitting air to the inside vacuum).

Figure 7-5 shows the type of air system socket often used for the glass envelope power tubes. In this case, the holes in the base of the tube admit air from beneath the chassis (cooling the seals), and allow it to pass over the hot glass envelope, leaving at the glass chimney at the top. This tube also uses a heat-dissipating plate connector. These provide the electrical connection to the anode and are designed to have fins that will dissipate the heat. This will keep the anode connection glass-metal seal intact.

The amount of air passed over the tube is also important. The manufacturer will specify the air flow rate in cubic feet per minute, as part of the normal tube specifications. Blowers in the 30 to 150 cu. ft./min. are easily obtained.

AMPLIFIER DESIGNS

Only a few different types of amplifier circuit are in use, despite the large number of power amplifier tubes available. Grounded-grid and grounded-cathode circuits are in abudance; few common-anode circuits are found in amateur work, however. You might find a common-anode power amplifier in some high-power broadcasting transmitter stations in which water cooling of the anode is used. These use a water plumbing system and radiator to dissipate the anode heat, so they may want to keep the plumbing system and radiator to dissipate the anode heat, so they may want to keep the plumbing at ground potential by turning the circuit "upside down," or operating with the plate grounded.

A Common-Cathode Design

An example of a 1000-watt common-cathode final amplifier is shown in Fig. 7-6. This circuit can be used with the 4-125, 4-250, 4-400, and 4-500 tubes. This series of tubes has the heat dissipate 400 watts. It is recommended that the 4-400 or 4-500 tubes be used if the full kilowatt is desired.

The cathode/filament circuit is fed by a filament transformer with a secondary center-tap to provide a balanced situation. The filament transformer with a secondary center-tap to provide a balanced situation. The filiments are bypassed by not less than three 0.005 μF capacitors—one from each side to ground and one across the filament terminals. These capacitors should be mounted as close as possible to the terminals on the tube socket, and the leads should be kept as short as possible. If the leads are too long, the capacitors may perform poorly because of lead inductance, and this could lead to instability in the power amplifier. Disc ceramic or mica capacitors should be used, and they must be of good quality.

The screen grid is similarly bypassed, but with 0.001 μF capacitors. Note that there are two screen connections on these tubes, and both are bypassed to ground. If an Eimac air-system socket is used, it is possible to buy the type with built-in screen bypass capacitors. The DC potential to the screen is supplied through 50-ohm, 4-watt, noninductive resistors. The best way to obtain this value (in noninductive) is to connect in parallel two 100-ohm, 2-watt carbon composition resistor. Do not use wirewound resistors, as their inductance could lead to unfortunate resonances. Again, it is necessary to make the leads of the capacitors as short as possible.

The input circuit to the control grid is tuned by a parallel resonant tank circuit C1L2. This tank is driven from the exciter through link coupling L1. This tank should be at a low impedance to ground for rf, but there must be a DC voltage for the C-bias to the tubes. Use bypass capacitor C2 for this purpose (470 pF silver mica). This bypass capacitor, and the 2.5 mH rf choke keep the rf signal from the exciter from getting into the c-bias supply. We would ordinarily want to use a high value for C2 as practical, but this capacitor is part of the neutralization network.

The neutralization of the tube is necessary because of the common-cathode configuration. A 15 pF variable capacitor is used for this purpose (see Chapter 9). Most of these capacitors are either vacuum variables or a simple pair of metal discs separated by an air space of a few centimeters. Details on the different types of neutralization circuits and their adjustment are given in Chapter 9.

There are three elements to the anode circuit of this final amplifier: parasitic suppression, the DC network, and the output matching network. The parasitic suppression networks are R1/L4 and R2/L5. These are used to provide a high series impedance to parasitic VHF oscillations that could interfere with other service. The inductors are 4 to 5 turns of No. 12 solid copper wire, while the resistors are 200-ohm, 2-watt composition types. Four resistors are required in parallel for each resistance R1 and R2.

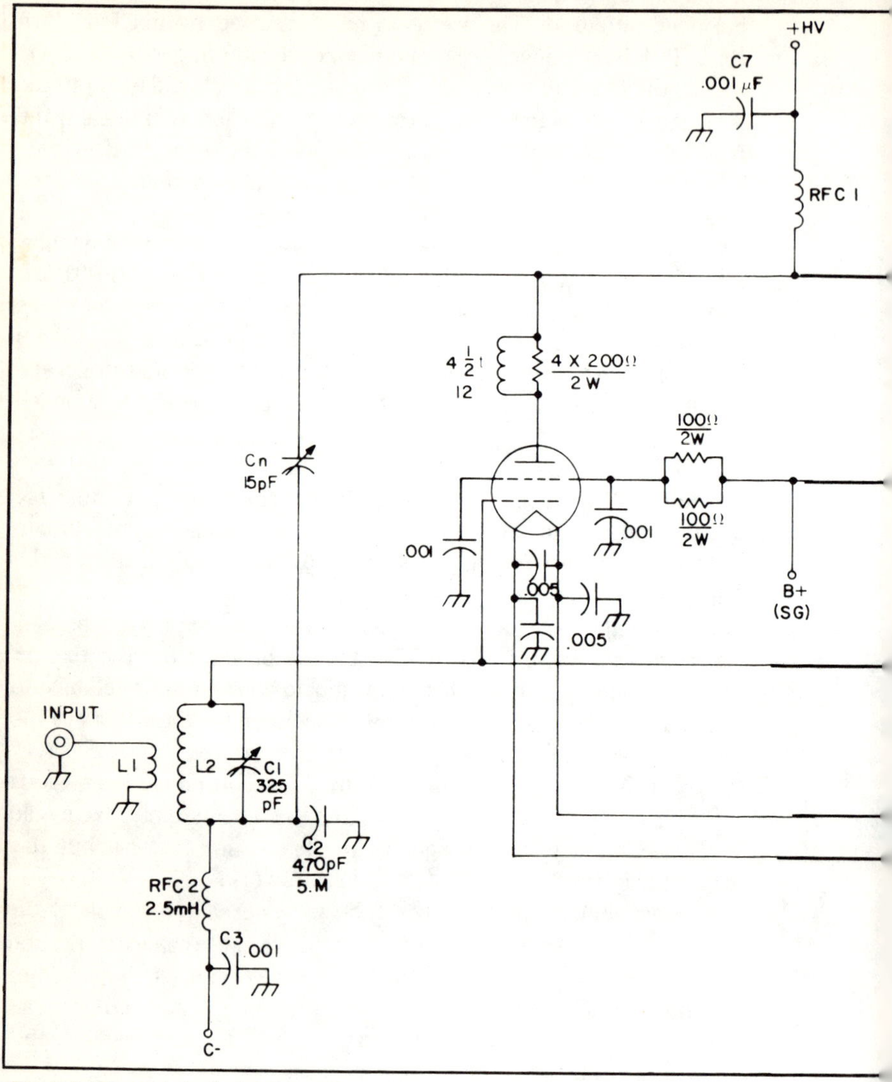

Fig. 7-6. Power tetrode final amplifier.

The DC network includes the rf choke (RFC1) and the power supply bypass capacitor, C7. The other fixed capacitor is C4, a DC blocking capacitor intended to keep the high-voltage DC from the output matching circuit. Both capacitors are 500 pF/20 kV "television" capacitors (two required in parallel for C7). These capacitors are available from Centralab and others, and they are sometimes called "doorknob" capacitors.

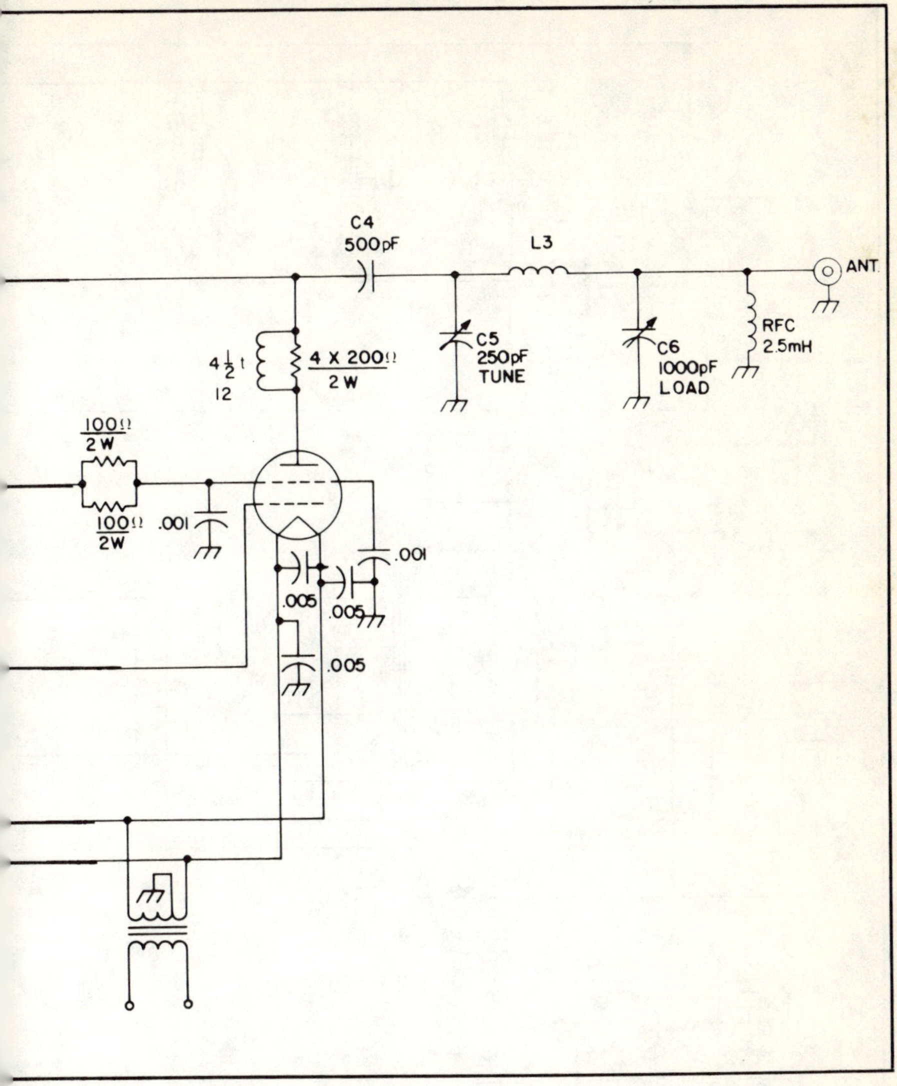

The output impedance matching network is the pi network, C5/C6/L3. This type of matching network was covered in Chapter 6. Note that B&W manufactures a premade pi network inductor with a built-on rf switch for use in amateur power amplifiers. The tuning capacitor, C5, must have a high-voltage rating of at least *twice* the DC voltage applied to the anode in linear amplifiers and *four* times in modulated class-C amplifiers.

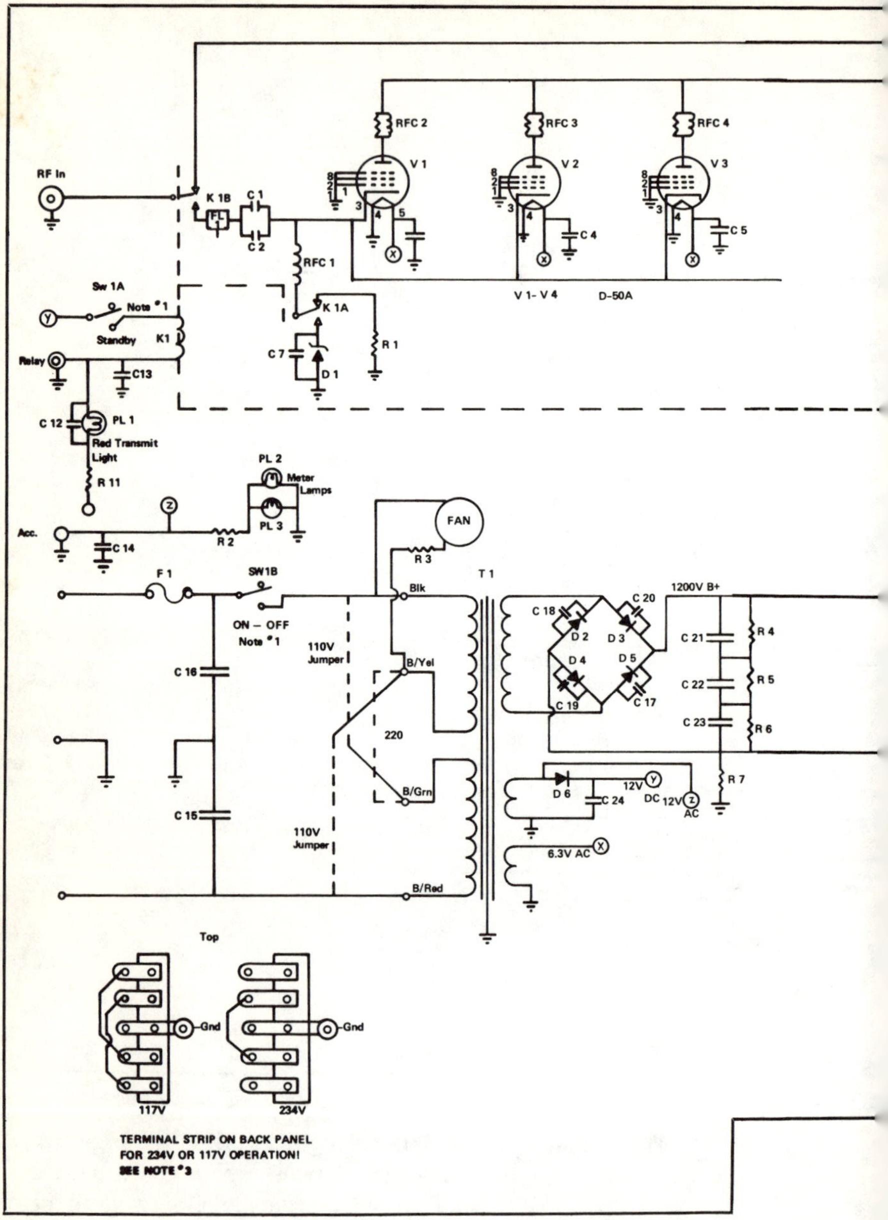

Fig. 7-7. Dentron GLA-1000 schematic.

A dangerous situation could occur if the DC blocking capacitor (C4) should become shorted: The lethal high-voltage DC on the

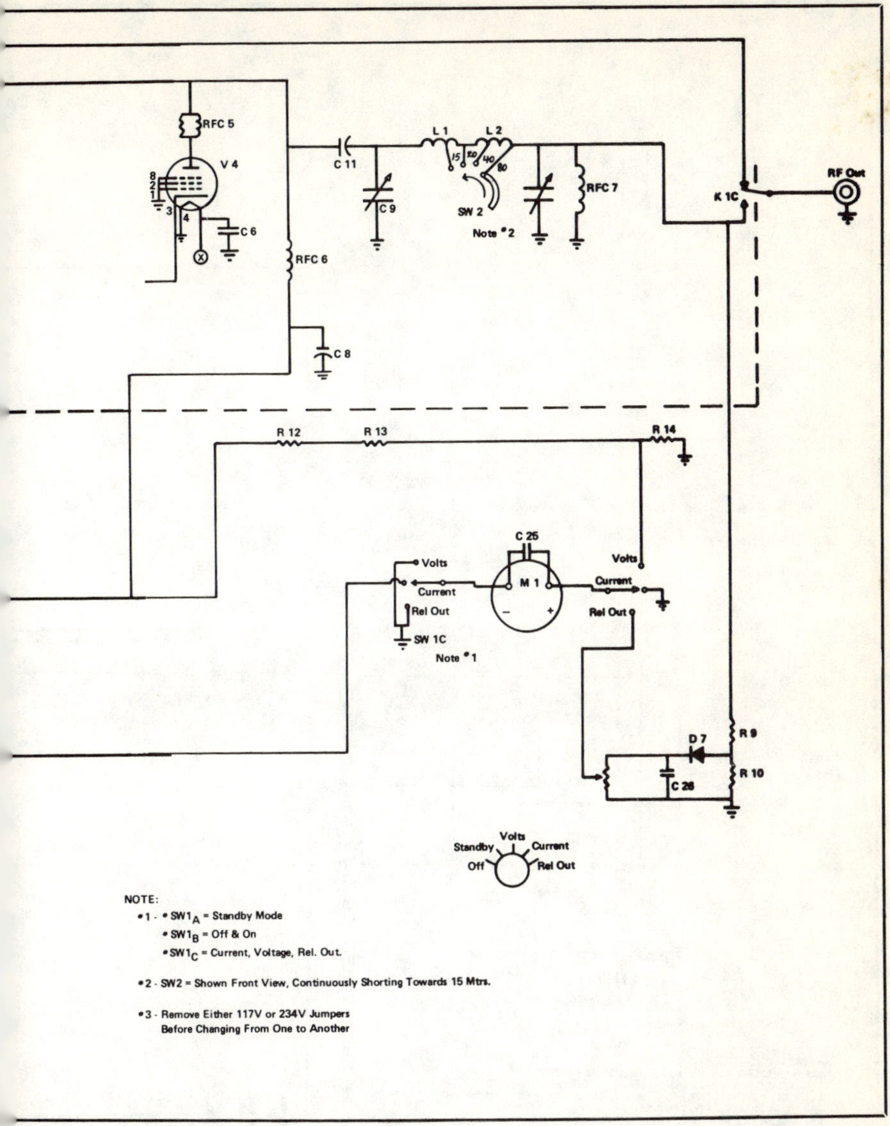

anode would exist across the antenna terminals and therefore on the antenna, itself. The rf choke across the antenna connector provides a low-resistance DC path to ground for the DC if the short occurs, yet maintains a high impedance to any rf. In informal operation, the choke is ignored by the amplifier, but in the vent of a failure, it protects you.

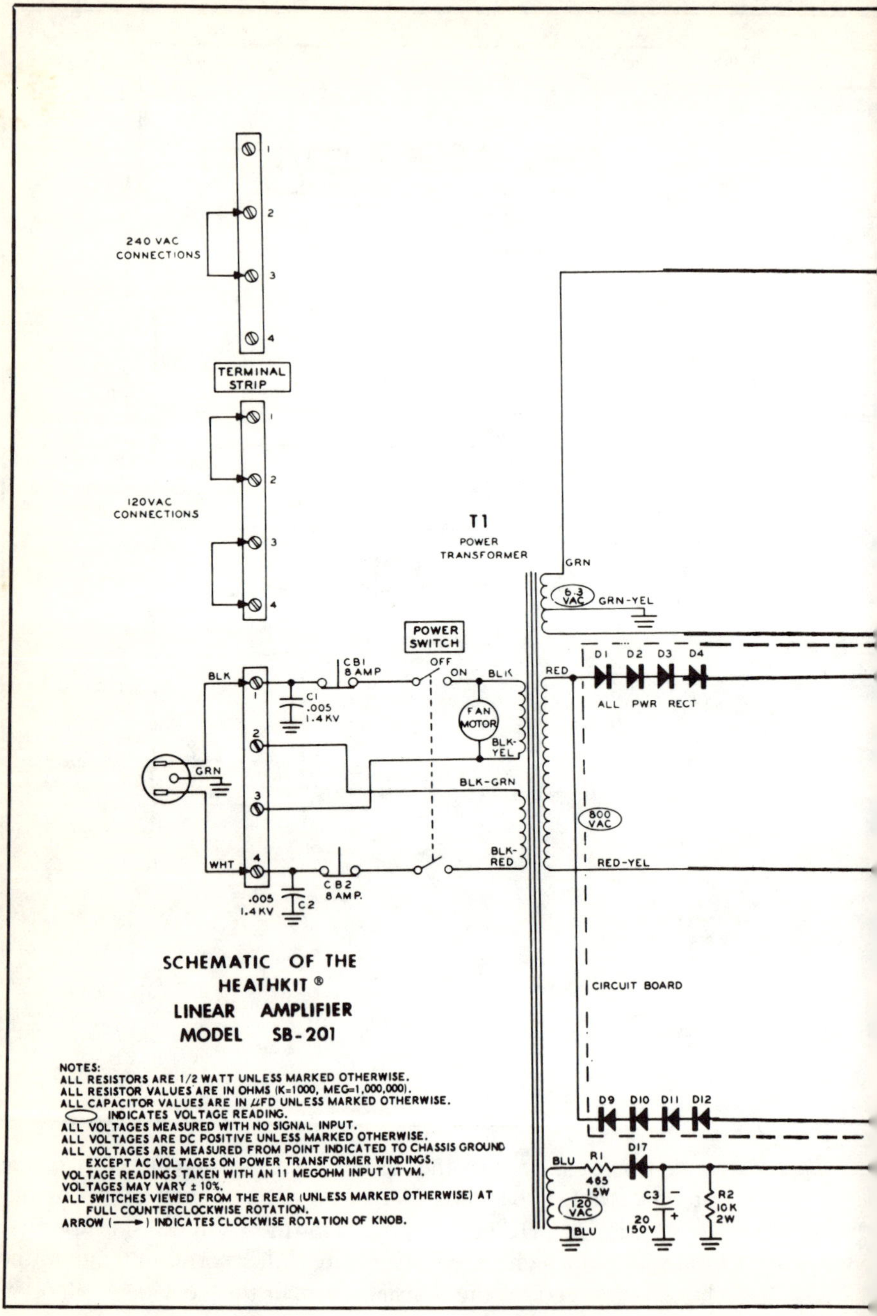

Fig. 7-8. Heath SB-201 schematic.

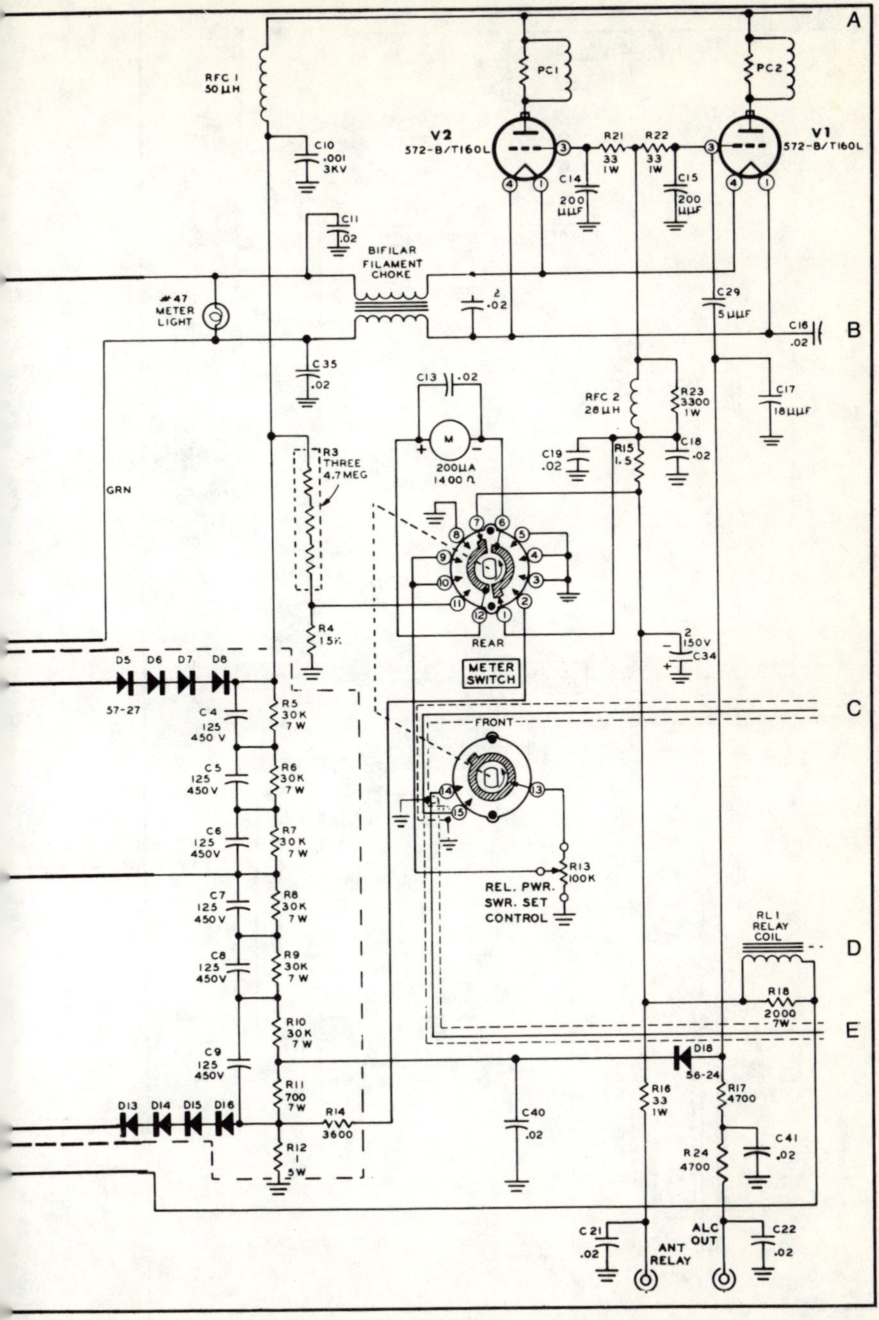
A
B
C
D
E
RFC 1 50 μH
C10 .001 3KV
V2 572-B/T160L
PC1
R21 33 1W
R22 33 1W
PC2
V1 572-B/T160L
C14 200 μμF
C15 200 μμF
C11 .02
BIFILAR FILAMENT CHOKE
#47 METER LIGHT
2 .02
C29 5 μμF
C16 .02
C35 .02
C13 .02
M
200μA 1400 Ω
RFC 2 28 μH
R23 3300 1W
C17 18μμF
C19 .02
R15 1.5
C18 .02
R3 THREE 4.7 MEG
GRN
R4 15K
REAR
METER SWITCH
2 150V C34
D5 D6 D7 D8
57-27
C4 125 450 V
C5 125 450V
C6 125 450V
C7 125 450 V
C8 125 450V
C9 125 450V
R5 30K 7 W
R6 30K 7 W
R7 30 K 7 W
R8 30K 7 W
R9 30K 7 W
R10 30K 7 W
R11 700 7W
R12 1 5W
FRONT
R13 100K
REL. PWR. SWR. SET CONTROL
RL1 RELAY COIL
R18 2000 7W
D18 56-24
R16 33 1W
R17 4700
C40 .02
C41 .02
R24 4700
D13 D14 D15 D16
R14 3600
C21 .02
ANT RELAY
ALC OUT
C22 .02

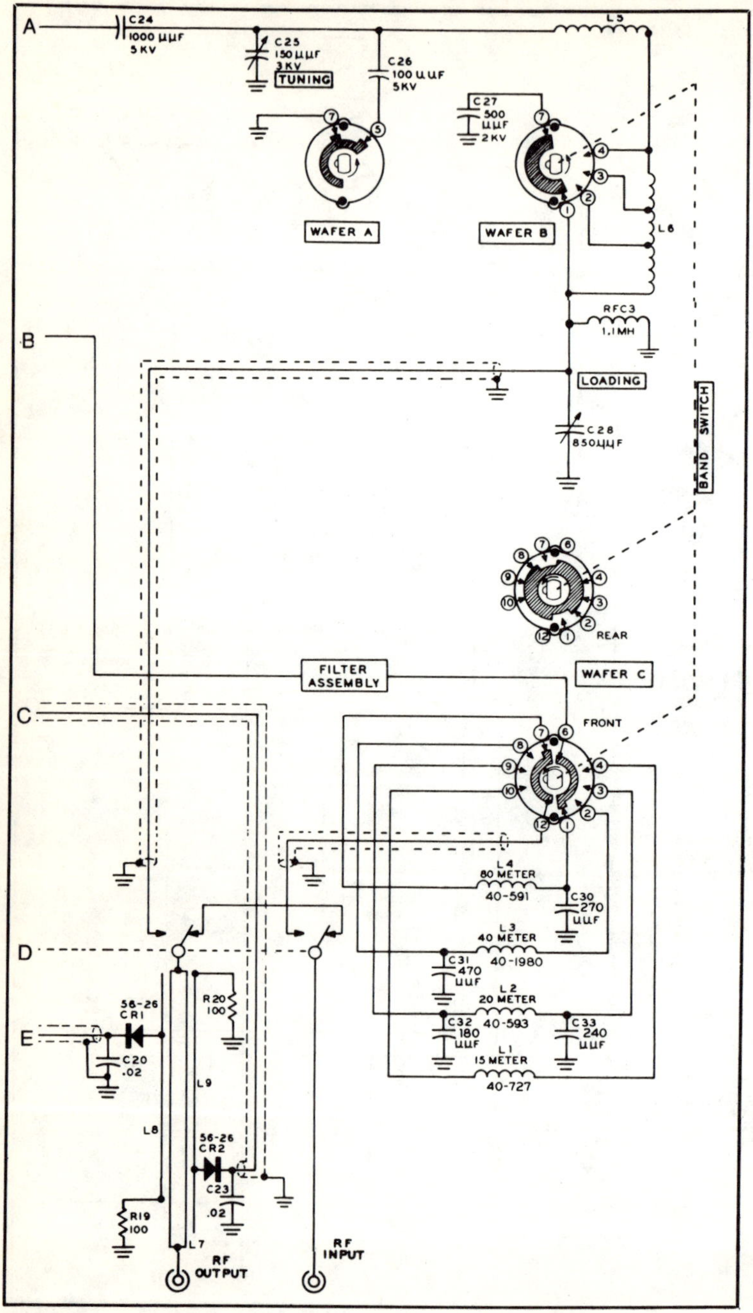

Fig. 7-8. Heath SB-201 schematic (continued from page 112).

Grounded-Grid Design

Grounded-grid operation is popular with HF and VHF power amplifier designers because the amplifier does not usually require neutralization, provided that good layout practice (isolation of input and output circuits) is followed. A number of power tubes are designed specifically for grounded-grid operation.

Figure 7-7 shows the circuit for the Dentron GLA-1000 power amplifier. This amplifier uses four power pentodes connected in parallel. In this case, the three girds from each tube are directly grounded.

The four parallel-connected cathodes from the final amplifier are driven by the exciter through a filter and DC blocking capacitors C1/C2. This circuit is cathode biased by R1 and Zener diode D1.

Another form of grounded-grid circuit is shown in Fig. 7-8. This one is based upon the 572-B/T160L triodes and is made in kit form by Heath (model SB-201). A negative grid bias is supplied to the grids through an isoltion network consisting of R21/R22 and the 200 pF bypass capacitors. These components isolate the grids from each other, while the additional components (RFC2, C18, C19, R15, and R23) isolate the DC power supply from the rf component that might exist on the bias line.

The plate circuit contains the same elements as in the circuit of Fig. 7-6, so it will not be repeated here. But the input circuit is a little different from either of the previously discussed circuits because the tubes in this amplifier use directly heated cathodes. We want to keep the filaments of V1 and V2 at a high rf impedance, while at the same time permitting the AC power supply to function. The 6.3-volt AC filament winding is balanced with respect to ground, and a bifilar filament choke is connected with one winding in series with each side of the transformer secondary. A *bifilar* winding is one in which two wires are wound in a parallel fashion, with the two conductors

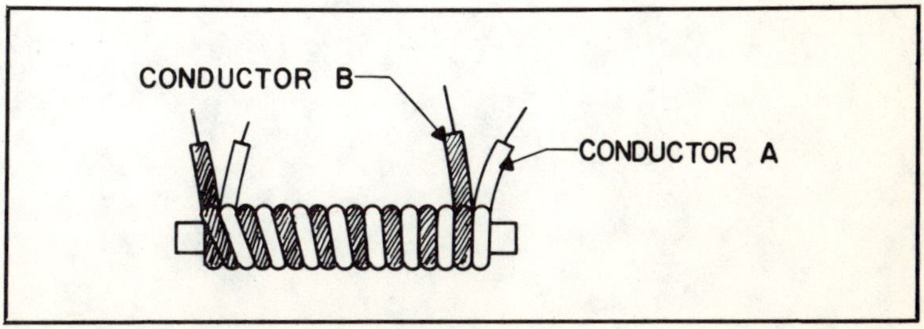

Fig. 7-9. Bifilar filament choke construction.

interleaved (Fig. 7-9). These coils are wound on a ferrite core. We must also try to keep the two sides of the filament at the same rf potential. Capacitor C12 is placed across the filament supply in order to facilitate this goal. The input signal is applied through a capacitor (C16). An input filter circuit will match the impedance of the input circuit to the output impedance of the exciter (50 ohms in most amateur radio amplifiers). These filters also aid in reducing harmonic content. This amplifier has a built-in VSWR/RF power meter. The sensor is monomatch elements L7 to L9.

Chapter 8

Solid-State RF Power Amplifiers

A decade ago rf power was always generated by vacuum tube circuits. Even though the transistor was well established and had supplanted the vacuum tube in almost all electronic applications, the generation of rf power was the sole province of the tube. There were a few transmitter projects being pushed, but all were QRP (low-power) circuits. It was simply too costly (and perhaps not technologically possible) to generate any substantial amount of rf power. But today, almost all currently sold rf power generators are solid-state. There are a few vacuum tube circuits of the 100-watt to 300-watt class still being sold, and even one new model. But by and large, the typical medium-power amateur radio transmitter uses solid-state rf power amplifiers.

Most of these circuits are wideband amplifiers. The basic circuit will operate over some wide range of frequencies. A lot of the high-frequency circuits, for example, operate from 3 to 30 MHz. The transmitter designers use low-pass filters in the output network to attenuate the harmonics that even linear amplifiers will inevitably generate.

Figure 8-1 shows a typical mobile power amplifier for the 2-meter band. This amplifier is a 40-watt model and requires only a few cubic inches of space. The heat sink mounted on the chassis is attached directly to the transistors. The circuit for this amplifier is shown in Fig. 8-2.

Fig. 8-1. Typical VHF power amplifier.

This circuit uses a pair of power transistors (Fig. 8-3), which are either 2N5591 or CTC 25-12s, connected in parallel. The output power from each half of the amplifier is summed at the input of the output impedance matching network; i.e., the junction of L5 and L6.

The impedance of the summing junction (point A) is approximately 5 ohms, and this must be matched to the 50-ohm output of the exciter stage. Capacitors C1 and C2 and inductor L1 form the impedance matching network. The power is then split into two paths from the summing junction to the bases of the two transistors. Each side of the amplifier contains an L-section network (L2/C4/C5/C6 and L3C7/C8/C9) that matches the input impedance of the transistor bases to the 5-ohm impedance of the summing junction. Note that three capacitors are used in parallel in each input network (C4 through C6 and C7 through C9). This is done not so much to obtain the 141-pF value as to distribute the current load in the capacitors, thereby reducing the losses. Capacitors are not simple devices at VHF but become complex RLC networks in their own right.

Note that there are two 10-ohm resistors connected between the two transistors. Resistor R1 is connected from base-to-base, while R2 is connected from collector-to-collector. The purpose of these resistors is to reduce, or damp out, the spurious low-frequency oscillations called *squeeging*.

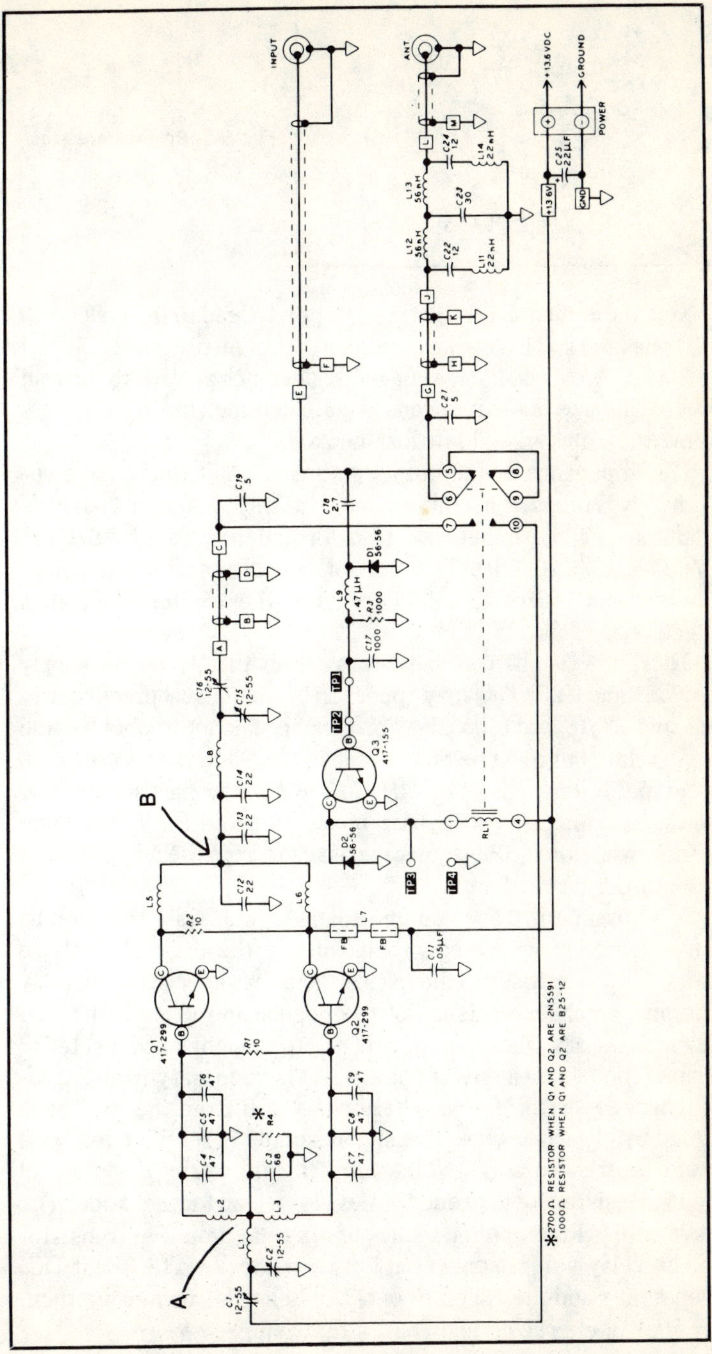

Fig. 8-2. Schematic of a power amplifier.

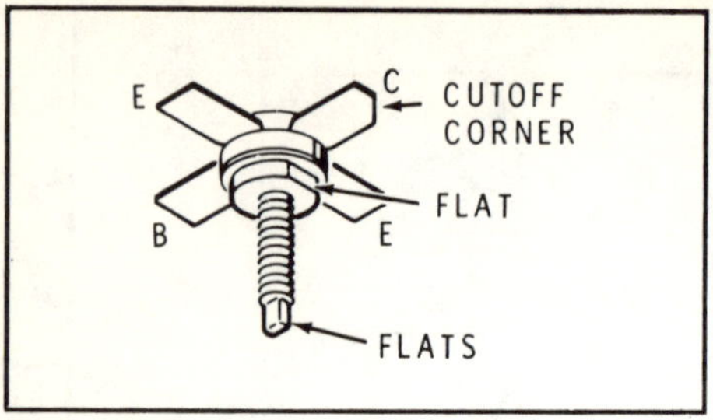

Fig. 8-3. Power transistor.

Notice the rf choke (L7) in the DC power lead to the collectors of the transistors. This choke is really a pair of ferrite beads slipped over the DC wire, and it has sufficient inductance at VHF to form an rf choke. The use of ferrite beads is quite common in VHF circuits, because they make good low-loss chokes.

The impedance at the output summing point (point B) is approximately 5 ohms. This impedance is raised to 50 ohms to drive the antenna by the impedance transformation network L8/C12/C13/C14/C15/C16/C19. This network is adjusted for a midpoint frequency and is not tuned every time the exciter channel is changed.

There are two bypass capacitors across the DC power supply leads: C11 and C25. This may appear to be an unusual practice, i.e. paralleling 22 μF and 0.05 μF. The idea here is not to obtain some specific value, but most be done because the electrolytic across the power input will not pass 144-MHz to 148-MHz signals adequately. This capacitor is used to bypass noise from outside (power line filtering), while the 0.05 μF unit bypasses rf from the amplifier that gets onto the power line.

It is possible in base station situations to operate the antenna changeover relay with a contact closure in the exciter. In those circuits, a switch closes in the exciter, and that energizes the relay in the final amplifier. This is not too popular in mobile situations, though, especially where a variety of exciters might be connected to the same model amplifier. In this case, it is necessary to automatically changeover the relay. The input signal from the exciter is sampled by capacitor C18. The signal sample is rectified by diode D1 and filtered to a DC voltage by the network consisting of L9/C17. Resistor R3 is needed to develop a load for the diode. The voltage across R3 is applied to the base of relay amplifier transistor Q3. The relay coil is connected between the V+ (13.8-volt DC) power supply and the collector of Q3. When Q3 is turned on, then, relay RL1 is energized and causes the changeover.

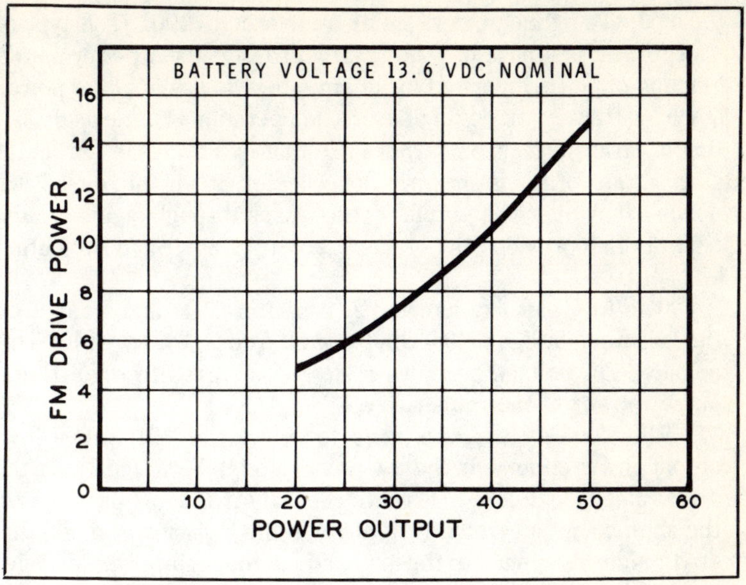

Fig. 8-4. Power transfer curve.

A low-pass filter is in the antenna network. The purpose of this network (L12/L13/L11/L14/C22/C23/C24) is to attenuate harmonics of the amplifier on transmit and incoming out-of-band signals during receive.

The rf output power of this amplifier is directly dependent upon the drive power applied to the input of the amplifier. If the input is 5 watts, the output will be 20 watts, but if the input power is 15 watts, the amplifier will output 50 watts. The amplifier is nominally rated at 40 watts output. The input-vs.-output power transfer characteristic is shown in Fig. 8-4.

HF 100-WATT POWER AMPLIFIER

The real change in transceiver design over the past decade is the use of solid-state power amplifiers in the 100-watt to 300-watt class. This is the power range of the typical amateur radio set. In this section, we consider the design of a typical solid-state power amplifier—in this case, the Heath SB-104A (transmitter section only, because the SB-104A is a high-frequency transceiver).

Figure 8-5 shows the driver amplifier for the SB-104A. This stage uses two transistors in push-pull and operates class-AB. The single-ended input signal is applied through a matching network to the input of a hybrid power splitter transformer, L901. This type of transformer is typical of solid-state HF designs and performs the function of phase inversion for driving the two bases of the power amplifier transistors out of phase. The operation is not exactly like that of a transformer, but is more like that of a transmission line. A similar transformer is shown in the collector circuit and is used as a power combiner. The output signal passes through an interesting network that provides an (ALC) DC control voltage and two output levels.

The ALC signal is sampled by capacitor C912 and rectified by the half-wave bridge D905/D906 and filtered to DC by C913. This negative DC voltage becomes a kind of agc used to regulate the output power of the transmitter.

DC diode switching is used to select high-output and low-output power options. In the low-power mode, the rf signal from the driver amplifier (Fig. 8-5) is routed directly to the low-pass filter in the antenna output circuit (Fig. 8-7). In the high-power mode, the driver signal is applied to the input side of the high-power amplifier stage (Fig. 8-6). The routing of the signal from the driver is controlled by DC potentials and two diodes (D903/D904). When a diode is heavily biased with a DC current, it will pass AC almost as if it were a piece of wire—no rectification occurs as would if the diode were not biased. DC voltages applied to either terminal 5 or terminal 6 select low-power and high-power options, respectively.

The high-power final amplifier for the SB-104 is shown in Fig. 8-6. This circuit is one of the best examples of high-frequency circuits for power amplifiers because it combines push-pull and parallel operation into one circuit. This circuit contains essentially *two* rf push-pull power amplifiers (Q951/Q952 and Q953/Q954). Each of these push-pull amplifiers operates in a manner similar to the driver amplifier described previously. The outputs of these two power amplifiers are combined in a hybrid combiner of a type described in Chapter 6. The network consists of the secondaries of L956 and L957, transformer L958, resistor R959, and the impedance of the antenna circuit. Note that the resistor has a value equal to twice the load resistance a slightly different arrangement from the Chapter 6 circuit. The combined output power is fed to the antenna from point E on L958.

The automatic load control (ALC) and output low-pass filter circuit is shown in Fig. 8-7. The low-pass filter is needed to reduce the harmonic content of the output signal. Both the driver and the final power amplifier operate in a linear class-AB mode, which is

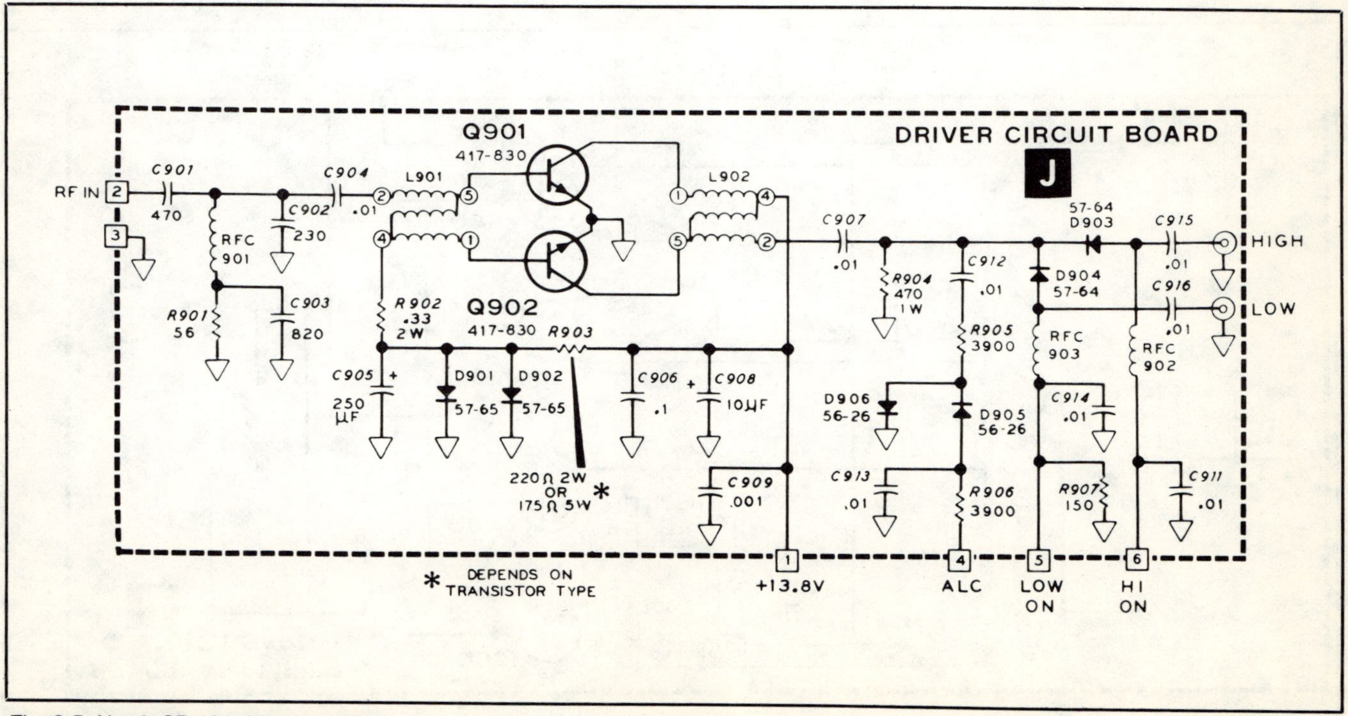

Fig. 8-5. Heath SB-104 driver amplifier.

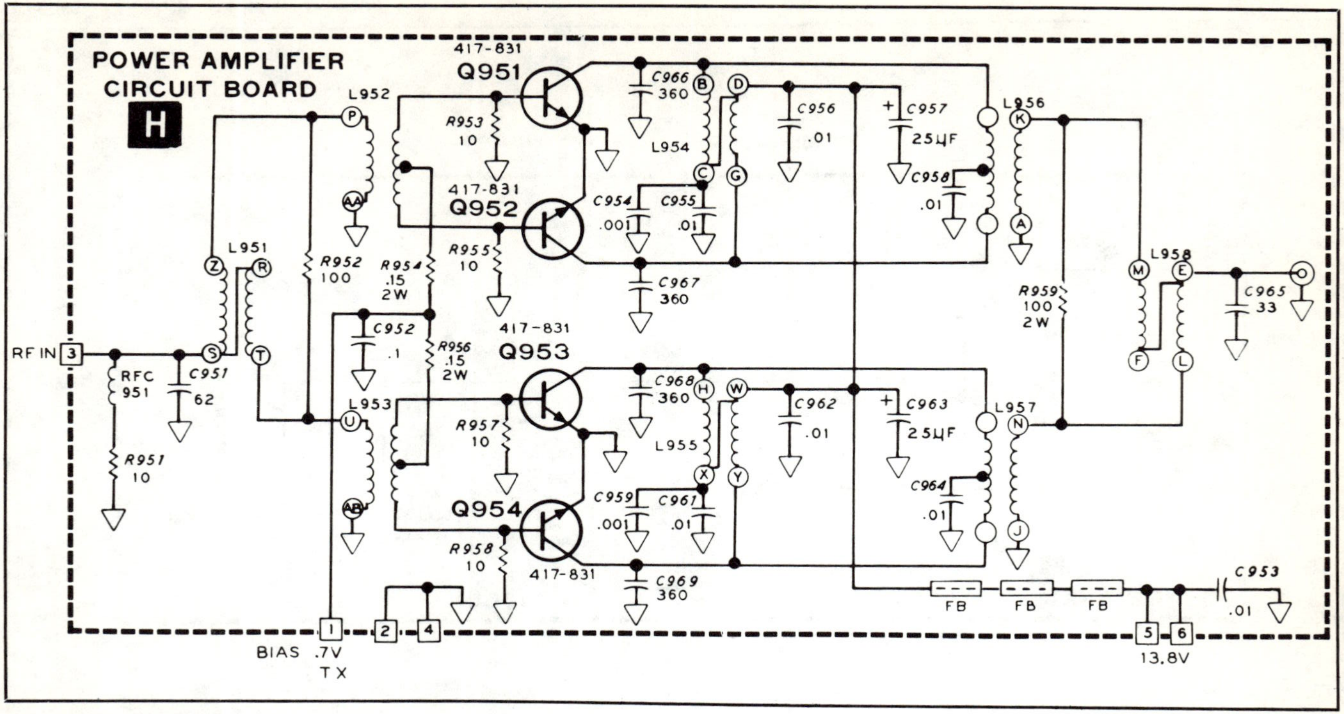

Fig. 8-6. Heath SB-104 power amplifier.

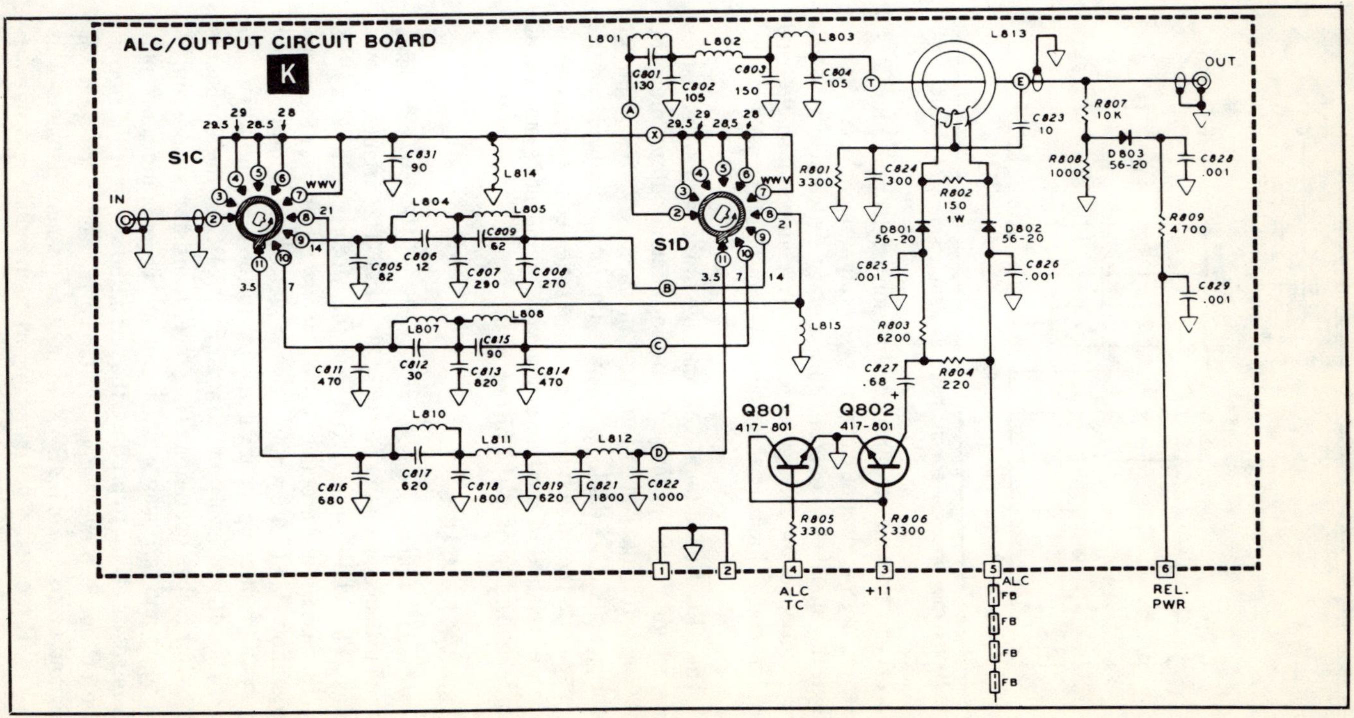

Fig. 8-7. Heath SB-104 ALC circuit.

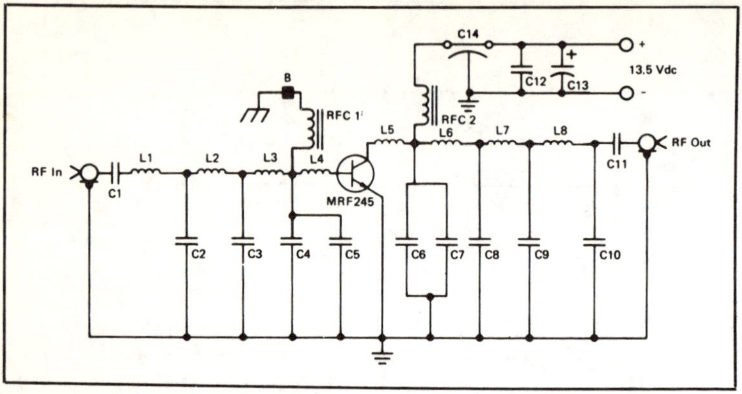

Fig. 8-8. VHF power amplifier schematic diagram (courtesy of Motorola).

theoretically not able to produce harmonics. In real amplifiers, however, a great deal of harmonic generation might occur, especially if the amplifier is overdriven. A rotary switch selects which filter is needed for any given band. Of course, a single-band amplifier would eliminate the switch, but the filter is needed anyway.

The ALC is used to prevent damage to the output transistors when a high VSWR is present on the antenna output line. The sensor is a toroidal coil in the same manner as VSWR/rf power meter pickoffs. One diode will sample the forward power and the other samples the reflected power. If the reflected power becomes too large, relative to the forward power, the negative output DC voltage is used to drop the power of the signal.

VHF 80-WATT AMPLIFIER

In this section we are going to describe a VHF power amplifier that will produce an output power of 80 watts. The circuit is shown in Figs. 8-8 and 8-9. This circuit uses the Motorola MRF245 VHF power transistor and produces 9.4 dB of power gain over the range of operation. There are actually two ranges of operation requiring slightly different component values: 143 to 156 MHz and 155 to 175 MHz. The MRF245 will operate at 12.5 volts DC, so the amplifier is suitable for mobile operation. The power transistor is more rugged than some and is readily suitable for mobile operation, not only because of its ability to produce the output power over the normal range of mobile power supply voltages, but also because it will withstand up to a ¼0:1 VSWR without permanent damage. This high-VSWR protection becomes important in mobile operation because many antennas have a higher VSWR than they should, especially after corrosion, wear, tear, and the tender mercies of vandals take their toll. The efficiency of this amplifier is approximately 60 percent, and the second harmonic should be at least 40 dB down from the fundamental without the use of an output low-pass filter.

The input matching circuit is a multiple section network (L1/L2/L3/L4/C1/C2/C3/C4/C5) that matches the 50-ohm input to the base impedance needed by the MRF245. Similarly, the output impedance of the transistor is matched to the 50-ohm impedance of the antenna by the collector network (L5/L6/L7/L8/C6/C7/C8/C9/C10/C11). The capacitors in both networks are the VHF chip capacitors (UNELCO). The placement of the capacitors is not supercritical, but make it as close as possible to the ends of the inductor lengths as possible. The inductors are not coils, but strips of printed circuit board material called *airline* inductors. The parts list shows these lengths for the 143 - 156 MHz band.

Assembly

The value of the inductors must be maintained. At the same time, it is necessary to heat sink the transistor. To maintain both of

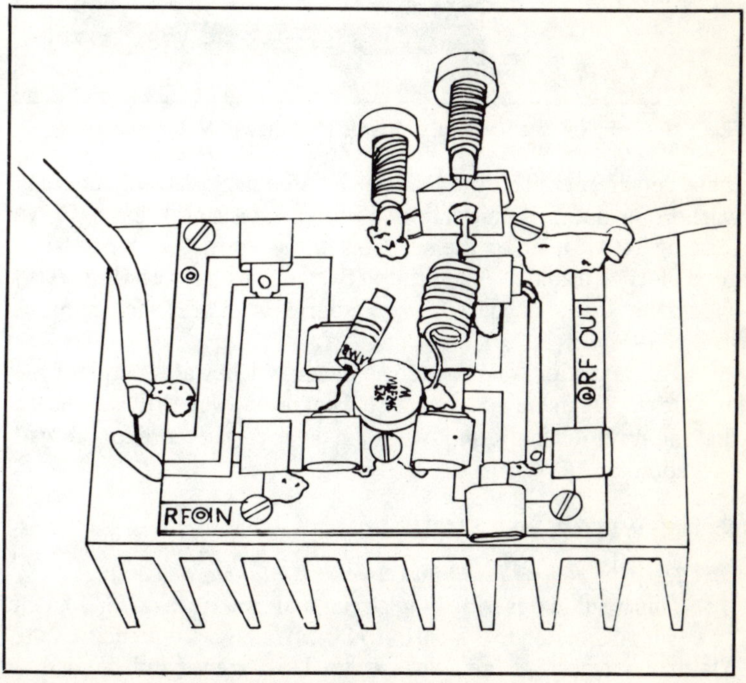

Fig. 8-9. VHF power amplifier (courtesy of Motorola).

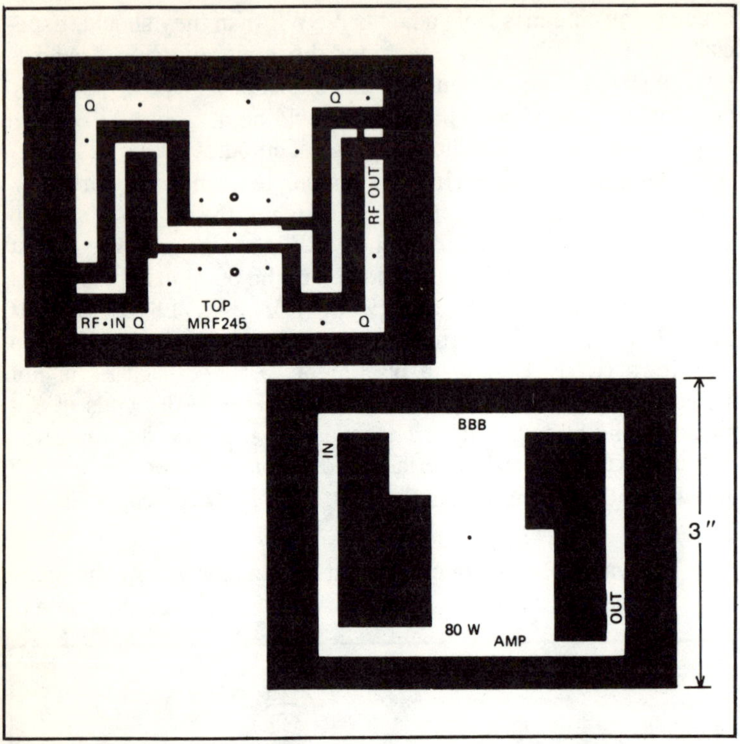

Fig. 8-10. VHF power amplifier PC boards (courtesy of Motorola).

these requirements lift the PC board off the heatsink by 2 mm (use washers or nuts). A thin film of a proper heat sink grease (Down Corning 340) is used on the heat sink of the transistor to maintain a good thermal transfer. The printed circuit board pattern (Fig. 8-10) should be made of G10 Epoxy/Fiberglass. Parts placement is shown in Fig. 8-11.

This amplifier will produce as much as 100 watts output if the drive level is high enough, but should not be allowed to do so. Insure that the drive level is kept lower than the level required for 80-watt operation.

HF 140-WATT POWER AMPLIFIER

The use of a linear amplifier on 11-meters (CB) is illegal in the U.S.

This section describes the circuit and construction for a linear power amplifier for the 2-MHz to 30-MHz high-frequency range. This frequency range encompasses the US citizens band, but operation on that band is totally illegal.

The amplifier will produce a 140-watt PEP output. It is linear and is an ideal amplifier for single sideband operation. Because the amplifier will operate from 12.5 volts DC (compatible with mobile power supplies in American automobiles), it makes a decent power booster for low-power rigs intended specifically for mobile operation.

The circuit of the 140-watt linear amplifier is shown in Fig. 8-12. Circuit boards and parts placement are shown in Figs. 8-13 and 8-14. The quiescent bias is set to approximately 500 milliamperes in order to prevent crossover distortion at the higher output power that this amplifier can produce.

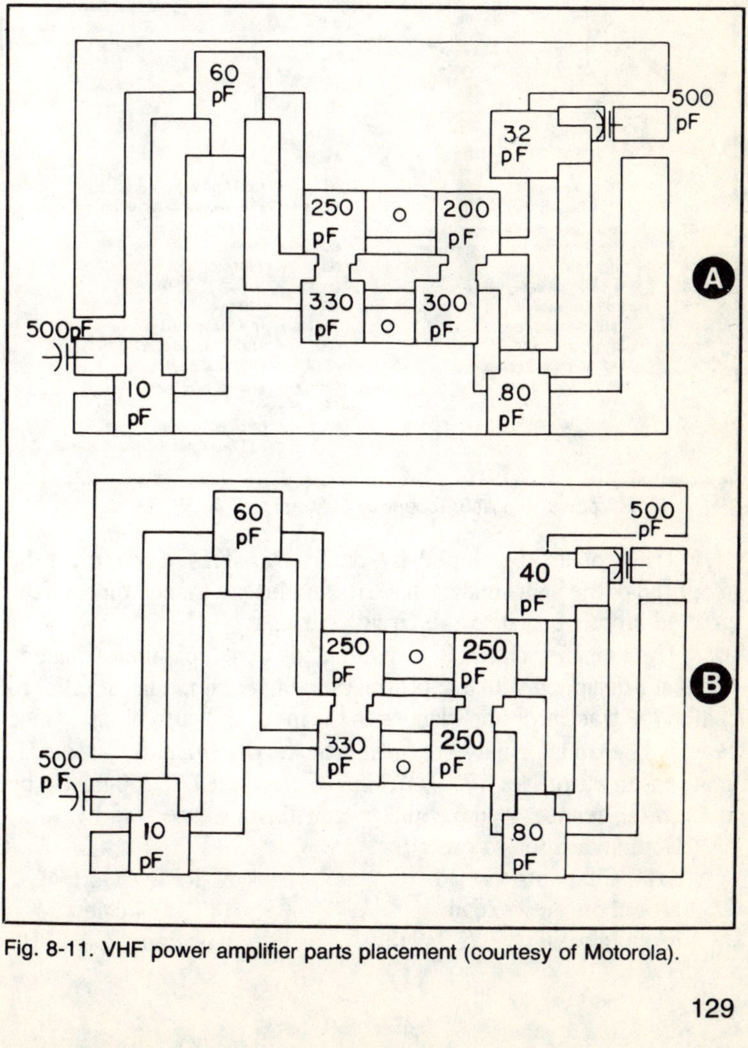

Fig. 8-11. VHF power amplifier parts placement (courtesy of Motorola).

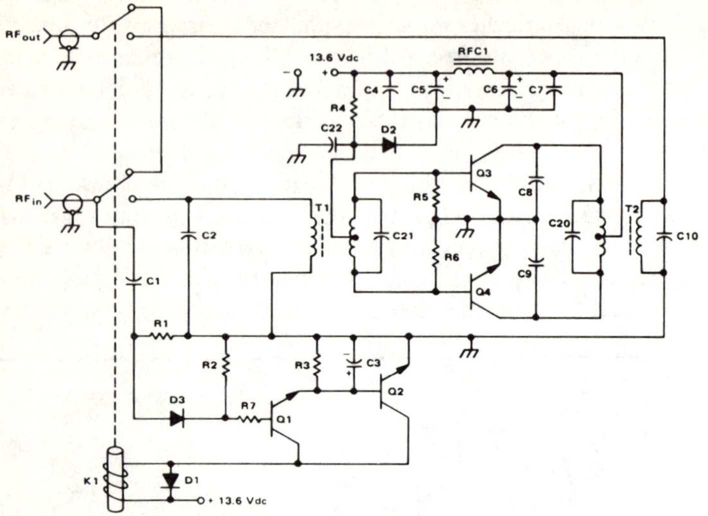

C1	=	33 pF Dipped Mica	R7	=	100 Ω 1/4 W Resistor
C2	=	18 pF Dipped Mica	RFC1	=	9 Ferroxcube Beads on #18 AWG Wire
C3	=	10 μF 35 Vdc for AM operation, 100 μF 35 Vdc for SSB operation.	D1	=	1N4001
			D2	=	1N4997
C4	=	.1 μF Erie	D3	=	1N914
C5	=	10 μF 35 Vdc Electrolytic	Q1, Q2	=	2N4401
C6	=	1 μF Tantalum	Q3, 4	=	MRF454
C7	=	.001 μF Erie Disc	T1, T2	=	16:1 Transformers
C8, 9	=	330 pF Dipped Mica	C20	=	910 pF Dipped Mica
R1	=	100 kΩ 1/4 W Resistor	C21	=	1100 pF Dipped Mica
R2, 3	=	10 kΩ 1/4 W Resistor	C10	=	24 pF Dipped Mica
R4	=	33 Ω 5 W Wire Wound Resistor	C22	=	500 μF 3 Vdc Electrolytic
R5, 6	=	10 Ω 1/2 W Resistor	K1	=	Potter & Brumfield KT11A 12 Vdc Relay or Equivalent

Fig. 8-12. HF power amplifier (courtesy of Motorola).

The bias is developed by diode D2. This diode must be mounted to the heat sink and have its anode lead passed through the printed circuit board to the circuit wiring.

The amplifier circuit is a push-pull design. The input transformers are designed with a 4:1 turns ratio (a 16:1 impedance ratio) to match the transistor impedances to the input and output impedance levels. The transformers are made from ferrite (FairRite Products, material mixture No. 77, 0.375-in. OD by 0.187/0.200-in. ID by 0.44-in. L, ferrite beads). Similar transformers are sold by:

Communications Power Inc.
2407 Charleston Road
Mountain View Road
Mountain View, CA 94043

and

M-RED - C&L
1475 Oakdale
Pasadena, CA 91106

The input driving power can be anything in the 1-watt to 7-watt range. The design is simple and straightforward. Antenna changeover switching is provided by a carrier-operated exchange (SOX) circuit consisting of transistors Q1 and Q2. A sample of the rf input driver signal is taken through capacitor C1. This sample is rectified and applied to the base of transistor Q1. This transistor is normally unbiased, so when the input signal is present, it will become forward biased. When Q1 is forward biased, its emitter current will cause a voltage drop across resistor R3, and this turns on relay driver transistor Q2. When Q2 is turned on, antenna changeover relay K1 is energized, and the transfer is completed.

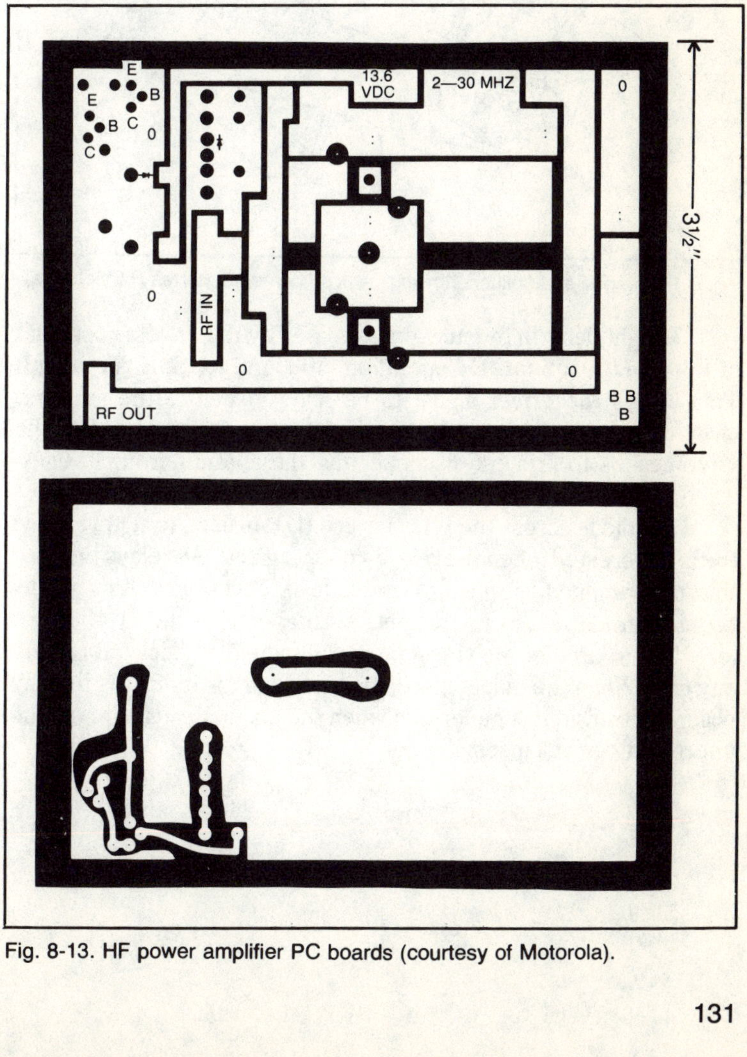

Fig. 8-13. HF power amplifier PC boards (courtesy of Motorola).

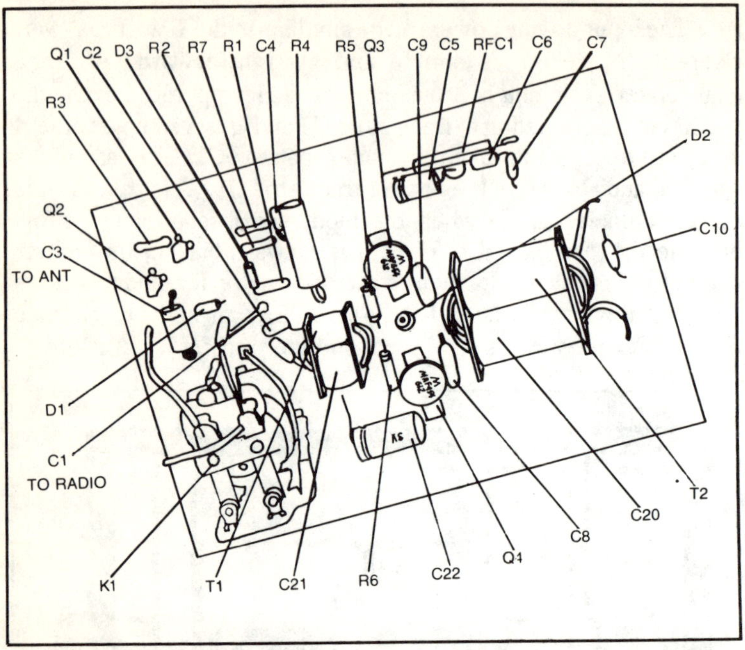

Fig. 8-14. HF power amplifier showing parts placement (courtesy of Motorola).

A slight delay in the turn-on is created by the RC time constant of R3/C3 (100 mS for AM operation, 1000 mS for SSB operation). This allows the driver signal to be fully on before the amplifier changes over, and keeps the amplifier on for a brief time after the drive ceases to precent disconnecting the antenna from an energized power amplifier.

The diode across the K1 relay coil (D1) is used to suppress the transient created when the relay is deenergized. All relays produce this phenomenon, so using this diode is necessary. Almost any rectifier-grade diode is sufficient, such as one of the 1N4000 ser ies). The source of the transient is the inductive "kick" normal to any coil. When the inductance is high enough, as it often is in relay coils, the transient is generated when the magnetic field around the energized coil collapses.

Chapter 9

Neutralization

Rf power amplifiers are necessary to produce the amounts of radio frequency energy needed for communications; however, a stability problem accompanies the construction of rf power amplifiers. If tuning the plate and grid circuits of a power amplifier stage causes any feedback problems, the circuit becomes essentially a tuned-grid—tuned-plate power oscillator at the resonant frequency of the tank circuits. This is especially true of triode amplifiers in the common cathode configuration, although tetrodes and pentodes may possibly also oscillate.

Also, the amplifier might oscillate at some VHF frequency. These oscillations are called *paristic oscillations*. All oscillations, regardless of source will reduce the ability of the amplifier to do its job and will probably cause interference with other established services. Amateur high-frequency power amplifiers, for example, frequently produce parasitics in the band occupied by the aviation radio service. These oscillations must be suppressed; even though amateur radio operators are licensed to produce a 1000 watts on the amateur bands, they are not licensed to produce so much as a single watt on any other band.

There are three principle sources of VHF oscillations: stray capacitance and inductance in the circuit wiring, loss of effectiveness of the screen grid at VHF frequencies, and ineffectiveness of screen grid bypassing at VHF frequencies. The first of these, stray LC, is a matter of layout. We must use short lead lengths. These reduce the stray inductance and keep the input and output circuits isolated from each other. This means shielding, such as placing the output circuit on the top of the chassis and the input circuitry on the bottom side. It also means that we should keep the mechanical and electrical design of the power amplifier within the KISS principle (Keep It Simple, Stupid).

The second of these is a fact of power tube construction. In many tubes, the structure of the screen grid is such that it becomes ineffective—usually too inductive—at VHF frequencies. Even though the screen grid might present a low impedance to ground for HF frequencies, its effectiveness is limited for VHF frequencies.

The last of these principal causes is the loss of effectiveness of the screen grid bypassing at VHF frequencies. Part of the problem is that the bypass capacitor may well be a relatively simple *circuit* at HF (a low capacitive reactance to ground), but it becomes a complex impedance at VHF (including RLC elements). This will place the screen grid at a higher VHF impedance from ground than the HF signal. There are several ways to reduce the effect of this problem. One is to insure that you use only disc ceramic capacitors with short leads. It is also necessary to know what ground points are used. Ground is ground is ground—*at DC only*. As the AC frequency gets into the rf range, however, various hot and cold spots exist on the chassis. Some points selected as ground might well be at a high impedance to true ground, even though on a low-resistance (DC) chassis. It is generally best to select a single-point ground, where all of the bypass capacitors connect to the same ground point. If one of the Eimac power tubes are used, it might be wise to spend the money for one of Eimac's air-system sockets that include the screen bypassing capacitors built into the socket.

The instabilities of the amplifier are caused by feedback between the output and input circuits. We can limit the external feedback by isolation of the output and input circuits. We can limit the external feedback by isolation of the output and input stages, per the discussion above. But there is still the matter of the *interelectrode capacitances* of the tube! These capacitances C_{gp} and C_{gc} form an AC voltage divider that will induce a small feedback voltage across the grid-cathode. This signal is then amplified by the tube to reinforce the feedback signal. The oscillation then builds and becomes sustained.

METHODS

There are several ways to eliminate this problem, and all of them involve creating a counterfeedback voltage that cancels the feedback voltage from the interelectrode capacitances. The major methods for *neutralization* of the feedback are:

- ☐ Losser resistor
- ☐ Plate neutralization
- ☐ Grid neturalization
- ☐ Direct neutralization
- ☐ Inductive neutralization.

A parasitic choke in the anode lead (see Chapter 7 on vacuum tube rf power amplifiers) can be used, too. These chokes are placed in series with the anode of each tube in the final, regardless of whether parallel or push-pull is used. The parasitic suppressor consists of a 50-ohm to 100-ohm carbon resistor, shunted with 3 to 6 turns of wire. The coil is wound directly over the resistor body in most cases. As the power level increases, it may be necessary to use several 2-watt resistors in parallel in order to achieve the 50-ohm to 100-ohm values. In low-power applications a 1-watt resistor is used, and in medium-power a 2-watt resistor is used. In no circumstances should a wirewound or other noninductive resistor be used.

The losser resistor is a small-value resistor connected in series with the grid of the power amplifier tube. The resistor reduces the gain slightly, but not sufficiently to make much difference. The losser resistor has a value between 100 and 1000 ohms, and must be noninductive (carbon). When a signal is applied to the losser resistor, an AC voltage is generated across the resistor in a polarity that will tend to be *degenerative*. This is therefore negative feedback, which will reduce the tendency of the amplifier to oscillate at its operating frequency. Recall that the amplifier may tend to act as a TPTG oscillator at its *operating frequency*. Although some stabilization occurs for off-resonance oscillation, the primary use of the losser resistor is to reduce the tendency to oscillate at the operating frequency. It is also helpful when the amplifier tries to oscillate while being tuned and on modulation peaks.

Figure 9-1A shows the plate neutralization system, called the Hazeltine circuit for its inventor. The interelectrode capacitances of the vacuum tube are C_{pg} (plate-grid) and C_{gk} (grid-cathode). These capacitances form an rf voltage divider, producing signal voltage V_f across the grid in the form of a feedback signal. Because this signal is proportional to the anode current, it will be in phase with the plate signal and will therefore be regenerative feedback.

To implement the Hazeltine system, a feedback capacitor C_n and a center-tapped inductor are needed for the plate tank circuit.

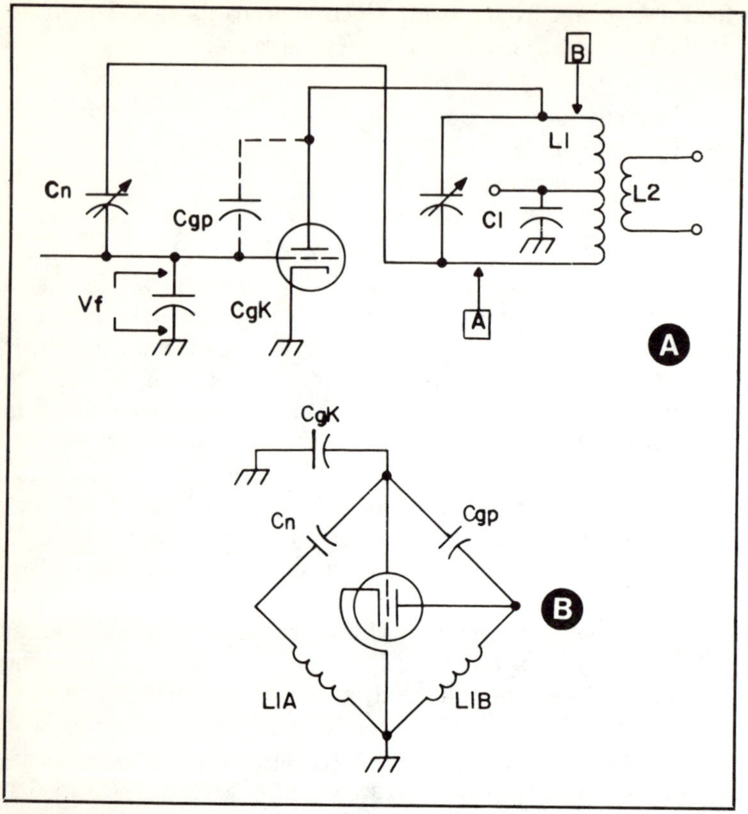

Fig. 9-1. Hazeltine neutralization at A. The bridge at B is needed to implement the circuit.

These components form a bridge (Fig. 9-1B) with the interelectrode capacitances. The center tap on the inductor is bypassed to ground via capacitor C1, so is at a zero volts rf potential. This means that the rf voltages at either end of the inductor will have equal magnitude but opposite phase—they are 180 degrees out of phase. The signal at the "B" end of L1 creates V_f across the grid through capacitance C_{pg}. The signal passing through C_n from the "A" end of the inductor forms an equal, but opposite polarity, signal at the grid $-V_f$. When these two are added, they null or cancel to zero, stabilizing the amplifier. The value of C_n, therefore, must be equal to the interelectrode capacitance of the tube C_{pg}.

A variation on the Hazeltine circuit is shown in Fig. 9-2. This circuit is essentially the same as the previous circuit, except that the plate tank circuit is tapped via a capacitor (C1) instead of the inductor. Balance is preserved by also tapping the inductor for the DC power, but this point is kept at a high rf impedance by rf choke RFC1. The zero reference point is created by grounding the junction of C1A and C1B. This capacitor is a dual-section split-stator type in which the rotor is normally grounded when the capacitor is mounted. Other than this modification, the circuit works the same as the inductive version shown in Fig. 9-1A.

The grid neturalization system, shown in Fig. 9-3, is also called the Rice neutralization system, again for the inventor. This circuit, perhaps, is the most common one used in amateur power amplifiers. Actually, two versions of the Rice system exist. One, not shown here, uses a tapped grid circuit inductor and operates in a manner not unlike the Hazeltine system of Fig. 9-1A.

The feedback capacitor C_n can be a vacuum variable or a pair of 1.5 to 3.0 pF ceramic discs counted parallel to each other. Neutralization capacitors use the former method of construction. In small amplifiers, such as those that use the 6146, the neutralization capacitance may be a small length of hook-up wire, 3 cm to 10 cm in length, connected at one end ot the grid circuit. It is then dressed so that it is parallel to the tube, forming a capacitance with the anode assembly inside the envelope. The "capacitor" is adjusted by varying the position of the length of wire (cheap, but effective).

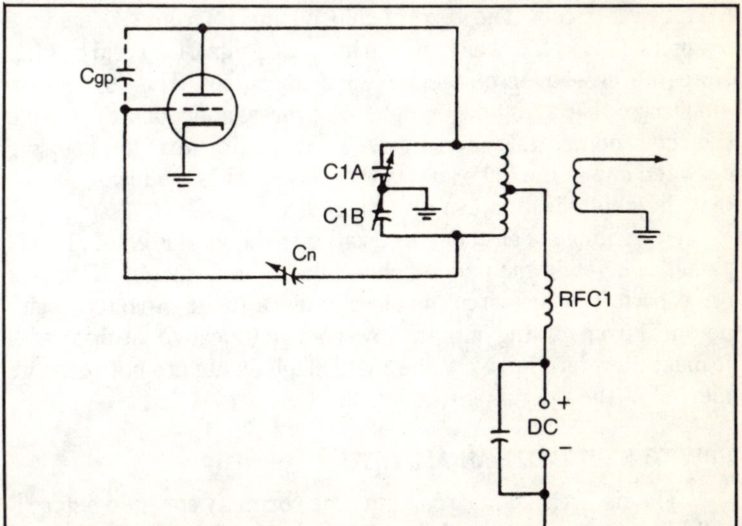

Fig. 9-2. Hazeltine neutralization.

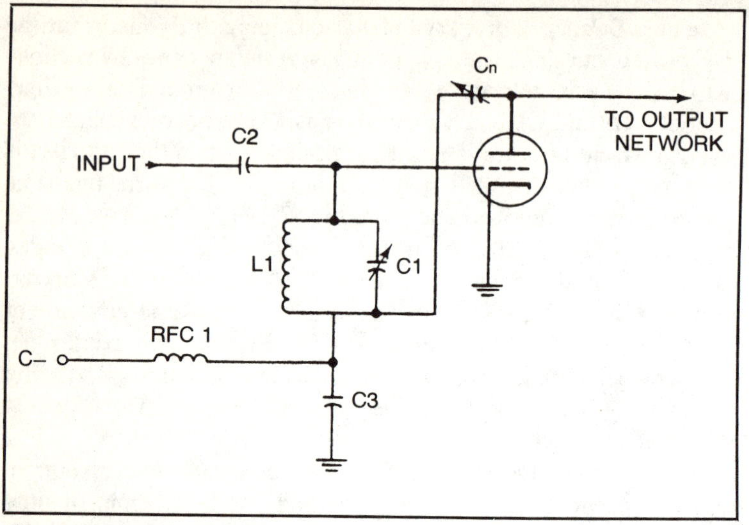

Fig. 9-3. Rice neutralization.

The circuit of Fig. 9-3 uses a capacitive voltage divider consisting of feedback capacitor C3. This circuit is used primarily in the neutralization of power tetrodes at the frequency of operation.

A circuit for inductive neutralization is shown in Fig. 9-4. In this circuit, the output rf signal is sampled by a link coupled to the plate circuit tank circuit. The current induced into L3 from L4 sets up a magnetic field in L2, which in turn induces a signal into grid coil L1. Note the crossover of the L2/L3 connections. This is done to emphasize that a 180-degree phase reversal is needed to provide the required neutralization signal. In actual practice, the coupling between one of the links and its associated coil is made variable so that the neutralization can be adjusted.

An example of direct neutralization is shown in Fig. 9-5. In this circuit, we obtain the needed phase shift in an inductor. Note that the capacitor in this circuit is used to block the high-voltage plate potential from getting onto the low-voltage (negative, at that) grid. It must therefore have a value that is high enough to not resonate the coil at the operating frequency.

HOW TO NEUTRALIZE AN AMPLIFIER

The neutralization circuit must be correctly adjusted before it will provide the degree of protection sought. Two basic methods prevail: one uses some form of rf output indicator, and the other uses the grid current of the final amplifier stage.

In both methods, it is necessary to turn off the high-voltage supply to the final amplifier anode. The job of adjusting the neutralization cannot be accomplished if the high voltage is turned on. There would be too much interference from the signal that is supposed to be present—the desired signal.

The rf output indicator method requires us to couple some sort of rf output indicator to the plate circuit. Some have been successful using one of the small VSWR/rf power meters now sold to radio amateurs at a cost low enough to make one wonder about the advisability of homebrewing one. In still other cases, use a loop of wire connected to a passive wavemeter or simply connected to an small flashlight bulb. Or if an oscilloscope is available, the loop can be connected to the vertical inputs of the oscilloscope (assuming that the vertical amplifier has sufficient bandwidth).

The grid circuit must be energized. This can be detected by the presence of grid current, or by assumption that turning on the transmitter will, in fact, excite the grid circuit (simple, huh?). Next, set the neutralization capacitor to one end of its range (either maximum or minimum capacitance). The rf output indicator should show some rf output, part of which is due to the neutralization capacitor, part due to the interelectrode capacitances of the tube. Next adjust the neutralization capacitance until the rf output drops

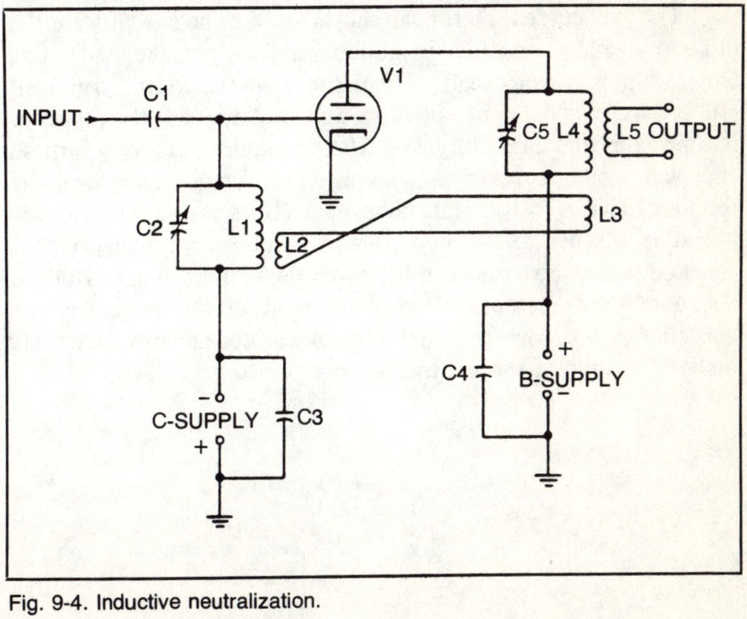

Fig. 9-4. Inductive neutralization.

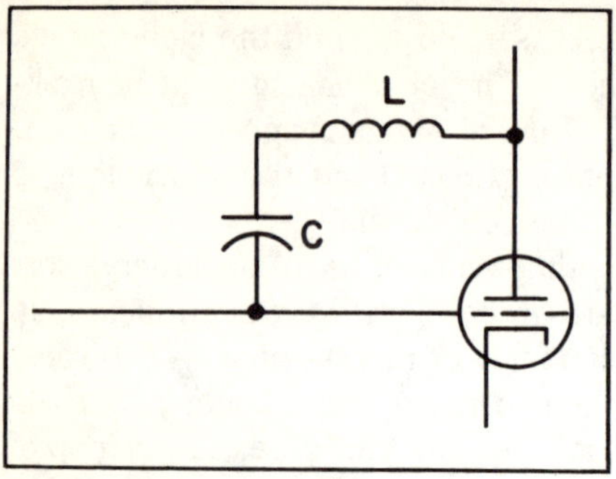

Fig. 9-5. Direct neutralization.

to zero. It may be found that two different rf detectors result in the best job. Use one of the less sensitive detectors, such as the lamp, to rough in the adjustment. Finish with one of the more sensitive types (VSWR/rf power meter) to fine tune the adjustment. Of course, an oscilloscope will probably work for all ranges.

The oscilloscope is recommended mostly because it lets you see the output waveform, and this will help detect problem situations, such as parasitic oscillation only on the rf peaks under modulation (AM). Also, make sure that the neutralization holds at *all* settings of the plate-tuning and grid-tuning capacitors. These adjustments should not be interactive when the neutralization is correct, but in recircuits there is frequently some problem interaction.

The grid current meter can also be used as an indicator for the rf neutralization condition. In an unneutralized amplifier, adjusting the plate tank-tuning capacitor (with the high voltage DC turned off-will cause a brief deflection of the grid current meter as the capacitor is tuned through a critical value. The procedure is then to turn off the plate voltage. Next simultaneously and vigorously vary the tuning capacitor in the plate tank through its range while slowly adjusting the neutralization capacitor. The correct adjustment is reached when the grid current meter deflections cease. Both of these methods require slow adjustment of the neutralization capacitor, as the correct point is critically placed and valid over only a short rotation of the neutralization capacitor.

Chapter 10

Power Supplies

At one time, several decades ago, radio transmitters operated directly from alternating current (AC). In the earliest days of radio, transmitters were nothing more than high voltage AC spark gaps with a tuned resonant tank circuit to cause oscillation at one frequency. In somewhat later designs, when vacuum tubes came into more general use, some people still operated transmitters with raw (unrectified) AC on the anodes of the tubes. Regardless of which system was used, however, the raw AC produced a raspy, wideband signal that was difficult to copy and took up much spectrum space. But in those less crowded bands, using only radiotelegraph, it was not an unworkable situation.

As time passed, however, things had to change. More and more users came on the air, and wideband signals were no longer tolerable. For many years, the FCC regulations required DC operation of all transmitters in the VLF through HF bands. Only VHF and UHF transmitters were permitted to use raw AC on the anodes. But today, pure DC is used throughout the spectrum. Those old jokes by VHFers about the "DC bands" (below 30 MHz) no longer have meaning; *all* of the bands are DC bands. The FCC doesn't appear to have a specific requirement as to the ripple component allowed, but they do examine certain other parameters of the transmitter (distortion, carrier-shift, etc.) that are affected by the amount of AC component on the power supply.

In this chapter we are going to consider the types of circuits used in transmitter power supplies, and their basic operating principles. Some of the material will be a review of basic electrical fundamentals, but it is highly recommended for all but the most seasoned professionals. A little review of basics never hurt anyone, especially since all higher technology is derived from basics.

BUT FIRST—SAFETY!

In this era of low-voltage, solid-state devices, we are apt to sometimes overlook the fact that electricity can be *lethal!* It can and will kill you under the correct circumstances. Those circumstances are very often too easy to unwittingly create.

Also, do not be too quick to believe two insidious rumors often passed on by both professionals and amateurs alike: voltage doesn't kill, it's the current; and 110 volts AC from the residential power mains is not terribly dangerous—after all, they use over a kilovolt in the electric chairs, don't they? Neither of these rumors are true! Both are lies from the pits of hell. I have even heard a medical doctor pass along the first rumor, and he really believed it! He had learned (quite correctly) that it is the *current density* in the heart that kills, but they apparently do not teach Ohm's law in medical schools. Remember that current applied to the heart is a function of the resistance of your body (as little as 50 ohms through a wound, or as much as 100K ohms through intact, dry skin) and the *applied voltage*. Consider the arithmetic for a minute. Currents of less than 100 mA have been known to cause death. Assume that the resistance of your body is 1000 ohms which is a realistic value when you are sweaty or wet. Then the current will be 110V/1000 ohms, or 0.110 amperes. That 100 mA is greater than the current known to be fatal.

Consider also the retrospective evidence. *Most* of the fatal accidents with electricity involve 110 volts AC residential power. This potential can cause as much as 300 mA of current flow through intact skin.

TODAY'S POWER SUPPLIES

At one time, a chapter on radio transmitter power supplies might well have included only high-voltage circuits. The only use for low-voltage supplies was in the grid bias section, if used. Today, things are quite different. Radio transmitter might operate at voltages as low as 6 volts DC up to potentials more conveniently measured in kilovolts. The range of operating voltages is now quite wide. The subject has expanded considerably, to say the least. We might have DC-only circuits, as in mobile and portable operation, where the power source is a battery. In those cases, the transmitter circuits operate directly from the battery voltage. Still others are AC-operated power supplies intended to derive DC either high or low voltage) from the AC power mains. In still other cases, DC-to-DC power supplies convert low-voltage DC from batteries to high-voltage DC for use in tube-operated transmitters. This latter circuit was commonly employed in mobile transmitters of any power level over a couple of watts. In fact, until the late '60s, even 5-watt CB mobile rigs had to have a DC-to-DC converter circuit on board.

DC-ONLY POWER SUPPLIES

Most modern mobile rigs use all solid-state circuitry; only a few models still use vacuum tube final amplifiers—a pair of 6146s or a pair of sweep-tubes from TV service. All-solid-state transmitters usually have circuits that will operate directly from the vehicle battery system. In most US-made cars, this will be 13.5 to 14.4 volts DC, with the negative terminal of the battery grounded. Note that most so-called "12-volt" electrical systems are actually 13.5 to 14.4 volt systems when the engine is running. Figure 10-1A shows a typical DC-only power for a solid-state mobile rig of moderate power output level.

It is not necessarily a direct shot from the vehicle battery to the transmitter circuits. This situation arises from the fact that vehicle power systems are very noisy. The pure DC from the battery may have riding on it a 2000+ Hz signal from the AC alternator used in the charging system, high-amplitude spikes from the ignition system, and miscellaneous high-frequency trash from assorted small motors (the heater blower), voltage regulator, and even the gasoline gauge sending unit in the fuel tank. Most of the components shown in Fig. 10-1A are for the suppression of this noise.

The primary section of the power supply consists of the components common in the high current side of the circuit; i.e., the line feeding the power amplifier. The components of the primary section are fuse F1; capacitors C1, C2, and C3; diode D1; and inductor L1.

Fuse F1 will usually be external to the rig, except in low-power models. In higher power transmitters, the fuse is likely to have a value in the 25-ampere to 30-ampere range, while for low-power CB and 2-meter FM rigs, it might be just a few amperes. In any event, it is usually considered good practice to use an external fuse to protect the vehicle wiring in case of a transmitter power line short that is connected to the internal fuse. Sometimes, when the power is high enough to warrant, a direct connection to the battery, a fuse block is mounted under the hood for the transmitter. In other cases, fuses up to 20 to 30 amperes can be mounted in the normal fuse block of the automobile. Most such fuse blocks will have at least one extra space for fuses to accessories not actually present on that specific vehicle. It is also possible to install fuses to 30 amperes in some in-line fuse holders.

The main DC power lead into the transmitter or receiver is usually called an *A-lead*. This is a throwback to early radios in which the low-voltage A battery was used to power the filaments. The terminology persists today, however, and is common in mobile applications.

Three different types of noise bypass capacitors are used in the primary section of the power supply. Capacitor C1 is a "spark plate;" C2 is a disc ceramic; and C3 is an electrolytic capacitor. The spark plate is not so common anymore, except in auto broadcast receivers, but it is seen occasionally. It consists of two pieces of copper foil glued, one to each side, to a paper insulator. This forms a crude capacitor that has an area of a few square inches. Most spark plate capacitors have a capacitance of 50 pF to 300 pF. The purpose is to attenuate high-amplitude noise pulses originating in the car ignition system spark coil (hence, the name spark plate). The effectiveness of C1 depends strongly upon its placement. It should be placed immediately inside the chassis at the point where the A-lead enters the rig. Otherwise, radiation from the A-lead to the other circuits can cause severe problems. In receivers, the low-amplitude volume control signals often pick up spark and alternator noise because the A-lead is connected to the potentiometer power switch. Many service bulletins from equipment manufacturers over the years have called for redressing either the A-lead, the volume control leads, or both.

The combination C2/C3 may look a little strange on first glance. Why put 0.1 μF in parallel with 1000 μF, especially since the tolerance normal to high value electrolytics is −20 percent to +100 percent? The goal is not the total capacitance obtained, however. The problem is that the electrolytic is ineffective at high frequencies. So C2, the 0.1 μF unit, takes out the high-frequency components, while C3 takes out the low-frequency components.

Diode D1 has two main functions: It protects the transmitter from reverse polarity connection to the power source, and it pro-

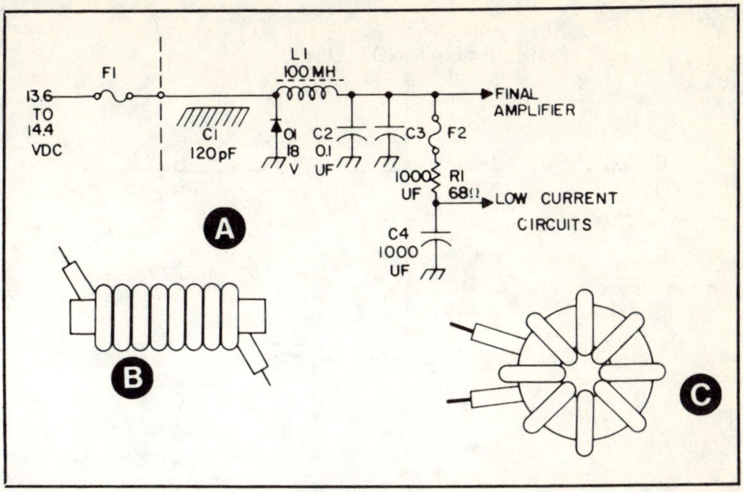

Fig. 10-1. Low-current DC mobile supply input at A wtih construction techniques at B and C.

tects the circuit from excessive input voltages. Both of these faults are common in mobile installations. Reverse polarity results from reverse installation of a new battery. Mistakes occur, despite the clear marking and *different terminal sizes*. The mechanic may not notice the problem because the car still runs. Even some solid-state ignition systems actually float the circuit and then use internal steering diodes to allow bipolar operation. But your radio equipment is terribly polarity sensitive. Audio output stages from receivers, and the final power amplifiers in transmitters, may easily burn out if the DC polarity is backward. The overvoltage problem occurs when the regulator in the battery charging system is defective or misadjusted. At high engine RPM, the alternator output voltage can exceed 18 volts. If the regulator is defective, then this voltage is applied to the electrical system of the car. Most of the devices on a car can tolerate overvoltage, even though their operating life is shortened. It may not be noticed in any form other than a slightly higher lamp replacement rate. Apply that 18 volts to a power stage using solid-state devices, however, and it will cause burnout.

Diode D1 will protect against reverse polarity because it will become forward biased when the positive terminal of the battery is grounded. This diode is usually a 50-watt zener. Overvoltage protection occurs because the zener potential selected is 16 to 18 volts (18V being most common).

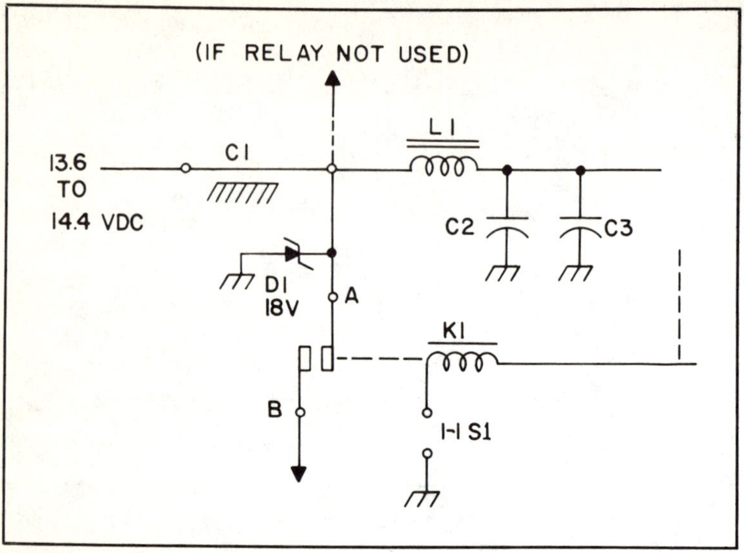

Fig. 10-2. High-current DC mobile supply input.

Inductor L1 serves two functions: It reduces the high frequency noise from outside of the rig, and it attenuates the rf on the power circuit wiring before it leaves the rig. Some of the electronics used in automobiles are not happy with high-power radio interference. We used to be amused at the stories circulating about a certain high-dollar European sports car that used a computerized fuel injection system. Everytime the ham owner would key his 20-meter transceiver, the car would cut off or otherwise misbehave. This is now a severe problem, though, due to the larger amount of auto electronics present in cars.

One principal requirement placed upon L1 is a *low DC resistance*. If a transmitter draws 18 amperes on *transmit*, and L1 has a DC resistance of, say, 0.2 ohms, then the DC voltage drop across the coil is 18A × 0.2 ohms, or 3.6 volts. If the circuits are designed to operate at 13.6 volts, then there will only be 10 volts left after the drop across the power supply choke. A substantial loss in output could be expected, and is the least of the possible problems. The rig might not even work at all. We meet the low-resistance requirements, while providing sufficient inductance by winding L1 on a ferrite core using heavy guage wire—No. 14 through No. 10. In general, it is wisest to use enamelled wire for L1.

Two methods for constructing L1 are shown in Figs. 10-1B and 10-1C. The solenoid-wound version in Fig. 10-1B uses a ferrite rod such as ordinarily is found as a loopstick antenna in portable AM broadcast radios. This type of core is also available from ferrite manufacturers in the form of rods. Amidon Associates of California, for example, sells such a core in the form of a kit to make the input choke for the filament circuit of grounded-grid power amplifiers. The toroidal version of Fig. 9-1C offers the advantage of a geometry that tends to concentrate its own magnetic field, which is an important consideration given the need for noise radiation suppression in a mobile rig. It is also easier to mount such a choke, as we can pass an insulated bolt through the center hole of the toroid for fastening to the chassis.

The circuit of Fig. 10-1A is really useful in low to intermediate power level rigs. At levels greater than 50 to 100 watts, however, the voltage drop across L1 is too great. In higher power rigs, one of the circuits in Figs. 10-2 and 10-3 will be used instead. In Fig. 10-2, an electromechanical relay is used instead of a simple switch. The high-current section of the power supply is fed from the battery through the high-current relay contacts. In some cases, the designers of mobile rigs connect the high current A-lead directly to the collectors of the final amplifier transistors, but this is possible only on those types of transmitters in which the current drain in the final amplifier is near zero under no-signal conditions. Many of the linear amplifiers used in SSB transmitters will not meet this requirement, so for them, the relay is required. Relay K1 is activated on transmit via the VOX (voice-operated exchange) circuit or the push-to-talk

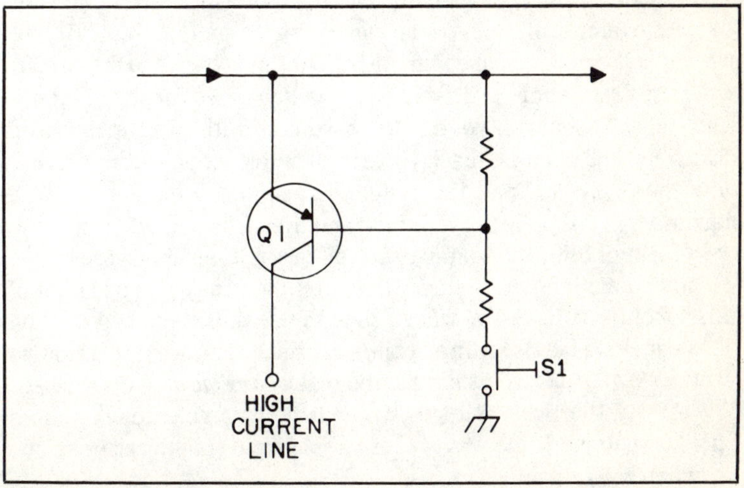

Fig. 10-3. Solid-state switch.

switch. When the relay is energized, contacts A and B close, and this applies V+ to the final amplifier stage.

A solid-state switch in Fig. 10-3. The PNP transistor, Q1 , is normally turned off; no current will flow. But when the push-to-talk or transmit switch is closed the transistor is forward biased to the point of saturation. There are several requirements placed upon Q1: I_c rating greater than the current demand of the transmitter, adequate P_d rating, and a low $V_{ce(sat)}$. The last requirement is usually interpreted to mean less than 0.9 volts.

The power supply circuits discussed thus far cover only the simplest case: a transmitter in which all circuits operate from the battery voltage. Most rigs, however, can operate from the AC power mains, even if an external AC-to-DC power pack must be used. Two types of AC supplies are used: low voltage/high current and high voltage/low current. In the next section, we will consider AC power supplies of both types.

AC POWER SUPPLY FUNDAMENTALS

A DC power supply that operates from the AC power mains must scale the 110 volts AC up or down to the proper level, rectify it to produce pulsating DC, and then filter it to produce nearly smooth DC output for the transmitter. Additionally, in some cases, it will also be necessary to *regulate* the DC output voltage. The major components of a DC supply, then, are: *transformer, rectifier, filter circuit,* and sometimes a *regulator*. Let us discuss each of these sections separately, in their turn.

An example of the AC sine wave is shown in Fig. 10-4. Before we can make reasonable comments about the voltages in the circuit, we must first denote precisely which voltages we mean. The pure sinusoid will begin at zero, rise to a maximum with current flow in on direction, and then fall back to zero. The direction of the current flow then *reverses*, and the action is repeated; i.e., it rises to a maximum in the opposite direction and then falls back to zero. The waveform of both the voltage and the current is sinusoidal, so we can find instantaneous voltage (the potential at any given instant of time) from $e = E_p \sin \theta$, where E_p is the peak voltage and θ is the phase angle. There are three points of principal interest to us in this waveform are the two peaks and the *root mean square* (rms) values. We specify the peak voltage, E_p, and the rms voltage, E_{rms}. In a pure, undistorted sine wave, there are certain simple relationships between these values:

$$E_{rms} = 0.707E_p$$

$$E_p = 1.414E_{rms}$$
$$E_{p\text{-}p} = 2E_p = 2.82E_{rms}$$

These values are practical simplifications based on more basic principles that apply to all periodic waveforms. As such, they are valid only to sine waves.

Definitions of the various waveform points (E_p, $E_{p\text{-}p}$ and E_{rms}) are necessary before one can apply them to fully understanding the operation of the various circuits. The definition of peak voltage is almost obvious when Fig. 10-4 is examined. It is the potential difference between zero and the maximum voltage in one direction. Putting number on the definition shows that E_p is the difference in potential between 0 degrees and 90 degrees on the sinusiodal waveform. The peak-to-peak voltage ($E_{p\text{-}p}$) is the potential difference between the maximum positive (90-degree point) and maximum negative (270-degree point). In a perfect sine wave, the positive and negative excursions are identical, so the peak-to-peak voltage is *twice* the peak voltage.

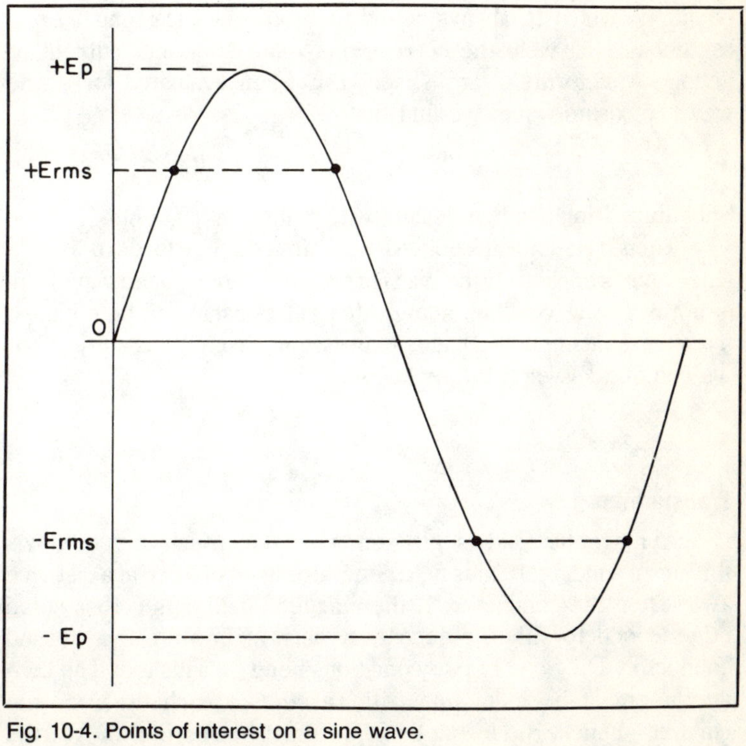

Fig. 10-4. Points of interest on a sine wave.

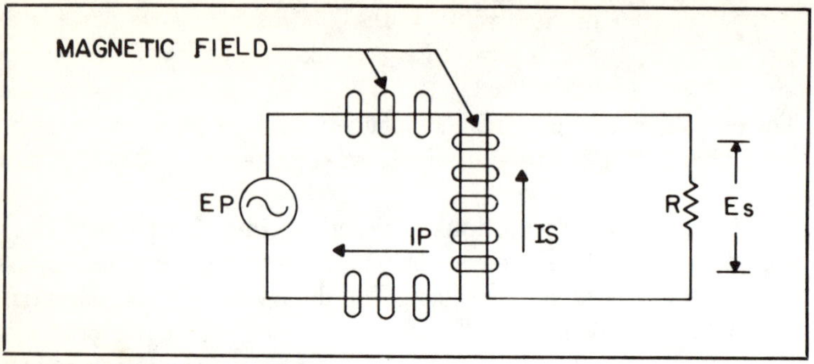

Fig. 10-5. Transformer action.

The rms voltage is a little more difficult to define. It is the so-called *effective* voltage and is numerically the same as a DC potential that causes the same amount of heating in a resistive load, or the *work* that is performed by the AC waveform. In fact, one method for measuring rms is to compare the unknown AC with a known DC in heating identical resistive loads. Mathemetically, E_{rms} and I_{rms} are expressed by an equation using integral calculus. That equation, which is always found in basic electrical engineering textbooks, will yield the correct rms value for either current or voltage waveforms of any shape. But when evaluated for a sine wave or cosine wave, we find that:

$$E_{rms} = E_p/(2)^{1/2} = E_p/1.414 = 0.707E_p$$

Substitute I for E when dealing with sinusiodal currents.

These relationships hold true only for resistive loads to which a sine wave is applied. If the load is reactive or the applied waveform is not a sine wave, then some other relationship will be obtained from the fundamental calculus expression. From the last equation, we can find E_p when E_{rms} is known:

$$E_p = 1.414E_{rms}$$

Transformers

A magnetic field is created when electrical current flows through a conductor. This field generates lines of force at a distance away from the conductor. If the magnetic field cuts across some other conductor, then a secondary current flows in that second conductor. Only one of two conditions need be satisfied: The conductors must be in motion with respect to each other or the current—hence, the magnetic field—must be varying. In AC, the current is constantly varying in a sinusoidal manner, so the magnetic field also varies. A transformer is a device in which two conductors (Fig. 10-5) are placed in close proximity to each other. One, called the *primary*, is energized from the AC source, so a current I_p is caused to flow in it. The magnetic field from the primary conductor cuts across the secondary conductor, causing secondary current I_s to flow.

Real transformers use many turns of wire, as in Fig. 10-6, not just one turn. In addition, most transformers and all that operate at low frequencies have their windings wound over a core material that makes the coupling between primary and secondary more efficient. At AC power line frequencies (60 Hz in the US), the core will be made of soft iron. But at radio frequencies, the core will be ferrite; i.e., a powdered iron material in an adhesion matrix. At VHF (over 30 MHz), the core may well be air or some nonferrous powdered metallic matrix.

Transformers are from 95 to 99+ percent efficient. Because of this pleasant fact, we may use simple expressions for the relationship between E_p, E_s, I_p, and I_s that contain no loss terms:

$$\frac{E_p}{E_s} = \frac{I_s}{I_p}$$

or alternatively:

$$E_s I_s = E_p I_p$$

☐**Example:**

A transformer with its primary connected to 110 volts AC produces a secondary potential of 6.3 VAC. Find the primary current, I_p, and the secondary current, I_s, when a 10-ohm resistive load is connected to the secondary.

Secondary current—

$$\begin{aligned} I_s &= E_s/R \\ &= (6.3)/(10) \\ &= 0.63 \text{ amperes} \end{aligned}$$

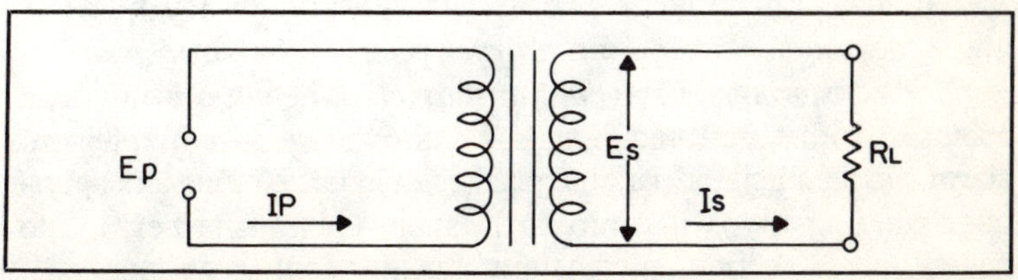

Fig. 10-6. Transfer action.

Primary current—

$$E_p/E_s = I_s/I_p$$
$$(110)/(6.3) = (0.63)/I_p$$
$$I_p = (0.63)(6.3)/(110)$$
$$I_p = 0.036 \text{ amperes}$$

From these two equations we see a relationship that is very important. It is especially clear in the last equation: The product of the primary voltage and current is equal to the product of the secondary voltage and current. This assumes, of course, lossless transfer between windings, which is almost true.

One rating of transformers is the VA rating, sometimes called the *primary VA ranging*. This is used because the primary is usually the innermost winding and cannot dissipate heat as rapidly as the secondary winding. The VA rating is not necessarily *watts* even though the product V × A is watts. In reactive circuits, a *power factor* equal to the cosine of the phase angle between voltage and current occurs. It is imperative that the VA rating be heeded; it must not be exceeded. If the VA rating is exceeded, the transformer may possibly be damaged. In any event, the manufacturer's guarantees are void if the transformer is placed in a circuit that causes the VA rating to be exceeded.

Transformers can be classified according to the ratio of the primary and secondary voltages. If $E_p = E_s$, then the *turns ratio* is 1:1. This means that the number of turns in the primary is the same as the number of turns in the secondary. Most 1:1 transformers are called *isolation transformers*. They are used to make ground-referenced AC power mains to floating, nongrounded AC power. Such is often necessary to insure safety in workbench situations.

A transformer in which E_p is less than E_s is a *step-up transformer*. Such transformers are used in radio transmitters to supply the high voltage for vacuum tube anodes. They might be billed under the terms *plate transformers* or *power transformers*.

The opposite type of transformer is the *step-down transformer*. A step-down transformer is one in which E_p is greater than E_s. They might be billed as *filament transformers, control transformers, rectifier transformers,* or *signal transformers*. The latter is not to be confused with electronic signals—audio voltages—but refers to alarm and annunciator panel signals used in large office and apartment buildings. We will return to the subject of transformers in due course in the section on transmitter power control wiring.

Rectifiers

Alternating current flows first in one direction, then reverses direction to flow in the other direction for one half of a cycle. AC, then, is *bidirectional*. Electronic circuits, especially radio transmitters because of FCC mandate to good engineering practices, require DC—direct current. A rectifier is a device that converts bidirectional AC to a unidirectional pulsating DC. It is not *pure* DC but is pulsating DC. There are several different types of rectifying devices, and not all of them are electronic. Examples of mechanical rectifiers are the *rotary rectifiers* and *synchronous vibrators*. Mechanical rectifiers are not used anymore, except in older equipment that ought to be replaced.

Both semiconductor and vacuum tube rectifiers are used in radio transmitters. We will briefly discuss vacuum tube types, but the remainder of the chapter will use semiconductor examples. If you want to plug these concepts into vacuum tube circuits, then simply replace the solid-state device with a vacuum tube.

Figure 10-7 shows a half-wave vacuum tube rectifier; it is a simple diode. The AC is applied to the anode. When the anode is positive with respect to the cathode, current will flow from the cathode to the anode. But on alternate half cycles, when the anode is negative with respect to the cathode, the tube is cut off, and no current can flow. The output waveform is as shown.

There are two basic types of tube rectifier commonly used in transmitter power supplies. For low-power and medium-power levels, *high-vacuum* rectifier tubes are used. Examples of these are the 5U4GB, 5R4GTB, and so forth. A higher power version is the

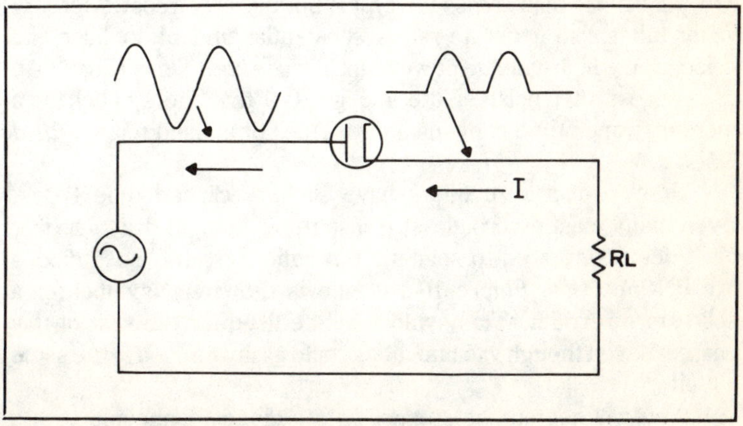

Fig. 10-7. Rectification in a vacuum tube diode.

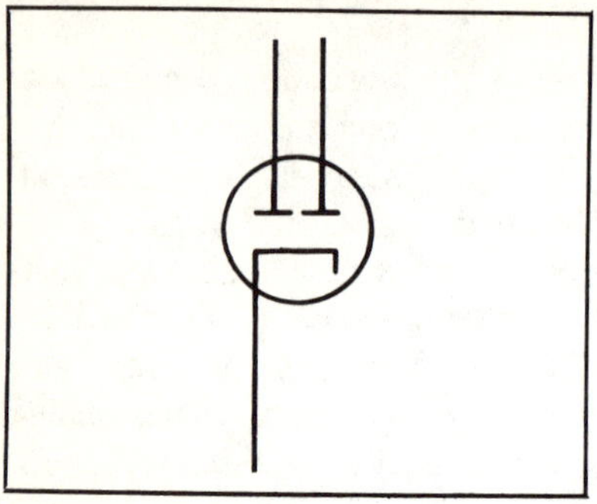

Fig. 10-8. Full-wave tube rectifier.

3B28. All of these are vacuum tube diodes. The 5U4 and 5R4 are full-wave devices, so they will have two independent anodes served by a common cathode (Fig. 10-8).

High power levels are usually handled by mercury vapor rectifiers. These are called "vacuum tubes," even though there is not a true vacuum inside of the glass envelope. The inside of the tube is evacuated of air, but contains a mercury vapor shortly after the filaments heat up. When the plate is made positive with respect to the cathode, that vapor will inoize, giving off a soft blue glow. Examples of mercury vapor tubes include the 866A, 966, and 872 devices. These tubes are considered obsolete, and even many applications formerly given the 866/966 devices are now solid state. In some cases, amateurs have replaced the 866 tubes in existing transmitters with 3B28 high-vacuum rectifiers.

The impedance of mercury vapor tubes is lower than that of high-vacuum types. This means that the voltage drop across the MV will be less than across HV types, but this is at a cost. Mercury vapor tubes also generate rather substantial amounts of hash-like noise. Some high-voltage power supplies use *hash filters* (Fig. 10-9) to suppress that noise. Note in Fig. 10-9 that the symbol for a mercury vapor tube contains a dot. This dot is used for *gas-filled tubes*.

Semiconductor rectifiers have all but eclipsed tube types. Even many existing broadcast transmitters, long in daily service, have been converted to solid-state rectifier circuits. Such diodes are PN junctions. Figure 10-10A shows the circuit symbol for a solid-state rectifier. This symbol will be used for the rest of this chapter, even though vacuum tubes could easily be used in the same circuits.

You will sometimes see solid-state diodes connected with a small series resistor (Fig. 10-10B). This is especially common in

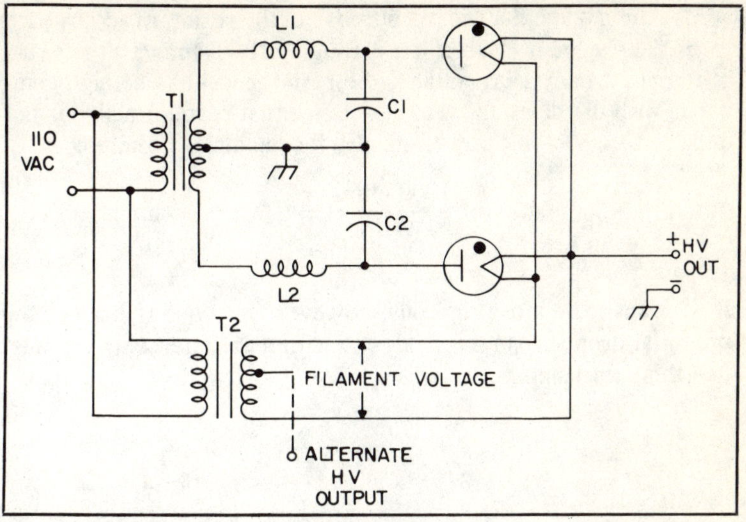

Fig. 10-9. Mercury vapor tube power supply.

older circuits. The purpose of this resistor is to limit the current surge and prevent it from damaging the diode. A large surge exists when some power supplies are first turned on. Most such resistors have a value between 1 ohm and 20 ohms.

Most high-voltage power supplies use several diodes in series in order to obtain the needed voltage rating (Fig. 10-10C). Each diode in the stack should be identical to the others. The voltage rating of the stack is the sum of the individual ratings. The current rating however, is not changed. But it is not good practice to simply connect the diodes in series. We must equalize the voltage drop

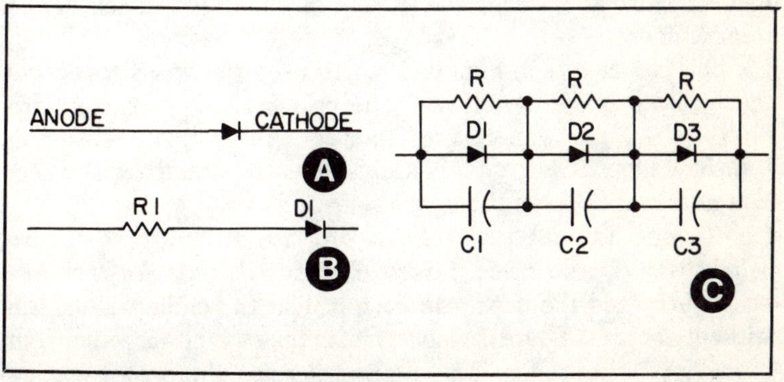

Fig. 10-10. Solid-state diode rectifiers with a simple circuit at A, a more complex one at B, and an even more complex one at C.

across the diodes. That is the purpose of the resistors across each diode. The value of the resistor is roughly 1000 ohms/volt as a rule of thumb. At any high voltage, the resistor value is so high compared with the diode forward resistance that rectifier action is not impaired. A 500-volt power supply, for example, requires:

$$500 \text{ volts} \times \frac{1000 \text{ ohms}}{\text{volt}} = 500{,}000 \text{ ohms}$$

If we want to be more rigorous, however, and wish to balance the potential drop across the diodes to within ±10 percent, we must use this expression:

$$R_{sh} = \frac{E_{piv}}{2I_r}$$

where R_{sh} is the shunt resistance in ohms, E_{piv} is the peak inverse voltage rating of the diode in volts, and I_r is the reverse leakage current in amperes.

□**Example:**

Find the resistance required shunted across each diode in an 800-volt power supply. Assume that three diodes are used to obtain a PIV rating of 2400 volts. Each diode, then, has a PIV of 800 volts. The reverse current of typical silicone diodes is 0.35 mA.

$$\begin{aligned} R_{sh} &= E_{piv}/\ (2 \times I_r) \\ &= (800)/((2)(0.00035) \\ &= 1.15 \text{ megohms} \end{aligned}$$

R_{sh} is a ***maximum*** value, so our rule of thumb would satisfy the requirement.

The capacitors in Fig. 10-10C are used to bypass transients that could damage the diodes. The values of the capacitors are 0.001 to 0.05 μF. There are other techniques that are used in transient suppression, and these will be dealt with in another section nearer the end of the chapter.

Diodes are rated by the *forward current* and *peak reverse voltage* (PRV), also called the *peak inverse voltage* (PIV). The forward current is the maximum current that the diode will sustain without damage. There is also an instantaneous, or surge, current rating for most diodes. This rating refers to the peak transient current that can be handled without damage. A transient, for the purposes of the surge current rating, is an event that lasts not longer than one complete cycle of the applied sine wave, or 1/60 of a second in US power systems.

The peak inverse voltage is the maximum reverse bias voltage that will not cause breakdown of the diode junction. The PIV rating is critical and must not be exceeded. The forward voltage rating is not nearly as important.

Rectifier Circuits

We have established that rectifiers convert bidirectional current such as AC into a unidirectional current such as pulsating DC. The raw output of a rectifier is not suitable for direct use in most types of electronic circuits, but is considered a step along the way to creating nearly pure DC.

Figure 10-11A shows a simple half-wave rectifier circuit. On the positive alternation of the AC cycle, the diode rectifier anode is positive with respect to the cathode, so current flows. On negative alternations, the diode is reverse biased, so current cannot flow (Fig. 10-11B). The output waveform (Fig. 10-12) is unidirectional, but it is not pure DC. It is pulsating DC. Because of the missing half cycles, the *average* voltage at the output of a half-wave rectifier is only 0.45 times the applied rms voltage. To deliver any given level of voltage and current with this circuit requires a transformer that has a primary VA rating at least 40 percent higher than would be required for full-wave circuits. The half-wave rectifier is not too efficient, and as such, is less popular in all but the lowest power applications. Rectifiers that use the full AC cycle are a lot more efficient. Examples of these full-wave rectifiers are shown in Figs. 10-13 and 10-14.

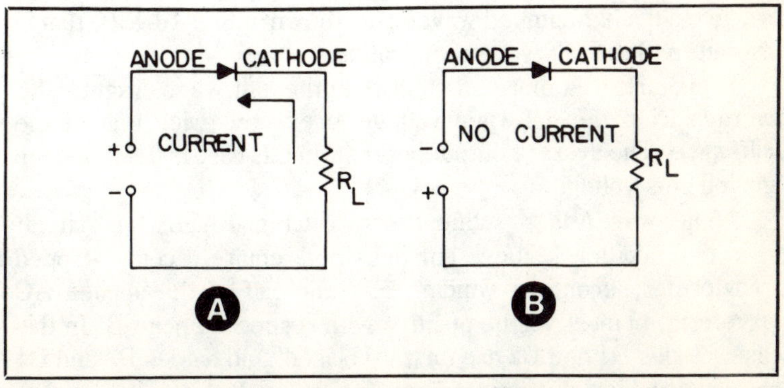

Fig. 10-11. Rectification in solid-state power supplies. Current is passing at A, and no current is passing at B.

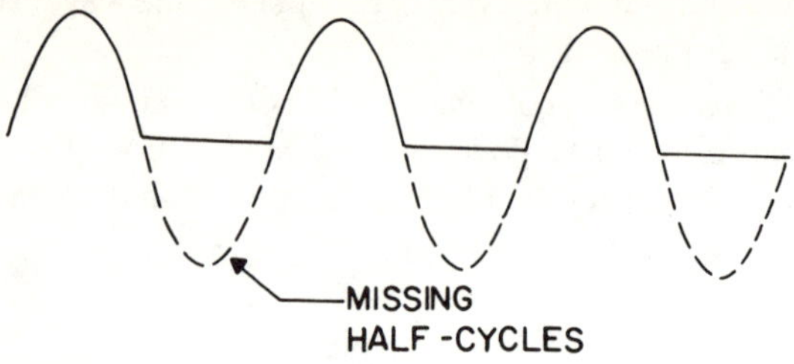

Fig. 10-12. Half-wave rectification.

Figure 10-13A shows a simple full-wave circuit using two rectifier diodes and a special transformer that has a center-tapped secondary winding. The center tap is used as the zero potential reference point from which all of the circuit voltages are measured. One end of the secondary will be positive with respect to the center tap, while the other end is negative with respect to the center tap. On one-half of the AC cycle, point A will be positive with respect to the center tap, and point B will be negative. In this case, diode D1 will be forward biased and diode D2 will be reverse biased. Current flows from the center tap through load resistor R_L, and diode D2 back to the transformer at point A.

On the second half cycle, the polarity across the secondary of the transformer is reversed. Point A becomes negative with respect to the center tap, and point B becomes positive with respect to the center tap. In this case, diode D1 is reversed biased and D2 is forward biased. Current flows from the center tap through load resistor R_L and diode D2 back to the transformer at point B.

It is important to note that the current flow through the load is in the same direction on both halves of the AC cycle. This will produce a double-humped waveform, shown in Fig. 10-13B, that is characteristic of full-wave rectification.

This circuit is more efficient than the half-wave circuit. The average from the full-wave voltage is exactly twice that of the half-wave. The average output potential in this case is 0.9 times the applied rms voltage.

A full-wave *bridge* rectifier is shown in Fig. 10-14A. This circuit uses two additional diodes but does not require a center-tapped transformer secondary winding. On one-half of the applied AC waveform, point A will be positive with respect to point B. In this case, diodes D1 and D2 are forward biased, and diodes D3 and D4 are reverse biased. Current flows from point B through diode D2,

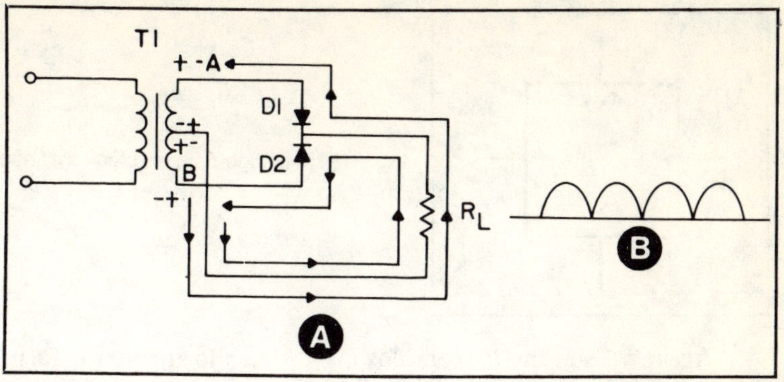

Fig. 10-13. Full-wave rectifier with center-tapped transformer at A and waveform at B.

resistor RL, and diode D1 to point A. One the second half of the AC cycle, the polarities of the transformer secondary reverse. Point A will become negative with respect to point B. Diodes D3 and D4 are forward biased, and D1/D2 are reverse biased. Current will flow from point A through D3 and RL to D4 and then back to the transformer at point B.

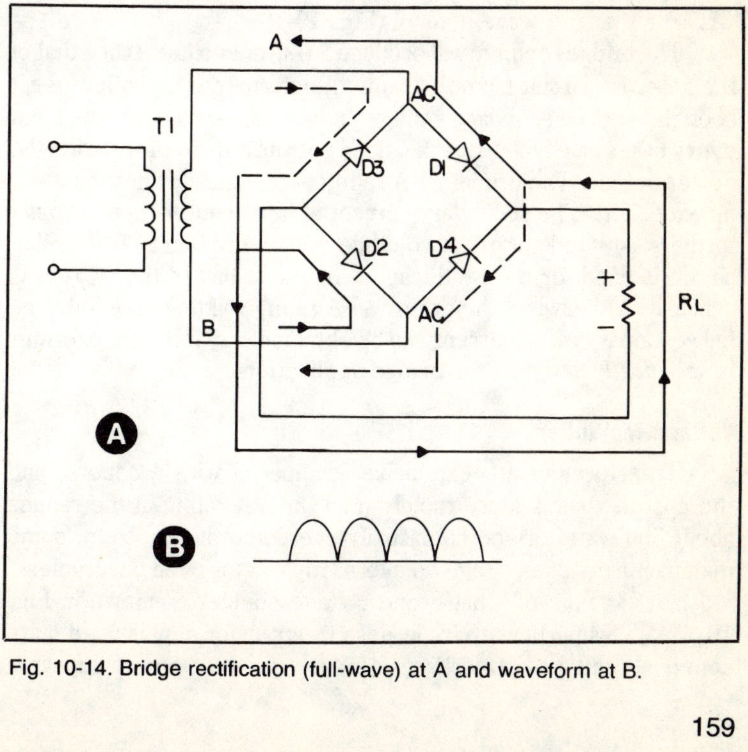

Fig. 10-14. Bridge rectification (full-wave) at A and waveform at B.

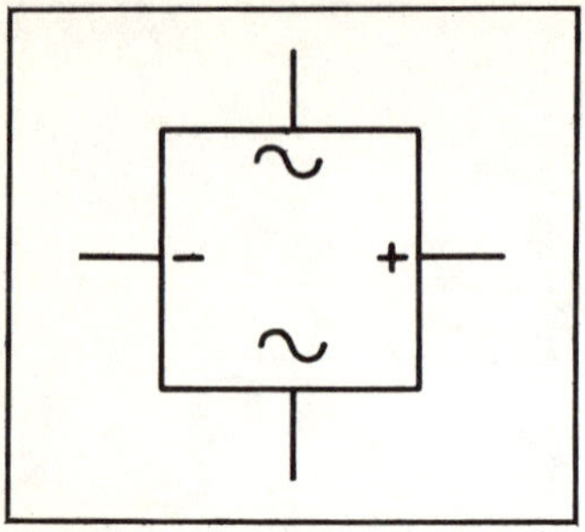

Fig. 10-15. Bridge rectifier block symbol.

Again we see the current flowing through load resistor R_L in the same direction on both halves of the AC cycle. The output waveform (Fig. 10-14B) is that characteristic of full-wave rectification. Once again the average output voltage is 0.9 times the applied rms voltage.

The transformer used with bridge circuits need not be center-tapped. The zero potential reference point is designated as the junction of the anodes of diodes D2 and D3. This point is labelled −, while the junction of D1 and D4 is labelled +. Some prepackaged bridge rectifiers have the DC terminals marked with the positive and negative signs and the AC terminals either with the letters "AC" or the sine wave symbol (Fig. 10-15).

The bridge rectifier will produce an output voltage twice that of the full-wave circuit from the same transformer. This is because it sees the entire secondary voltage on each half cycle instead of just every other half cycle. But this does not allow us to draw twice the power output. The primary VA rating of the transformer must not be exceeded. The secondary current of most center-tapped transformers is rated for the two-diode full-wave circuit. If a bridge circuit is used for the rectification, only one-half of the current is available. Otherwise, the primary-VA rating will be exceeded. The full secondary rated current can be obtained only if the transformer is *specifically designed* for bridge applications.

Voltage Multipliers

Transformers are expensive, compared with rectifiers, and the cost increases more rapidly than the VA rating in the ranges about 500 watts or so. To partially overcome this problem, some manufacturers use voltage doublers, triplers, or even quadruplers.

An example of a half-wave voltage doubler is shown in Fig. 10-16A. This particular circuit, less the transformer, is used quite commonly in transformerless television receiver power supplies.

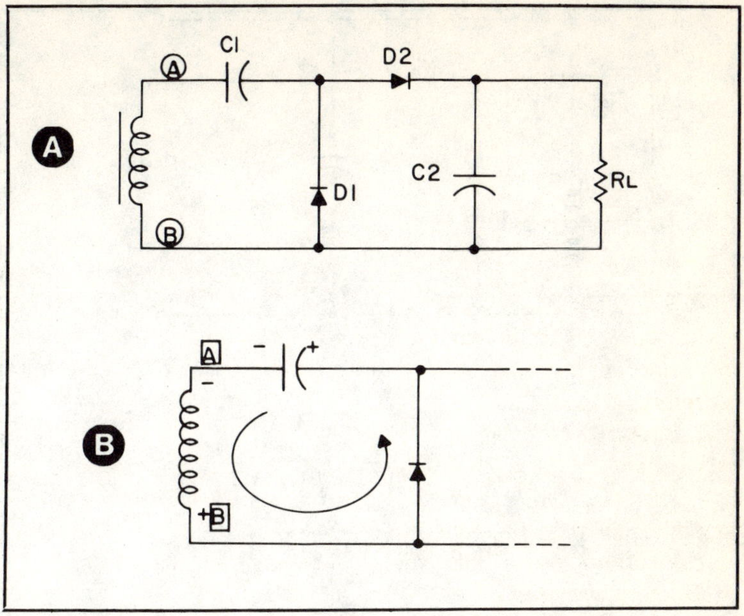

Fig. 10-16. Half-wave voltage doubler at A and simple action B.

One salient feature of this circuit is the half-wave bridge (D1/D2) fed from the AC source through capacitor C1.

On one-half of the AC cycle, point A will be negative with respect to point B (Fig. 10-16B). This condition will cause diode D1 to be forward biased and D2 to be reverse biased. Capacitor C1 will charge to approximately E_p (current flows through the transformer secondary and D1).

On the second half of the AC cycle (Fig. 10-17), the transformer polarity reverses, and point A is now positive with respect to

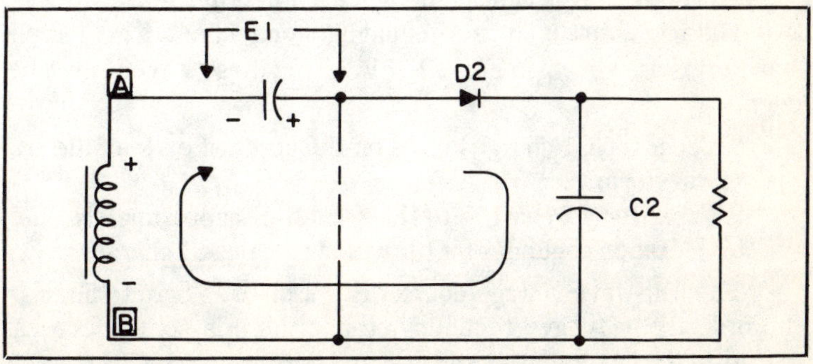

Fig. 10-17. Full action of a half-wave voltage doubler.

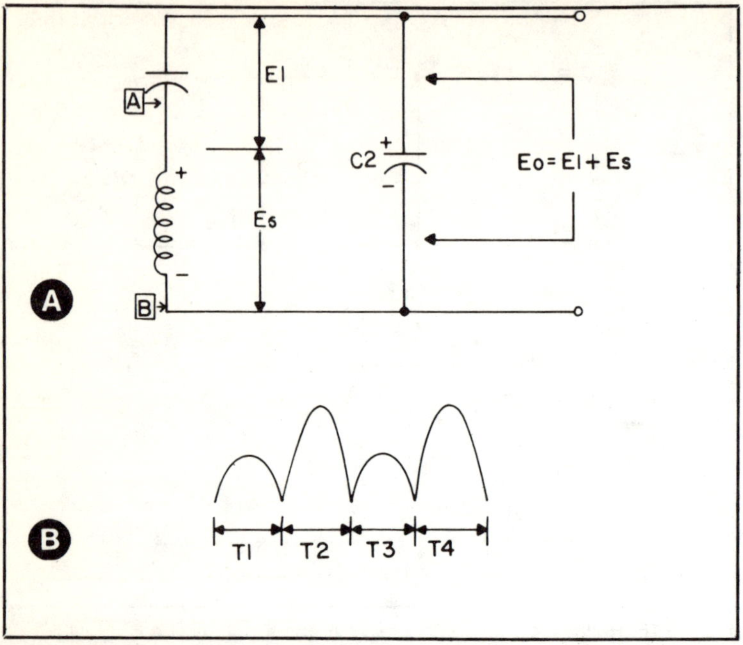

Fig. 10-18. Half-wave voltage doubler at A and waveform at B.

point B. This condition will reverse bias D1 and forward bias D2. Capacitor C2 will now charge to a potential of approximately $2E_p$. One-half of the potential is the transformer secondary voltage, while the other half is the charge placed on C1 during the first half cycle of operation.

$$E_o = E_{C1} + E_s$$
$$E_o \approx (E_p) + (E_p) \approx 2 \times E_p$$

Figure 10-18A is simplified somewhat in order to show these two voltage sources in series-aiding connection. The circuit waveforms are shown in Fig. 10-18B. Using these waveforms, the events are:

- □Capacitor C1 charges to E_p on the first half cycle of the AC waveform
- □Capacitor C2 charges to $(E_s + E_{C1})$, or approximately $2E_p$.
- □Operation continues for the next cycle; i.e. C1 charges to E_p.

The half-wave voltage doubler is not quite as hard to filter as the ordinary half-wave rectifier because there is some electrical energy in the circuit during the first and third half cycles. But it still does not make full use of the AC cycle, so it is less efficient than a full-wave circuit.

The full-wave voltage doubler of Fig. 10-19A, however, does use the entire AC cycle. On one-half of the AC cycle, point A will be positive with respect to point B (Fig. 10-19B). Diode D1 is reverse biased, and D1 is forward biased. Current will flow through D1 to charge capacitor C1 to approximately E_p.

On the second half of the AC cycle, the transformer polarity reverse, so point A is now negative with respect to point B. Diode D1 is now in the reverse bias state, and D2 is forward biased. Current will now flow through D2 to charge capacitor C2 to approximately E_p.

The output voltage E_o will be the sum of the voltages across capacitors C1 and C2 (Fig. 10-20).

$$\begin{aligned} E_o &= E_{C1} + E_{C2} \\ &= (E_p) + (E_p) \\ &= 2E_p \end{aligned}$$

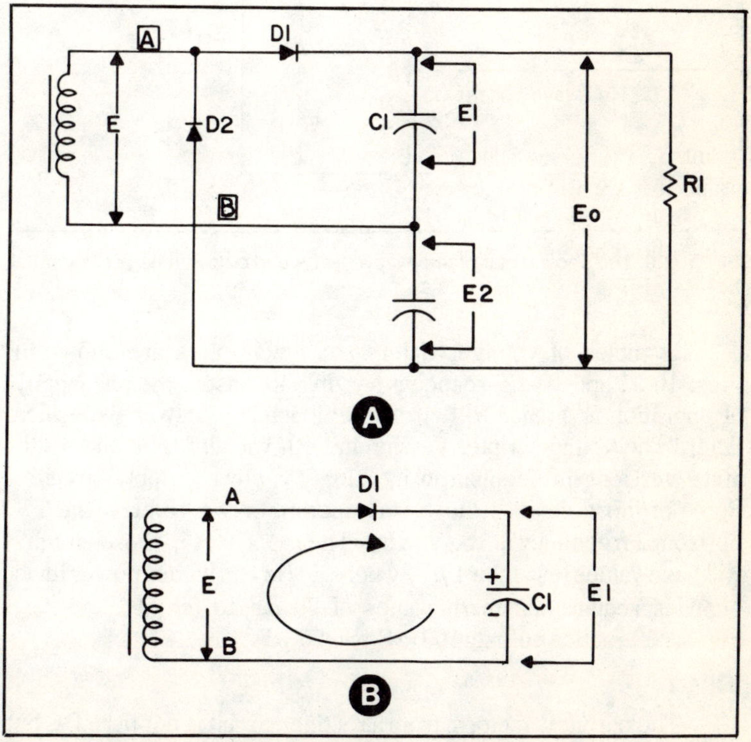

Fig. 10-19. Full-wave voltage doubler at A and its action at B.

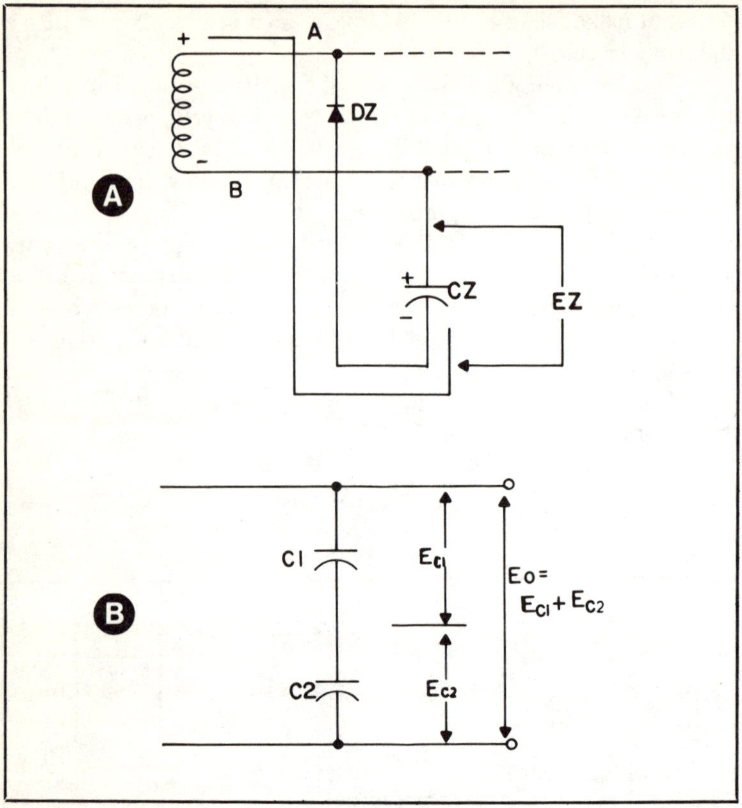

Fig. 10-20. The second half of the AC cycle at A and output voltage diagram at B.

Examples of voltage triplers and quadruplers are shown in Figs. 10-21 and 10-22, respectively. In both cases, the philosophy of operation is similar to that of the doublers—only the specifics differ. The voltage tripler version in both vacuum tube and solid-state versions is seen often in color TV power-supply circuits. Some of those devices, however, operate only a 15 kHz (the TV horizontal frequency is 15,734 Hz). The capacitors in those circuits will have values less than 1 μF, where 60 Hz and higher power level supplies require minimum values of 30 μF to 50 μF, and it is standard practice to use 100 μF to 250 μF.

Filters

The output waveform from a rectifier circuit is not pure DC but pulsating DC. True, it is unidirectional, but it has a half-sinusoidal

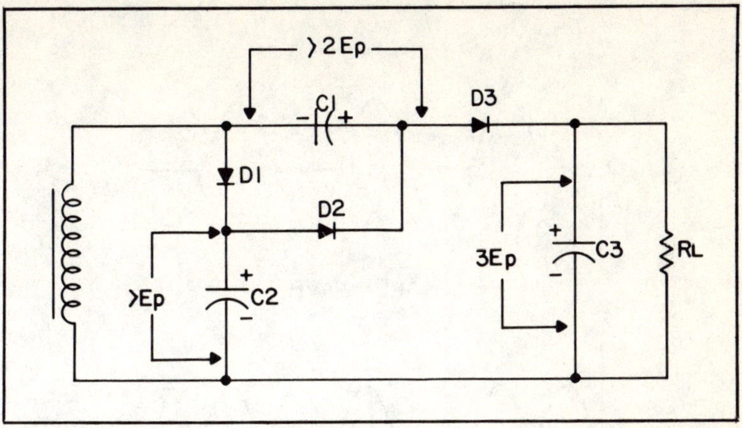

Fig. 10-21. Tripler.

waveshape. Figure 10-23 shows the waveforms for a half-wave and full-wave rectification, filtered pulsating DC, and nearly pure DC. Electronic equipment is happiest with pure DC, but will usually tolerate well-filtered pulsating DC.

The purpose of the filter circuit is to smooth the ripple output of the rectifier circuit and make it approximate pure DC. Various types of filter circuits exist, but they all involve capacitors and sometimes inductors and resistors. Different types of filters have different applications.

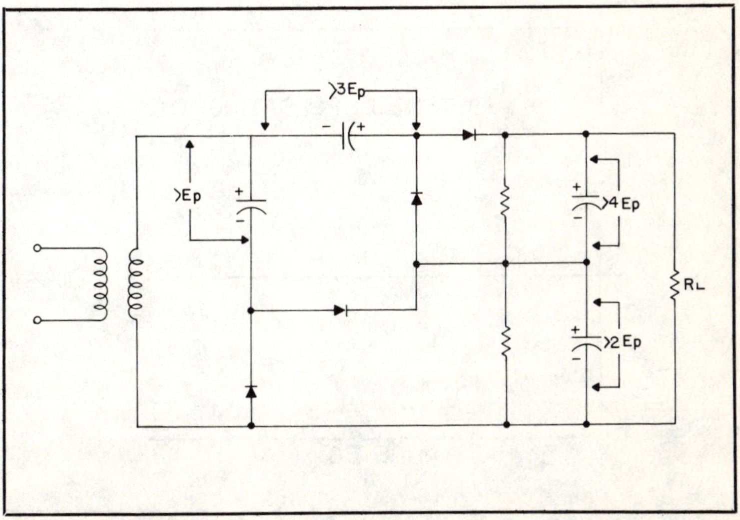

Fig. 10-22. Quadrupler.

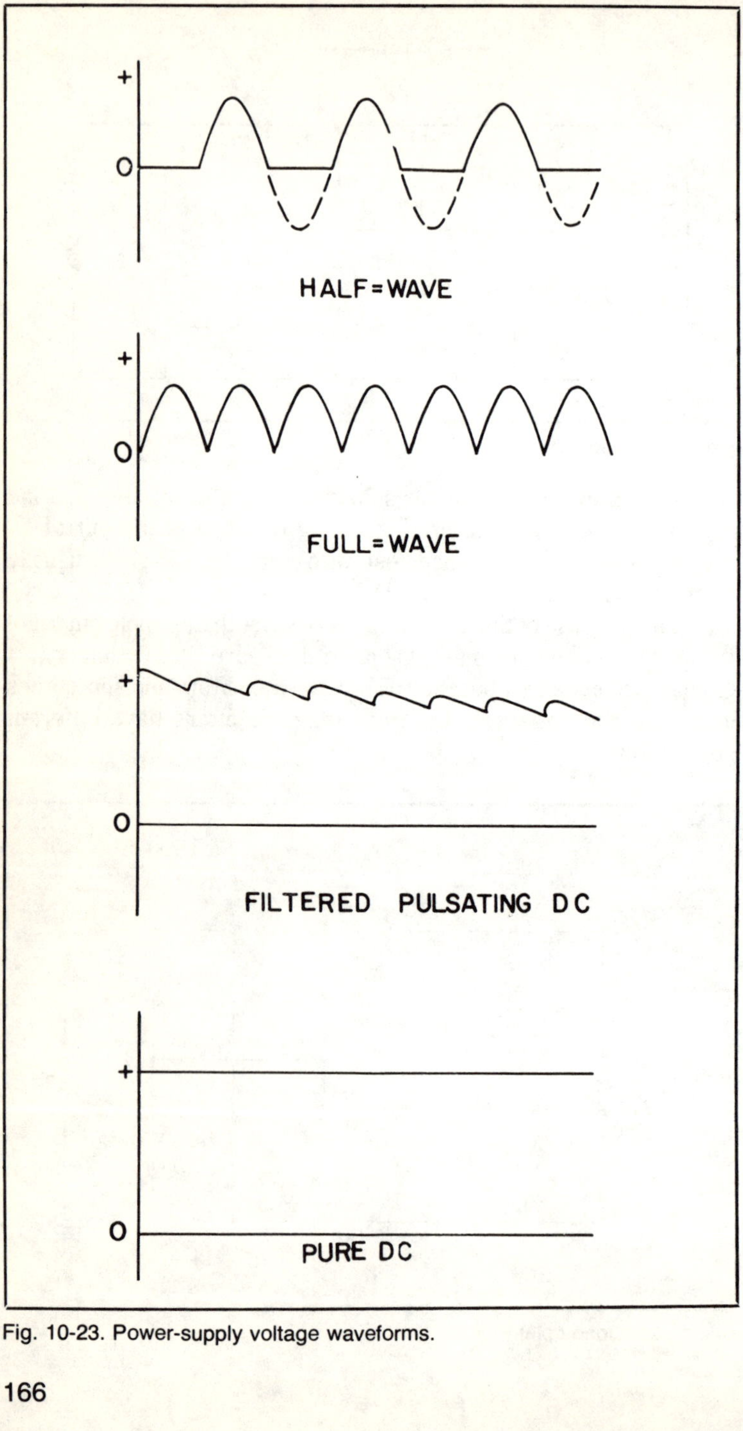

Fig. 10-23. Power-supply voltage waveforms.

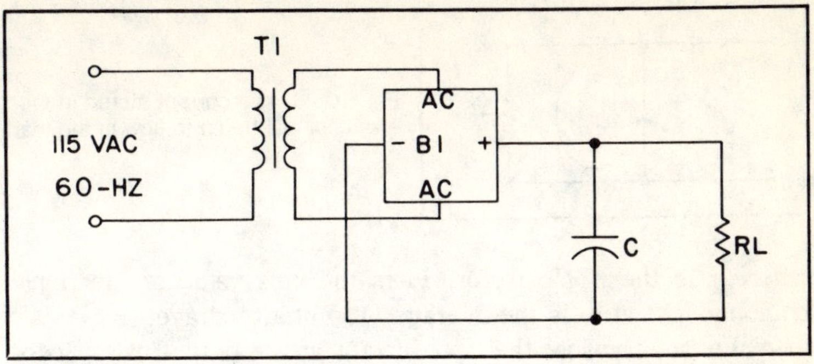

Fig. 10-24. Brute force filter circuit.

Capacitor "Brute Force" Filters. The simplest form of filter is the "brute force" filter of Fig. 10-24. The filter consists of a capacitor in parallel with the load resistor (Fig. 10-25). When the pulsating DC waveform is increasing, the rectifier output current serves to charge capacitor C and supply the load current. The capacitor will charge to approximately E_p during this period. After the peak has passed, however, the capacitor voltage is greater than the rectifier output voltage. This means that the current stored in the capacitor will begin to flow in the load (Fig. 10-26). The total load current will be partially due to the rectifier current and partially due to the capacitor current. But the contribution of the rectifier current to the total load current becomes less and less, diminishing to zero at the conclusion of the half cycle. The process repeats itself as the next half cycle begins (Fig. 10-27).

The filter tends to smooth out the pulsations of the rectifier output, but it does not reduce it to zero. There is still a certain ripple factor (r) present. The ripple factor can be defined as the ratio of the ripple component to the average DC voltage:

$$r = \frac{E_r}{E_a}$$

Fig. 10-25. Brute force filter is a capacitor and resistor in parallel.

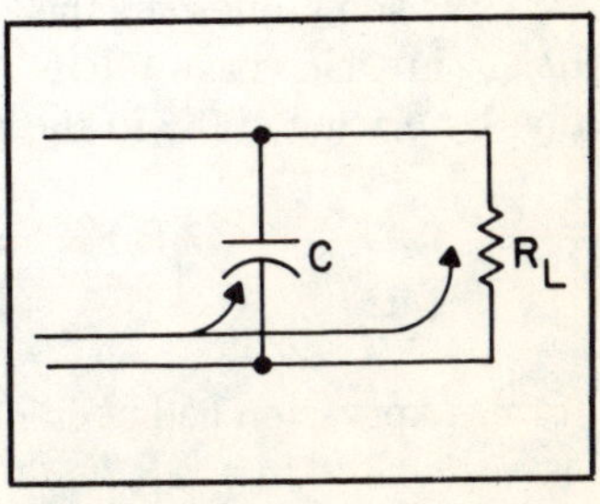

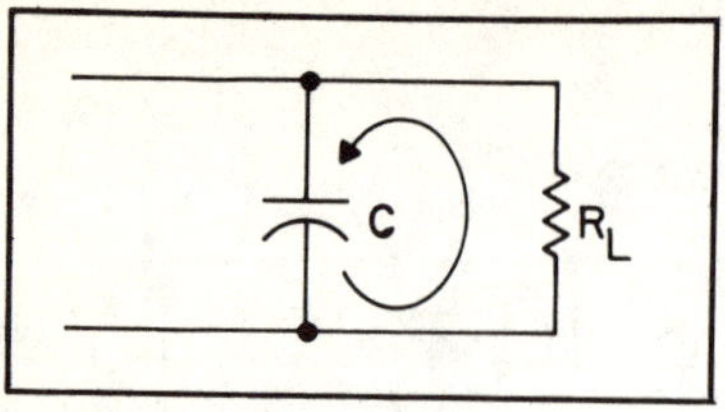

Fig. 10-26. The current stored in the capacitor will begin to flow in the load.

where r is the ripple factor, E_r is the rms value of the ripple component, and E_a is the average DC output voltage.

We can calculate the ripple factor at the output of capacitor filters from the expression:

$$r = \frac{1}{2\sqrt{3}\, f\, C\, R_L}$$

where r is the ripple factor, f is the ripple frequency (60 Hz in half-wave supplies and 120 Hz in full-wave supplies), C is the filter capacitance in farads, and R_L is the load resistance in ohms ($R_L = E_{DC}/I_{out}$).

☐**Example:**

Calculate the ripple factor if a 1500-volt full-wave power supply delivers 500 mA. The filter capacitance is 16 μF.

$$r = \frac{1}{2\sqrt{3}\, f\, C\, R_L}$$

$$= \frac{1}{(2)(3)^{1/2}(120\text{ Hz})(1.6 \times 10^{-5}\text{F})(1500\text{V}/0.5\text{A})}$$

$$= 1/19.95 = 0.05$$

The ripple factor is reduced by increasing the capacitance. For your own satisfaction, you might want to calculate the ripple factor of this same power supply with 4 μF and 50 μF filter capacitors.

We sometimes see this basic equation in the form of *percentage ripple*, in which case %R.F. = (100%)(r), so modify the equation by placing 100% in the numerator:

$$\%\text{R.F.} = \frac{100\%}{2(3)^{1/2} f C R_L}$$

If this expression had been used in the previous example, it would have resulted in a %R.F. of 5% for r = 0.05.

In most practical applications, we will want to rearrange these expressions to solve for the capacitance required to achieve a specified ripple factor. Regulated power supplies, for example, frequenctly specify input ripple factor of 10 percent or less. Radio transmitters using unregulated power supplies usually ask for a ripple factor of one percent or less with 0.1 percent being considered by many as "good engineering practice." The corrected expression to find the minimum capacitance required in any given power supply is:

$$C_{farads} = \frac{1}{2\sqrt{3}\, f\, r\, R_L}$$

☐**Example:**

Find the minimum required capacitance to achieve a 0.9 percent ripple factor from a 13.6-volt DC power supply using a brute force filter when 10 amperes is drawn (%R.F. of .9% is r = 0.009).

$$C = \frac{1}{2(3)^{1/2}\, f\, r\, R_L}$$

$$= \frac{1}{(2)(3)^{1/2}(120\text{ Hz})(0.009)(13.6\text{V}/10\text{A})}$$

$$= 1/7.6 = 0.196 \text{ farads} = 196{,}500\ \mu\text{F}$$

The above example gives a result that is terribly large: a 196,500 μF capacitor? But this result is totally reasonable, and the desired capacitance is actually within reach. Many manufacturers actually offer low-voltage (15 to 30 WVDC) capacitors in the 15,000 to 150,000 μF range.

We can simplify the expression even further by consolidating the constants (1,2, and $(3)^{1/2}$), assuming that only full-wave rectifiers are used (making the frequency 120 Hz) and expressing the resultant capacitance in the more useful microfarads:

$$C_{\mu F} = \frac{2405}{rR_L}$$

where $C_{\mu F}$ is the filter capacitance in μF, r is the ripple factor, R_L is the load resistance.

☐**Example:**

Find the minimum capacitance required for a 13.6-volt DC, 10-ampere power supply to achieve a ripple factor of 0.9 percent (r = 0.009).

$$
\begin{aligned}
C_{\mu F} &= 2405/rR_L \\
&= 2405/(0.009)(13.6/10) \\
&= (2405)/(0.0122) \quad = \quad 196{,}500\ \mu F
\end{aligned}
$$

This equation is the correct expression for designing brute force filter power supplies. Keep in mind that it is desirable to keep the ripple factor reduced as low as possible. The ripple factor often considered good practice in transmitters is 0.1 percent ($r = 0.001$). We could modify this equation even further to account for this r-value. In that case, $C_{\mu F} = 2.4 \times 10^6/R_L$.

There is a common rule of thumb that is heard around amateur circles from time to time. It is:

$$C_{\mu F} = \frac{100}{\omega R_L}$$

or

$$C_{\mu F} = \frac{0.265}{R_L}$$

where $C_{\mu F}$ is the minimum required capacitance in μF, R_L is the load resistance in ohms, and ω is the term $2\pi f$ (377 at 60 Hz), where f is the AC line frequency, not the ripple frequency.

The use of this expression assumes a value for r of 0.01. As a rule, it produces the same result as the previous expression, assuming $r = 0.01$.

RC Filters. In some cases, the simple brute force filter results in capacitance values that are simply too high for the required level or ripple reduction. This is especially true in some low-level signal applications, such as receiver front ends and audio microphone preamplifiers where the ripple would tend to modulate the signal being processed in the electronic circuitry.

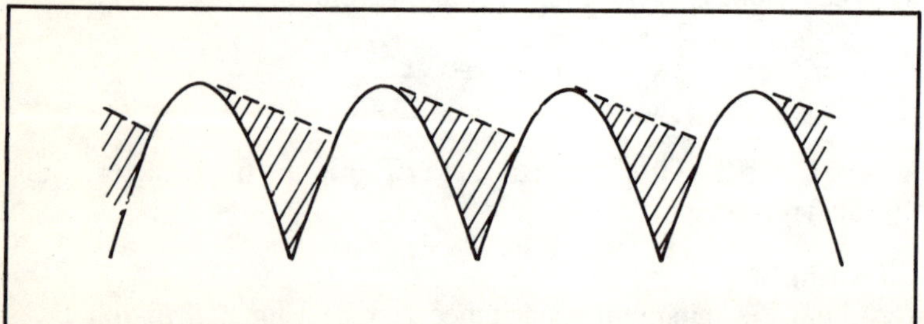

Fig. 10-27. The contribution of the rectifier current to the total load current drops to zero at the end of the half cycle.

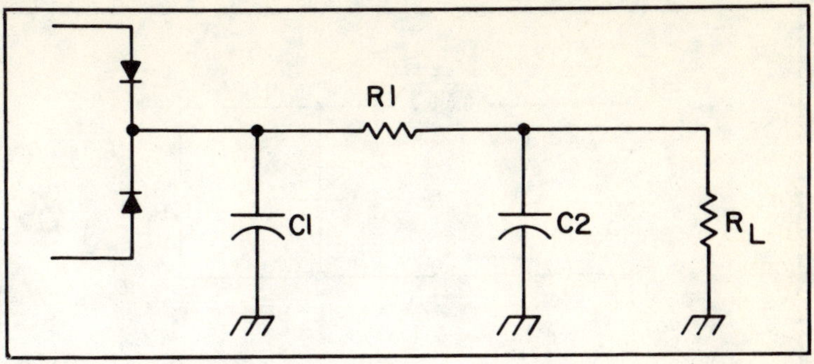

Fig. 10-28. Pi-section RC filter circuit.

Figure 10-28 shows a pi-section RC filter that will cause a substantial reduction in output ripple. Unfortunately, this is only at the expense of increasing the series resistance by the amount of R1, thereby limiting this circuit to relatively low-current applications. In addition, the current drain should be kept constant or at least reasonably so. Such demands make the principal applications of this circuit nonpower-amplifier stages (audio preamplifiers, i-f/rf amplifiers in receivers, etc.). The ripple factor from the pi-section filter is found from:

Half-wave circuits—

$$r = \frac{1}{C_1C_2R_1R_L}$$

Full-wave circuits—

$$r = \frac{2.5 \times 10^{-6}}{C_1\ C_2\ R_1\ R_L}$$

where r is the ripple factor, and C_1 is the capacitance of C1 in farads, C_2 is the capacitance of C2 in farads, R_1 is the resistance of R1 in ohms, and R_L is the load resistance in ohms.

□**Example:**

Calculate the ripple factor at the output of a 250-volt, 100-mA, full-wave power supply if C1 = 16 μF, C2 = 8 μF, and R1 = 560 ohms.

$$r = (2.5 \times 10^{-6})/(C_1C_2R_1R_L)$$

$$= \frac{(2.5 \times 10^{-6})}{(1.6 \times 10^{-5}\text{F})(8 \times 10^{-6}\text{F})(560\text{ ohms})(250\text{V}/0.1\text{A})}$$

$$= 0.014$$

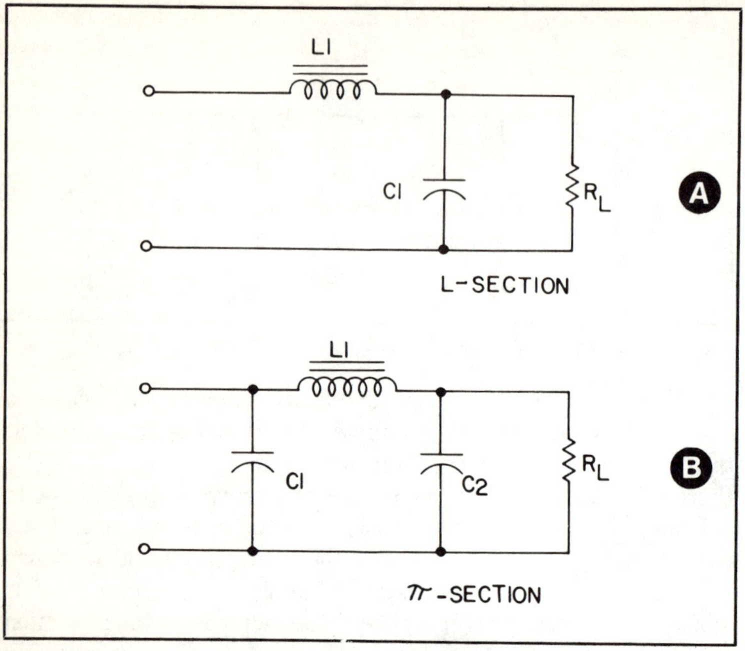

Fig. 10-29. L-section and pi-section inductive filters.

Inductive Filters. High-power, vacuum tube transmitters frequently employ inductive filters in the DC power supply. Inductive filters might be one section or several sections. They might have capacitor input or inductor input. As a general rule: do not use any form of inductor-input filter with a half-wave rectifier. Inductors respond to the rate of change of current. The voltage across an inductor is proportional to the inductance and the rate of change of the current:

$$E_L = L \times \frac{\Delta I}{\Delta t}$$

where E_L is the voltage across the inductor, and $\Delta I/\Delta t$ is the rate of change of the inductor current with respect to time.

The rate of change of the current from a half-wave rectifier becomes very large close to the zero crossings (when I is near zero). This situation results in a very high value of E_L, so high in fact as to exceed the peak inverse voltage rating of the rectifiers. Although one might tend to wonder why anyone would want to use a half-wave rectifier in an application that required an inductor filter, it has been done.

Figures 10-29 through 10-31 show several examples of inductor power-supply filters. The two basic configurations are shown in Fig. 10-29. The circuit of Fig. 10-29A is a simple L-section filter; it is an inductor-input filter circuit. It gets its name from the fact that the LC portion resembles the letter "L." The pi-section filter of Fig. 10-29B is the capacitor-input version. The name is derived from the resemblence of the C1-L1-C2 circuit to the Greek letter "π." The ripple factors for these filters are:

L-section, full-wave—

$$r = \frac{8.3 \times 10^{-7}}{C\ L}$$

Pi-section, half-wave—

$$r = \frac{2.6 \times 10^{-8}}{C_1\ C_2\ L_1\ R_L}$$

Pi-section, full-wave—

$$r = \frac{3.3 \times 10^{-9}}{C_1\ C_2\ L_1\ R_L}$$

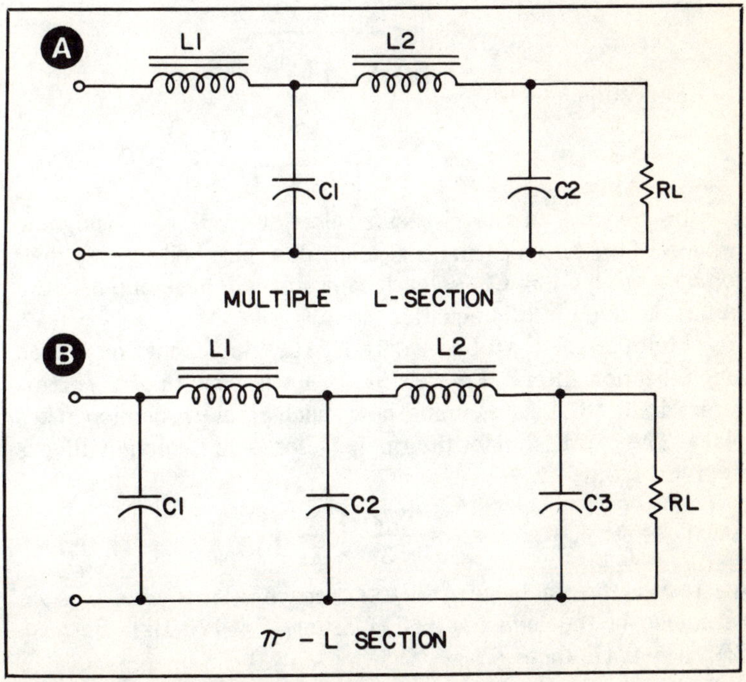

Fig. 10-30. Multiple L-section and pi-L-section inductive filters.

As noted earlier, an inductor tends to oppose changes in the current flowing. Consequently, the inductor in a power-supply filter will tend to keep the output voltage more constant as the current level changes.

By examining the above expressions, we can learn something about the characteristics of these filters. Note that the R_L term appears for the pi-section filter, but does not appear in the expression for the L-section filter. This fact tells us that the ripple factor is independent of the load current, because $R_L = E_o/I_L$.

As a general rule, it is standard practice to select a value for the inductor according to a rule of thumb, and then calculate the minimum values of capacitance required to achieve the desired ripple factor. The inductor should have a value of at least:

$$L = \frac{R_L}{1100}$$

We may rewrite the previous expressions to select the capacitance (fullwave case only):

L-section

$$C = \frac{8.3 \times 10^{-7}}{rL}$$

Pi-section

$$C_1C_2 = \frac{3.3 \times 10^{-9}}{r\ L_1\ R_L}$$

In the case of the pi-section filter, we are calculating the product of the two capacitances. Many designers will simplify their job by setting C1 = C2, so each capacitor will have one-half the result obtained in this equation.

Multiple-section filters will reduce the ripple much more than single-section filters. Figure 10-30 shows a multiple L-section filter circuit. Let us examine how much ripple reduction takes place. The expression for the ripple factor of an L-section filter is derived from

$$r = \frac{\sqrt{2}}{3}\left[\frac{X_c}{X_L}\right]$$

where X_c is the reactance of the filter capacitor $\frac{1}{2}\pi fC$, and X_L is the reactance of the inductor $2\pi fL$ (assume f = 120 Hz). Because $(2)^{1/2}/3 = 0.47$, then:

$$r = 0.47(X_c/X_L)$$

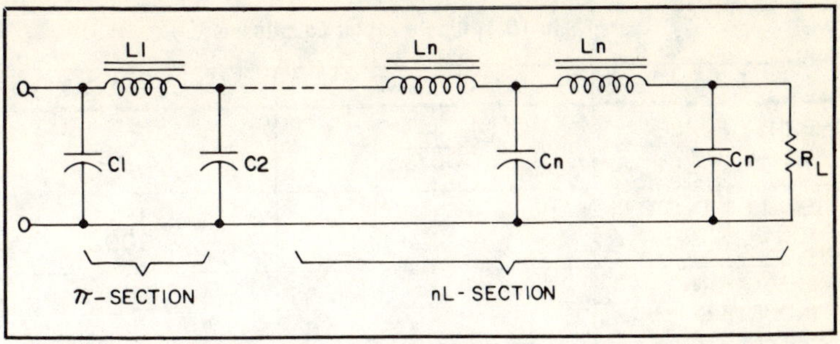

Fig. 10-31. Pi section-multiple L-section filter.

When this expression is evaluated, the result is the expression given earlier. But in this form, we see that the ripple is reduced by the factor (X_c/X_L) for each L-section filter connected in series. So for a two-section filter, it is:

$$r = 0.47 \left[\frac{X_c}{X_L}\right]\left[\frac{X_c}{X_L}\right] = 0.47 \left[\frac{X_c}{X_L}\right]^2$$

and for three sections:

$$r = 0.47 \left[\frac{X_c}{X_L}\right]\left[\frac{X_c}{X_L}\right]\left[\frac{X_c}{X_L}\right] = 0.47 \left[\frac{X_c}{X_L}\right]^3$$

But being practical minded, we simplify the expressions using certain assumptions, such as 120-Hz ripple frequency, and come to the conclusion that:

$$r = 0.47 \left[\frac{1.76 \times 10^{-6}}{LC}\right]^N$$

where r is the ripple factor, L is the inductance in henrys, C is the capacitance in farads, and N is the number of identical sections.

Similar reasoning will yield the ripple factor expressions for the other multi-section filter circuits. All of the ripple factor expressions are summarized in Table 10-1.

Filter Output Voltage. The output voltage that may be obtained from any given filter depends upon the type of filter used. Table 10-2 lists the types of filters and the mathematic expressions that will yield the approximate output voltage. These figures assume more or less ideal components, especially in the case of the inductor circuits. If the inductor has any appreciable DC resistance for the level of current being drawn, there will be a voltage drop

Table 10-1. Ripple Factor Equations.

TYPE	HALF-WAVE	FULL-WAVE
BRUTE FORCE CAPACITOR	$\frac{1}{208 R_L}$	$\frac{1}{416 R_L}$
SINGLE INDUCTOR (NO CAPACITOR)	X	$\frac{R_L}{1600L}$
L-SECTION INDUCTIVE (LC)	X	$\frac{8.3 \times 10^{-7}}{LC}$
π-SECTION INDUCTIVE	$\frac{2.6 \times 10^{-8}}{C_1 C_2 L_1 R_L}$	$\frac{3.3 \times 10^{-9}}{C_1 C_2 L_1 R_L}$
MULTIPLE-L	X	$0.47 \times \left[\frac{1.76 \times 10^{-6}}{LC}\right]^N$
π-MULTIPLE-L	$\frac{1.22 \times 10^{-8}}{C_1 C_2 L_1 R_L} \times \left[\frac{7 \times 10^{-6}}{C_N L_N}\right]^N$	$\frac{1.4 \times 10^{-9}}{C_1 C_2 L_1 R_L} \times \left[\frac{1.76 \times 10^{-6}}{C_N L_N}\right]^N$
π-SECTION RC	$\frac{10^{-5}}{C_1 C_2 R_1 R_L}$	$\frac{2.5 \times 10^{-6}}{C_1 C_2 R_1 R_L}$

across the choke coil. In the cast of an L-section choke, for example, the ideal components result in an output voltage of approximately $0.636E_p$. But if there is any DC resistance in the choke, the correct expression is $E_o = 0.636E_p - I_oR_o$, where R_o is the resistance of the choke winding.

Swinging and Smoothing Chokes. There are two types of filter chokes used in multi-section power supply filters. The smoothing choke is an ordinary inductor and is used as the first choke in the two-section filter. When there is only one section or in two-section filters in which varying current demands are noted (when powering a class-B power amplifier), however, it might be wise to use a *swinging choke*. The equation, $L \geq R_L/1100$, gives us the *critical value* of inductance, or the minimum required for proper filtering. But note in this equation that the load resistance is in the numerator. This means that the minimum required value of inductance will vary as the current values. The swinging choke is designed with a gap in its magnetic core, such that the inductance will vary more or less inversely porportional to the current passing through the winding. This means that reducing the current will increase the inductance, and increasing the current (the same as reducing the output load resistor) will decrease the inductance. Swinging chokes will have a dual rating; for example, they will be rated as 5/20 henrys. This means that the inductance will vary from

5 to 20 henrys as the current varies. The use of this type of choke effects some savings in the cost of the power supply for those applications where the current may vary considerably.

Bleeder Resistors. There are two main uses for the bleeder resistor in a power supply: load regulation and safety. The load will vary considerably, so when inductor filters are used, we have to insure that the load is never greater than some maximum amount. Otherwise, a different value of critical inductance results. The next equation gives the maximum value of the bleeder resistor, or a resistor that is connected across the power supply output in parallel with R_L, that will insure the proper operation of the inductor filter:

$$R_b = \frac{R_{bm} R_L}{R_L - R_{bm}}$$

where R_b is the bleeder resistance maximum value in ohms, R_L is the maximum load resistance (not including R_b), and R_{bm} is the load resistance required to satisfy the equation, $L \geq \frac{R_L}{1100}$.

The second reason for using a bleeder resistor is a matter is pure safety. The filter capacitors will store a lethal charge. If you go to work on a power supply, and the charge is still on the filters, then

Table 10-2. Output Voltage for Supplies With Different Filters.

TYPE	HALF-WAVE	FULL-WAVE
BRUTE FORCE CAPACITOR	$EP - \frac{Io}{120\,C}$	$EP - \frac{Io}{240\,C}$
SINGLE INDUCTOR (NO CAPACITOR)	X	0.636 Ep
L-SECTION (INDUCTIVE)	X	0.636 Ep
π-SECTION (INDUCTIVE)	$Ep - \frac{Io}{240\,C_1}$	$Ep - \frac{Io}{120\,C_1}$
π-SECTION (RESISTOR)	$Ep - Io\left[\frac{1}{120\,C_1} + R\right]$	$Ep - Io\left[\frac{1}{240\,C_1} + Ro\right]$
MULTIPLE-L	X	0.636 Ep
π-MULTIPLE-L	$Ep - \frac{Io}{120\,C_1}$	$Ep - \frac{Io}{240\,C_1}$

contacting those filters can kill you. The purpose of the bleeder is to drain off the charge after the power supply is turned off. If the circuit is an inductor filter, then use this last equation to calculate the bleeder. When voltage regulation and ripple stabilization is unimportant, however, use a rule of thumb that sets the bleeder resistor not higher than 100 ohms per volt. This means (100 ohms/V) × 3000V, or 300K ohms maximum for a kilowatt power supply. The wattage rating of the actual resistor selected for the bleeder should be very conservative. In the case above, let us say we ordered a "standard" value resistor of 250K ohms for our 3000-volt power supply. The wattage required would be

$$\begin{aligned} p &= E^2/R \\ &= (3000)^2/(250{,}000) \\ &= 36 \text{ watts} \end{aligned}$$

The next higher standard wattage rating would be 50 watts, but it is usually recommended that you use the next wattage higher than *twice* the actual wattage. And this assumes that you use the entire resistor for the bleeder. Many power resistors are open, so that you can tap off the correct value with a special metal clamp. The wattage rating, however, assumes that the entire end-to-end resistance is being used. If you must tap down on the body of the resistor in order to obtain the maximum value of bleeder by either method, then it is necessary to *derate* the power specification. Since it is not necessarily a linear derating that it used, it is recommended that you consult the resistor manufacturer or buy another resistor that will work without tapping down.

The bleeder resistor is one area where you don't want to play around! It is a critical safety device. Having a failed bleeder is worse than not including one in the first place. An open bleeder resistor can give you a lethal, false sense of security.

Always discharge all filter capacitors when working on a power supply. This is done with a screwdriver across the terminals or by connecting a known good bleeder resistor across the capacitor. Of these two, the first is most common while the second has most common sense behind it. Keep a second bleeder handy to serve this function. Mount it in an insulated box and then connect it with insulated alligator clip leads, but mind that most wire is insulated only to 600 volts! Always connect the ground end first and then attach the hot end. Wear dry gloves and *never* touch the exposed conductors in either the bleeder box or the power supply until you *know* that the capacitors are safely discharged. Use a voltmeter across the capacitor to tell you when this occurs.

Repeat the process several times! A phenomenon known as *dielectric absorption* prevents a capacitor from discharging all of its charge on the first try. The practical result of this is that you will discharge the filter, and then get "bitten" a short while later from a capacitor you *assumed* was cold.

People have accused me of flailing the safety issue "too much." Perhaps they are right, but I have seen one man killed by a transmitter power source and don't want to repeat the experience. Besides, anyone good enough to pay money for my book ought to get the conservative, safe advice that will tend to keep him breathing.

Voltage Regulation

A perfect (ideal) voltage source power supply will produce the same output voltage regardless of the load current. But real power supplied may show some variation in the output voltage with changes in the output load current. In general, the voltage will drop as the load current increases.

The cause of the problem is the power-supply internal impedance, which is usually called *output impedance, source impedance,* or *internal impedance*. We shall use the term, source impedance (R_s), but be aware when reading other books that the other terms are synonyms. The value of R_s is:

$$R_s = \frac{E}{I_{o(max)}}$$

where R_s is the source impedance in ohms, E is the open-terminal output voltage, and $I_{o\,(max)}$ is the short circuit output current.

Open-terminal output voltage is the voltage that is produced by the DC supply when the load is disconnected. The short-circuit output current is exactly that: the current that will flow in an output short. Don't actually measure this value, as few real-world power supplies can withstand a direct short across the output for long.

Figure 10-32 shows a model circuit of a typical DC power supply. Voltage E is the open-circuit potential, I_o is the output current, E_o is the *loaded* output voltage ($R_L \neq 0$) and R_L is the load resistance E_o/I_o. Some part of E will be dropped across R_s, so is not available to the load R_L. R_s and R_L then form a resistor voltage divider, which is shown in the redrawn version of the circuit in Fig. 10-33. The actual output voltage E_o is:

$$E_o = \frac{E\,R_L}{R_s + R_L}$$

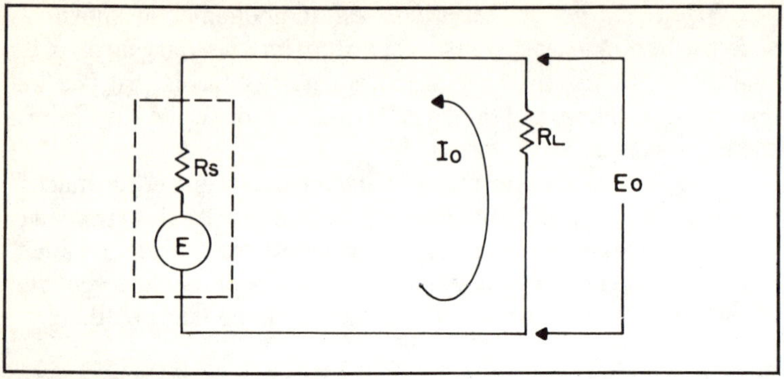

Fig. 10-32. Mechanism of voltage regulation problems.

□Example:

A DC power supply for a mobile two-meter rig has an open-terminal voltage of 14.0 volts DC. Its internal resistance is 1.5 ohms, and the load resistance on *transmit* is 3 ohms. Calculate the voltage available to operate the rig. Also, if the rig requires 13.6 VDC ± 15 percent, is this voltage sufficient?

$$E_o = ER_L/(R_s + R_L)$$
$$= (14)(3)/(1.5 + 3)$$
$$= (42)/(4.5) = 9.3 \text{ volts}$$

and "no," this voltage is not sufficient.

The problem in power supplies comes from the fact that R_L depends, in part, on the load current, I_o. If I_o varies, then R_L varies also, and this causes E_o to vary.

Another problem that may cause the output voltage to vary is changes in the AC line voltage applied to the primary of the power transformer. The AC voltage is *nominally* 110 volts, but will be within the normal range if it is anything between 105 and 120 volts. Sometimes, potentials may rise to 125 volts or drop to 95 volts temporarily. It has been as low as 90 volts during intentional brown-outs.

Regardless of the root cause, variations in power supply output voltage can be a real disaster in electronic circuits. The oscillator that keeps the transmitter on the correct frequency, for example, will vary in frequency with changes in supply potential. The measure of how well a power supply works is given by its *percentage of regulation*, which is defined as:

$$\% \text{ Regulation} = \frac{E - E_o}{E} \times 100\%$$

□**Example:**

A DC power supply is designed to deliver 600 mA. The open-terminal voltage is 800 VDC, while the output under a load of 600 mA is 725 volts. Find the percentage of regulation.

$$\begin{aligned} \% \text{ Reg.} &= (E - E_o)(100\%)/(E) \\ &= (800 - 725)(100\%)/(800) \\ &= 9.4\% \end{aligned}$$

□**Example:**

Calculate the percentage of regulation for the power supply in the example of the 14-volt power supply for the 2-meter rig given earlier. In that example, the open-terminal voltage was 14 volts, and the voltage on transmit was 9.3 volts.

$$\begin{aligned} \% \text{ Reg.} &= (E - E_o)(100\%)/(E) \\ &= (14\text{-}9.3)(100\%)/(14) \\ &= (4.7)(100\%)/(14) \quad = \quad 34\% \end{aligned}$$

A voltage regulator circuit is designed to stabilize the output voltage of a DC power supply. There are several basic types, but only a couple of them are considered here. Also missing are some of the antique regulator circuits which use vacuum tubes. Today, almost any voltage required in a radio transmitter can be supplied by a solid-state circuit.

Zener diodes. The zener (zen-ner) diode is a specialized PN junction device that will provide some voltage regulation. Figure 10-34 shows the transfer characteristic of the zener diode. In the forward bias region, it will behave the same as any ordinary PN

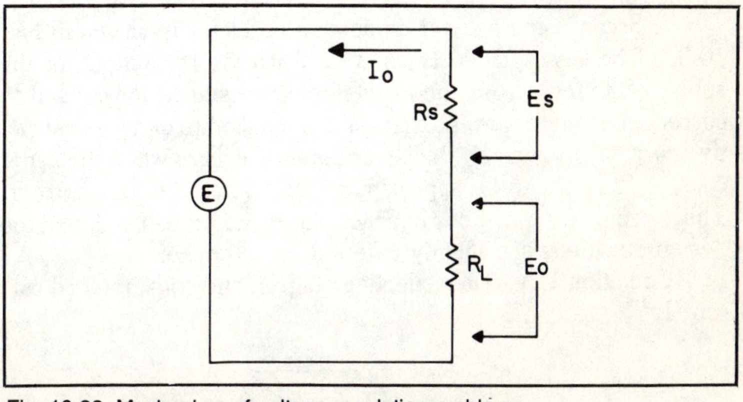

Fig. 10-33. Mechanism of voltage regulation problems.

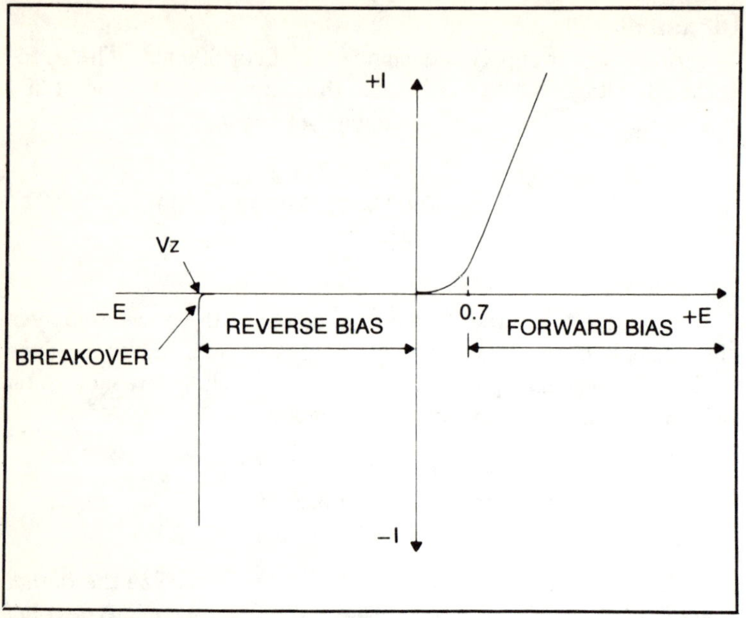

Fig. 10-34. Zener diode regulator curve.

junction diode. The current is nearly zero for voltages less than a certain threshold (about 0.6 to 0.7 volts for silicon devices), but will increase rapidly in a nearly linear manner for voltages greater than 0.7 volts. In the reverse direction, however, things are different. For potentials between 0 and some reverse bias V_z, the diode is like any diode. The only reverse current flow will be a minute leakage current. but at V_z the diode will breakover (avalanche). The zener diode will regulate the applied potential to this potential V_z.

A circuit using a zener diode as a regulator is shown in Fig. 10-35. The zener diode is connected across the output of the supply, shunted across the load. Series resistance R1 is used to current limit the circuit, so that D1 is not asked to carry a destructive current. In general, the best operation occurs when the zener current I2 is set to be approximately 10 times the load current. There are three conditions in which zeners might be used, and the design equations are slightly different for all three:

Condition 1: Variable supply voltage E and constant load current I.

$$R_1 = \frac{E_{min} - V_z}{1.1\ I_3}$$

Condition 2: Constant supply voltage E and variable load current I.

$$R_1 = \frac{E - V_z}{1.1\ I_{3(max)}}$$

Condition 3: Both supply voltage E and load current I are variable.

$$R_1 = \frac{E_{min} - V_z}{1.1\ I_{3(max)}}$$

The power dissipation expressions for all three cases are:

$$P_{D1} = \frac{E_{max} - V_z}{R_1} - (I_3 V_z)$$

and

$$P_{R1} = P_{D1} + (I_3\ V_z)$$

The simple zener diode regulator suffers from several problems. One of these is that the zener current is much higher than the load current. This means that a 50-watt zener is not too useful. They are used in stacks of several zeners to regulate the screen potentials in some linear amplifiers, and they find some application in low-current stages, but we must look to one of the other circuits for high current levels.

Series-Pass. Figure 10-36 shows a method for using a series-pass transistor to boost current capacitor of a zener diode.

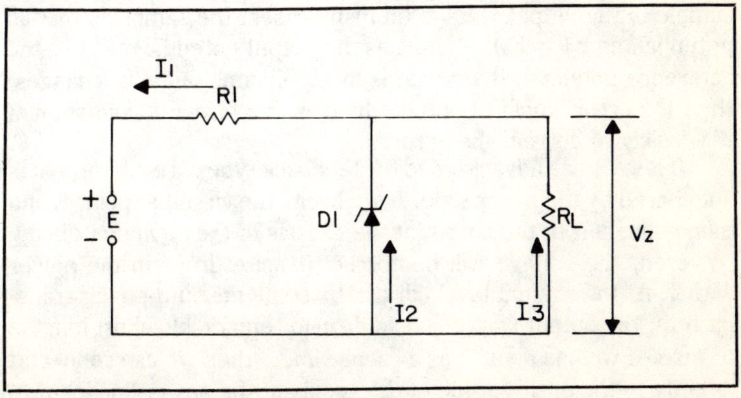

Fig. 10-35. Zener diode regulator.

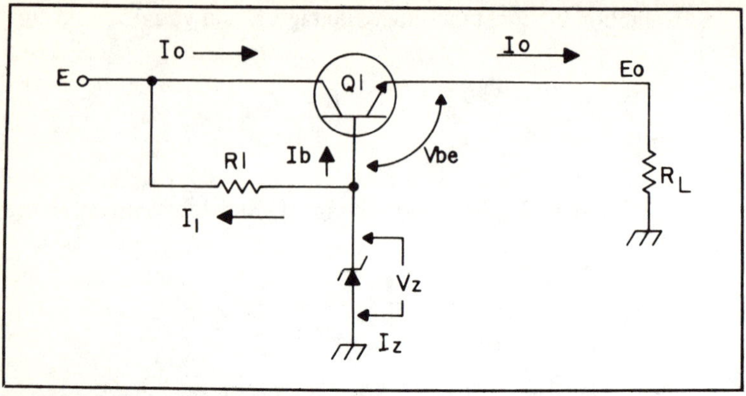

Fig. 10-36. Series-pass Zener regulator.

The transistor regulates the output voltage by acting as an electronically variable resistor. The base voltage on Q1 is held constant by the zener diode D1. The diode is selected such that I_b can be 10 percent of I_z. The maximum output current will be 4 $H_{fe} \times I_b$. In other words, the load current is amplified by the beta of the series-pass transistor. The output voltage is approximately $V_z - V_{be}$, which for silicon transistors, is approximately 0.7 volts below the zener potential V_z.

The circuit shown in Fig. 10-36 is but one type of series-pass regulator. The circuit shown in Fig. 10-37 is the other major type. In this circuit, the base voltage on the series-pass transistor is controlled not by the zener but by a differential *error amplifier*. There are two inputs to this amplifier. The noninverting input is connected to a reference potential, E_{ref}, while the inverting input is connected to a sample of the output voltage (in many cases, the sample is the full potential, so $E1 = E_o$). As long as the output potential matches the reference potential, the circuit is in equilibium. But if E_o changes, then the error amplifier output changes in a direction and amount necessary to correct the error.

There is one advantage to the feedback type of regulator that is not shared by the zener type. If we break the circuit at point A and then connect it to the output at the chassis of the equipment being powered, the voltage will be correct despite drops in the power wiring. At the current levels drawn by some medium-power transmitters, the output voltage could drop in remarkable short lengths of wire. If we use point A as a "sense line," then we can connect it directly to the positive side of the supply at the power line input to the transmitter.

POWER/SUPPLY PROTECTION

There are two demons that require exorcising in all regulated power supplies: overvoltage and overcurrent. The former will destroy the equipment being powered, while the latter will destroy the power supply. Consider two scenarios. In one case, you are powering your kilobuck transceiver with a series-pass regulated power supply. The series transistor shorts, applying +18 volts from the rectifier directly to the transceiver (smoke gets in your eyes). Alternatively, the power line connector to the transmitter (or the final amplifier transistors shorts to ground. This draws a mammoth current from your supply, shorts the series-pass transistor, and blows out the rectifiers. If you are lucky, it might spare the power transformer, but don't count on it.

Overvoltage protection is provided by the circuit shown in Fig. 10-38. This is an SCR crowbar, in testimony to the brute force method of providing protection. Zener diode D1 has a V_z that is at the point where you consider the equipment in danger (17 to 18 volts in most mobile rigs). Q1 is the series-pass transistor in the voltage regulator. When Q1 shorts out, it places the rectifier voltage on the E_o output line. This will cause D1 to breakover and produce a current in R1 that is capable of turning on the silicon-controlled rectifier, D1. When the SCR is turned on, it shorts out the power-supply line and blows the fuse. In some cases, D2 has sufficient rating to simply sink the total current.

Overcurrent protection is provided by the circuit of Fig. 10-39. The protection components are Q2 and R1, while Q1 is the series-pass regulator transistor. Transistor Q2 is a silicon unit, so will be

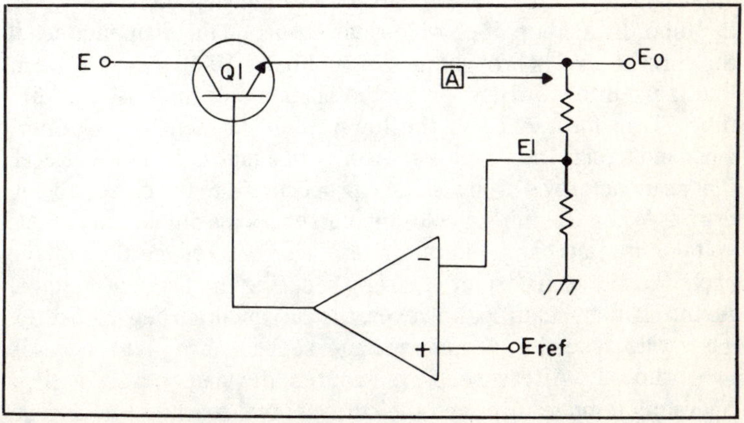

Fig. 10-37. Feedback amplifier/series-pass regulator.

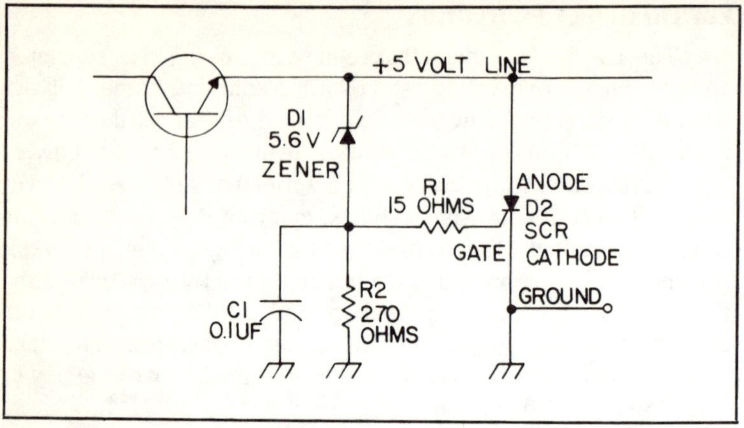

Fig. 10-38. SCR crowbar protection.

turned on by a bias voltage E_b of 0.6 volts. As long as the output current I_o is less than $0.6/R_1$, the circuit will operate as a normal power supply. If I_o is less than $0.6/R_1$, the circuit will operate as a normal power supply. If I_o exceeds this Ohm's law value, however, then Q2 becomes forward biased and will short exists, placing the emitter of Q2 at near-ground potential. The circuit will gently turn off, instead of becoming a "silicon-to-carbon converter."

FILAMENT STARTING

The resistance of a cold transmitter tube filament might be as low as 10 percent of its hot resistance. As the power level of the transmitter rises to the 1000 watts and above level, it becomes important to preheat the filaments prior to applying anode voltage to the tube. In addition, it provides less stress on the filament itself if the current level is brought up slowly. Figure 10-40 shows a circuit that is used to slowly warm up the filaments of the final amplifier tube. Transformer T1 is the low-voltage filament transformer, while inductor L1 is a series reactor with a high inductance. Recall that an inductor will drop a voltage proportional to the current being drawn. When the tube is cold, the current is maximum, so the AC voltage drop across L1 is maximum. This will reduce the voltage appled to the transformer, thereby reducing the filament voltage. As the filament heats up, however, the current drain begins to drop. This will cause the filament voltage supplied by T1 to increase proportionally. After a couple of minutes, the filaments are at their operating temperature, and the current level reaches equilibrium. The voltage drop across L1 then drops to near zero.

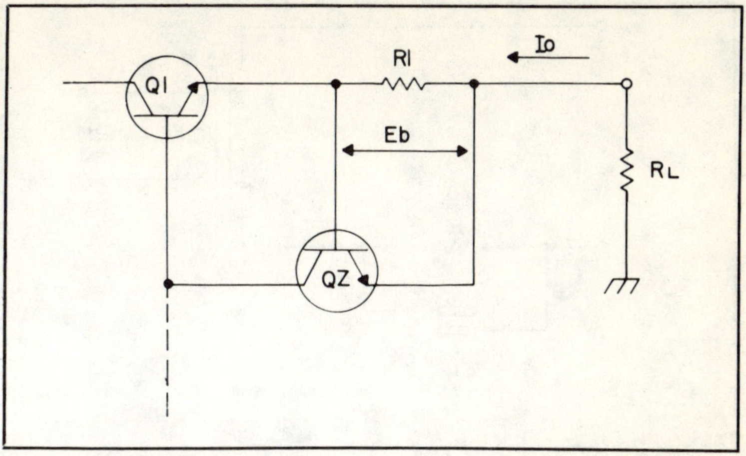

Fig. 10-39. Output current limiter.

A second approach is shown in Fig. 10-41. In this case, a resistor is connected in series with the transformer primary. This will reduce the filament voltage for a period while the tube is cold. A timer will not allow relay K1 to turn on for 1 or 2 minutes. But when the time expires, then K1 is turned on and its contacts short out the series resistor. *The filaments then go to full heat.*

An alternative to the second circuit, often favored by radio amateurs, is to use separate on/off switches for filament and anode power. The operator then manually performs the function of the timer. Some transformers used for the filament voltage will also perform the function of the inductor in Fig. 10-40. These transformers are high-reactance devices, especially designed for this application.

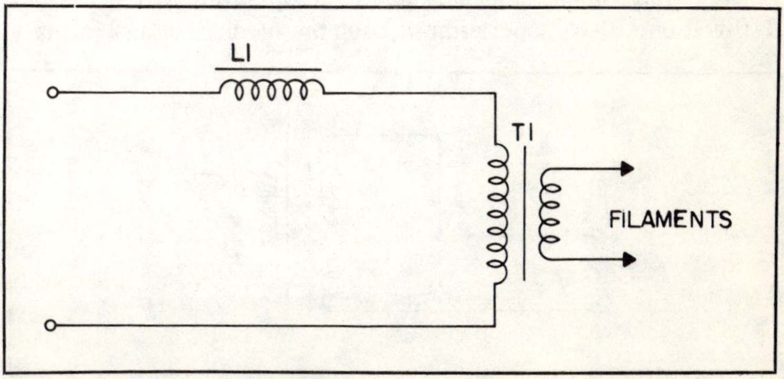

Fig. 10-40. Filament starting circuit.

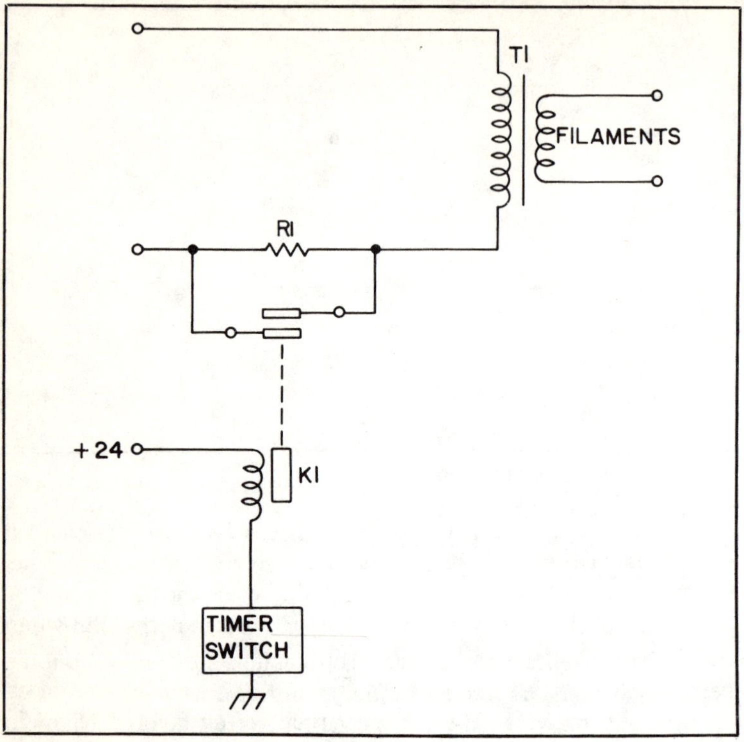

Fig. 10-41. Another filament starting circuit.

110/220-VOLT OPERATION

Moderate power level transmitters usually draw power from the ordinary 110-volt AC residential service. But for rigs in the kilowatt class, it becomes necessary to go to 220-volt wiring. Most amateur radio linear amplifiers allow the operator to select either 110-volt or 220-volt operation through the medium of dual primary

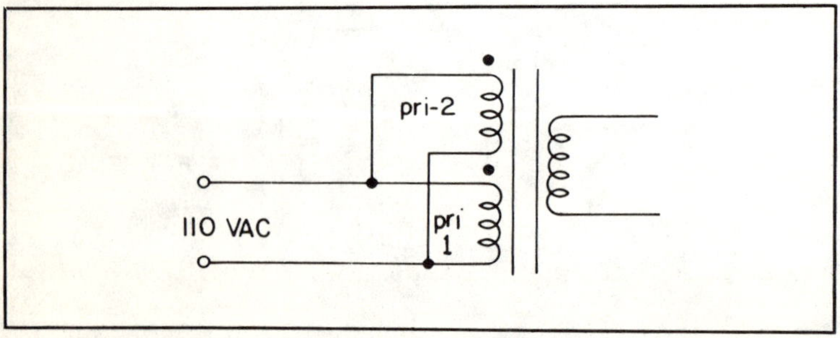

Fig. 10-42. Primary wiring for 110 volts.

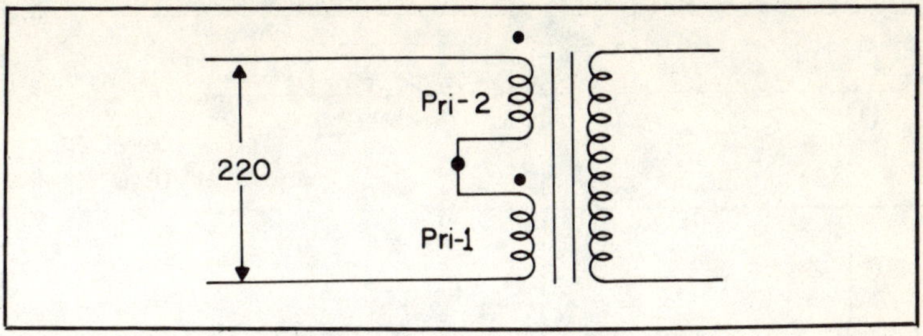

Fig. 10-43. Primary wiring for 220 volts.

power supplies, as in Figs. 10-42 and 10-43. When the primaries are connected in parallel, the equipment is wired for 110-volt operation (the dots are to show the sense of the windings, so that they will not be reverse phased, cancelling the output voltage). On the other hand, if the primary windings are series connected, as in Fig. 10-43, the equipment is wired for 220-volt operation.

AC POWER CONTROL

In low-power and medium-power transmitters, the control of the AC side of the power supply is simple. The AC line is interrupted by a simple switch, and it contains a single fuse. At high power levels, though, it becomes necessary to use a little bit more protection. Figures 10-44 and 10-45 two typical circuits.

The circuit in Fig. 10-44 shows two double-switches and a double-fuse. These are all matters of safety. The double-switch, S1, is a standard AC wall switch for most amateur-level transmitters. Such switches will handle up to 15 to 20 amperes of current, depending upon rating. The main safety switch, S2, should be located away from the transmitter, at a point where a rescuer can

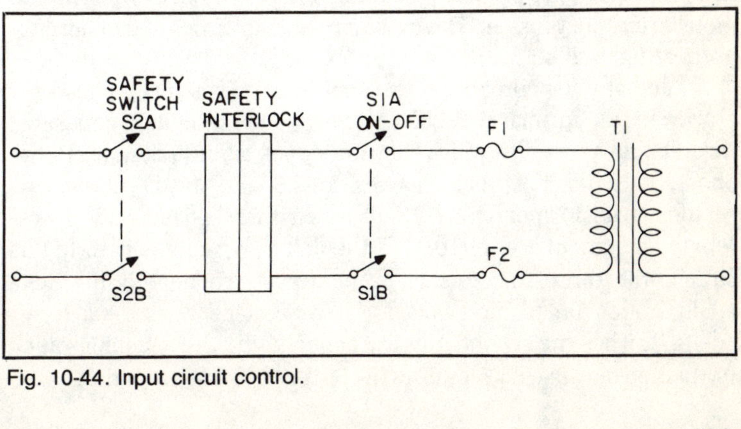

Fig. 10-44. Input circuit control.

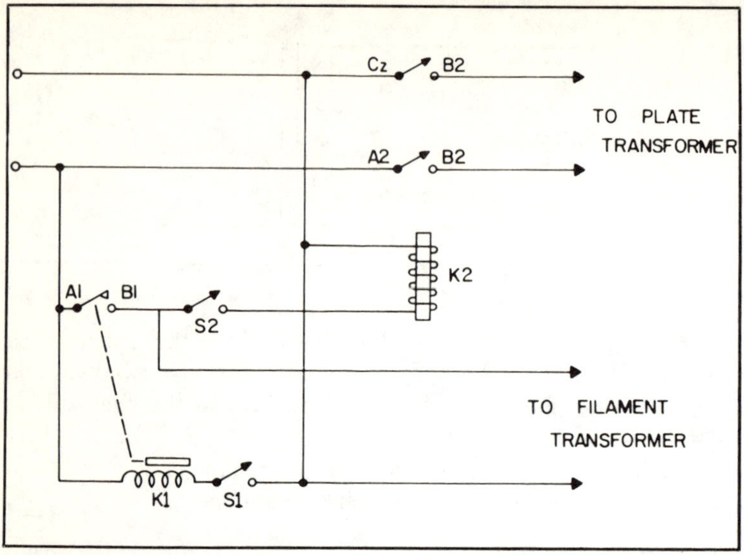

Fig. 10-45. Using a relay to power the primary circuit.

turn off power to the transmitter at no risk to himself. This is usually an electrical box switch with a pull lever.

Figure 10-45 shows the use of a relay to power the primary circuit. Relay K1 is energized by closing S1. This will apply power to the filament transformer and will allow K2 to be energized. Plate power to the transmitter is not available until K2 is energized by closing switch S2. Relay K2 should have contacts that are capable of carrying the entire primary current, which could be 25 to 35 amperes. In some cases, this circuit will be incorporated with one of the filament starting circuits. If mercury vapor rectifiers are used, it is also necessary to provide a warm-up time for them to vaporize the metallic mercury inside. The timer circuit, allowing 1 to 2 minutes of preheating, will suffice.

Almost every broadcast transmitter requires three-phase AC power in the primary circuits. The power company generates three-phase power, in which the sine waves are 120 degrees out of phase with each other. Figures 10-46 and 10-47 show two rectifier/transformer circuits for use in three-phase power systems. The circuit in Fig. 10-46 is the delta-wye connection. In this circuit, the three-transformer primaries are connected in a delta configuration (so named because it resembles the Greek letter "delta"), while the secondaries are connected in a wye configuration (so named because it resembles the letter "Y"). Rectifier diodes are

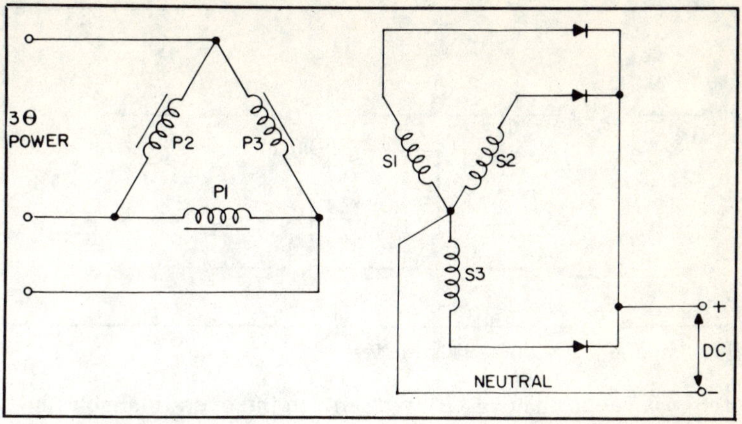

Fig. 10-46. Three-phase AC.

connected to each of the three secondary phases, and their outputs are combined in a single DC line.

A delta-delta connection is shown in Fig. 10-47. In this circuit both primary and secondary windings are connected in the delta configuration. Note the absence of the neutral line. This requires us to provide an artificial common connection by bridge wiring the diode rectifiers. Note that the secondary voltages are not necessarily equal in all three windings.

TRANSIENT PROTECTION

High voltage transients existing on the power lines can destroy the rectifier diodes and regulator components in some power supplies. The average residential electrical power distribution sys-

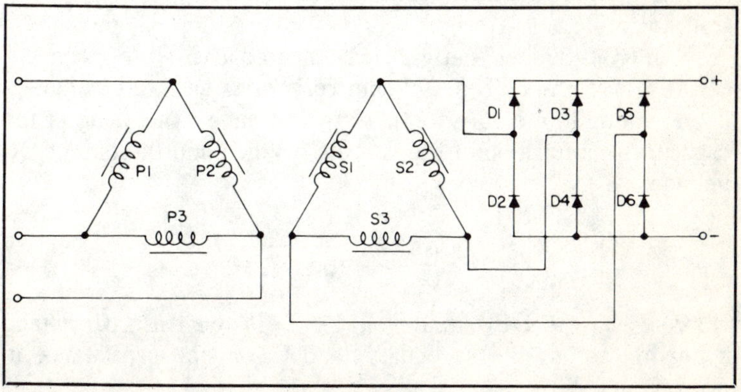

Fig. 10-47. Three-phase AC power supply.

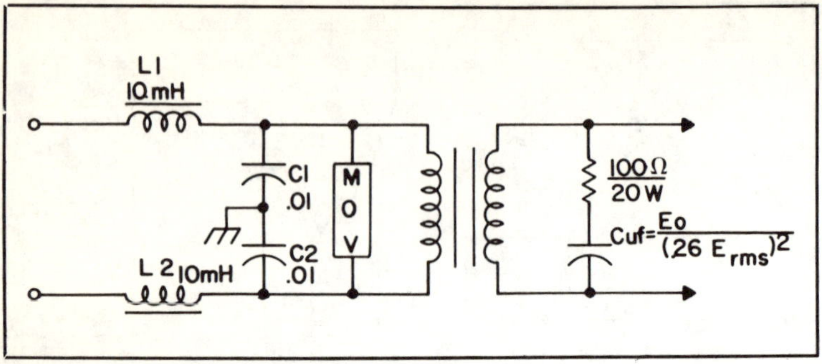

Fig. 10-48. Noise and transient suppression.

tem will see as many as 10 transients in the over-1000-volt range every day. When there is an electrical storm in the vicinity, the number rises even greater. These transients should be suppressed, and several different techniques are available.

Figure 10-48 shows a transformer circuit with several different types of protection. The primary winding uses a brute force rf filter and a metal oxide varister (MOV) device. The rf filter serves to reduce the amplitude of the impulses in the line because these transients are primarily high-frequency in content. The filter also serves to keep the signal inside the transmitter from getting out via any route other than the antenna terminal.

The MOV device, of which General Electric is a principal supplier, will have a high impedance at normal AC power line voltages. If a high voltage transient comes along, however, it will take on a sudden drop in impedance and shunt the transient harmlessly across the primary. The device should be selected according to the manufacturer's recommendations for high-power transmitters.

An RC "snubber" network is connected across the secondary of the transformer. The 100-ohm resistor is selected to have a power rating in the 20-watt to 100-watt range. The value of the capacitor is a function of the rms AC voltage and the output DC voltage:

$$C_{\mu F} = \frac{E_o}{(0.26\ E_e)^2}$$

where E_o is the DC output voltage, E_r is the rms AC voltage appearing across the secondary, and $C_{\mu F}$ is the capacitance in microfarads.

In power supplies with an inductive filter circuit, it might be advisable to use a snubber network across the filter choke(s). This is shown in Fig. 10-49. This practice is most valuable in transmitters which have sudden increases in output current drain. A radiotelegraph transmitter, with its on and off keying is a prime example. This circuit will also snub out transients caused when the DC high-voltage supply is turned on as the transmit switch is pressed. The sudden increase in current will create a very high transient voltage across the inductor, and this is capable of puncturing the rectifier diodes in the reverse direction. The resistor should have a value of 100 ohms at 20 to 100 watts, while the required capacitor value is:

$$C_{\mu F} = \frac{0.1\, L\, I^2}{E_o^2}$$

where $C_{\mu F}$ is the capacitance in microfarads; L is the inductance of the choke in henrys; I is the current drain (constant, after turn-on), or the normal load current; and E_o is the DC output voltage.

SOME PRACTICAL CIRCUITS

In this section we will consider some practical circuit examples that can be constructed. We will show examples of low-voltage—low-current power supplies, low-voltage—high-current models, and high-voltage power supplies. Also, we will give an example of a variable voltage bench power supply that can be used to test new circuits, troubleshoot small projects, and provide a general power source for your electronics work.

Low-Voltage—Low-Current Supply

Small electronic projects, including QRP transmitters, require a small DC power supply. As often as not, this supply must be

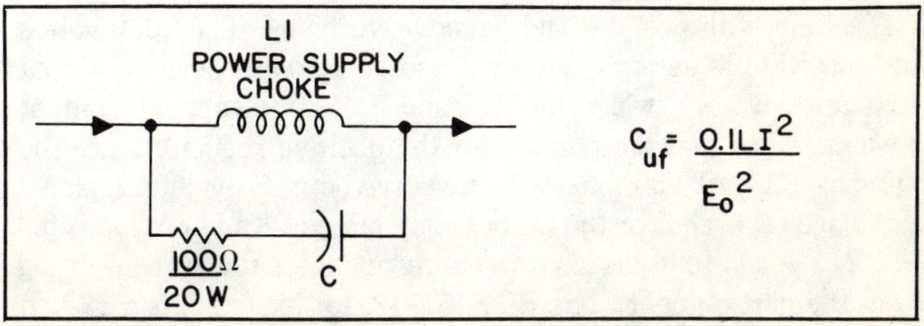

Fig. 10-49. Using a snubber network across the filter choke(s).

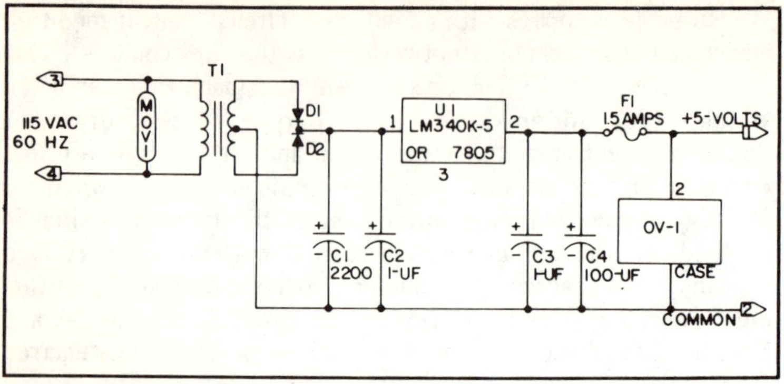

Fig. 10-50. Low-voltage regulated DC supply.

regulated to a high degree. Figure 10-50 shows the basic low-current power supply, and this circuit may be used for any potential between 5 volts and 24 volts DC. The particular example shown is for 5 volts at 1 ampere.

The regulator used in these circuits is the key to design simplicity. Regulator U1 is a three-terminal IC regulator. These ICs are packaged in transistor cases and have three leads: input, output, and ground. We can tell the package (and by inference, often the current rating) and the voltage from the type number. Several basic series of these devices are available. In 5-volt-only devices we have the LM309 and LM323. The LM309 is available in three package styles. The H-package is a TO-5 transistor case, and the LM309H will pass 100 mA. The 1-ampere model is the LM309K. The K-style package is the TO-3 transistor case. The third case style is the T-package, which is the TO-220 plastic (also called the P-66 package by some) power transistor case. T-style devices will pass 750 mA in free air and 1 ampere if properly heat sinked. The LM323 device is a K-package (TO-3) but is capable of passing up to 3 amperes of current at 5 volts.

There are two general lines of 750 mA to 1A regulators available in both positive and negative versions. The 7000 series and the LM300 series. The 78xx and LM340 are positive output voltage regulators, while the 7900 and LM320 are negative output devices. Note that the pinouts for the positive regulators are the same as in Fig. 10-50, but in the negative output regulators, pins 2 and 3 are reversed, or the output is on pin No. 3 and ground is pin No. 2. We can read the voltage rating and infer the current rating from the part number. The LM340T-12, for example, is a 12-volt (−12), plastic package (T) regulator. We can infer from the package style that the device will dissipate 750 mA in free air and 1000 mA when heat sinked. The LM340K-5, is a 5-volt, 1-ampere device in the TO-3 package. As such, it is directly compatible with the LM309. Standard voltages from 5 to 24 volts are available in both case styles.

The 7800 series and 7900 series are similar to the LM340 and LM320, respectively. In fact, although Motorola devices tend to have different case styles that are really 78xx and 79xx devices, many are available in TO-3 and TO-220 styles. Perhaps, we should label these as LM340 or LM320, but the common industry practice is to use the 78xx and 79xx designations. The voltage rating is given by replacing the "xx" in the type number with the voltage rating; i.e., 7905 is a 5-volt, negative output, 1-ampere voltage regulator, while 7805 and 7815 are 5-volt and 15-volt regulators, respectively (both positive output).

The circuit in Fig. 10-50 can be used for any of these three-terminal IC regulators. It is, however, necessary to make certain circuit value changes for other than 5-volt operation. Input capacitor C1 should have a value of 2000 μF/ampere, while output capacitor C4 should have a value of 100 μF/ampere. If one of the higher current devices are used, then you must raise the value of these capacitors accordingly. If the 100 mA regulator is used, however, do not reduce C1 to less than 500 μF.

Capacitors C2 and C3 are for improved noise immunity. These capacitors can have any value between 0.1 μF and 2 μF, regardless

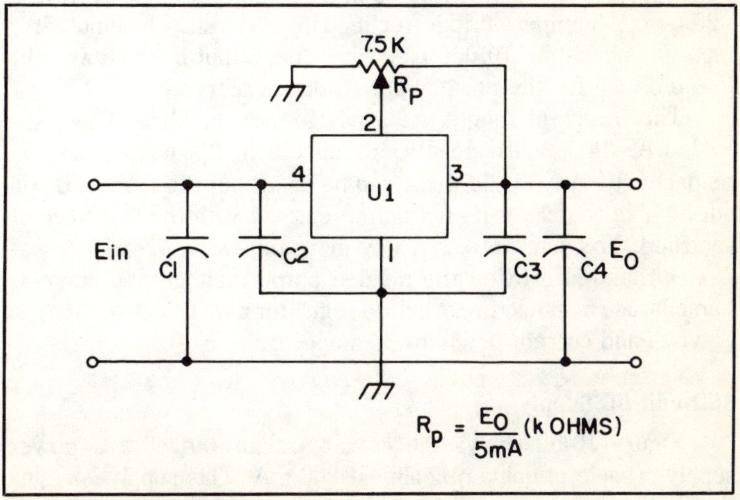

Fig. 10-51. Adjustable output low-voltage DC supply.

of the current rating. Device OV-1 is an overvoltage protection circuit, such as the SCR crowbar described earlier in this chapter. You could also use one of Lambda Electronics overvoltage crowbar modules. These are devices packaged in either TO-3 transistor cases or special hybrid packages that include all the circuitry needed for the SCR crowbar. They are two-terminal devices.

The transformer must be selected to provide a DC input voltage to the regulator that is 2 to 3 volts higher than the regulator output voltage. For the 5-volt supply shown, then, a 6.3-volt filament transformer is sufficient. For a 13.6-volt power supply, we can usually get away with using a 12.6-volt filament transformer. These will output approximately 17.5 volts DC if the capacitance across the rectifier is sufficient. In both of these cases, incidentally, a bridge rectifier is assumed instead of the center-tapped transformer shown.

Note that Lambda Electronics manufactures a series of three-terminal voltage regulators in current levels up to 8 amperes. One of these is the LAS-19CB, which will provide the 13.6 volts at 5 amperes needed to power small mobile rigs on AC power. Most small 2-meter and other VHF/UHF mobile transceivers will operate with this supply or its newer 8-ampere brother. Of course, a 10,000 mF capacitor is needed at the rectifier output.

Adjustable Bench Supply

Figure 10-51 shows the circuit for a 4-volt to 30-volt regulated power supply that can be adjusted with a simple potentiometer. Use a 25.6-volt, 5-ampere bridge-rectified input circuit. The input filter capacitor should be 10,000 μF, while the output filter is 500 μF. Use 0.33 μF for the noise suppression capacitors.

Three current ranges are available from Lambda. The LAS-15U, LAS-14U, and LAS-19U produce 1.5, 3, and 5 amperes, respectively. Any of the three can be used, but the LAS-19U will not be able to deliver its full output current with the transformer specified. You can, however, buy more expensive *rectifier transformer* that will provide the needed output voltage and current. Lambda also manufacturers hybrid regulator modules at voltages to 15 volts and current levels to 30 amperes.

800-Volt DC Supply

Figure 10-52 shows the circuit for an 800-volt DC power supply capable of delivering almost 200 mA. This supply contains the principal voltage needed for most 100-watt output SSB trans-

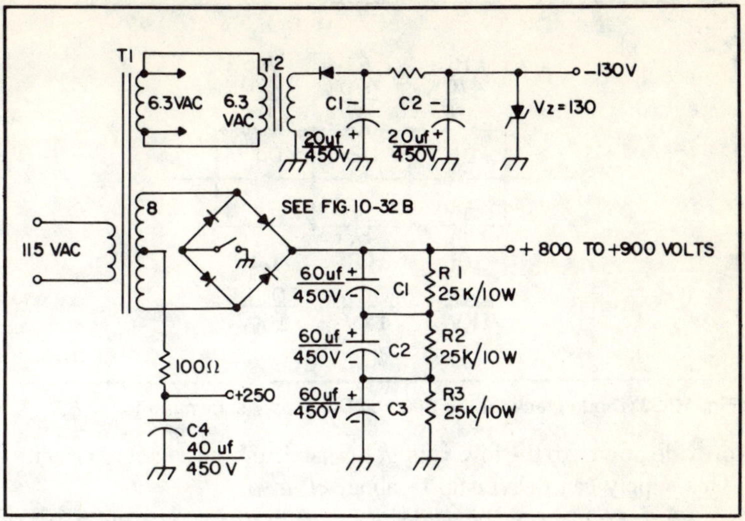

Fig. 10-52. An 800-volt DC supply.

ceivers. The transformer is a Stancor PC-821w, which is rated at 800 VCT at 250 mA. This transformer also has a 6.3-volt filament winding to power the heaters in the transceiver.

The transformer output is bridge rectified. Each arm of the bridge consists of at least three 1000-volt (PIV), 2.5-ampere, rectifier diodes. These diodes are standard color TV diodes and should be available from any electronic wholesaler that deals with service technicians. Alternatively, you can use the 1N4007 diodes rated at 1000 volts PIV at 1 ampere. Each diode is shunted with a 470K ohm, 1-watt resistor for equalization of the diode voltage drops (Fig. 10-53). Also shunting the diodes are 0.001 μF, 1000-volt capacitors for transient suppression.

The diode rating is one area where some amateur transceivers have been lacking. One well-known transceiver had a habit of popping the rectifier diodes on a regular basis. The problem was that only two 1-kV diodes were used in each arm of the bridge. But the inverse voltage applied to these diodes, when a filter is used, is $2.82 \times E_{rms}$, which in that case was 2300 volts! It required at least three 1-kV diodes in series to make this supply work safely.

The filter stack is a brute force capacitor filter made from a series of electrolytic capacitors. The overall capacitance of this stack is 20 μF, but the working voltage is 1350 WVDC. The parallel resistor across each capacitor is to equalize the DC voltage drop.

A +250-volt supply is also provided. This supply is used to

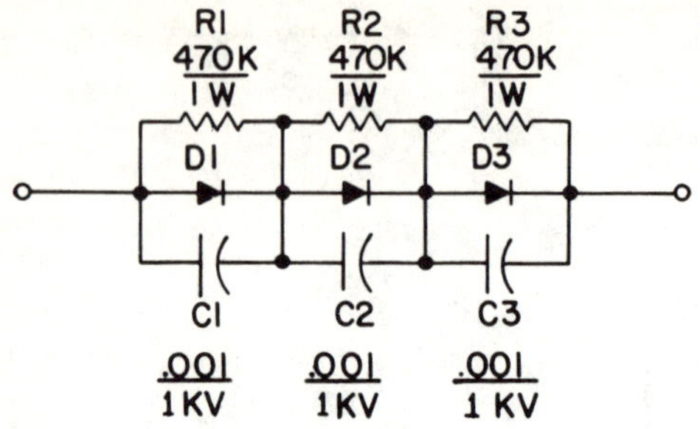

Fig. 10-53. Diode stack for the 800-volt DC supply shown in Fig. 10-52.

provide power to the low-voltage transmitter and receiver circuits. This supply can deliver up to about 60 mA.

A bias supply can be made by connecting a low-current 6.3-volt filament transformer (T2) as shown. We turn this transformer around so that the 6.3 VAC winding becomes the primary, and the 115-volt winding is the secondary. The primary is connected to the 6.3-volt filament winding of T1. The rectifier is reversed so that the output is negative with respect to ground. This will, of course, also require that the filter capacitors must also be reversed; i.e., the positive terminal is grounded in this case. The 130 volt, 5-watt zener diode provides regulation of this bias supply.

3000-Volt High-Voltage Supply

Figure 10-54 shows an elementary 3000-volt HV supply for a kilowatt linear amplifier. The transformer is rated at 2250 to 2400 volts DC at 500 mA or more. Such a transformer may not be rated in a catalog with the secondary current rating, but instead with the VA rating. A 2250-volt, 1200-VA transformer is needed. The Peter Dahl Company of El Paso, Texas, can supply transformers of various voltage ratings for amateur service transmitter. Other companies also make such transformers, but many are not geared to serve the unit-quantities of the amateur market.

The rectifier stack is made from the same 1000-volt PIV at 2.5-ampere color TV diodes as in the 800-volt supply. In this case, however, at least eight are needed, and 10 would not be unjustified, for each arm of the bridge. Again 470K ohm at 1-watt resistors and 0.001 μF/1000-volt disc ceramic capacitors are shunted across each diode in each stack (Fig. 10-55).

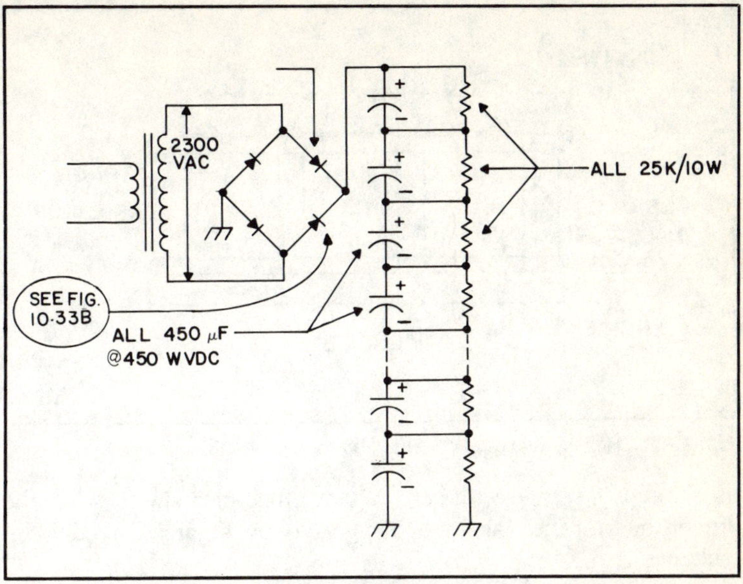

Fig. 10-54. A 3000-volt high-voltage supply.

The filter stack is the same as in the previous example, but there are 10 450 μF/450-volt, computer-grade, electrolytic capacitors. These capacitors and the diodes are at high voltage, even on the cases. As a result, it is mandatory that high-voltage insulation be provided. An example of a method for mounting the capacitors (only three are shown) is given in Fig. 10-56. A half-inch sheet of Lucite or plexiglass is used as the base. Two quarter-inch sheets actually hold the capacitors. The smaller sheets have the holes drilled in them to accept the capacitor bodies. The lower sheet

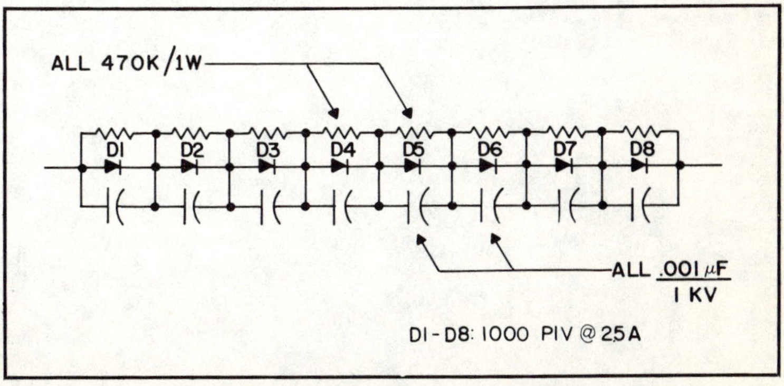

Fig. 10-55. Diode stack for the high-voltage supply shown in Fig. 10-54.

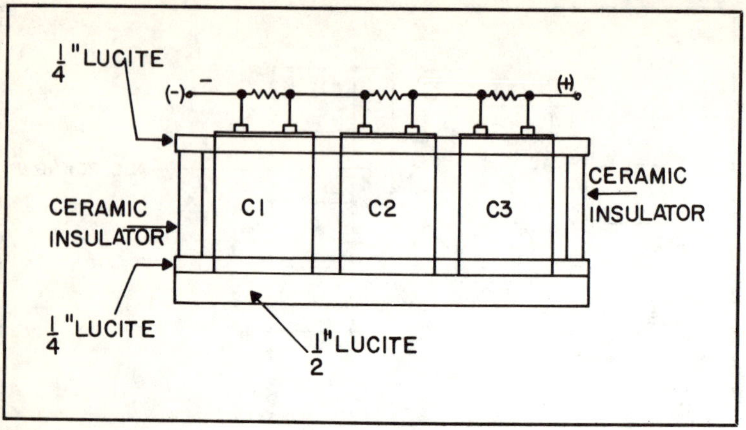

Fig. 10-56. Typical insulated construction for filter bank.

is glued or fastened with screws to the half-inch sheet, while the upper quarter-inch sheet is supported by ceramic standoff insulators.

The rectifiers must be similarly mounted, well away from the chassis. It is all right to use perfboard, only if the board selected is rated for high-voltage properties. Much of the hobby perfboard is not suitable for high voltage applications, though. The perfboard or lucite holding the rectifiers must be mounted off the chassis by two to three inches with ceramic insulators.

A more complete treatment of power supply construction techniques, as well as many practical circuits, are given in TAB book No. 806, *The Power Supply Handbook*. This book is highly recommended by this author.

Chapter 11

Radiotelegraph Transmitters

The radiotelegraph transmitter was the first type to be used in "wireless" communications and can be implemented with simple on-off keying of the carrier. The radiotelegraph transmitter is the simplest form used, consisting only of an oscillator and a power amplifier and some method for turning the amplifier on and off with a telegraph key.

Most radiotelegraph communications take place with the international Morse code of dots and dashes. Note that this code is not the code made famous by Samuel F.B. Morse, but is an international adapation of the original code. The original code is now called American Morse code. The international Morse code used for radiotelegraph communications is shown in the Appendix.

There is a pernicious rumor that radiotelegraphy (called CW for *continuous wave*) is a dead art. In fact, our own FCC has contributed to this feeling with proposals for "codeless" amateur radio licenses, which may say would simply be technical CB licenses. But listen to the lower portions of the HF amateur bands and to the marine radiotelegraph bands. CW is very much alive in those parts of the spectrum. I asked one Navy radio operator chief petty officer if the service would remove the radiotelegraph code in the near future. His reply was a definite no! The reasons were that there are still ports and ships on the seas that carry only radiotelegraphy equipment (at least for long-distance communications); and when the ship is sinking and you are in the life boat, and the complicated radioteletype and radiotelephone equipment is on the fritz, it is still possible to make communications with CW by beating two wires together!

Morse code to at least the slow speeds required of the US Novice and Technician amateur radio license classes is simple for most people to learn. At those speeds, you can almost count the dots and dashes and decode the characters from a written chart!

The General and Advanced US amateur radio license classes require a code speed of 13 words per minute. A word is defined as five alphabet characters, with punctuation and numerals given extra weighting (1½ characters/2 characters, etc.). The speed required is a little difficult for some to reach, but most eventually reach that level with practice. It is not until you reach for the 20 words per minute (wpm) level that some natural ability comes into play. But even then the natural ability is not intelligence so much as it is "stick-to-itiveness."

Keying the transmitter requires some sort of system that will turn the carrier on and off. But there are a few problems in this regard: *chirp* and *keyclicks* of assorted types.

CHIRP

Chirp is caused when the oscillator frequency is pulled slightly as the transmitter is keyed. The slight change in frequency is readily noticed by the operator at the other end.

One cause of chirp is poor isolation between the oscillator and amplifier circuits. This can be corrected most readily when the transmitter is designed, and is effected by providing good shielding between the two stages.

The second cause of chirp, and probably the more common, is poor regulation of the DC power supply feeding the oscillatior. This is especially likely if the B+ potential for the oscillator is derived from the high voltage of the final amplifier, a very common and economical practice in low power CW transmitters. Changes in the oscillator plate/collector voltage cause the frequency shift. The correction is to regulate the oscillator potential. In most vacuum tube designs, gas-filled voltage regulator (VR) tubes are used to stabilize the oscillator voltage. Alternatively, a stack of Zener diodes equal to the oscillator B+ voltage will provide adequate regulation. A separate, low-current, DC power supply could be provided for this purpose. In transistor oscillators, any of the regulators shown in Chapter 10 should suffice, although some feedback circuits do not like rf.

Chirp can be caused by poor filtering in the DC power supply. Inadequate filtering is a primary cause of poor regulation in homebrew power supplies, and should be examined when a new homebrew transmitter shows excessive chirp. In addition, a transmitter that suddenly develops a chirp where it had none before might have an open power-supply filter capacitor. These filters not only contribute to the ripple reduction and regulation but also to the decoupling of the amplifier and oscillator circuits.

KEY CLICKS

Key clicks are hard snapping sounds caused by spurious frequency components either side of the carrier frequency. There are several causes of key clicks. One is poor neutralization. If the amplifier is poorly neutralized, the amplifier turn-on will cause a modulation product that is manifest as a key click. The solution here is to neutralize the amplifier, which should have been done on any event, prior to placing the transmitter on the air!

Another cause of key clicks is *power-supply transients*. These are especially common in transmitters that use an inductive filter for the transmitter B+ power supply. The voltage across an inductor is proportional to L(dI/dt), which means the rate of change of the current in the inductor. When the amplifier turns on suddenly, the current through the inductor is suddenly increased, and that causes an excessive potential spike across the inductor. This transient will add to or subtract from the power-supply B+ output, causing the click in the output. The solution to this problem is to use a different type of filter circuit or place an RC snubber network in shunt with the inductor (see Chapter 10).

I can recall an American Radio Relay League Sweepstakes in which an East Coast radio amateur used an AM/CW transmitter. The modulation transformer was normally connected in series with the B+ lead of the final amplifier during AM operation, and was shorted out for CW operation. During this particular on-the-air contest, the switch failed to short out the transformer secondary. That placed an unsnubbed inductance in series with the B+ path. Every time the key was depressed, a terrific transient was generated, and this created key clicks over *tens* of megahertz. The key click effectively amplitude modulates the final amplifier, and the *modulating frequency* is the rise time of the final plate current. This means that the particular amateur in question generated clicks every 4 or 5 kHz througout the high-frequency portion of the

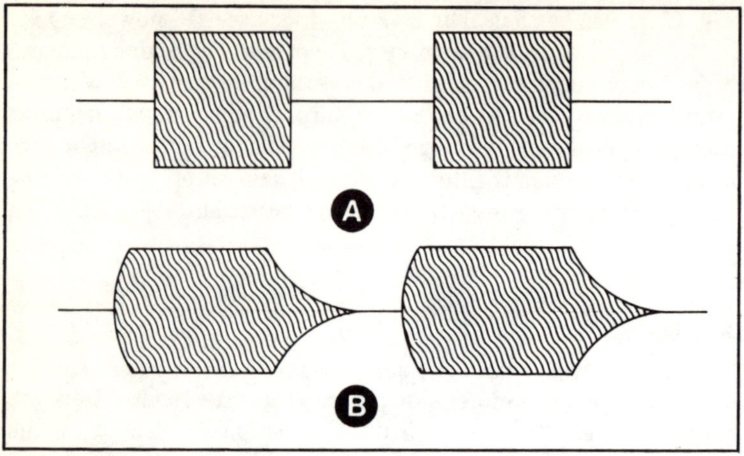

Fig. 11-1. Keying waveforms.

spectrum. Fortunately, he was using only 100 watts of rf power, so the problem seemed confined to the middle Atlantic states of the East Coast.

The last cause of key clicks considered here is the rise time of the transmitted signal. Figure 11-1 shows the output waveform of a CW transmitter sending a string of *dots* (di di di di di . . .). The waveform in Fig. 11-1A has a fast rise time and will produce a substantial quantity of key clicks. The second waveform (11-1B) is softer and produces fewer clicks, because the rise and fall times are slow. The waveshaping can be a function of the keying circuit. But first, determine that the final amplifier is not being *overdriven* from the oscillator or any previous driver/buffer amplifier stages.

The simplest form of keying is shown in Fig. 11-2. In this method, a telegraph key or the contacts of a keying relay interrupts the cathode circuit of the final amplifier and possibly the oscillator as well. There is a danger in this circuit, incidentally. If the key is used to directly switch the cathode, the plate high voltage, which is limited only by the plate resistance of the tube, will appear across the open key terminals. This can be very dangerous if a high-voltage vacuum tube final amplifier is used! In that case, it is heartily recommended that you use a keying relay. The relay contacts will actually switch the amplifier/oscillator cathode circuit while a low-voltage (i.e. 6-24 volts D.C.) power supply energizes the relay. The relay coil, power supply, and telegraph key are connected in series.

If the rise time of the amplifier output waveform is too fast, this is equivalent to modulating the carrier with a sine wave having a frequency equal to the reciprocal of the rise time. Recall that frequency is the reciprocal of period (f = 1/t). A 10 μS rise time, this is equivalent to a sinewave frequency of approximately $1/10^{-5}$, or 100 kHz. The fast rise time will create key click products out to 100 kHz on either side of the carrier. This is illegal on any amateur radio band, even if the clicks fall entirely inside of the bands.

The rise time can be shaped in a *key click filter* circuit. The goal is to create a rise time of not much faster than would be required to pass the third harmonic of the keying frequency. This is based upon a word length of five characters. At a keying frequency of 100 words per minute—100 wpm is *very* rapid with the world receiving record at less than 75 wpm—a rise time of approximately 5 milliseconds is needed, which corresponds to a bandwidth of 200 Hz. The actual bandwidth of a CW signal will be proportional to the code speed being transmitted: The faster the speed is, the greater the bandwidth will be.

Figure 11-3 shows some examples of key click filter circuits. The switch in each of these cases will be the contacts that actually turn the amplifier on and off; these are the key contacts themselves in directly keyed circuits or the relay contacts in indirectly keyed circuits.

Placing an inductance in series with the key in a cathode-keyed amplifier (such as Fig. 11-2) will control the rise time of the leading edge, while a capacitor across the key (Fig. 11-3A) will control the fall time of the trailing edge of the output signal. The capacitor will have a value in the 0.001 μF to 0.01 μF range, depending upon the characteristics desired. The sparking at the key contacts (relay contacts) can be reduced by connecting a small resistance in series with the capacitor, such as shown in Fig. 11-3B. This resistor will

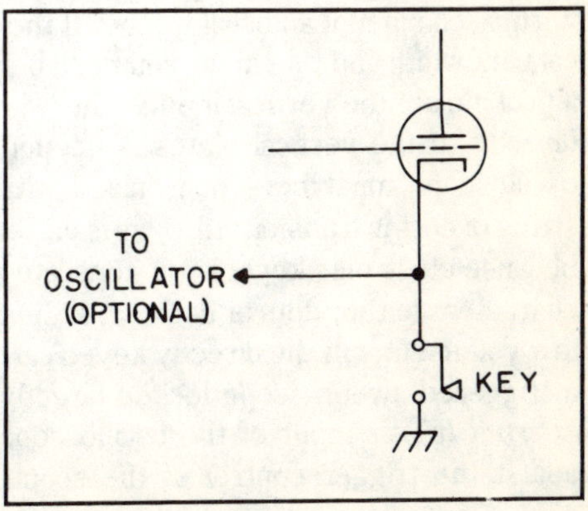

Fig. 11-2. Cathode keying.

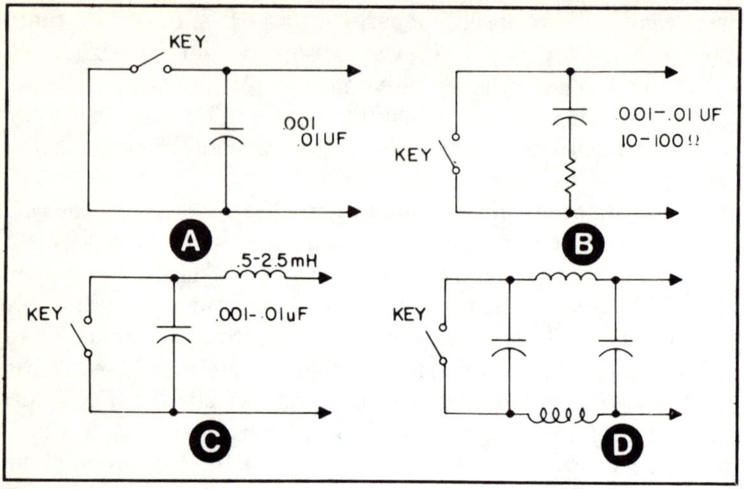

Fig. 11-3. Key click filters.

have a value in the 10-ohm to 100-ohm range with 47 ohms being very common.

Figure 11-3C shows the full key click filter that will control both the rise and fall times of the output waveform. This circuit contains both a parallel capacitor and a series inductor. As in the previous case, the capacitor value is 0.001 μf to 0.01 μF. The value of the inductor will be 0.5 mH to 2.5 mH, also depending upon the keying characteristics desired. One last circuit is the double filter of Fig. 11-3D. This is essentially an example of the LC filter above, but with the inductance and capacitance split into two components.

An oscilloscope is necessary to check the waveform of the final amplifier stage. Using techniques discussed in Chapter 21, connect an output sample of the signal produced by the transmitter to the vertical channel of an oscilloscope. If the scope has adequate vertical bandwidth and rise time, you need only connect a loop of wire to a probe from the vertical input; otherwise, a tuning unit connected directly to the vertical plates is needed. The sweep time of the oscilloscope must be synchronized with the code speed when a string of dots is transmitted. This is easier to do now than in the past because electronic keyers will generate a constant-speed string of characters (either dots or dashes). If the transmitter is keyed with a low voltage (*not* in the directly keyed cathode circuits), you can use a triggered sweep scope locked directly to the key. Connect the *external trigger* input of the oscilloscope, turn on the keyer, and adjust the trigger control of the scope until a stable pattern of several waveforms is displayed on the CRT screen. If the time base of the oscilloscope is calibrated, you can measure the rise time of the waveforms directly or just rough it in, looking for a shape similar to Fig. 11-1B.

BLOCKED-GRID KEYING

The cathode keying of the previous example is not one of the best systems. Requiring the switching of the final amplifier cathode current is fraught with difficulty in any event. The blocked-grid keying system of Fig. 11-4 is a little bit better. In this system, the grid circuit of the power amplifier sees a high negative bias when the key is up. The negative voltage is sufficient to bias the final amplifier tube to cut off. When the key is closed, the blocking voltage is shorted to ground, causing the grid bias to reduce enough to turn on the power amplifier. In some cases, there might be a separate C-bias supply which supplies protective bias to the final amplifier. In other cases the grid return resistor R3 is merely grounded. The typical design calls for R1 and R2 to have very high

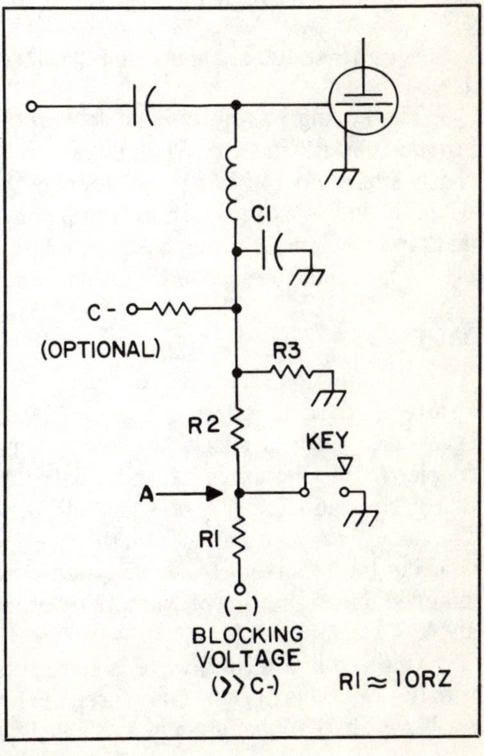

Fig. 11-4. Simple grid block keying.

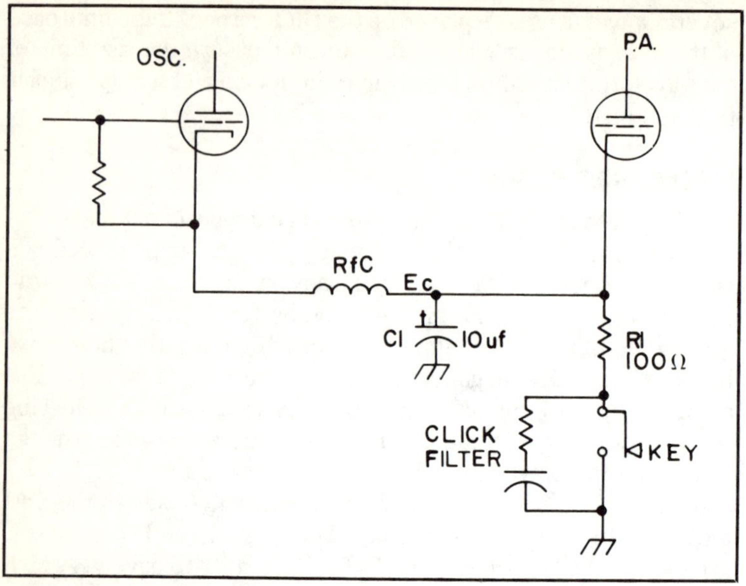

Fig. 11-5. Differential keying.

values (at least 400-K ohms), and that R1 be 10 times the value of R2.

The keying waveform is controlled by the RC time constant produced by R2 and C1. When the key is up (open), capacitor C1 charges to the cutoff bias point, keeping the tube turned off. But, when the key is closed, it shorts out point A, and this causes C1 to discharge to ground through R2. It will discharge to the bias voltage, which could be supplied by either grid leak or fixed bias supply.

BACKWAVE

If the final amplifier is keyed, and the oscillator is left running in an attempt to reduce chirp, it is possible that some energy from the oscillator, which is applied to the power amplifier grid, will be coupled to the antenna circuit and radiated. FCC regulations require that this radiation be suppressed 40 dB or more below the carrier. The name *backwave* comes from the fact that the oscillator signal is heard in the receiver of a distant station when the transmitter is unkeyed. Neutralization problems serve to increase backwave radiation.

One solution to backwave is to turn the *oscillator* on and off with the key. This must be done carefully, however, especially if the oscillator shares a common power supply with the final amplifier.

Any regulation problems in the power supply during keying will only exacerbate chirp problems.

Figure 11-5 shows two methods for curing the chirp problem in transmitters where both oscillator and amplifier are keyed simultaneously. Both of these circuits cause the oscillator to turn on a fraction of a second ahead of the final amplifier, giving it time to produce a stable frequency. Such a system is sometimes called *differential keying*, perhaps alluding to the turn-on time differential between the oscillator and amplifier.

The circuit in Fig. 11-5 uses an RC time constant but no active components. Note that the grid leak bias for the oscillator is returned to the cathode, instead of ground. Furthermore, note that the final amplifier grid leak is returned to ground. These stages will come to full operating level when the cathode-grid potential becomes sufficiently small. In this circuit, make the cathode more positive than the grid. The cathode voltage is developed across capacitor C1. With the key open, E_C will be high enough to turn off the oscillator. When the key is closed, this capacitor will discharge through resistor R1. When the voltage across C1 reduces a little bit, the oscillator will turn on. But it takes a larger reduction in order to turn on the amplifier, thereby inducing the time differential.

A more common form of circuit is shown in Fig. 11-6. In this circuit, a high negative voltage is applied to the grids of both the final

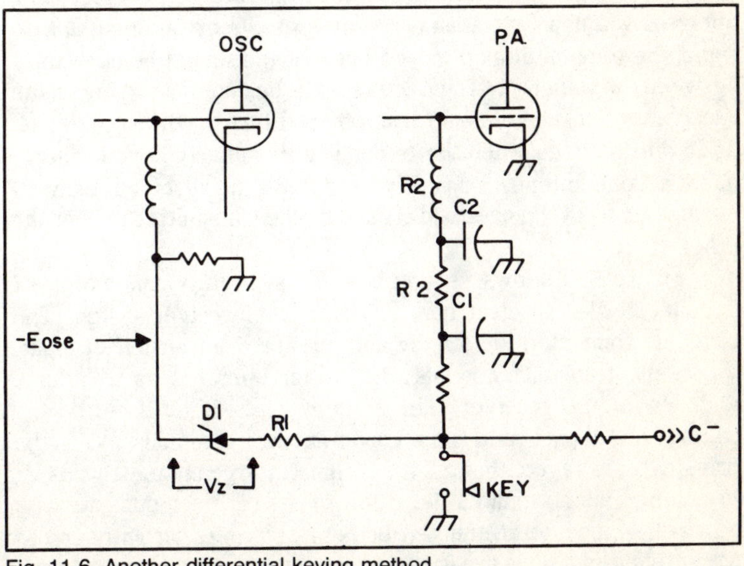

Fig. 11-6. Another differential keying method.

amplifier and the oscillator stages. The voltage must be high enough to turn on the zener diode and bias the oscillator to cutoff. When the key is closed, the zener immediately shuts off, because the potential is no longer greater than V_z. This will turn on the oscillator immediately, but the amplifier sees an RC time constant in its grid circuit. The capacitors in the grid block circuit have been charged to a voltage high enough to cut off the amplifier tube. These must discharge through the resistors and the key contacts. The exponentially rising waveform of Fig. 11-1B is obtained because these capacitors will discharge according to the exponential manner of capacitor-resistor circuits.

The differential keying system will produce some chirp if the time constants are not correctly adjusted. The solution to this problem lies in checking for operation when the zener diode is removed from the circuit (open). If there is no chirp then, *probably* no chirp will exist in the transmitter output with D1 in the circuit.

BREAK-IN OPERATION

High-speed (don't laugh, data transmission freaks) CW operations require the ability for the receiving operator to interrupt the transmitting operator. This is done to get a reread of missed portions of the message, or for other purposes known to operators who do such things. Reception must be possible to detect the break-in request between Morse code characters. The breaking operator will usually transmit a couple of *dits* or one long *dah* to signal the transmitting operator. These signals must be detectable between the transmitted characters. This be done by turning on the receiver when the transmitter is off, and turning the receiver off again during times when the transmitter is actually keyed. Simply use a second antenna for receive and hope the receiver recovers sufficiently from the overload caused by the transmitter to hear the other station.

Figure 11-7 shows a simple break-in system. Vacuum tube V1 is connected as an electronic TR (transmit/receive) switch. The switch is connected to both the antenna and the transmitter input. When the transmitter is off, the circuit operates as a cathode follower for the receiver. The antenna signal is available on the cathode and is passed to the receiver antenna terminals. When the transmitter is keyed, though, the rf signal is large enough to bias V1 to cutoff. This attenuates the signal to the receiver considerably; however, it must be attenuated further by placing a capacitor across the antenna line. This capacitor is connected to the circuit by a pair

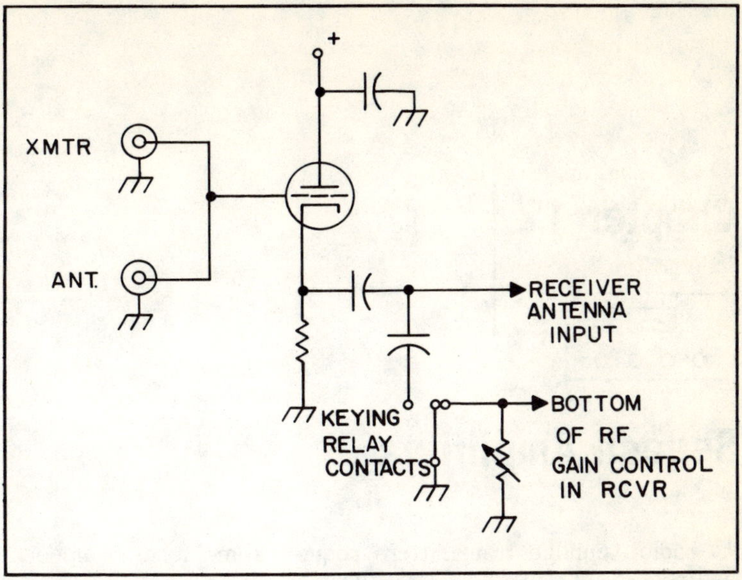

Fig. 11-7. Electronic T-R switch for break-in operation.

of terminals on the transmitter keying relay. In the receive mode, these terminals serve to ground the rf gain control inside of the receiver. To modify a receiver for full break-in requires us to lift one end of the rf gain control and bring it out to an extra potentiometer (R_b). Most receivers of the tube variety use the cathode bias of the rf amplifier as the source of the rf gain. When the relay contacts are shorted, the bias is normal and is controlled by the rf gain control. On transmit, R_b is shorted, inserting an additional resistance in series with the rf gain control, and reducing the rf amplifier sensitivity.

Chapter 12

Speech Amplifiers

All radiotelephone transmitters require some form of speech amplifier ahead of the modulator. The speech preamplifier circuits of this chapter are applicable to any of the modulation/modulator chapters to follow. The speech amplification system contains several different components that must work together to produce the audio signal at the input of the modulator section: the microphone, amplifier stages, frequency response shaping, and in some cases, clipper/limiter stages.

The frequency response of the audio amplifier will affect the output signal considerably. In amplitude modulation, and its single sideband cousin, the bandwidth of the signal is directly related to the frequency response of the modulator/preamplifier stages. If the audio sections of the transmitter have a frequency response of 2.5 kHz, the bandwidth of the AM transmitter will be 2 × 2.5 kHz, or 5 kHz. Similarly, the bandwidth of the SSB transmitter emission will be 2.5 kHz. In FM two-way communications, it is the usual practice to limit the bandwidth of the audio signal to the 300 to 3000 Hz range. But in FM broadcasting, the audio bandwidth is 15 kHz (more if the transmitter is stereo, because it has to account for L-R subcarrier modulation to 53 kHz). Broadcasters in the AM band are required to trim the audio bandwidth to 5 kHz, allowing a 10-kHz channel width.

MICROPHONES

The microphone is an important link in a two-way communications system. Radio amateurs are sometimes a little haphazard when they select microphones for their transmitters. Several different types of microphone are available and these have different properties.

Frequency Response

One property of all microphones is the frequency response. For high-fidelity broadcasting this becomes a little bit more critical, and some broadcast-grade mikes cost proportional to their frequency response. In two-way communications work, though, a microphone needs only be flat (or at least relatively so) between 300 and 3000 Hz. Almost all available commercial microphones will meet this specification.

Directivity

Another property is the directivity of the microphone. In some application, one wants an *omnidirectional* (all directions) or *cardoid* (heart-shaped pattern) microphone. These will pick up sounds coming from a wide angle. This could prove to be a problem in communications work. The operator might not want to pick up any other audio sources than his voice. For this reason, directional or unidirectional microphones are desirable for two-way work.

Some communications microphones are designed to eliminate extraneous noises from the environment. For instance, mobile operation and aircraft cockpit operation tend to have high ambient noise levels. Two methods can be used to desensitize the microphone to local noise. One of these is to make the microphone hyperdirectional, meaning that it will only pick up the sound that enters directly into a small slit or collection of tiny holes in the speaking face of the microphone. Sound from other directions is highly attenuated. The other method is to select an internal microphone element that is not as highly sensitive as normal. In that case, the output component due to the environmental noise is less than the operator's voice signal. In both cases, the operator is instructed to speak directly into the microphone, or "swallow" the mike.

Several different types of microphone are commonly used in electronic communications. As classified by the element style, these are *carbon, dynamic, crystal* (piezoelectric), and *electrotet* (condenser).

Carbon Microphones

The carbon microphone was one of the earliest types used in communications. Although it has been largely eclipsed by other designs in modern equipment, a lot of carbon microphones are still in service and are still being sold. The telephone mouthpiece was a carbon button for many decades. The construction of a typical carbon microphone is shown in Fig. 12-1A. The actual microphone element consists of a small pot full of carbon granules, loosely packed. When an acoustical (audio) wave hits the diaphragm, the granules are compressed and released in step with the variations of the audio wave. This creates a varying electrical resistance across the element that follows the acoustical wave.

Making a carbon microphone work properly requires an external circuit consisting of a small power source (battery E) and either a resistor/capacitor combination as shown or an audio transformer. When no sound is hitting the element, its electrical resistance is constant. This causes a constant current to flow in the series circuit made by the element with resistor R1. The voltage drop across R1 is the constant under that condition. When an acoustical wave hits the element and causes the element resistance to vary, the current through the element and the resistor will vary also. This produces a varying voltage drop across R1 that follows the variations of the acoustical signal. It's an audio signal made from the acoustical wave hitting the element. Capacitor C1 is used to keep the DC component across R1 from getting into the microphone preamplifier circuits. In many cases, the resistor was replaced with a small interstage audio transformer. In those cases, the varying current produced by the element resistance creates a varying magnetic field in the transformer. That in turn is reflected as the output voltage changes at the transformer secondary.

The carbon microphone has several advantages. One is that it is more rugged than some of the other microphone types. It produces a relatively high output voltage, so its output signal requires little amplification before being applied to the modulator. The carbon microphone also has its share of problems. The carbon granules sometimes become packed and will not compress anymore when acoustical wave strikes the diaphragm. This reduces the audio output to an almost negligible level. Two-way radio operators who use these microphones soon learn the trick of smacking the microphone against the dashboard or against the side of a desk to loosen the granules inside the microphone element. Repair technicians might wince at that techniques, but it works! Another problem is that the carbon element sometimes produces a frying sound when the DC bias current pass through it. This hash has to be filtered in the modulator stages of the transmitter to prevent it from being radiated.

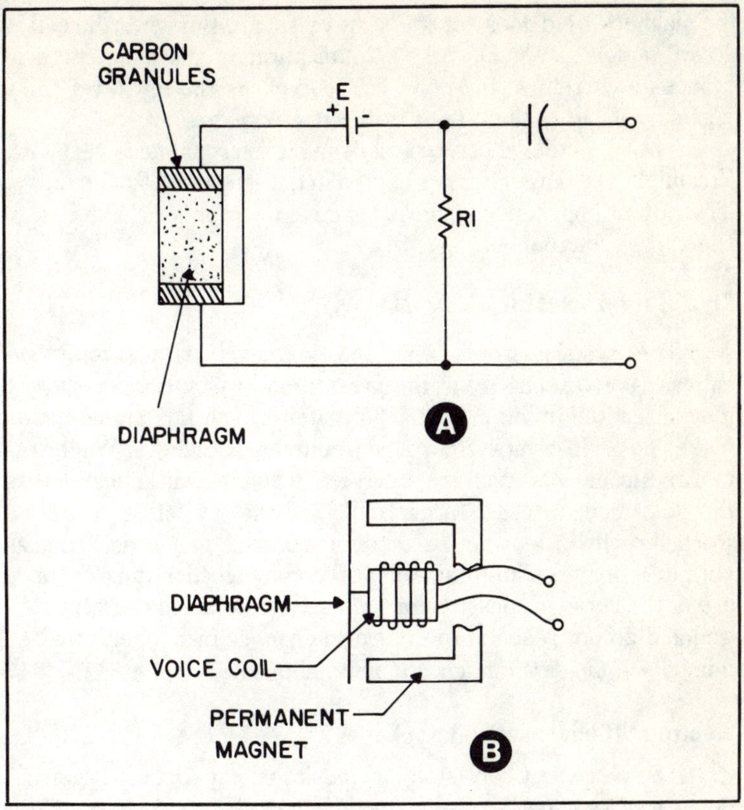

Fig. 12-1. Carbon microphone at A and dynamic microphone at B.

Dynamic Microphones

The dynamic microphone requires no power supply, but it is generally less rugged than the carbon microphone. The dynamic element is very much like a permanent-magnet loudspeaker operated in reverse. In fact, if you connect a loudspeaker to the input of an audio amplifier, allowing an impedance matching transformer to boost the voltage output from the loudspeaker coil, you will find out just how good of a microphone can be made from a loudspeaker. Most intercoms, and not a few two-way hand-held portable radio transceivers use a loudspeaker for both "talk" (transmit) and "listen" (receive).

Figure 12-1B shows the basic configuration for a dynamic microphone. A lightweight coil of wire is mounted to the back of a diaphragm. The coil is positioned inside the field of a permanent magnet. When the coil is stationary, as when there is no sound, the magnetic field does not affect the coil at all. But when the coil is *set in motion* as sound strikes the diaphragm, the magnetic field induces a current in the coil. This current is the low-level audio signal available at the output of the microphone.

The output level of a typical dynamic microphone is very low. You might see output figures quoted in the –50 to –60 dBm range. The output impedance of most dynamic microphones, like that of the carbon microphone, is low.

Crystal (Piezoelectric) Microphones

The crystal microphone is used extensively in radio communications. Piezoelectricity is the generation of electrical energy by mechanical deformation of a solid material, and this phenomenon makes possible crystal-controlled frequency oscillators. When certain crystalline materials are deformed, a voltage will appear across the deformed surfaces. In microphones, the crystals used will be Rochelle salts or a man-made ceramic containing materials such as barium–, or zirconium-titanate. Most communications microphones are of the ceramic construction. These microphones are very rugged and do not respond too much to changes in temperature and humidity (humidity kills carbon microphones).

Electrotet (Condenser) Microphones

The electrotet consists of an insulator that contains a permanent (well, not *really* all that permanent) electrical charge. The diaphragm forms a pair of capacitor plates with the insulator as the dielectric. When the diaphragm is distended by the acoustical wave, the capacitance changes. In any capacitor, because the voltage across the capacitor is proportional of the charge (fixed in the electrotet insulator) and the capacitance (varies with the sound), the output voltage will vary with the sound wave. Most electrotet microphones contain a small battery to power a JFET preamplifier. The typical output impedance is the source resistance of the preamplifier, or about 1000 ohms. Most electrotet microphones are not rugged enough for mobile operations. Both heat and humidity can affect the electrotet element, but such microphones *are* suitable for base station operation.

SPEECH PREAMPLIFIERS

The output of most micrphones is too low for direct use in the modulator of a transmitter. A high-level broadcast modulator might want to "see" anywhere from 1 volt to tens of volts at the input. The preamplifier stage must develop the output of the microphone to a point where it will drive the modulator.

Let's consider a hypothetical situation. A radio transmitter requires a signal of 0 VU at the input to the modulator. A VU is a *volume unit*. It is a power decibel measurement in which 0 VU is 1 milliwatt dissipated in a 600-ohm load. The VU can be calculated for any given power level using the well-worn expression:

$$dB_{VU} = 10 \log_{10} \left[\frac{P}{0.001 \text{ W}} \right]$$

If the power level, P, is less than 1 mW, the VU rating is negative. If P is greater than 1 mW, the VU is positive. A typical dynamic microphone might have a rating of −55 dBm (decibel related to 1 mW), so a gain of 55 dB must be provided from the preamplifier.

Figure 12-2 shows the circuits for bipolar transistor preamplifiers. The circuit shown in Fig. 12-2 uses a PNP transistor in the bias circuit configuration for negative-ground operation. This means that the circuit is compatible with mobile battery power supplies.

The bias network consisting of resistors R3 and R4 sets the operating conditions of the transistor (Q1). For speech amplifier

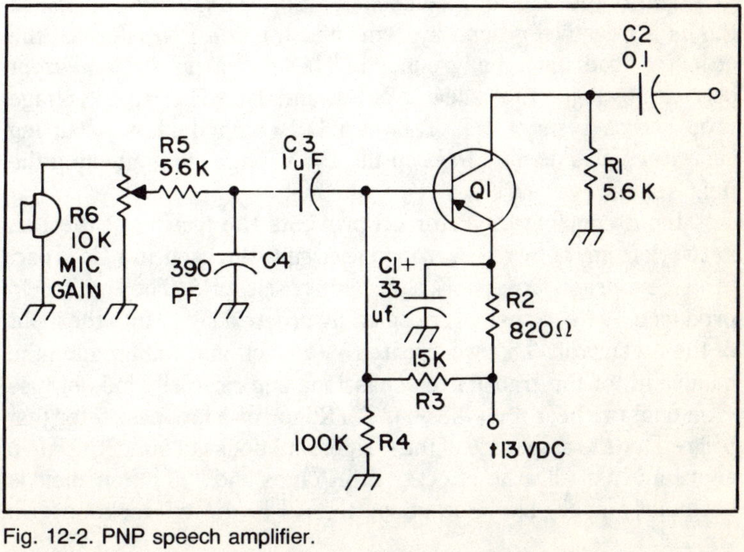

Fig. 12-2. PNP speech amplifier.

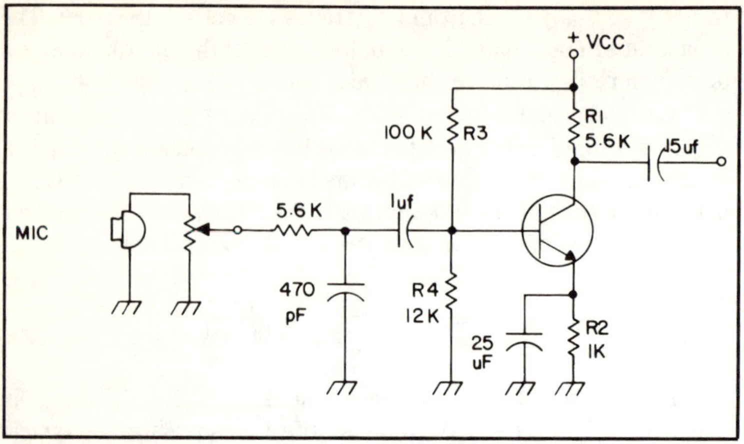

Fig. 12-3. NPN speech amplifier.

purposes, the transistor is biased for class-A operation. Resistor R2 is for thermal stabilization of the transistor. The emitter bypass capacitor (C1) is used to set the low-frequency response of the preamplifier. Select a value for C1 which will provide a reactance that is one-tenth of the resistance of R2. In this case, it's 82 ohms. It is important in single sideband and AM transmitters to prevent the low-frequency information from modulating the transmitter. This becomes a matter of the frequency/power distribution in the modulated envelope.

The output signal is developed across resistor R1. Because this is a negative ground system, and Q1 is a PNP device, the collector load must be grounded. There is a normal quiescent current flowing in the collector of Q1, and this will create a voltage drop across resistor R1. This is a DC component, so coupling capacitor C2 is needed to keep the DC voltage from affecting the next stage.

Input coupling capacitor C3 prevents the loading of the bias network from Q1 by the microphone circuit. But capacitor C4 is part of an rf suppression network along with resistor R5. The high rf field produced by the transmitter can easily create a signal into the input of the modulator. This will create rf feedback that will broaden the bandwidth of the transmitter emissions and raise all kinds of fuss (you ought to hear it on the air). C3/R5 forms a low-pass filter that will reduce the amplitude of the rf signal but does not affect the audio signal much at all. The effect of 5.6K ohms and 390 pF on audio is essentially negligible. Potentiometer R6 is the microphone gain control.

An NPN version of the same circuit is shown in Fig. 12-2. Most of the components in this circuit have counterparts in the previous circuit. Note that due to the reversed polarity, the collector load goes to the positive side of the power supply. In the negative-ground situations considered here, the hot side of the power supply (if such is even a valid concept) is positive.

The transistors used for either Figs. 12-2 or 12-3 can be any medium-beta silicon device. The original transistors for these amplifiers are 2N3904/2N3906 devices for NPN/PNP, but there should be little trouble experienced with hobbyist/service-replacement line transistors.

A small collection of field-effect transistor audio preamplifiers is shown in Figs. 12-4 through 12-6. The circuit in Fig. 12-4 uses a simple junction field-effect transistor (JFET). This particular circuit is an N-channel, which is the most commonly employed in negative-ground operation. The JFET is more like a vacuum tube in its characteristics, so the circuit looks much like a vacuum tube preamplifier. The microphone signal is applied directly to the gate of the JFET through a potentiometer that serves as the *microphone gain* control. This potentiometer should have a value of 100K ohms for low-impedance (Z) microphones and 1 megohm for high-Z mikes. Note that the 1 megohm value will also work with the low-Z microphone, so there's good cause to use 1 megohm to universalize the circuit.

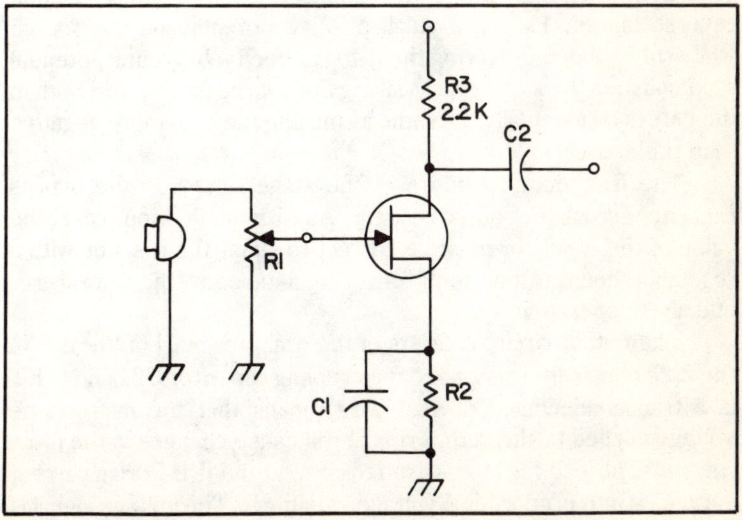

Fig. 12-4. JFET speech amplifiers.

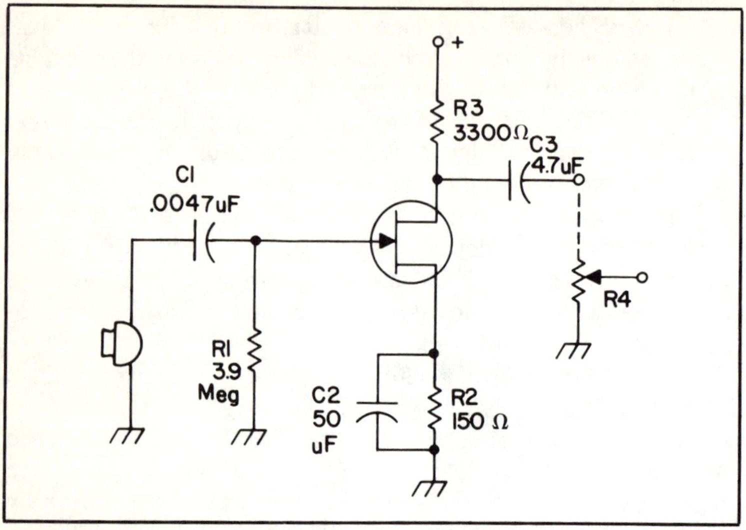

Fig. 12-5. JFET speech amplifiers.

The bias on the JFET is provided by a resistor between the source and the ground. This is analogous to the cathode bias generally used in low-level audio voltage amplifier vacuum tube circuits. The channel current flowing in the JFET produces a voltage drop across the source resistor. This voltage drop becomes the bias voltage for the gate-source of the JFET. The goal is to keep the gate more negative than the channel to make the gate junction reverse biased. Placing a small positive potential on the source while simultaneously placing the gate at effectively ground potential accomplishes the same goal. Making the source more positive than the gate is conceptually the same as making the gate more negative than the source.

The frequency response of the stage is set by the bypass capacitor across the source resistor. As with the previous case, the value of the capacitor must be correct to shunt the resistor with a reactance equal to one-tenth of the resistance at the lowest frequency of operation.

The output circuit consists of the drain terminal of the JFET, the 2.2k ohm load resistor and the coupling capacitor C2. The JFET is a transconductance device. This means that the microphone voltage applied to the gate terminal will cause changes in the drain *current*. The load resistor is used to convert the JFET drain current variations into proportional voltage variations. The voltage signal is coupled through C2 to the stages that follow.

A slightly different JFET circuit is shown in Fig. 12-5. In this circuit, however, no microphone gain control is in the input; the gain of the stage is fixed. You might connect a potentiometer either in the output (as shown in the optional circuitry) or in a following stage. Both of these circuits will work with any of the garden-variety JFET devices, such as the popular MPF102.

The last JFET circuit is the MOSFET (metal-oxide semiconductor field-effect transistor) circuit of Fig. 12-6. Most of the elements of this circuit are analogous to Fig. 12-4. There is, however, a second gate in this circuit. The second gate is biased on by a resistor voltage divider network, but it can also be used as an automatic gain control (agc) terminal. This agc is sometimes used in radio transmitter speech amplifiers to keep the modulation level constant, thereby compensating for variations in the radio operator's voice level. In those circuits, the audio in some later stage would be rectified to provide a DC control voltage. This voltage will be applied to gate G2 in a polarity that will increase the gain of the stage when the audio level drops and will decrease the gain when the audio level rises.

Integrated circuits play a large role in all areas of electronic technology. Large strides have been made in so many areas that it is doubtful that any form of radio equipment will escape using an IC. Some IC preamplifier circuits will be discussed here. After all,

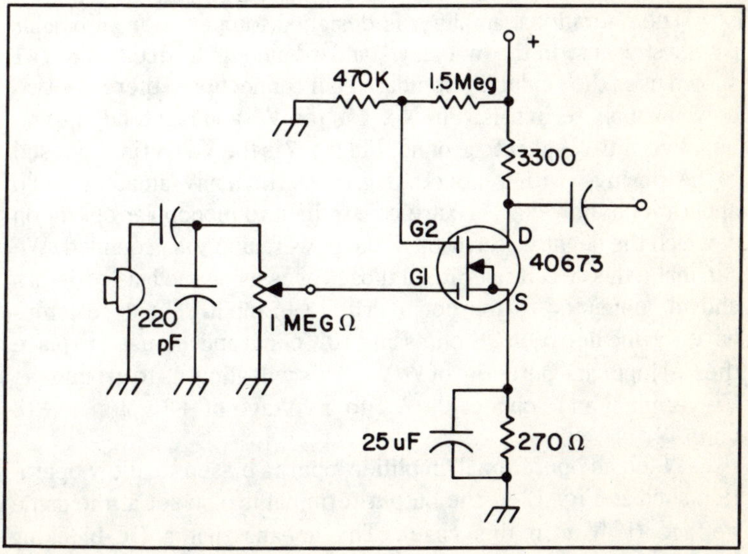

Fig. 12-6. JFET speech amplifiers .

somebody will soon be manufacturing complete transmitter audio system ICs containing all needed stages for any type of transmitter.

The use of an operational amplifier (op amp) as a speech amplifier is shown in Fig. 12-7. The operational amplifier is a wonderful little device that allows almost everyone to become a solid-state circuit designer. This little device has a very high open-loop (all feedback resistors open) voltage gain. The theoretical operational amplifier has an *infinite* open-loop, and practical op amps have gain figures in the 20,000 to 2,000,000 range. The gain of the closed-loop stages is controlled solely by the ratio of the feedback to the input resistances. In the circuit shown, the gain is set by the ratio of (R1 + R2)/R3.

The operational amplifier is differential. This means that it has two inputs, each of which has an equal but opposite-polarity effect on the output signal. The inverting input produces an output signal that is 180 degrees out of phase with the input signal. The noninverting input produces an output signal that is in phase with the input signal. The circuit configuration selected here is the inverting amplifier (the signal is applied to the – input), so the output signal will be 180 degrees out of phase with the input. We do not have the space here to go any deeper into operational amplifiers so for those who desire a deeper understanding, let me recommend TAB book No. 787, *Op amp Circuit Design & Applications* and TAB book No. 856, *Master Op-AMP Applications Handbook* by Harry Fox.

The operational amplifier is designed to operate from bipolar power supplies. In the industry standard pinouts devices—the 741 shown uses the "industry standard" pin connection—there are two power-supply terminals. Pin No. 4 is the V_{ee} and is intended to be negative with respect to ground. Pin No. 7 is the V_{cc} and is supposed to be positive with respect to ground. In many amateur radio applications, however, we are constrained to monopolar operation in which the negative terminal of the power supply is grounded. We can make this circuit operate in those cases using the bias resistors shown connected to the noninverting (+) input. These resistors have a value between 2K ohms and 10K ohms and are used to place the (+) input at a potential of ½-V_{cc}. This will allow us to ground the V_{ee} terminal and connect the V_{cc} to a voltage of +8 volts to +18 volts.

When an operational amplifier input is biased to allow operation such as Fig. 12-7, the output terminal is also set at the same voltage (½-V_{cc}, in this case). This means that a DC-blocking capacitor is needed in the output (C3); otherwise, the potential

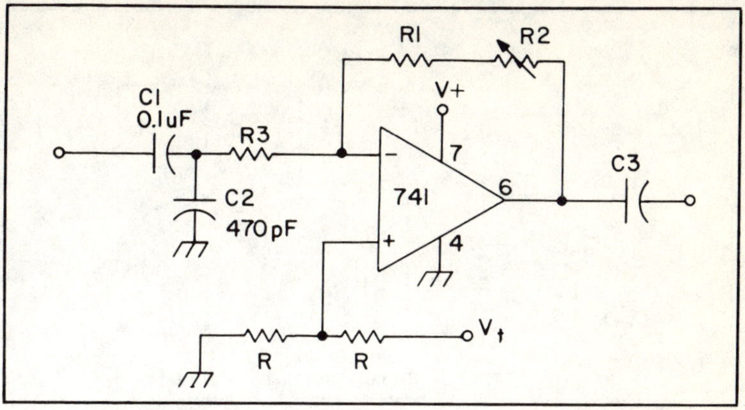

Fig. 12-7. Op amp speech amplifier.

applied to the next stage will contain a 4 to 9 volt DC component.

The 741 family of operational amplifiers is designed to be unconditionally stable. What this means in lay language is that they "kick the pants off" the frequency response. A 741 that operates much over 8 kHz is a rarity. There are op amps that operate in the higher frequency ranges, but these must be compensated with external capacitors or RC networks, in order to narrow the frequency response. Since we are concerned solely with a frequency response under 3 kHz, the 741 is adequate for our purposes.

Theoretically any gain up to the open-loop gain of the device can be obtained; however, there are practical limitations to this circuit caused by problems in the mass production of operational amplifiers—or any IC for that matter. Limit the practical gain from the circuit to under × 200. If you want to obtain more gain, consult the books listed earlier to tricks and techniques in operational circuit design, especially the sections on offset nulling.

A noninverting amplifier circuit is shown in Fig. 12-8. This circuit is intended for bipolar operation and is therefore suitable for mobile applications. One very well respected magazine in the amateur radio area insists on publishing all op amp circuits in the monopolar configuration, using the lame excuse that the bipolar circuits won't operate in mobile applications. But that is just plain stupid once you consider that 99.9999 percent of all mobile rigs are commercially made and the operational amplifier works best in bipolar circuits, and monopolar circuits just cannot get the most out of the op amp. Besides, the bipolar circuit produces no DC offset in the output. When there is zero signal at the input, the output voltage is zero.

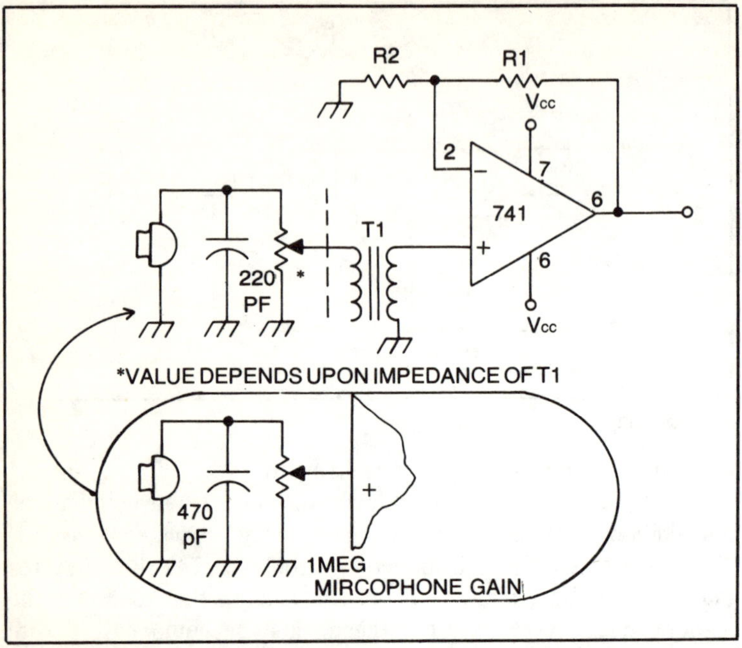

Fig. 12-8. Op-amp speech amplifier.

The gain of this circuit is set by the same method as in the inverting case: the ratio of the two resistors. But in the noninverting amplifiers, the gain is (R1/R2) + 1. At high gains normally expected of microphone preamplifiers, the "+1" factor becomes almost negligible, but at lower gains a significant error will result if the "1" is dropped.

The signal applied to the noninverting input of this preamplifier is from the secondary of the interstage transformer. If this transformer provides any voltage increase because of its turns ratio, the overall gain of the circuit will be higher by the turns ratio factor. When the microphone gain control is at maximum, the gain is:

$$A_v = \left[\frac{N_p}{N_s}\right]\left[\frac{R1}{R2} + 1\right]$$

The transformer can be deleted (see Fig. 12-8 inset), but only if the overall gain of the stage is small or the operational amplifier is a premium type in which the input bias currents are small. Those bias currents will create a DC offset voltage when they flow through a resistance. The transformer secondary has a low DC resistance so this effect is of little importance. In the nontransformer version, though, the resistance of the microphone gain control can be significant, and a DC offset might exist. Of course, this offset can be handled, and you are referred to the previous references.

Microphone preamplifiers are usually located inside the transmitter in amateur applications. Some CBers and not a few radio amateurs, however, are prone to using power microphones. One has to doubt the wisdom of that course of action, when he considers that the manufacturer of the transmitter in all likelihood provided sufficient preamplification to 100 percent modulate the transmitter using any standard microphone. The use of additional microphone gain would overmodulate the transmitter. With CB rigs, the manufacturer normally includes a modulation limiter stage to counteract the use of these devices.

There is one type of operation, however, in which an external microphone preamplifier is justified, and that is when the microphone and operating position are remotely located from the actual transmitter. In that type of radio station (typical of broadcasting and commercial base stations) the level of the microphone signal needs boosting prior to sending it over a long line to the transmitters. All microphone lines wil pick up a certain amount of 60-Hz hum from the power lines. In short runs, the microphone signal is usually so much higher in level that no one can hear the hum. In long runs, however, the signal is attenuated, while the hum pickup is *greatly increased*. In those cases, we want to boost the audio signal to overcome line losses and overcome the hum pickup.

CLIPPERS

Many transmitters include a clipper circuit (Fig. 12-9) to prevent overmodulation. The microphone preamplifier (A1) output is fed to a pair of diodes D1 and D2. These diodes will clamp the line to 0.6 volts, assuming they are silicon, (0.2 volts if germanium diodes

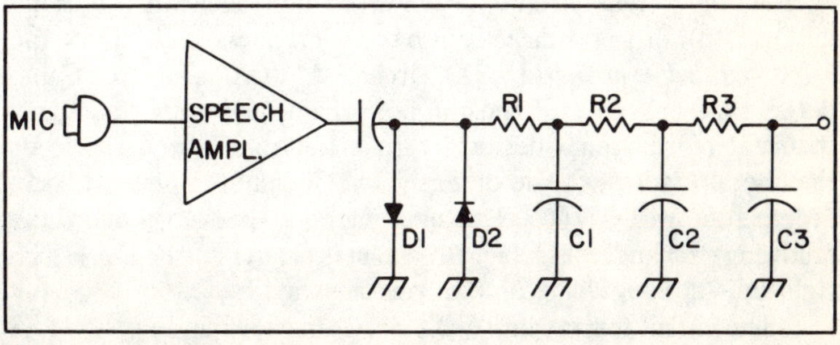

Fig. 12-9. Clipper/filter circuit.

are used) when forward biased. Diode D1 will clip the positive peaks of the audio signal, while D2 clips the negative peaks. The purpose of these diodes is to assure a maximum audio signal level is not exceeded, thereby insuring that the transmitter is not overmodulated.

But this is not for free. When you clip a sine wave, you generate a lot of harmonics. These harmonics distort the signal fed to the modulator and will cause a great deal of splatter in the emitted signal produced by the transmitter. The RC network that follows the clipper is used to remove these harmonics. The RC circuit is a passive low-pass filter. All RC low-pass filters produce a severe attenuation of the output signal. This signal must be again built up in fixed gain stages that follow the filter.

Figure 12-10 shows the circuit for a unity-gain (no loss) filter circuit that could be used to replace the RC network in modulation limiter stages. This circuit is an operational amplifier active filter. The active filter should be used in bipolar power-supply circuits, which is not unreasonable in fixed station equipment. The filter components are R1/R2/C1/C2. The 1 μF input capacitor is used to block the DC level that might exist at the output of the preamplifier or the level across the diodes. In many cases, this capacitor is deleted. Similarly, the purpose of the 100K ohm resistance to ground is to form a resistance across which to generate the audio signal from the capacitor. If the capacitor is deleted, then so is the resistor. Active filter circuits can be built with gain. This is done by removing the short between the output and the inverting input and replacing it with a resistor network. But the gain filters become a little bit more complicated.

VOX CIRCUITS

There are three ways to turn on a radio transmitter: operate the front panel *transmit* switch, press a *push-to-talk* button on the microphone (often merely a switch wired in parallel with the transmitt switch), or speak directly into the microphone and actuate the *voice-operated exchange* (VOX) circuit. A VOX is a circuit that detects the presence of audio in the preamplifier circuit and uses that neat information to decide to turn on the transmitter. This frees the operator's hands to do other things. Of course, one has to be a little careful with a VOC. This morning on 20 meters, I heard the following exchange: "Herbie! Breakfast is ready!" "OK, hon. I'll be right up—nothing doing on the band anyway!"

The circuit for a typical VOX circuit is shown in Fig. 12-11. A three-stage transistor amplifier is used to turn on the transmit-

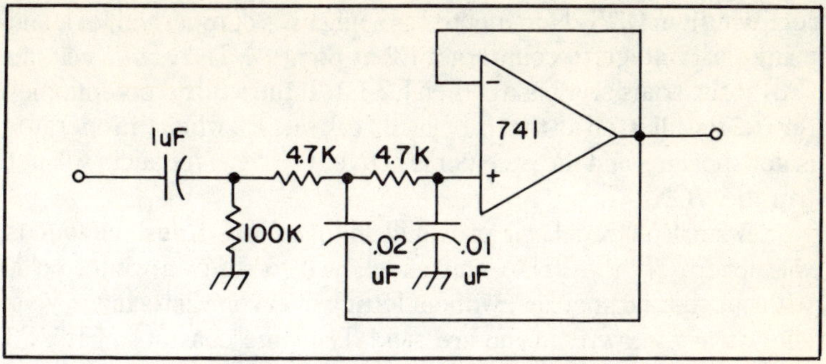

Fig. 12-10. Active low-pass filter for 3-kHz bandwidth.

receive relay (K1). In the normal receive mode, transistor Q1 is cut off, Q2 is turned on to saturation, and Q3 is turned off. Let us first consider only the VOX-trip circuit. We will isolate the antitrip circuit by pretending that point B is grounded. This will allow us to consider only the effects of the trip circuit. The audio signal is sampled through capacitor C1, and this sample is rectified by D1/D2. The pulsating DC output of the rectifer is applied to the VOX adjustment potentiometer, R2. The DC voltage at point A is then proportional to the audio signal. Transistor Q1, the relay trigger, needs to have 0.6 to 0.7 volts at the base base before it will conduct. The VOX potentiometer is adjusted so that this voltage (E3) is applied to the base of Q1 when the operator speaks into the microphone, but not when lower level background noise is present.

When the operator speaks into the microphone, then, E3 goes greater than 0.7 volts, turning on transistor Q1. This forces the collector of Q1 to ground. Transistor Q2 is normally fully turned on by the voltage across capacitor C3. When the operator speaks, however, the Q1 collector provides a path to ground for the charge in C3, so the capacitor begins to discharge. When it has discharged below 0.6 volts, Q2 turns off and its collector goes HIGH. When the Q2 collector goes HIGH, it will forward bias relay switch transistor Q3, and this energizes the relay.

Audio from the receiver loudspeaker should not trip the VOX. An antitrip voltage must be inserted to counteract the sound from the loudspeaker. This sound will *ordinarily* create a voltage output from the trip circuit. As the loudspeaker operates, therefore, sound enters the microphone, is amplified in the mike preamp, and is then rectified to form a contribution to voltage E3. If the loudspeaker output is high enough, the VOX will trip on itself! We counteract this by taking a sample of the signal across the speaker voice coil, rectifying it in D3/D4 (connected as a negative output rectifier), and using this voltage to counteract E2 at point A. The actual voltage (E3) at the transistor base is then E2-E1. If the antitrip potentiometer (R3) is adjusted so that E1 exactly cancels E2 when the operator is not speaking and the receiver is active, the receiver audio will not trip the VOX.

A small VOX delay is provided so that short, transient sounds will not trigger the VOX. This allows you to suffer a winter cold, with coughs and sneezes, without letting everyone listening to your kilowatt rig know that you are sick! The time constant of R5/C3 gives us a little delay in the actuation of the VOX. If the transient sound drops out before the RC network times out, then the VOX will fail to fire. If the sound is sustained, the RC network will time out and the VOX fires normally. Of course, there has to be a trade-off in this respect. We want the time constant delay long enough to keep the rig silent during inappropriate transient sounds in the shack, yet trigger it on the first syllable of the operator's transmission. We also don't want the rig to go back to receive just because the operator pauses momentarily; otherwise, the operator would have to go uhhhhhhhh between thoughts.

An IC VOX circuit is shown in Fig 12-12. The job of comparing the amplitude of the trip and antitrip signal is performed, interestingly enough, by a *comparator* IC. A voltage comparator will examine the voltages at the two inputs and issue an output that indicates whether they are equal or not. Again, let us assume that E3 is grounded, and the antitrip is out of the circuit. Voltage E_{ref} is a reference potential that is analogous to the 0.6-volt bias needed to trip the transistor in Fig. 12-11. The comparator output will be LOW (point A is at ground potential) if E2 is less than or equal to E_{ref}. This prevents normal ambient background noise from tripping

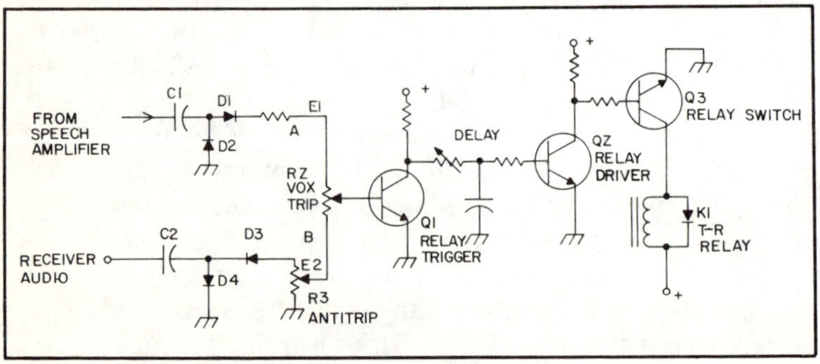

Fig. 12-11. VOX (voice-operated exchange).

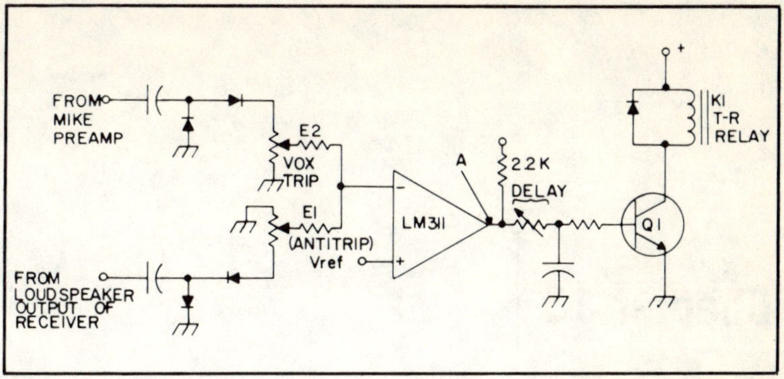

Fig. 12-12. VOX using an IC.

the VOX. When the rectified audio signal is greater than E_{ref}, the comparator output (point A) goes HIGH and turns on the relay switch transistor. The RC network in the base circuit of the transistor provides the delay function in a manner similar to the previous discussion. The antitrip signal subtracts from the voltage applied to the – input of the comparator. It is adjusted to exactly cancel the contribution of the loudspeaker sound to E2.

Chapter 13

Amplitude Modulation

There are four ways that a radio signal can be varied to convey information. The first of these is by on-off keying. The information is contained in the existence or nonexistence of the signal. The carrier amplitude, frequency or phase can be modulated with the signal containing the information. In this chapter we will discuss amplitude modulation of the radio carrier signal.

Modulation is the superposition of information on a radio signal which is called the carrier. In radiotelephoney, the modulating signal will be speech—the operator's voice. There is a classification system for the different types of modulation, and now is as good a time as any to introduce it. See Table 13-1.

In this chapter we will deal with the variety of modulation called amplitude modulation, with full carrier and both sidebands (A3). This form of modulation is the type used in the AM broadcast (how about that!), international broadcasting, and on much of the citizens band. Few radio amateurs still use AM, however, preferring single sideband for normal high-frequency operations and FM/PM for VHF radiotelephone operations.

AM BASICS

Figure 13-1 shows just what we mean when we call this system amplitude modulation. The waveform in Fig. 13-1A is the unmodulated radio signal, or the rf carrier. In order to simply our arithmetic, we are going to state that the frequency of the carrier is 1000

Table 13-1. Emission Types.

DESIGNATION	MODULATION
A0	UNMODULATED CARRIER
A1	RADIOTELEGRAPHY, ON-OFF CARRIER CW
A2	AUDIO TONE MODULATED CW (MCW)
A3	FULL AMPLITUDE MODULATION WITH THE CARRIER & BOTH SIDEBANDS
A3a	AM, WITH REDUCED CARRIER AND BOTH SIDEBANDS
A3b	INDEPENDENT SIDEBAND AM
A3h	SSB, FULL CARRIER
A3j	SINGLE SIDEBAND, SUPPRESSED CARRIER (SSBSC)
A4	SLOW SCAN TELEVISION
A5c	VESTIGIAL-SIDEBAND (AM) TELEVISION
F1	FREQUENCY-SHIFT KEYED TELEGRAPHY
F2	FM TONE TELEGRAPHY
F3	FREQUENCY-MODULATED OR PHASE-MODULATED RADIOTELEPHONY
F4	FM FACSIMILE
F5	FREQUENCY-MODULATED TV SOUND CARRIER
P0	PULSE RADAR

kHz. This is right in the center of the AM broadcast band, so don't try to tune up there unless the FCC gives you a license to do so. We want to impress a 1000-Hz (1 kHz) sine wave (Fig. 13-1B) onto this signal by amplitude modulation. These two signals are superimposed to form a waveform such as shown in Fig. 13-1C. The 1-kHz sine wave causes the *amplitude* of the rf carrier to vary in accordance with the waveshape of the modulating signal. The audio modulating signal algebraically adds with the rf carrier.

We can measure the amplitude modulation of the carrier with the *percentage of modulation*. We define 100 percent modulation as the condition in which the positive peak of the modulating sine wave causes the rf carrier to drop to exactly zero. At percentages of modulation that are less than 100 percent, the addition of the two signals is not so complete, so the positive peak does not quite double and the negative peak does not quite cause the rf envelope to drop to zero.

We use sine waves in our discussions because they obey nice, clean, simple laws that are easy to consider. The math works out to

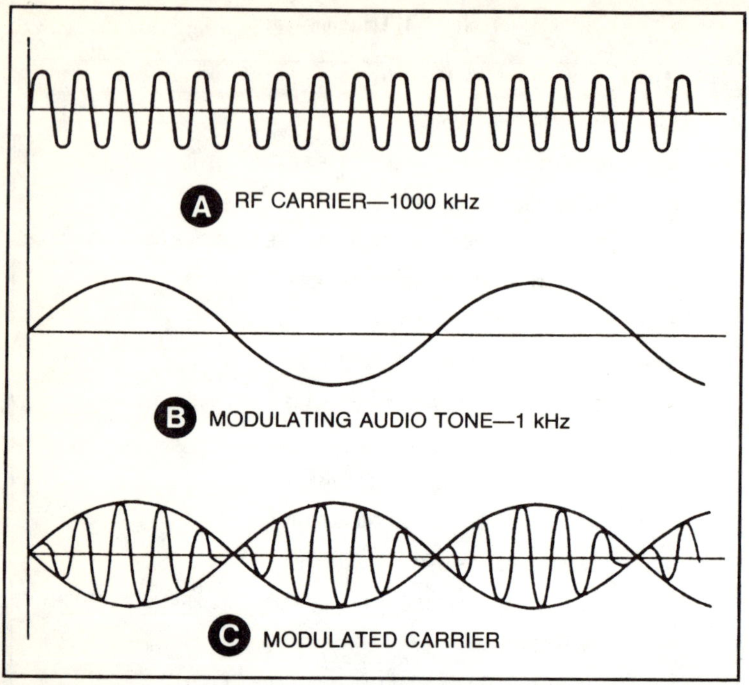

Fig. 13-1. Amplitude modulation of a carrier with carrier at A, audio tone at B, and modulated carrier at C.

be easier, and it allows us to make some standard simplifications. Figure 13-2 shows a modulated rf signal. The unmodulated portion of the waveform is section A, and this is the amplitude of the rf signal if there is no audio signal applied to the modulator. The section marked B is the point when the negative peak of the audio sine wave reduces the carrier from its unmodulated value. At 100 percent modulation, this point would kiss the zero baseline on both sides (section B would reduce to zero width). The third section, marked C, is the amplitude when the positive peak of the modulating sinewave causes the rf carrier to increase. At 100 percent modulation, this peak would be exactly twice its amplitude in section A.

The pattern shown in Fig. 13-2 is representative of what we would see if we examined a amplitude-modulated rf carrier with an oscilloscope (see Chapter 21). The carrier signal is sampled (never applied directly to the scope!), and the sample is applied to the vertical deflection amplifiers (or directly to the vertical plates) of the oscilloscope. The carrier is modulated with an appropriate sine wave, and the scope time base is adjusted to show one or two cycles of the *modulating* frequency. The percentage of modulation can be measured from the relationship:

$$\% \text{ Mod.} = \frac{C - B}{C + B} \times 100\%$$

where C and B are as defined in Fig. 13-2.

☐ **Example:**

We examine the modulation envelope of a transmitter using an oscilloscope. It is found that the amplitude of C is exactly 7.5 cm, and the amplitude of B is 1.4 cm. What is the precentage of modulation?

$$\begin{aligned}\%\text{Mod.} &= (C-B)(100\%)/(C+B)\\ &= (7.5 - 1.4)\ (100\%)/(7.5 + 1.4)\\ &= (6.1)\ (100\%)/(8.9)\\ &= 68.5 \text{ percent}\end{aligned}$$

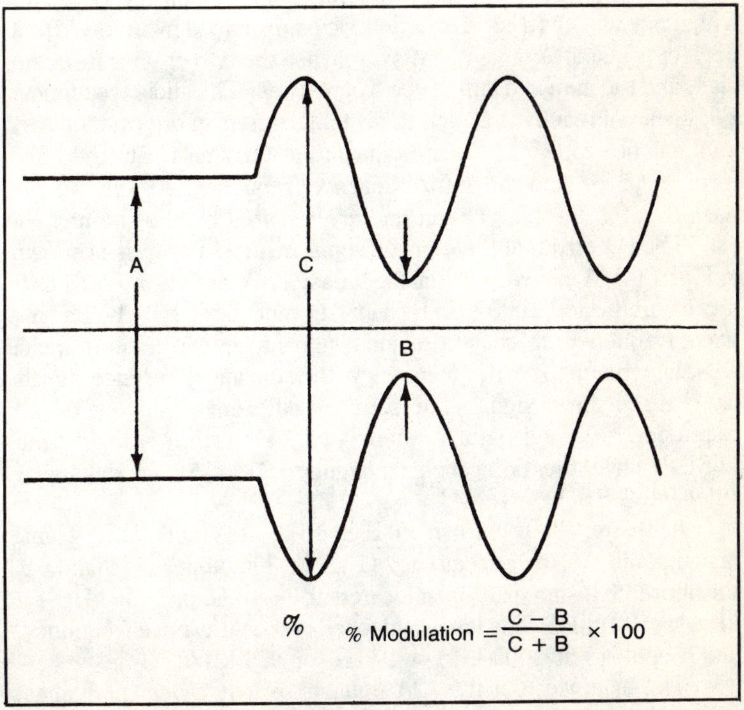

Fig. 13-2. AM Carrier with relationships.

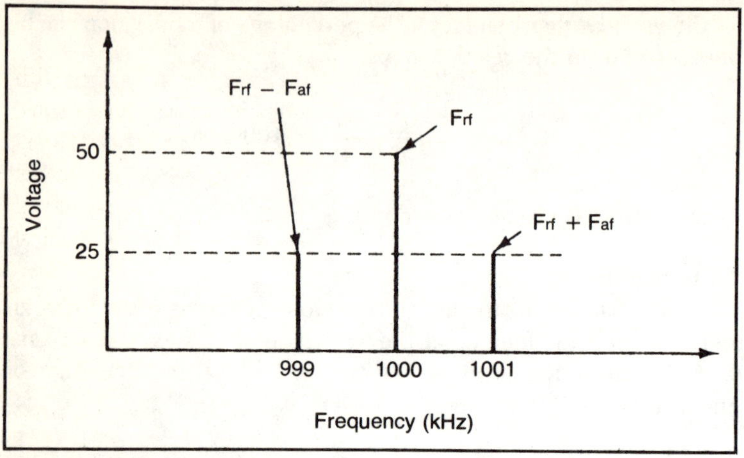

Fig. 13-3. Spectrum of an AM signal.

WHAT ARE SIDEBANDS?

Amplitude modulation is a multiplication process, like heterodyning, and this means that frequencies other than the rf carrier frequency are emitted. The frequencies produced in an ideal (that perfect transmitter again) AM system are the two original frequencies, and the sum and difference frequencies. The audio frequency, F_{af}, does not make it through the rf tank circuits in the transmitter, so it will not appear at the antenna output terminals. But the radio frequency, F_{rf}, and the sum/difference frequencies ($F_{rf} \pm F_{af}$) will pass through the tank circuit and are radiated by the antenna.

The spectrum of the emitted signal in an AM system is shown in Fig. 13-3. Again we are making it easy on ourselves by considering a hypotherical 1000-kHz transmitter modulated by a 1-kHz sine wave frequency. Because the audio signal frequency is very much less than the radio carrier frequency, the sum and difference signals lay close to the rf signal. The sum and difference frequencies are called *sidebands*. The sum frequency, $F_{rf} + F_{af}$, is the *upper sideband* (USB), while the difference frequency, $F_{rf} - F_{af}$, is the lower sideband (LSB).

In the example shown in Fig. 13-3, the modulating signal contains only a single frequency (1 kHz). The upper sideband will therefore be a single rf signal located at $F_{rf} + F_{af}$, or 1000 kHz + 1 kHz, or 1001 kHz. The lower sideband is the difference frequency, and it appears at 1000 kHz − 1 kHz, or 999 kHz.

The bandwidth of the AM signal is exactly twice the highest modulating frequency. In the simple case shown here, the modulating frequency is a pure, 1-kHz sine wave, so the bandwidth of the signal is 2 kHz.

Speech is not a simple sine wave, but a complex waveform with many different frequencies present. Even the most monotonic voice is actually a complex mixture of fundamental tones and overtones (harmonics). This is the reason why we can identify the voices of different persons, for each contains its unique combination of resonances and overtones. It is also why two different musical instruments sound differently when the same note is played. Consider the note A above middle C (440 Hz). If you play a sine wave of 440 Hz, the tone is pure, and there is little coloration, if any. Sine waves from different sources, but at the same frequency, will always sound the same. But when a trumpet plays a 440-Hz note A, it sounds different between these instruments is the harmonics present. Each instrument will produce its own collection of harmonics, thereby its own timbre, or coloration.

Figure 13-4 shows what the spectrum of a speech radiotelephone signal will look like. The sidebands spread out from the carrier a distance dependent upon the frequency response of the audio modulator in the transmitter. In normal radiotelephony, the spectrum of the audio signal is 300-Hz to 3000-Hz, so the signal will spead out 3 kHz above and below the carrier frequency. Because the lower limit is 300 Hz (and not DC), however, there will seem to be a dip in the signal between the upper end of the LSB and the lower end of the USB, with a peak right at the carrier frequency. This can be observed on any communications receiver equipped with a narrow (500-Hz) filter, called a CW filter).

The bandwidth of an AM signal modulated with speech is twice the highest audio frequency in the modulating signal. In the AM broadcast band, the modulator is allowed to pass 5000 Hz, so the modulation bandwidth of the signal will be 2 × 5000 Hz, or 10,000 Hz. Amateur radio transmitters which use AM are generally limited

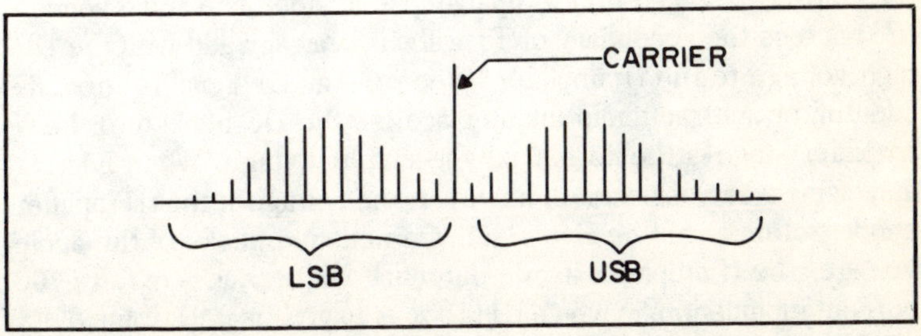

Fig. 13-4. Sidebands of an AM signal.

to 3-kHz audio bandwidth, so the bandwidth of the transmitted signal is double that, or 6 kHz.

MODULATION METHODS

To amplitude modulate an rf carrier, we must conspire somehow to change the amplitude of the rf carrier with the audio signal. It turns out that we can apply the modulating signal to almost any element of a vacuum tube rf power amplifier and achieve AM. Of course, these do not all result in equal performance, but they will produce at least some degree of amplitude modulation.

We can modulate any stage in the transmitter chain, except the oscillator. If we modulate the final amplifier, which is usually the best case, we call the system *high-level modulation*. If, on the other hand, we modulate some lower power stage, such as a buffer amplifier, we have *low-level modulation*. There are pros and cons to each system, but it is usually considered best to use high-level modulation, even though it requires a higher power modulator circuit. One problem with low-level modulation is that all rf power amplifier stages following the modulated stage must be linear, so we cannot take advantage of the high efficiency of class-B or class-C rf amplifiers in that case.

The plate, screen grid, control grid, suppressor grid, or cathode of the rf amplifier tube can be modulated. Figure 13-5 shows an example of high-level *plate modulation*. Yes, this is the final amplifier. This is probably the best form of modulation and is considered the most effective system.

The power amplifier portion of this circuit is the same as the power amplifiers discussed in Chapter 7. But note that the secondary of the audio modulator transformer is connected in series with the DC high-voltage lead to the plate of the rf amplifier. Transformer T1 is called a *modulation transformer*. The purpose of T1 is to match the impedance of the audio amplifier power output stage to the plate resistance of the rf final amplifier tube. The audio voltage (E1) across the secondary of T1 will algebraically add with the DC high voltage to the rf amplifier. When the audio signal is zero, no addition or subtraction is taking place, so the DC applied to the rf amplifier stage is the normal power-supply voltage. When E1 (the audio sine wave) is positive, however, the voltage at the rf amplifier anode potential will be $E_{DC} + E1$. On negative peaks of the audio voltage, the rf amplifier anode potential will be $E_{DC} - E1$. In 100 percent modulation, in which E1 is a sine wave, the peak value of E1 will be exactly equal to E_{DC}. Under this condition, the plate voltage

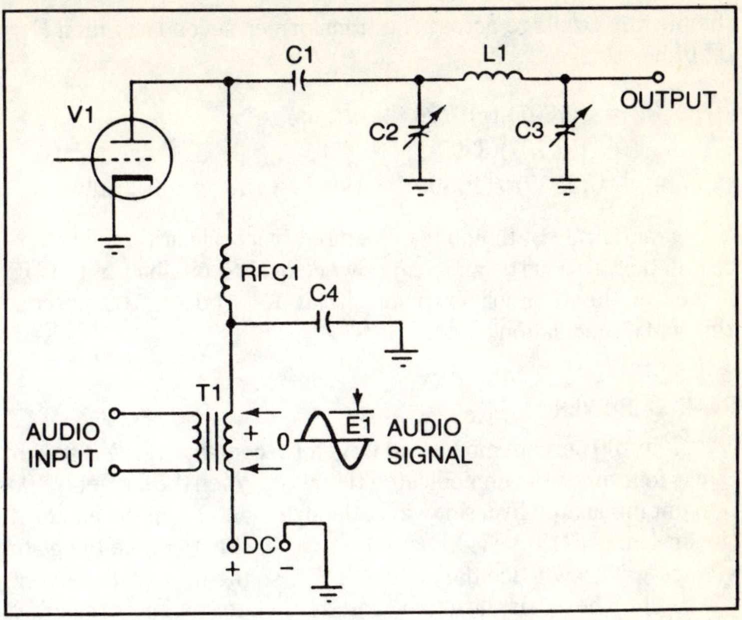

Fig. 13-5. Plate modulator.

on the rf amplifier will exactly double on audio peaks and drop exactly to zero on negative peaks. This voltage must not be allowed to go negative or go higher than $2E_{DC}$.

A certain amount of audio power is required to modulate the transmitter. We must know the plate resistance of the vacuum tube to develop an audio power that will create a voltage equal to E_{DC} across this resistance. The familiar expression $P = E^2/R$ is used.

☐ **Example:**

How much audio power is required of the plate modulator to produce 100 percent modulation of a 1-kW transmitter if the anode potential, E_{DC}, is 3300 volts and the anode current is 303 mA.

Find plate resistance R_p.

$$R_p = E_{DC}/I_{DC}$$
$$= (3300)/(0.303) = 10{,}890 \text{ ohms}$$

Calculate the amount of audio power needed to develop a peak voltage equal to E_{DC} across 10,890 ohms:

$$P_{af} = E^2/R_p$$

where P_{af} is the audio power in watts, E is the *rms* voltage across the secondary of the modulation transformer, and R_p is the plate resistance of the final amplifier tube. Note that since E_{DC} is equal to the peak AC voltage across the transformer secondary, then $E = 0.70E_{DC}$.

$$\begin{aligned} P_{af} &= (0.707E_{DC})^2/(10{,}890 \text{ ohms}) \\ &= [(0.707)(3300)]^2/(10{,}890) \\ &= (2333)^2/(10890) = 499.8 \text{ watts} = 500 \text{ watts} \end{aligned}$$

The general rule of thumb to remember from this arithmetic exercise is that it requires an audio power equal to one-half of the DC power in the rf amplifier plate circuit to produce 100 percent sinusoidal modulation.

PEAK RF POWER

The 100 percent modulated waveform contains a *peak* rf power that is four times the unmodulated rf power. When the carrier is 100 percent modulated by a sinewave, the peak voltage on the anode of the final amplifier tube is $2E_{DC}$ ($E1 + E_{DC} = 2E_{DC}$). Twice the plate voltage will draw twice the plate current, so the peak plate current will be $2I_p$, where I_p is the unmodulated DC plate current of the tube. By the ordinary power equation:

$$\begin{aligned} P &= IE \\ P &= (2I_p)(2E_{DC}) \\ P &= 4I_pE_{DC} \\ P_{mod} &= 4P_{unmod} \end{aligned}$$

This does not really tell us much about the average power in the waveform, however, and it is the average power that does the work. We can use the DC unmodulated power of the final amplifier to calculate the amount of audio power required to achieve any desired percentage of modulation:

$$P_{af} = \frac{M^2P_{DC}}{2}$$

where P_{af} is the audio power requirement in watts, M is the percentage of modulation desired, expressed in *decimal* (50% = 0.50), and P_{DC} is the DC input power to the plate of the modulated stage.

☐ **Example:**

How much audio power is needed to modulate a 5000-watt AM broadcast transmitter 85 percent?

$$\begin{aligned} P_{af} &= M^2P_{DC}/2 \\ &= (0.85)^2(5000)/2 \\ &= (0.73)(5000)/2 = 1{,}825 \text{ watts} \end{aligned}$$

CATHODE MODULATION

An example of a cathode modulator is shown in Fig. 13-6. In this circuit, the audio power amplifier is connected in series with the rf power amplifier DC path, such that the anode of the modulator tube is connected to the cathode of the rf power amplifier tube. An rf choke and a bypass capacitor are used to keep rf from the power amplifier tube from getting onto the plate of the modulator tube, and to maintain a low rf impedance to ground at the cathode of the rf amplifier tube. Cathode modulation is also called *series modulation* in some texts.

Modulation tube V2 is used as an electronically variable resistor. The resistance of its plate-cathode path is made to vary with the audio signal applied to the grid. Consider the situation when the modulating signal is a sine wave. On positive excursions of the wave, V2 will conduct harder, so its plate resistance increases. The plate resistance variations in the modulator tube serve to vary the plate current of the final amplifier tube. The plate milliammeter of the rf amplifier will not be affected because the plate resistance changes for V2 are approximately equal.

Cathode modulation can also be implemented with power transistors. If we replace modulating tube V2 with a transistor power amplifier capable of withstanding the voltages, we can then use solid-state audio amplifier circuits for the modulator. Like the vacuum tube version, it is necessary to operate the transistor in class-A.

The principal advantage of cathode modulation is low cost and low audio power requirements. A disadvantage is that there is a

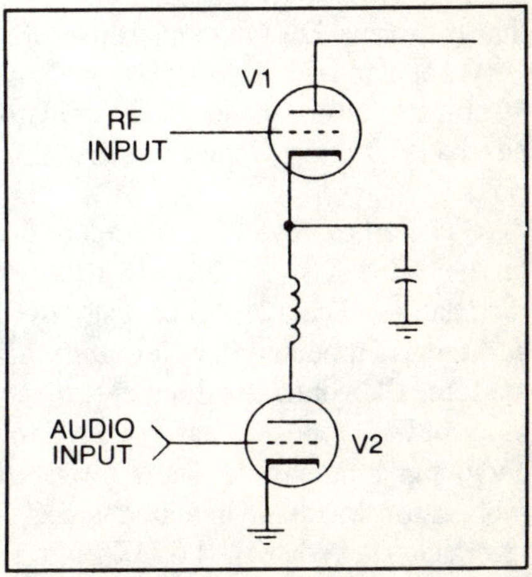

Fig. 13-6. Cathode modulator.

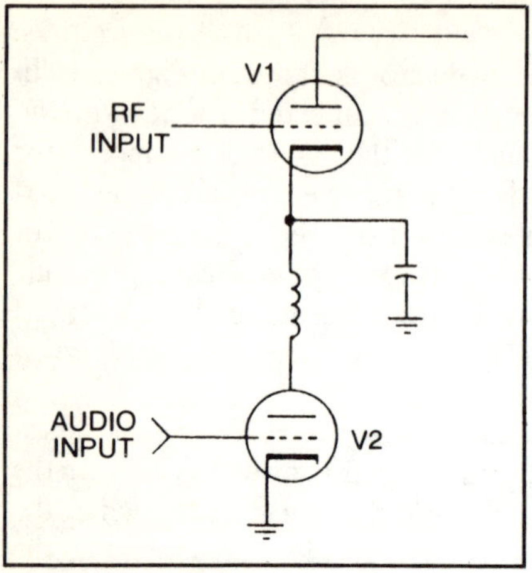

Fig. 13-7. Screen modulator.

lower rf efficiency in the power amplifier than could be obtained using the same tube in plate modulation. Also, in the cathode-modulated scheme, the DC power supply voltage divides between the modulator and rf amplifier tubes. This means that it is necessary to use a DC power-supply voltage of twice that required for the plate modulator.

SCREEN MODULATION

As shown in Fig. 13-7, an audio signal can be applied to the screen grid of the power amplifier. We need to apply a DC voltage to the grid that is approximately one-half of its normal unmodulated voltage, and then supply an audio modulating voltage that will cause the screen potential to peak up approximately 10 percent higher than its normal DC unmodulated value. Screen modulation was once very popular with in low-power "Novice" transmitters, and manufacturers often made the modulator an add-on accessory for purchase after the Novice licensee upgraded to at least the General class license.

The efficiency of the screen grid-modulated rf amplifier is low, on the order of 33 to 50 percent, depending upon the percentage of modulation. This can be partially overcome by using a version of screen grid modulation called *controlled carrier modulation*. In that system, DC control voltage is created from the audio signal at the input of the modulator, and is used to reduce the carrier when the audio signal amplitude has a low average amplitude. This form of modulation was used in a long series of Heath beginner's transmitters that started with the DX-20 in the mid-1950s.

Other forms of grid modulation include suppressor grid modulation and control grid modulation. The latter is sometimes called *efficiency modulation*.

HEISING MODULATION

The *Heising*, or *constant current, modulation* scheme is shown in Fig. 13-8. This circuit is a lower cost variation on the plate modulation scheme, and it cannot attain full 100 percent modulation of the rf carrier. This circuit operates on the principle of current division from a constant current source. The DC plate currents from rf final amplifier tube V1 and modulator tube V1 must both pass through a high-value inductor, L1, in the DC power-supply line. The inductance of this choke will tend to maintain a constant current when the current through it changes. The current in L1 (current I3) is the sum of two different currents: I1 (the V1 plate current) and I2 (the V2 plate current), or I3 = I1 + I2.

Vacuum tube V2 operates as a class-A audio power amplifier and is considered here as an electronically variable resistor. The plate resistance of V2 changes in proportion to the audio signal applied to its control grid. When the audio signal is making a positive

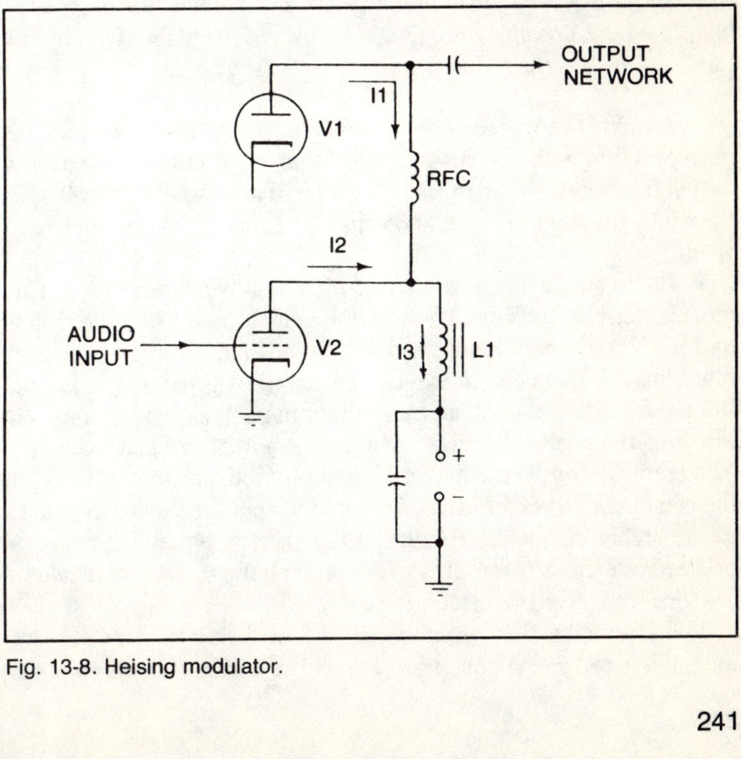

Fig. 13-8. Heising modulator.

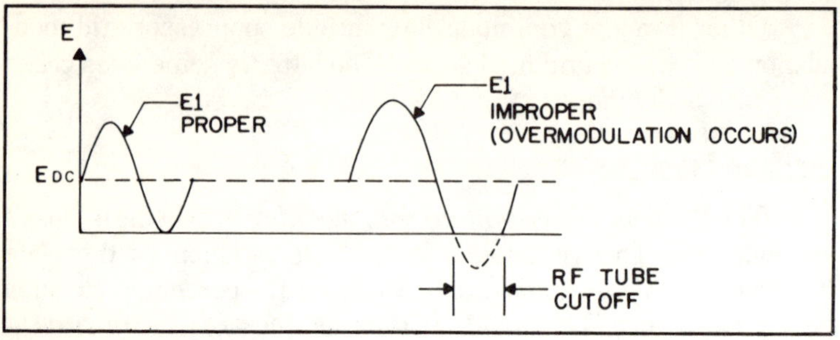

Fig. 13-9. Overmodulation.

excursion, the current in the anode of V2 increases. Because I3 is a constant, this action can occur only by taking some current away from I1. So I1 will drop when the audio signal is making its positive excursion. Similarly, when the audio signal is on a negative excursion, current I2 will drop, allowing I1 to take on a higher value. Current I1 will then increase on negative excursions of the audio signal and decrease on positive excursions.

OVERMODULATION AND UNDERMODULATION

Neither overmodulation nor gross undermodulation is desirable. The best possible situation is to always maintain 100 percent modulation. This is seldom possible in practical situations, however.

Perhaps the worse of these problems is overmodulation. The main problem with overmodulation is that it will generate spurious output frequencies that are called *splatter*. These hash rf signals spill over into the next channel and cause interference with other services.

The major cause of overmodulation is shown in Fig. 13-9. The modulating voltage should be exactly equal to E_{DC} when the transmitter is 100 percent modulated. When the modulator is driven into more than 100 percent, however, the voltage is greater than E_{DC} on the modulation peaks. This means that the actual plate voltage on the negative peak will be less than zero—a negative plate voltage. The rf amplifier tube is cut off during this period, and that will distort the rf modulated waveform and create the splatter. Some transmitters, notably CB models (overmodulation is rife on the CB because of the widespread use of power microphones), use modulation limiters to prevent overmodulation.

Undermodulation often occurs when trying to 100 percent modulate a tetrode rf power amplifier. The cure for this problem is

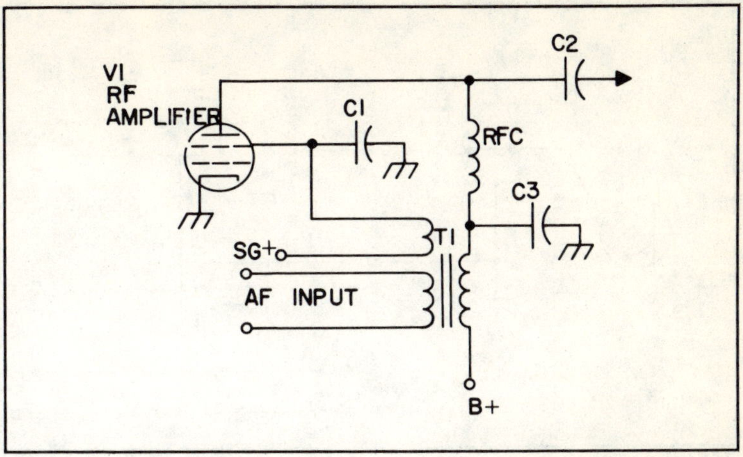

Fig. 13-10. Modulation of the screen grid in plate-modulated tetrodes.

to also modulate the screen grid, along with the plate. Figures 13-10 and 13-11 show two methods for accomplishing this goal. In Fig. 13-10, a tertiary winding is used on the modulation transformer. Most of the circuit is like the ordinary triode plate modulator discussed earlier. The secondary of the modulation transformer is connected in series with the B+ high voltage to the anode of the rf power amplifier. The operation thus far is exactly the same, but note that there is another secondary winding. This is the tertiary winding, which is connected in series with the screen grid SG+ DC power supply. Capacitor C1 is used to keep the screen grid at a low

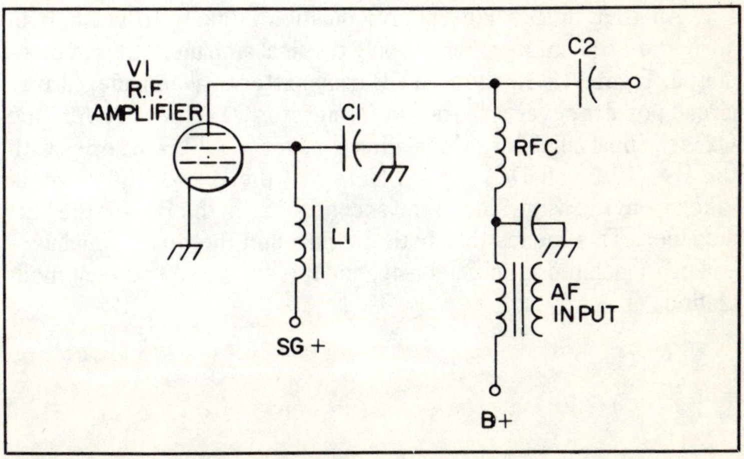

Fig. 13-11. Screen grid modulation.

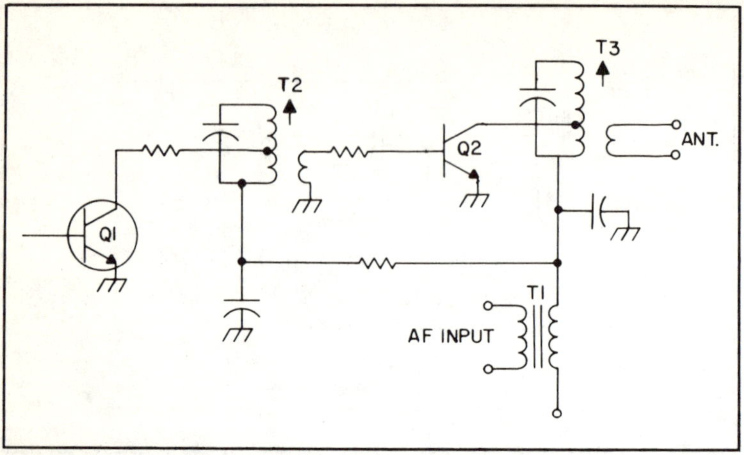

Fig. 13-12. Modulation of transistor power amplifiers.

rf impedance, as is normally required in tetrode power amplifiers. The added modulation of the screen grid will allow the amplifier to be 100 percent modulated.

A second approach is shown in Fig. 13-11. In this case, the SG+ DC power supply is passed through an iron core inductor, L1. Current variations that will normally occur in the screen circuit under plate modulation conditions create a varying voltage across L1 as the choke tries to maintain a constant current. These voltage variations are effectively in series with the SG+ supply and will therefore tend to modulate the screen grid. This circuit also allows full modulation of the tetrode vacuum tube.

Another time when undermodulation is likely to occur is in solid-state transmitters where only the final amplifier stage is modulated. Figure 13-12 shows a CB transmitter final amplifier/driver circuit popular several years ago. Transistor Q1 is the rf driver, and Q2 is the final amplifier. Modulation transformer T1 is in series with the B+ (12.5-volt DC) supply. Note that the B+ for the driver is taken from the same side of the secondary as is the B+ for the final amplifier. This means that both the final and the driver amplifiers will be modulated simultaneously, and this allows 100 percent modulation, or nearly so.

Chapter 14

Double And Single Sideband

Two decades ago the amateur radiotelephone subbands were alive with AM signals. An occasionally heard single sideband (SSB) operator would be met with some unkind comments from other users of the band. Today, the superiority of SSB has been given it the edge over AM by, shall we say, a larger margin. There is practically zero AM activity on the amateur bands, at least from radio amateurs (Radio Moscow and other international broadcasters still pollute the 40-meter amateur band). Commercial point-to-point radio stations have used SSB even longer than amateurs, and well over a decade ago the medium-wave (2 to 3 MHz) marine channels went SSB. In fact, it is no longer legal to operate an old AM transmitter in the medium-wave marine radiotelephone band. All pleasure boaters are required to use VHF FM for close-to-shore communications. They only qualify for an SSB license if they plan to go offshore any great distance.

The citizens band market is also heavily SSB, even though there seems to be a greater reluctance to "go Donald Duck" among CBers. This will possibly change as the cost of CB SSB rigs drops drastically. Now, however, the CB market is split between AM and SSB users. There have been some gentlemen's agreements regarding SSB and AM channel assignments, but these are not only unofficial, they are also disregarded by as many who follow them.

WHAT ARE DSB AND SSB?

In Chapter 13 we considered amplitude modulation (AM). SSB and DSB are merely specialized forms of AM. Consider Fig. 14-1.

The spectrum plot shown in Fig. 14-1A is for a normal AM transmitter output. The carrier signal has the greatest amplitude and is accompanied by two sets of *sidebands* (LSB and USB). Sidebands are mixing products between the carrier frequency and the modulating frequency, representing sum and difference components. For example, suppose we modulate a 14,300-kHz rf carrier with a 1 kHz sine wave. The output spectrum of the transmitter will contain three different frequencies: the carrier (14,300 kHz), the sum (13,301 kHz), and the difference (14,299 kHz). The sum and difference frequencies are the *sidebands*. When a complex speech waveform is used to modulate the carrier, the output spectrum would contain the carrier and one sum/difference product for each sine wave frequency present in the complex speech waveform. This is the situation shown in Fig. 14-1A. The sum products form the *upper sideband* (USB), while the difference products form the *lower* sideband (LSB).

There are two major problems with the AM signal. One is that it wastes needed spectrum space, and the other is that it wastes transmitter power. In the first of these, we must note that the bandwidth of the AM signal is twice the highest modulating frequency. If, in an amateur transmitter, the modulator passes frequencies as high as 3000 Hz, the bandwidth needed to transmit that signal is 2 × 3000 Hz, or 6000 Hz (6 kHz). But note that the same information is repeated in *both* sidebands. Only *one* of these sidebands is needed to convey all of the information present in the signal; transmitting two identical (but mirror image) sidebands is redundant.

The transmitting of the carrier and two sidebands is also a waste of transmitter final amplifier power. The carrier and one sideband are freeloaders—they are not needed to convey the information in the modulating signal. True, the carrier *is* needed to demodulate the signal at the receiver, but, provided that the carrier can be reinserted locally in the receiver, it can be eliminated at the output of the transmitter. For a transmitter that is 100 percent modulated with sinusoidal audio signal, the voltage vectors representing the upper and lower sidebands each have an amplitude of 50 percent of the carrier voltage vector. Power is proportional to the square of the voltage, so the total power contained in each sideband is only $(½)^2$, or ¼ of the power in the total output signal. Each sideband contains one-fourth of the total power, and the carrier consumes (gobbles up, is more like it) one-half of the total power. This tells us that only one-fourth of all

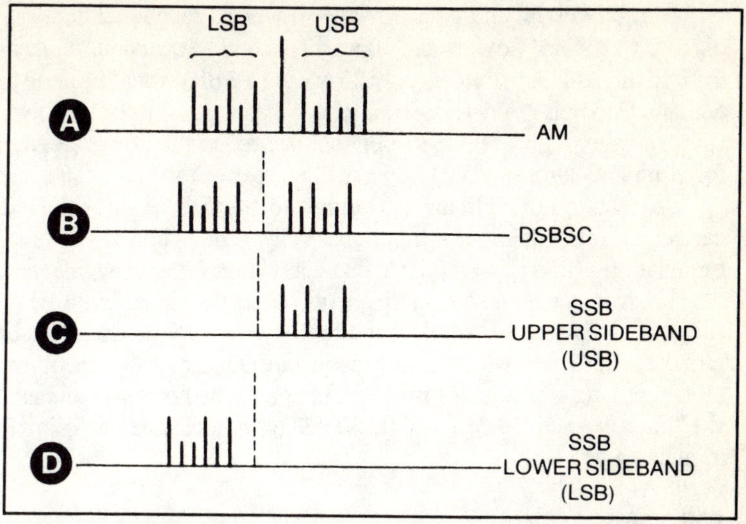

Fig. 14-1. AM, DSB, USB, and LSB spectrum.

power transmitted is used to send the information (the modulating signal); three-fourths are totally wasted.

We gain some of this power back, although none of the wasted spectrum space, by using a *double sideband suppressed carrier* (DSBSC) techniques, as shown in Fig. 14-1B. In this situation, the carrier is suppressed as much as possible, and only the remaining sidebands are transmitted. This means that one-half of the power in the rf output waveform is actually used send the information, and the other half is redundant.

The two *single sideband suppressed carrier* (SSBSC) signals of Figs. 14-1C and 14-1D go a long way toward completely solving the problem. These signals take care of the power waste and the spectrum waste. The bandwidth of each is only one-half that of the AM signal in Fig. 14-1A and the DSBSC of Fig. 14-1B.

There are actually several advantages to the SSB (the term SSBSC is correct, but is usually shortened to SSB in the jargon) signal. We have already noted the reduction in required spectrum space. This is the basis, incidentally, for the claims of some CB manufacturers to having 80 channels. They don't really have 80 channels; they have the regular 40. But it is *theoretically* possible for an LSB carrier to coexist on the same channel as a USB carrier. This is actually not always true in practice, however, and presumes that both transmitters have locked oscillator frequencies, and that the receivers have very *steep* filter passband characteristics.

Our power advantage is obvious. If we are capable of generating 100 watts of rf power, all 100 watts is used to transmit information. In the AM transmitter, only 25 watts (−6 dB) would be used to actually transmit needed information. We can either use the power that we have to greater effectiveness, or reduce the power needed for communications, which is what the letter of the law requires.

We get one additional advantage to the SSBSC signal: The carrier is not present, so the signal *disappears* when there is no modulation present. On today's crowded bands, this is significant. An AM transmitter produces a carrier all of the time, even when there is no speech present. This creates *heterodynes* when two or more carriers are within any reasonable distance of each other. Heterodyning accounts for the loud, unspeakable howl present all of the time (even at 3 AM) on CB; 20 million AM carriers are simply impenetrable!

DSB/SSB GENERATORS

The same circuit used to generate the DSB signal will also start the SSB signal on its way. The first job in both cases is to remove that darn carrier. There is one way to accomplish this noble goal: the *balance modulator*. Several different types of balanced modulator are in common use, but they all accomplish the same thing: They null the carrier out of the AM signal. If the transmitter is a DSBSC type, the output of the balanced modulator can be sent on its way to the output final amplifier. In fact, there were a few designs that used high-level balanced modulators that *were* the final amplifiers. If it is an SSBSC transmitter, however, some additional signal processing is needed (filtering or phasing) in order to rid ourselves of the unwanted sideband. The filtering and phasing methods are considered in the next chapter. In this chapter we will consider only the generation of the raw DSB signal that become the basis for the SSB signal.

BALANCED MODULATOR CIRCUITS

The goal of any balanced modulator is to remain in balance to the rf carrier signal when there is no audio present, and to then become unbalanced to the rf signal when audio is applied. There are many different ways that this is accomplished, but the most common are present here.

Many balanced modulator circuits operate as electronic switches that will toggle in step with the audio signal. When there is no audio signal present to toggle the electronic switches, then the rf

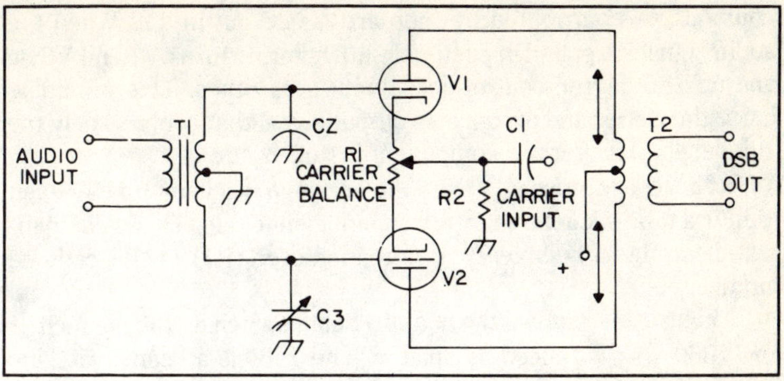

Fig. 14-2. Tube balanced modulator.

output is zero. Such switches can be placed either in series with, or in shunt across, the signal line between the oscillator and the stages that follow.

We will no longer long on vacuum tube circuits, but they must be covered at least in passing in sort of a genuflection to the older SSB transmitters still ambling around the used equipment and hamfest markets. Figures 14-2 and 14-3 shows two popular vacuum tube balanced modulator circuits. Figure 14-2 is based on a pair of triodes. In most rigs that used this system, the two triode tubes were separate sections of a dual triode, such as 12AU7, 12AX7, etc. The audio signal is applied to the primary of push-pull transformer T1 and drives the grids in push-pull (180 degrees out of phase). The plate circuit of the balanced modulator is also connected in push-pull, but in this case, transformer T2 is an rf transformer. To maintain the best balance, this transformer is usually constructed with bifilar windings for the primary. Each half of the primary is wound (interleaved) with the other half. In other words, turn No. 9 (as measured away from the center tap) of the top half of the primary is next to turn No. 9 of the bottom half of the primary.

The carrier oscillator signal is fed to the common connection of the tube cathodes. Capacitors C2 and C3 are selected to present a low impedance to the oscillator frequency and a high impedance to the modulating frequency. It is important that C2 = C3 and that they be matched closely. These capacitors place the grids of V1 and V2 at rf ground potential, so the oscillator thinks that it is looking into a grounded-grid push-pull stage. The oscillator signal generates two different plate signals that are in phase with each other. Because they combine from different ends of a push-pull transformer, however they cancel each other. As long as V1 and V2 are balanced, the components contributed by tube are nulled out in T2. When the audio signal is applied in push-pull, however, it turns V1 and V2 on and off 180 degrees out of phase with each other. This will unbalance the circuit and produce an output signal that contains only the sidebands. The carrier component is nulled out.

There are normal variations between vacuum tubes, even identical tube sections sharing a common envelope. This is compensated for the *carrier balance* potentiometer (R1) in the cathode circuit.

Figure 14-3 shows the use of a beam switching tubes, such as the 7360, as a balanced modulator. The 7360 is a beam switching tube because it uses a common cathode-grid structure to supply an electron stream to each of two different anodes. The beam is switched to one anode or another by the potential applied to the beam deflection plates. In the circuit shown in Fig. 14-3, plate No. P1 is controlled by beam deflection plate B1, and P2 is served by B2.

The audio signal is applied in push-push to the beam deflection plates, while the oscillator is applied to the control grid. In many circuits, the oscillator was designed to use the cathode-control grid-screen-grid portion of the 7360 as a triode oscillator circuit. In this particular circuit, however, the oscillator is external to the 7360, and is applied through a coupling capacitor to the control grid.

When the audio signal is zero, the potential on B1 is then equal to the potential on B2, so the electron beam is split equally between the two anodes. This sets up equal currents in the primary of transformer T2. This is a push-pull primary transformer, so current I1 is equal to I2. The net current current is I1-I2, so under zero audio conditions, the net current is zero. No signal will appear in the output circuit because of the cancellation of I1 and I2.

When an audio signal is applied, however, it drives the deflection plates in push-pull to create a 180-degree out of phase condition. The electron beam will be switched back and forth between the anodes, causing unequal currents to flow in the primary of T2. Because there is no longer total current cancellation, there will be an output signal from the secondary of T2. Although the circuit action is very similar to that of the triode case, it is still possible to obtain theoretically better suppression of the carrier because of the closer match in the characteristics of the two halves of the circuit.

Bipolar or field-effect transistors can also be used in the same kind of push-pull arrangement. A number of balanced modulators have been designed over the years, but all lacked some of the ability

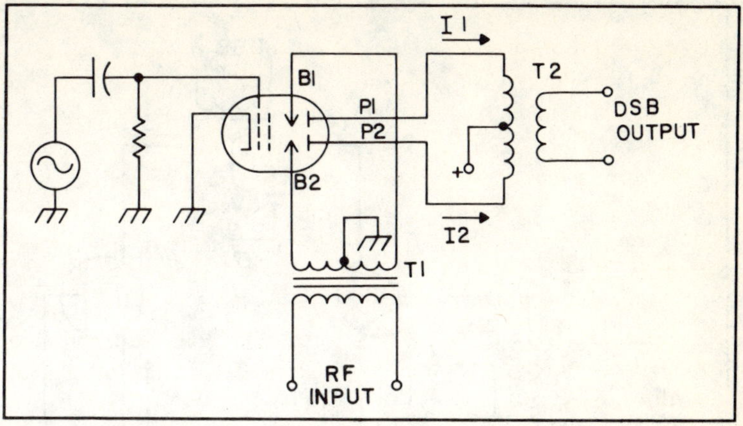

Fig. 14-3. Another tube balanced modulator.

to suppress as much of the carrier as the 7360 tube—until integrated circuits took over. The problem had been the difficulty in obtaining a really good balance with discrete transistors of either type. But this problem is solved when all transistors are formed at the same time from the same batch of silicon on the same *substrate*. Not only are the initial conditions better in a properly made IC, but two transistors in the push-pull amplifier will tend to track each other more closely when the environmental temperature changes. Both transistors share the same thermal environment, so they will see the same changes equally.

Figure 14-4 shows an IC balanced modulator circuit. The three transistors are formed into a *differential amplifier* configuration. This type of IC is very popular with rf designers and is made by Motorola, RCA, Fairchild, and others. An example might be the RCA CA3028 device. Transistor Q1 is used as a constant current source, so I1 remains constant for a fixed input voltage. Currents I2 and I3 are the emitter currents for the differential pair, Q2/Q3. The collector current for these transistors are I4 and I5, respectively. In a differential amplifier circuit, current I1 = I2 + I3. Because I1 has a constant peak value determined by the amplitude of the oscillator signal applied to the base of Q1, one of these currents can be increased only at the expense of reducing the other. Similarly, I4 and I5 can be increased only at the expense of decreasing the other current.

The audio signal is applied in push-pull to the bases of the differential pair Q2/Q3. When the audio signal is zero, these two transistors are equally biased, so their respective collector currents

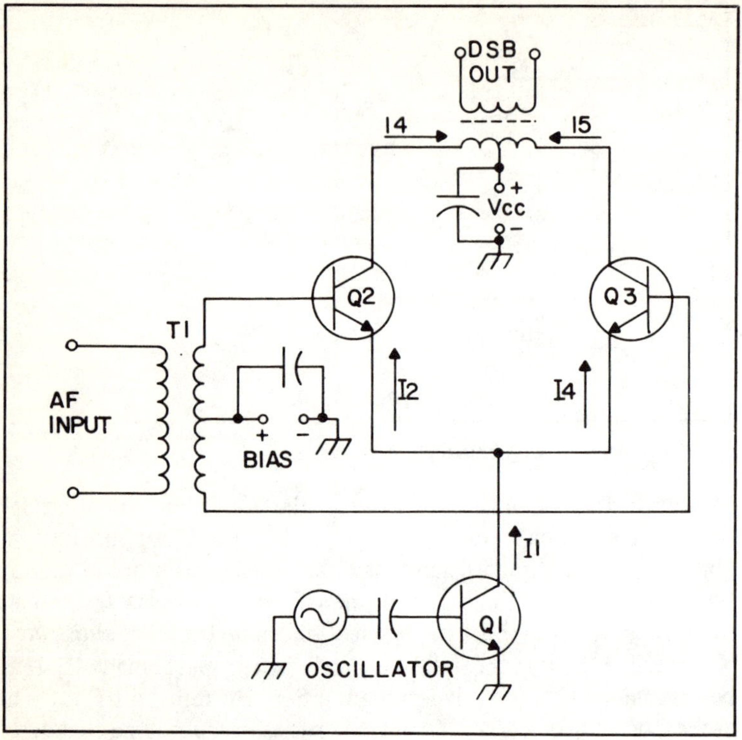

Fig. 14-4. IC differential amplifier balanced modulator.

are equal to each other. Because output transformer T2 has a push-pull primary, these currents cancel each other. The output is therefore zero. When the audio signal is not zero, however, one transistor turns on and the other turns off during each half of the AC signal cycle. The situation reverses on each successive half cycle. This action is analogous to the action in the triode circuit discussed earlier. The net result is that the output is nulled when only the carrier exists and will produce only a DSBSC output when the audio is applied to T1.

The balanced modulator is little more than a switching circuit, so diode switches can be pressed into service in balanced modulator circuits. Figure 14-4 shows two basic types of diode balanced modulators. The diodes used can be well matched discrete diodes or IC diodes. Several manufacturers make diode arrays for this type of service, but the most common is the RAC CA3019 device. All of the diodes needed for this circuit are contained in a single metal can

IC package. If discrete diodes are used, they must be well matched. For simple amateur experiments, matching of the forward and reverse resistances on an ohmmeter is sufficient, but the best performance will occur if they are matched on a curve tracer.

The circuit in Fig. 14-5 shows a simple series switch circuit. The circuit is set up so that no currents will flow in D1 or D2 when the audio signal is not present. The output is zero. When the audio is applied, however, these diodes will conduct out of phase with each other. On the positive peak of the audio signal, diode D2 is forward biased, so current will flow. Diode D1 is cut off on that half cycle, so I1 is zero. On the second half cycle, the audio signal is negative, so diode D2 is cut off and D1 is forward biased. In that case, I1 will flow in the primary of T2, and current I2 is zero.

Two controls aid in nulling the carrier in this circuit. Potentiometer R1 and trimmer capacitor C2 are adjusted together (they are often very interactive in practice) for the best obtainable null at the output under zero audio signal conditions. The carrier is monitored at the output of the transmitter or at the next stage in cascade with the balanced modulator.

The circuit in Fig. 14-6 uses four diodes and would make more use of the RCA CA3019 diode array. The rf carrier applied to this circuit has an amplitude that is approximately 10 times the amplitude of the audio signal. The oscillator signal is applied through transformer T3, while the audio signal is applied through transformer T1. The output signal is taken from transformer T2. The oscillator signal controls the switching action of the diodes because of its

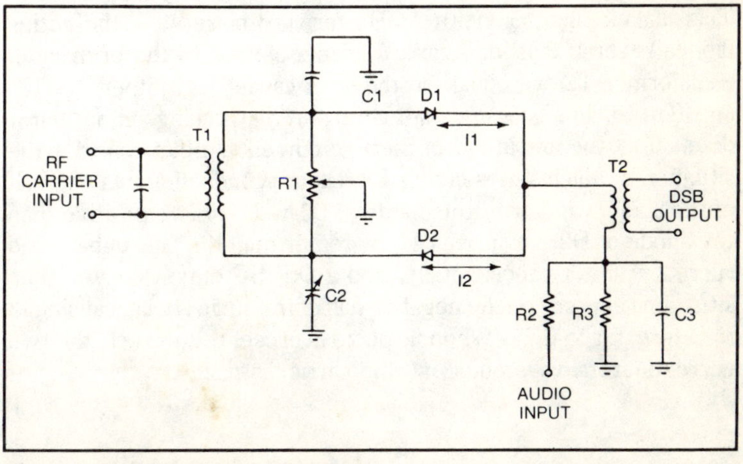

Fig. 14-5. Diode balanced modulator.

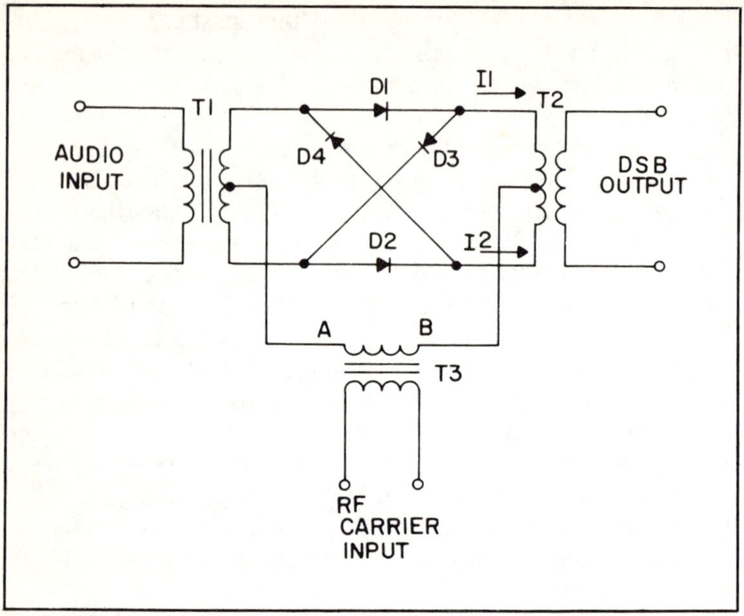

Fig. 14-6. Diode balanced modulator.

greater amplitude. On one-half of the rf carrier sine wave, point A will be positive with respect to point B. Under this condition, series diodes D1 and D2 are forward biased, causing equal currents to flow in the primary of transformer T2. These currents null each other out when there is no audio signal present. On the second half cycle, the polarity reverses, so point A is negative with respect to point B. This makes shunt diodes (D3/D4 forward biased and the series diodes reverse biased. Again, currents applied to the primary of transformer T2 are equal, so they will cancel each other.

The audio signal does not control the switching action, but it does affect the amplitude of the two currents. Let's consider the situation in which series diodes D1/D2 are conducting. On one-half of the audio waveform, the anode of D1 will be more positive than the anode of D2, so current I1 is greater than I2. This unbalanced current will not cancel totally, and a DSBSC output occurs. The situation reverses on the negative half of the audio signal, causing I2 to be greater than I1. When no audio is present, however, the two currents are always equal, so the carrier is nulled.

Chapter 15

SSB Transmitters

Radio amateurs use single sideband transmitters almost exclusively in the high-frequency range and use it extensively in the VHF/UHF region. There are two methods for generating an SSBSC signal: *filtering* and *phasing*. A balanced modulator is used to suppress the carrier (see Chapter 14). In the filtering method the unwanted sideband is sliced off with a sharp bandpass filter. In the phasing method the unwanted sideband is mulled to zero by producing it in two versions that are 180 degrees out of phase.

FILTERING METHOD

Figure 15-1 shows the block diagram to a filter-type single sideband transmitter. The carrier oscillator, speech amplifier, and balanced modulator are the same circuits as were discussed in Chapter 14. The DSBSC output is applied to the input of a very sharp bandpass filter that is designed so that only *one* of the sidebands will pass. The other sideband is attenuated. Following the filter is a bandpass buffer amplifier, sometimes called the *transmit i-f* amplifier. Standard practice is to design SSBSC transmitters such that the generation of the SSB signal occurs at one frequency, which will then be heterodyned to the operating frequency.

There are two methods for creating separate LSB and USB signals. Recall these signals appear on opposite sides of the carrier frequency, so they occupy slightly different frequency bands. One technique for separating which sideband is to be transmitted is to

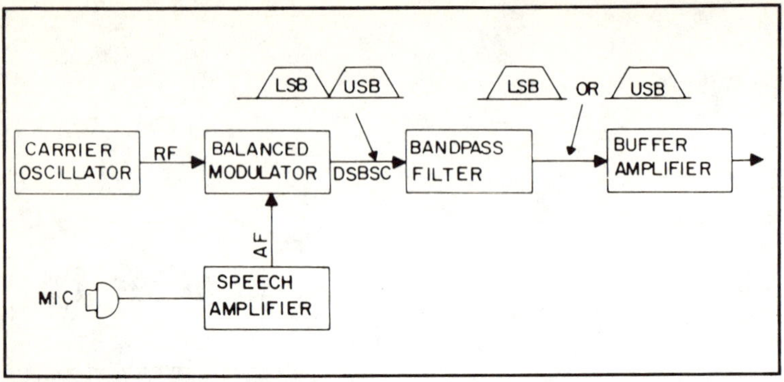

Fig. 15-1. Filtering method for SSB generation.

use filters of different frequency. The center frequency of the LSB filter will be approximately 1.5 to 1.8 kHz *below* the carrier oscillator frequency, while the center frequency of the USB filter is at a frequency the same distance *above* the carrier frequency.

But this method is *expensive*! The cost of good quality bandpass filters is quite high, so this method almost doubles the price of the filter system. It is used in some military and commerical transmitters, but is rarely if ever used in amateur radio transmitters.

In amateur radio transmitters, CB transmitters, and some other commercial/military transmitters, the *carrier* frequency is actually shifted above and below the passband of the single filter used. If the filter passband is 3 kHz, it will extend 1.5 kHz above and below the center frequency. Accounting for the 300-Hz lower limit of the audio signal, the two carrier frequencies would be 1.8 kHz above and below the filter center frequency. The LSB carrier oscillator crystal will be located 1.8 kHz *above* the center frequency of the filter, while the USB carrier oscillator frequency is 1.8 kHz *below* the filter frequency. It is then a simple matter of changing crystals in the carrier oscillator in order to select LSB or USB modes.

Filter Types

The principal characteristics required of the filter used in the single sideband exciter are relatively flat response inside of the passband an a very steep rolloff for frequencies outside of the passband. Three types of filters are commonly used: *LC, mechanical*, and *crystal*. The LC filters are normally used in SSB generators that operate in the 40-kHz to 500-kHz range. They consist of several tank circuit sections that shape the response.

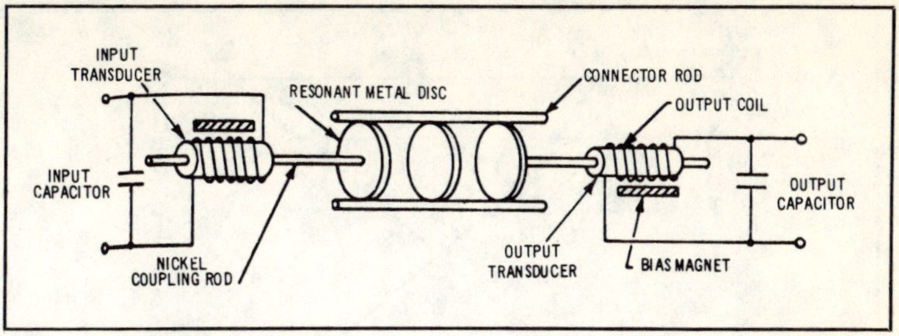

Fig. 15-2. Mechanical filter.

A simplified mechanical filter is shown in Fig. 15-2, and a representative frequency response characteristic is shown in Fig. 15-3. Most mechanical filters are made by Collins/Rockwell and operate in the 200-kHz to 500-kHz frequency range. Mechanical filters are smaller than LC filters, relatively rugged, and have a Q factor that is far greater than that obtained by LC filter sections. The major components of the mechanical filter shown in Fig. 15-2 are the input and output transducers (consisting of moving coils and bias magnets), resonant metal discs, nickel coupling rods, and input/output capacitors.

The transducers are used to convert electrical/mechanical energy. On the input side, the transducer converts electrical energy from the DSBSC source to mechanical energy that can vibrate the resonant discs. On the output side, the transducer converts this mechanical vibration back into electrical signals. The basic operation of the transducers is the same as that of the permanent magnet loudspeaker or the dynamic microphone. When a

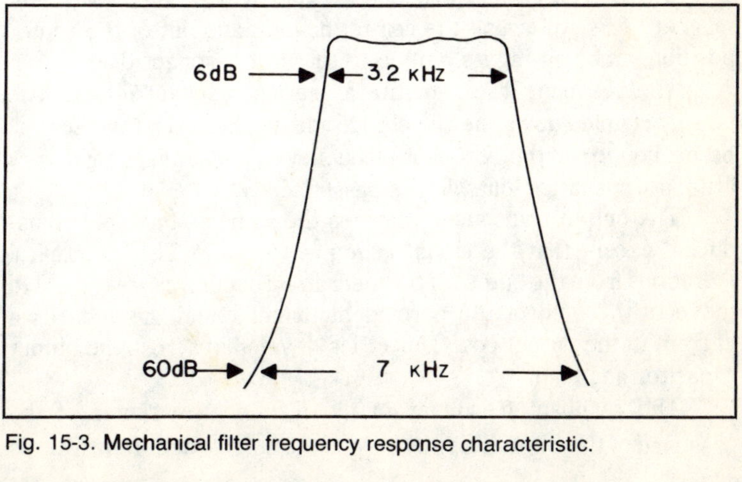

Fig. 15-3. Mechanical filter frequency response characteristic.

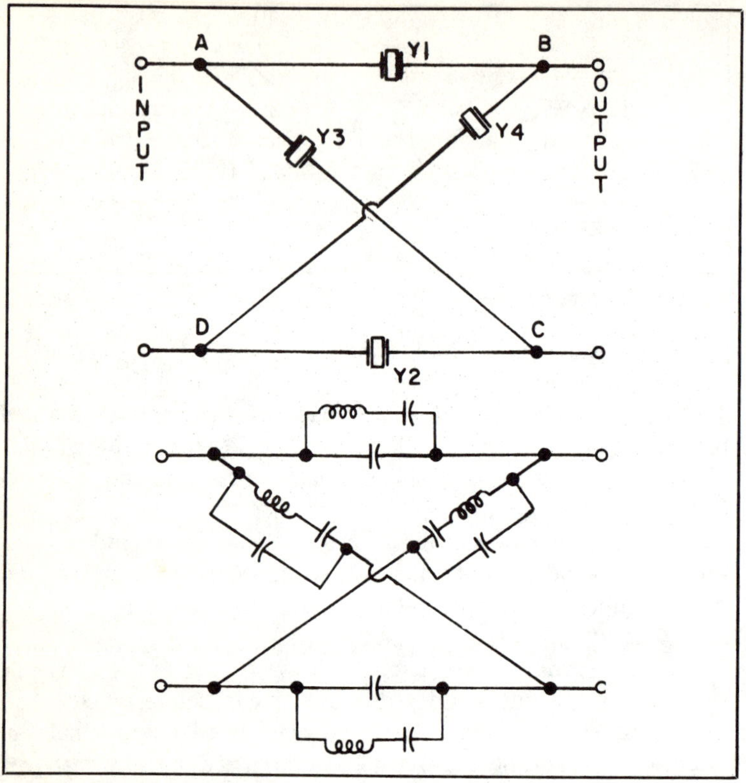

Fig. 15-4. Crystal lattice filter.

signal is applied to the input transducer, it sets up a magnetic field that will alternately aid and oppose the field of the permanent magnet. This will cause the coil to move in and out of the center position, transmitting its motion to the first resonant disc.

The resonant discs operate as series resonant filters. The energy coupled down the nickel rods and mechanical filter discs will be peaked up at the resonant frequency of these discs, and very little is transmitted outside the passband.

The output transducer operates the same as the input transducer, except that the circuit action is opposite. The mechanical vibrations from the output of the mechanical section cause the coil to move in the field of the permanent magnet, and this induces a current in the output coil. This coil is developed across the output capacitor and load.

The resonant frequency of the discs is the center of the passband of the filter. The measure of the selectivity of the filter is the *skirt factor*, which is defined as the ratio of the passband at –60 dB to the passband width at –6 dB. In the filter shown in Fig. 15-3, the –6 dB bandwidth is 3.2 kHz, and the –60 dB bandwidth is 7 kHz. The skirt factor, then, is 7/3.2, or 2.2. This is the normal skirt factor for a mechanical filter with six discs. For the seven-disc and eight-disc filters, expect skirt factors of 1.85 and 1,5, respectively.

The input and output load capacitors are used to resonate the coils in the input and output transducers to the frequency of the discs. In many circuits, these capacitors are variable, so that the passband can be peaked to various load and source capacitance circuits. There is no guarantee, for example, that the input of the amplifier following the output transducer will have a standard capacitance that will resonate the coil when a fixed capacitance is used.

Another popular filter is the piezoelectric crystal lattie filter, shown schematically in Fig. 15-4A. The equivalent circuit is shown in Fig. 15-4B. This is a single-section filter. Actual crystal filters used in SSB transmitters might use two or more sections. Actually, two sets of crystals are in this circuit. Crystals Y1 and Y2 form a series pair, which must be matched to the same frequency. A shunt pair Y3/Y4, must be matched to each other at a second frequency. The frequency difference between the series and shunt pair is the effective bandwidth of the filter. This can be visualized in Fig. 15-5, which graphs the effective reactances of the two pairs of crystals as a function of frequency. These curves are combined to form the passband characteristic of the filter which is shown in Fig. 15-6.

The graph in Fig. 15-6 is somewhat idealized. In actual practice, it is difficult to obtain matched pairs of quartz crystals that are as close as might be desired. Except for commercial filters of high quality, there is usually a lopsided passband characteristic that

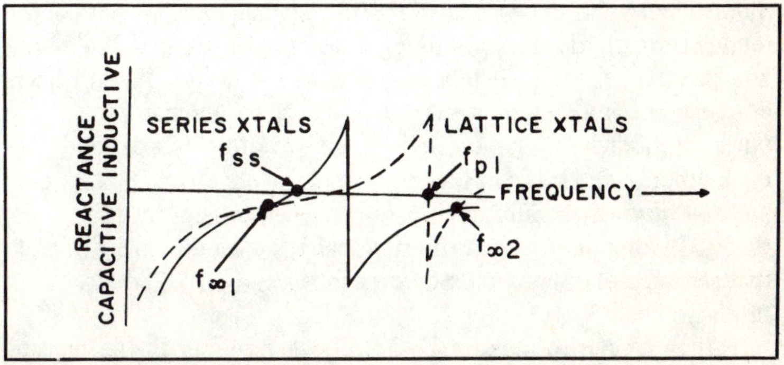

Fig. 15-5. Crystal filter properties.

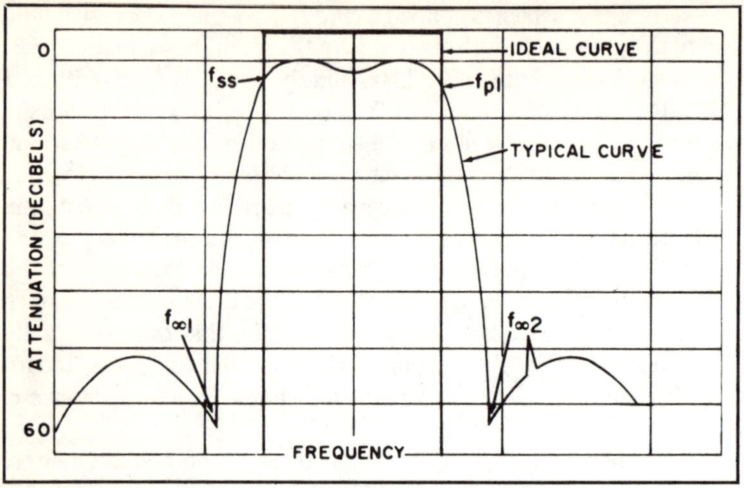

Fig. 15-6. Crystal filter frequency response characteristic.

should be idealized as much as possible. In some circuits, tuning inductances or capacitances are used to compensate for some of the problems.

PHASING METHOD

The phasing method is not quite as popular as it once was, because of a substantial decrease in the cost of filters and the relative simplicitity of the filter system. Phasing was the first sytem selected by many transmitter manufacturers, and is still seen occasionally. In the phasing system, as shown in Fig. 15-7, the audio and carrier signals must be split into two quadrature components each. Quadrature components are out of phase with each other by exactly 90 degrees. These signals are then combined in a pair of balanced modulators to produce a pair of DSBSC signals. The outputs of the modulators are then mixed with each other at the input of the output amplifier. Sideband selection is a matter of phasing. When LSB is desired, its components from the two balanced modulators will be in phase with each other, so they will aid—in other words, add to—each other. The USB components are out of phase by 180 degrees, so they *cancel out*. Similarly, when we want to generate the USB, the USB components from the two balanced modulators are additive, and the LSB components are subtractive, or 180 degrees out of phase.

It is simple to make a USB/LSB changeover in the phasing system. All that is needed is some means for switching the output

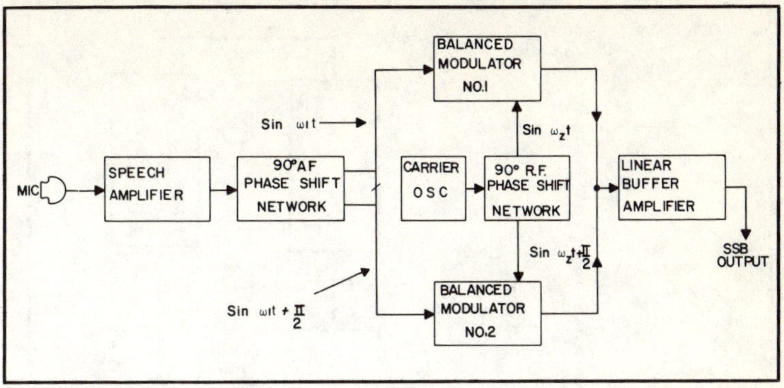

Fig. 15-7. Phasing method of SSB generation.

leads of *either* the audio or rf phase shift networks. Figure 15-8 shows an LSB/USB selector at the output of the audio phase shift circuit.

In early '60s and late '50s, some regular AM/CW amateur transmitters were adaptable to SSB by using an external SSB generator. This system generally used the phasing method and would take the audio and rf carrier from the transmitter, process them into SSB, and then feed the SSB signal back into the amplifier. Bias changes are generally required in the final amplifier, because SSB rf amplifiers must be linear, whereas it was common to use class-C amplifiers for CW and high-level modulated AM circuits. E.F. Johnson, B&W, and Heath made such adapters for their AM/CW 180-watt class transmitter. The only real problem with these systems was the unnerving need to return the phasing exciter whenever the VFO output of the transmitter was moved even slightly.

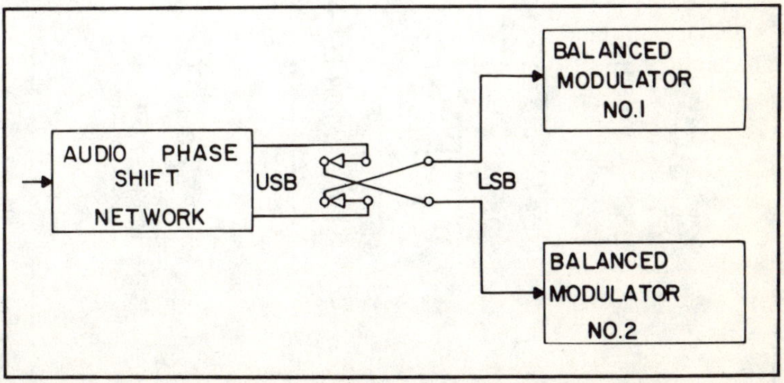

Fig. 15-8. Simplified diagram of phasing method of SSB generation.

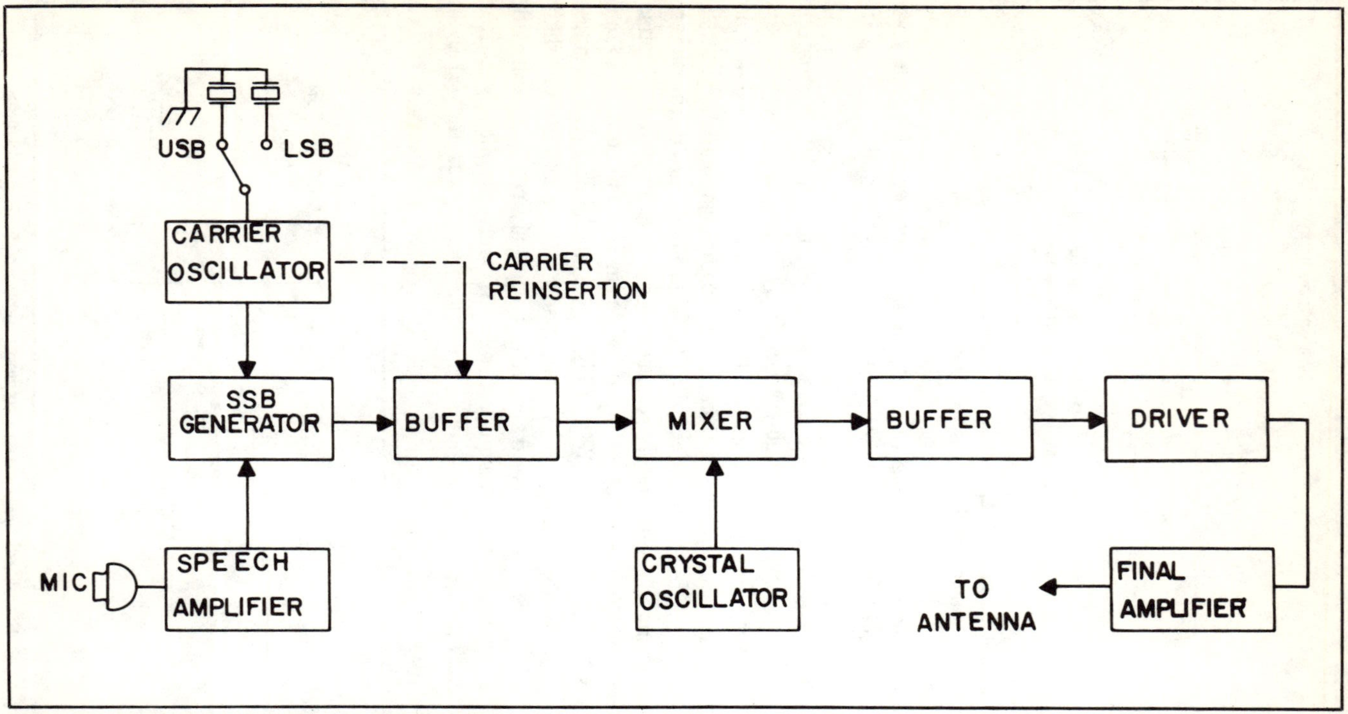

Fig. 15-9. Block diagram of a typical SSB transmitter.

A TYPICAL SSB TRANSMITTER

Figure 15-9 shows the block diagram of a simplified SSB transmitter. The SSB generator can be either the filtering or phasing type as discussed earlier.

In this particular transmitter, the sideband selection is by placing the carrier on either side of the filter passband. A sideband selector switch is used to determine which crystal becomes the carrier oscillator frequency control element.

The buffer amplifier at the output of the SSB generator is sometimes called the *transmit i-f*, especially in transceivers where both receiver and transmitter stages are used in the same circuit. If fact, as a matter of economy, many designs use the same stages for similar use in both receiver and transmitter. There seems to be little reason, for example, to have two i-f systems—each with its own expensive bandpass filter—so why not combine the two?

The SSB generator can only work at one frequency. It is common practice to specify this as the intermediate frequency of the transceiver; 455 kHz, 3350 kHz, and 9 MHz are common selections, although some manufacturers are now using i-fs in the 30-MHz to 50-MHz region. A crystal oscillator and mixer stage are used to heterodyne the frequency of the SSB generator to the operating frequency of the transmitter—the amateur bands, in the case of the ham transmitter. The buffer following the mixer is a tuned rf stage that is used to suppress the unwanted mixer products. Finally, the driver amplifier to full output.

Many new designs of amateur SSB transceivers no longer require extensive tune-up procedures. Most of the solid-state final amplifiers used today are actually broadband linear amplifiers (see Chapter 8), so they require no tuning. A fixed tuned low-pass filter will be at the output of the amplifier in order to eliminate any harmonics that pass through, or are generated by, the final amplifier stage.

Chapter 16

FM/PM Transmitters

Amateur radio operators have taken to frequency modulation in a big way. In the late 50s and early 60s, when I was first licensed, FM was something you learned about if you went to obtain a commerical license; taxi and police radios were typically FM, and a few FM stations were on the air. The only amateur radio FM transmitter that I can remember was a little 10-meter/6-meter job by Hallicrafters. Some people were also using converted 10-meter tank radios left over from World War II. But for the most part, amateur activity was on AM, with some going over to the new upstart, single sideband.

By the mid-60s, however, some West Coast amateurs began to get into VHF FM in a big way, using converted commercial/military equipment mixed with a blend of homebrew equipment. TAB Books author Ken Sessions was one of the first amateurs to go into FM in a big way, and built one of the earliest VHF FM repeaters used in the amateur 2-meter band. A repeater will simultaneously transmit and receive on different frequencies (usually 600 kHz apart). The receivers in the same system will transmit and receive on the opposite frequencies. For example, if the repeater transmits on frequency A and receives on frequency B, then the mobile stations in the system will receiver on A and transmit on B. This allows the repeater to pick up and demodulate the low-level mobile and portable signals and retransmit the audio on its output frequency. The advantage is wider coverage. The receiver in the

Fig. 16-1. Portable VHF transceiver.

repeater is usually well situated to pick up low-power signals from all over a region, and will also be heard over the same region. In contrast, two low-power mobiles would have to be a lot closer together in order to work each other, yet both could work the repeater. Therefore, they can work each other through the repeater (wonderful, huh?).

Figures 16-1 through 16-3 show several different types of VHF FM amateur radio transceivers. Most transmitters used by amateurs in these bands are, incidentally, in the form of transceivers. The equipment shown in Fig. 16-1 is a hand-held portable unit. Most of these instruments are called *walkie-talkies*, but this is in error because that term is actually a trademark. The hand-held portable shown here is a relatively sophisticated transceiver that contains a frequency synthesizer to set the operating frequency. A front panel keyboard allows the selection of the frequency. This instrument also has a *Touch-Tone®* capacility that allows it to interface with autopatch systems in some repeaters. Autopatch is automatic equipment that connects the repeater to the telephone lines so the user can make a telephone call while driving. The autopatch is one of the mainstays of amateur radio public service because it permits mobile and portable radio amateurs to call police, fire, rescue, and other public safety officers in the event of emergency. Amateur organizations never fail to let the public know about particularly heroic efforts in this vein, of which many exist.

The unit shown in Fig. 16-2 is a base station "all-mode" transceiver. These units permit operation on the 2-meter band with CW, classical AM, USB, LSB, and FM modes. The rig is VFO-controlled, which allows it to be set to almost any in-band frequen-

Fig. 16-2. Yaesu FM/SSB rig.

cy. But because of the nature of phase-locked loops, it is possible to store the preferred channel frequencies (local repeater channels) in a memory. A number of these devices on the market now are microprocessor-controlled.

Figure 16-3 shows a small frequency-synthesized mobile rig for the 2-meter FM band. Like the previous example, this rig is synthesized with a VFO and memory unit that allows the previous example, this rig is synthesized with a VFO and memory unit that allows setting the channel of operation almost at will. Mobile FM rigs for the 2-meter band tend to have output powers in the 10 to 25 watt range. Earlier solid-state rigs had power levels such as 1 to 5 watts, but most produce at least 10 watts of rf output power today. Some produce more.

BASICS OF FREQUENCY MODULATION

Frequency modulation, or FM, is the process by which a radio carrier is *varied* (modulated) in frequency by an applied audio signal. The amplitude of the rf carrier remains constant, but the frequency of the carrier is varied proportional to the applied modulating signal. FM reception problems and methods are discussed in the companion volume to this book, TAB book No. 1182, *The Complete Handbook of Radio Receivers*.

Figure 16-4 shows the graph of the FM signal as it varies with the audio signal. When the voltage of the modulating signal (in this case an audio sine wave) is zero, the carrier frequency is *F*. But as the amplitude of the modulating signal advances in the positive direction, the carrier frequency will begin to change until it reaches a maximum at F_h when the audio sine wave reaches its voltage peak. After the modulating signal passes its peak, however, the frequency of the carrier will begin to drop again until it is at *F* when the sine wave makes its zero crossing.

On the negative excursion of the sine wave, exactly the opposite occurs. The frequency will begin to *decrease* as the audio signal amplitude goes negative. The carrier frequency will bottom out at a minimum of F_L when the audio signal passes its negative peak and will then begin to increase in frequency (back towards *F*) as the audio signal amplitude becomes less negative. Over the time that the sine wave is applied to the modulator, the carrier frequency will swing back and forth between F_L and F_h in the manner described.

Several basic concepts are to be understood in FM systems, and these are often misunderstood because of their similarity in meaning. Among these are *modulation index* and *deviation ratio*.

Fig. 16-3. Yaesu mobile VHF FM rig.

First, though, let's define *deviation*. The deviation of an Fm carrier is the maximum frequency change between the unmodulated carrier frequency *F* and *either* F_L or F_h. In any given FM signal, there might be positive deviation ($F_h - F$) and negative deviation ($F - F_L$). If the transmitter is working properly, and the modulating signal is a perfect sine wave, then positive and negative deviation are identical. The units of deviation are frequency. For example, we may speak of 8-kHz deviation.

Frequency swing is the total swing from the lowermost frequency (F_L) to the uppermost frequency (F_h). In the case when the sine wave is perfect, the frequency swing will be exactly twice the deviation. But if the modulating signal is not symmetrical, the frequency swing will not be twice the deviation.

Modulation index and deviation ratio are often mixed up, as are deviation and frequency swing. The deviation ratio is defined as the ratio of the maximum *allowable* deviation to the maximum *allowable* modulating frequency. The modulation index, on the other hand, is the ratio of the maximum *allowable* deviation to the *actual* modulating frequency. The only time DI = MI is when the transmitter is actually being modulated 100 percent by the maximum allowable frequency. Consider the case of an FM broadcast signal, in which the deviation is 75 kHz and the maximum allowable audio frequency is 15 kHz. The deviation ratio is 75 kHz/15 kHz, or 5. The MI is dependent upon the actual frequency. If the transmitted audio is 15 kHz, the MI is 5, the same as the DI. But if the modulating frequency is, say, 1 kHz, the MI is 75. In FM systems, 100 percent modulation is set almost arbitrarily. In the AM mode, certain physical points could be called "100 percent," such as when the audio made the carrier drop to zero on negative peaks and double on positive peaks. But in FM, no such easily recognized points exist. In an FM broadcasting transmitter, 100 percent modulation is defined as a deviation of 75 kHz, while the FM used for broadcast television audio is 100 percent modulated with just 25 kHz of deviation. Commercial two-way transmitters can be considered 100 percent modulated with 5-kHz deviation, and some radio amateurs define 100 percent as approximately 8 kHz deviation, although some also use the 5-kHz commercial standard.

FM MODULATOR CIRCUITS

Only a couple of different classes of circuits are used to modulate an FM carrier. Before we discuss their operation, however, it is necessary to digress a moment or two and define two different forms of modulation that are generally lumped together under the rubric "FM." These are true FM and phase modulation (PM). Both are compatible forms of angular modulation, but only one is actually FM. Interestingly enough, both will work in the same receiver, so are considered compatible with each other in most cases.

True frequency modulation causes the frequency of the carrier to be changed with the audio signal. This mode is also called *direct FM*. Phase modulation causes the *phase* of the rf carrier to be varied with the audio, but the actual frequency remains the same. This is sometimes called *indirect FM*. The advantage of PM was more apparent some years ago when it was difficult to modulate the crystal oscillator needed for stability of the carrier frequency. In PM, the carrier signal could be generated in a rockbound crystal oscillator. Then a reactance modulator stage following the output of the oscillator was used in order to vary the phase of the signal with the modulating signal.

Note well that the audio modulating frequency controls how many times per second the rf carrier makes the sweep back and

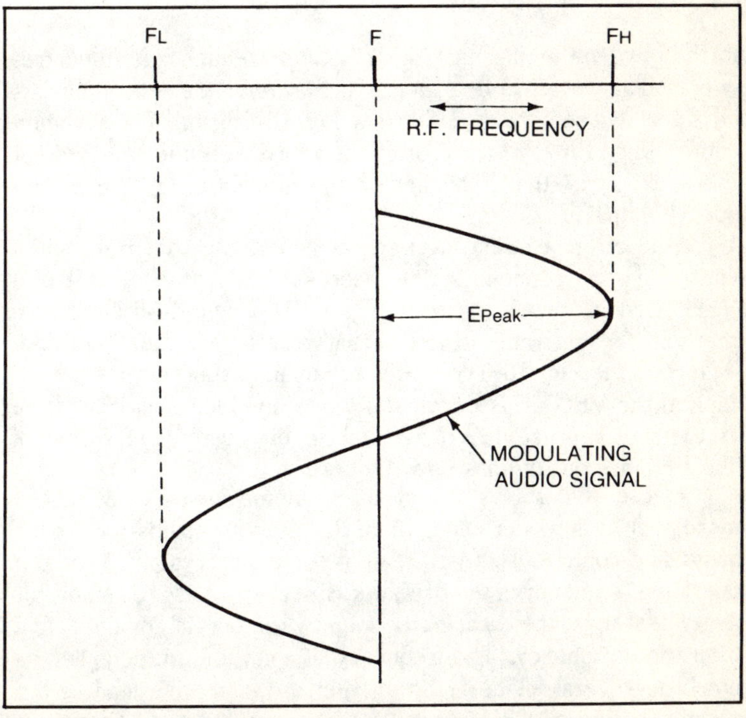

Fig. 16-4. FM carrier frequency-vs.-modulating signal curve.

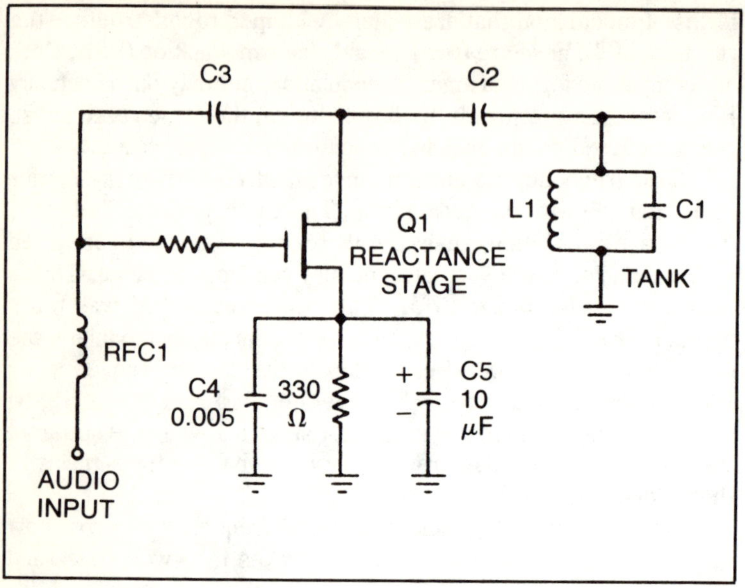

Fig. 16-5. Direct FM modulator.

forth between the maximum and minimum frequencies. Audio frequency does *not* affect deviation. It is the *amplitude* of the modulating signal that sets the deviation (see Fig. 16-4 again). The deviation control in an FM/PM modulator, therefore, sets the amplitude of the audio signal that is applied to the modulator input from the speech ampilfiers.

One of the earliest attempts at generating true FM used a capacitance microphone shunted across the LC tank circuit of a VFO. It didn't work too well. Figure 16-5 and 16-6 show two methods for generating true FM that seem a little more sensible. The circuit in Fig. 16-5 connects a reactance stage across the LC tank in the VFO. This circuit *stimulates* an electronically variable capacitance because the voltage and current vectors in the drain of the FET are out of phase by 90 degrees.

Figure 16-6 shows a modulator scheme for true FM that is more popular in modern equipment that uses crystal oscillators for frequency control. This circuit depends upon crystal Y1 for the operating frequency of the oscillator. But all crystals are somewhat dependent upon the *capacitance* seen by the crystal for the actual oscillation frequency. The frequency of a crystal can be pulled as much as several kilohertz by varying the capacitive load of the crystal. In the circuit of Fig. 16-6, the variable capacitance is a varactor diode. These diodes exhibit a PN junction capacitance that is a function of the voltage across the junction. If the diode is connected in a DC-isolated circuit or given a certain reverse bias in some cases, the capacitance seen by the crystal can be varied by applying the modulating signal to the diode. This will vary the operating frequency of the oscillator proportionally to the audio signal.

The variable-capacitance diode is linear only over a portion of its range, so it is not possible to obtain good linear FM with wide deviation in a single-stage oscillator. It is the usual practice in FM transmitters to generate a carrier at a lower frequency that is a subharmonic of the operating frequency. Frequency multiplier stages are then used to increase the oscillator frequency to the operating frequency. Fortunately, deviation is also increased in the same proportion. If we generate the carrier at a frequency that is 1/24 of the operating frequency, we need a deviation at the oscil-

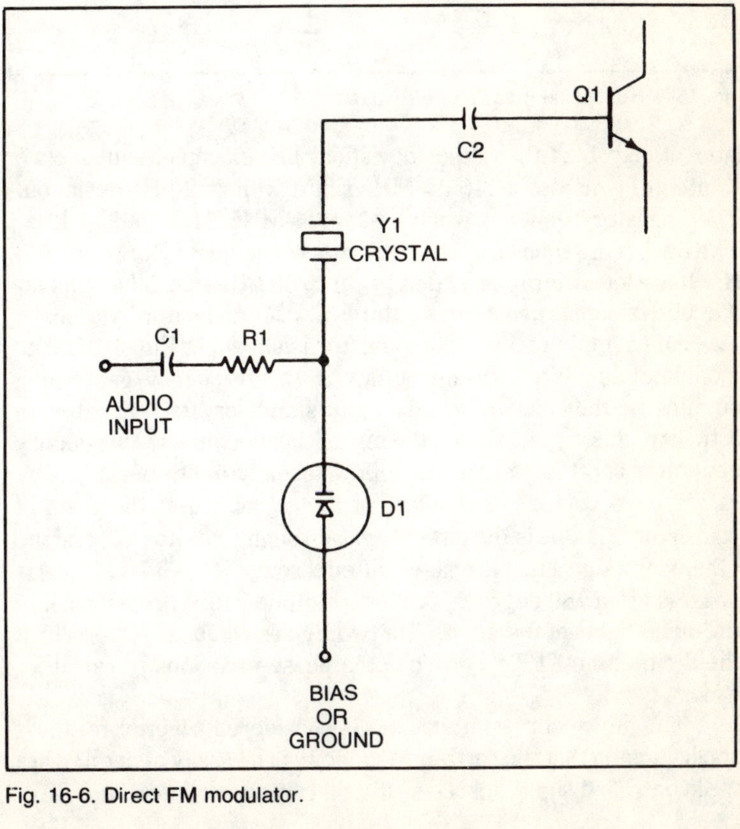

Fig. 16-6. Direct FM modulator.

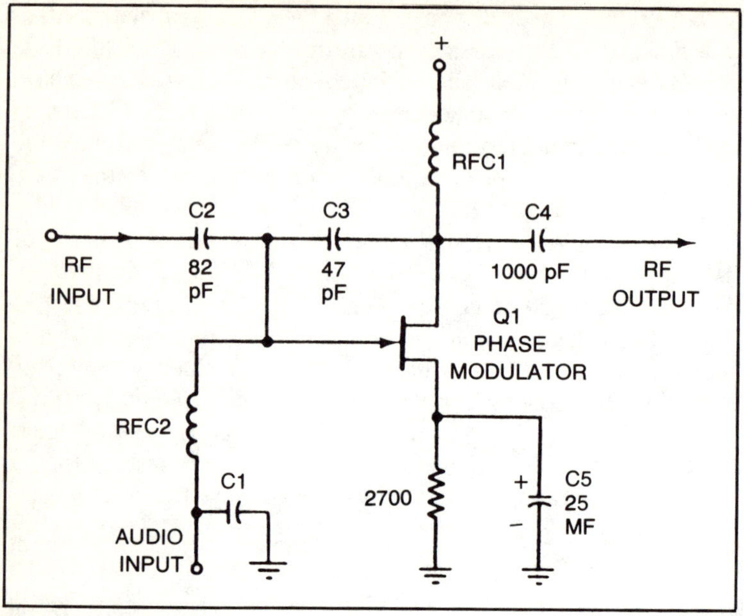

Fig. 16-7. Reactance modulator for indirect FM.

lator of only 1/24 the output deviation. For example, suppose we wanted to generate a 146.34-MHz signal with an 8-kHz deviation. The oscillator frequency would be 146.34-MHz/24, or 6097.5 kHz. The deviation required at the oscillator would be 8 kHz/24, or 333 Hz. It is a lot easier to provide a *linear* ± 333-Hz signal than a linear ±8000-Hz signal using *any* of the FM/PM modulator systems!

An example of a phase modulator is shown in Fig. 16-7. The modulator is placed at the output of the frequency generation circuits, be they phase-locked loop or simple crystal oscillator. In actuality, this circuit shunts the signal line between the frequency generator and the stages that follow (usually multipliers).

Two vector components of the signal are at the drain of transistor Q1. One is the direct signal through capacitor C3, and the other is the signal that is phase-shifted through Q1. The Q1 signal is phase-shifted 180 degrees, but its amplitude varies proportional to the audio signal at the input. The two signals combine vectorially at the drain of the FET to produce the phase variations required for PM.

The principal reason that this is considered superior for most applications is that the carrier frequency can be easily generated in a well controlled signal source such as a crystal oscillator, or a PLL.

The varactor oscillator method is useful but tends to produce deviation figures that are different for different crystals. This means that its major use is in single-frequency transmitters. The reactance modulator, on the other hand, yields more consistent results when frequency changes are desired.

A TYPICAL FM TRANSMITTTER

The block diagram of a typical FM/PM transmitter for use in the amateur and commercial land-mobile services is shown in Fig. 16-8. The circuits of each block have been discussed elsewhere in this book, so the material will not be repeated here. Instead refer back to the appropriate chapters.

The frequency generator stage will be one of several different circuits, including a single-frequency/multi-frequency crystal oscillator of the classical type, a crystal-controlled PLL, or a lockable PLL variable frequency oscillator. The first of these was popular some years ago but is not used much anymore. It requires the transmitter user to buy crystals for each frequency of operation. This also means limited frequency availability. While such is not a problem when you remain close to home—for some, at least—it becomes difficult when traveling.

The digital PLL became more popular several years ago. The operator was provided a thumbwheel or some other switch for setting the operating frequency by changing the "N-codes" of the PLL circuit (see Chapter 5). In the VFO models, the oscillator is a variable type that is locked to a reference by its PLL circuitry. Most newer designs have this type.

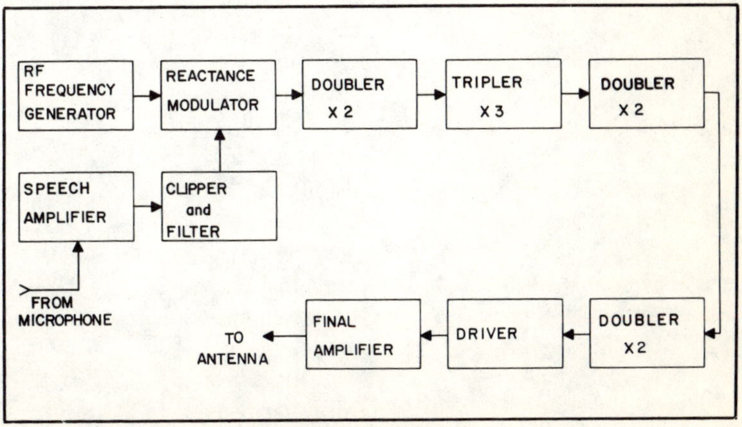

Fig. 16-8. Block diagram for typical FM transmitter.

The reactance modulator stage in Fig. 16-8 is the same type of stages shown in this chapter, although some active device other than MOSFETs (including an IC) might be used. The speech amplifier will amplify the microphone output to a level consistent with the input of the reactance modulator. But there must also be a clipper/filter circuit. The clipper prevents overmodulation on voice peaks, and the filter reestablishes the audio "purity" of the speech signal. This stage might also add some preemphasis, although PM naturally has a +6 dB/octave rising characteristic.

The doubler and tripler stages following the reactance modulator build up the frequency of the carrier and widen the deviation until, at the input of the driver, it has the frequency/deviation required at the output. It is not usually good practice to require frequency multiplication of the last stage (s) of a radio transmitter, as this might increase the harmonic output of the unit. Examples of frequency multipliers are also shown in Chapter 5.

The final amplifier and driver stages build up the power amplitude of the signal for useful radiation by the antenna. These are solid-state amplifiers in most cases. Bipolar predominate, although VMOS field-effect transistors are beginning to make their dent.

Chapter 17

Tuning and Using Transmitters

There are certain rules to follow when using transmitters. These are necessary to get the best out of a rig and to prevent interference with other services. I once knew a radio amateur of especially high technical competence, who claimed that he could tune up any commercial transmitter on the market sight unseen! This was not mere bragging, because the basic principles are the same for all types within any given class. In the days when that remark was made, the typical transmitter had only a few controls. Today, it might be wise to obtain the manufacturer's service manual, as many (perhaps most) of the adjustments will be internal. The modern solid-state transceiver transmitter section is pretty much a broad-banded affair that requires little no tuning.

CLASSICAL HF TRANSMITTERS

The *classical* high-frequency amateur transmitter has vacuum tube final amplifiers matched to the antenna with a pi-network. There are always at least three controls in such tuning: *bandswitch, final tune,* and *loading*. In addition, there will also be *driver tune* or *grid tune* and, sometimes, a *drive level* control on some transmitters. Figure 17-1 shows a Heath HW-101 transceiver that is so constructed. A line drawing of the front panel is shown in Fig. 17-2. Several aspects of this front panel need explanation.

Both final amplifier controls (tuning and loading) are concentric on the same control marked *final*. The center knob is the *tuning*

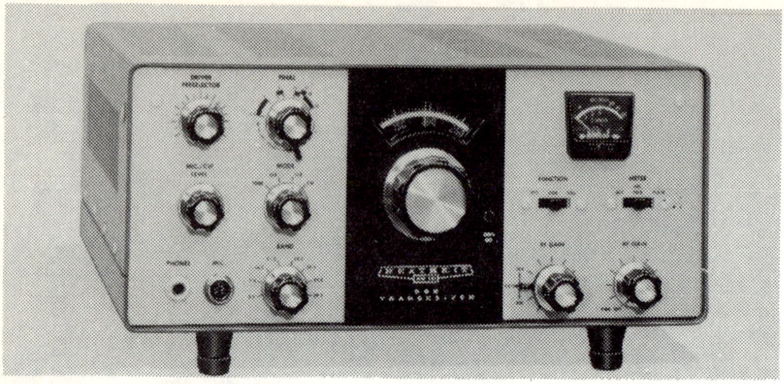

Fig. 17-1. Heath HW-101 tube-type HF SSB and CW rig.

capacitor control, while the outer knob is the *antenna loading* control. The driver tuning is connected to the same knob as the preselector tuning in the receiver section. This knob is immediately to the left of the final amplifier controls. This control is called *grid tuning* in some models.

Some transmitters also have a *grid drive level* control. This control is usually a potentiometer that controls the screen voltage of the driver stage (or some other technique) in order to control the amplitude of the drive signal to the grid of the final amplifier. In some solid-state transmitters, the drive level can be set by the collector voltage applied to the driver or predriver rf amplifier stages. Many rigs in this class have a high/low power switch (typical of some VHF FM rigs), in which a resistor is placed in series with the collector power supply of a stage preceding the final amplifier. This will reduce the drive level to the final when low power is desired.

Some transmitters also have a *tune* function built into the mode switch, as does the Heath HW-101 shown here. This will "throttle back" the final amplifier for initial tune-up. Several different methods are used to provide this feature. In some transmitters, the *tune* function grounds or reduces the SG+ to the screen grids in the final amplifiers. In still others, a higher DC negative bias is applied to the final amplifier grids to limit the plate current that can flow. In still others, the drive level is limited, but this is more common in transmitters, that use a *protective bias*, such as a fixed negative bias voltage on the final amplifier grids. Some transmitters use only *grid leak bias* in which the rf drive signal is used to derive the negative grid bias. In those transmitters, a loss of drive or reduction will cause the plate current of the final amplifier tubes to rise to a point where it can possibly damage the tubes.

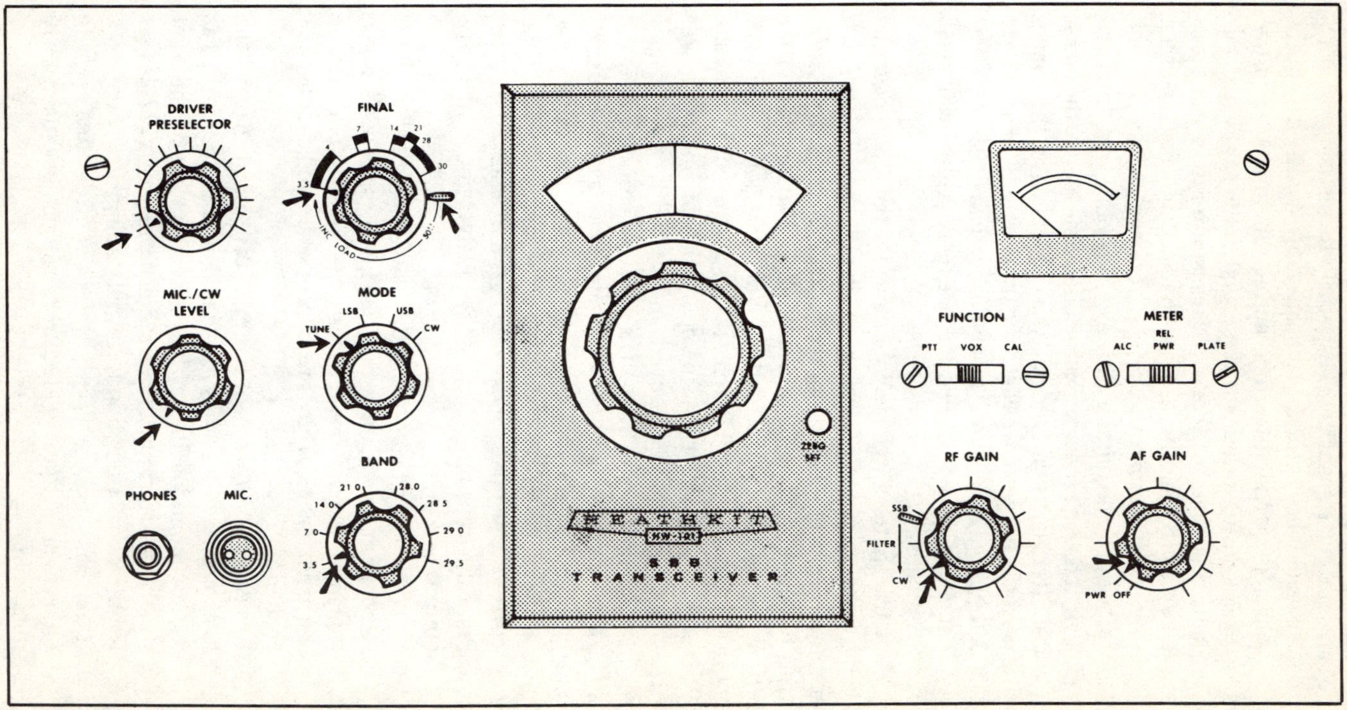

Fig. 17-2. Heath HW-101 front panel.

Many amateur transmitters, especially those using color-TV horizontal sweep tubes in the final amplifier, are damaged by keeping the amplifier energized too long during tune-up. At least one manufacturer (Yaesu) includes a 10-second timer to prevent this problem in one of its models. For a CW transmitter, it is possible to press the key down, quickly resonate the final amplifier, and then key the key up. By letting the key up between successive tune operations, we give the tubes a critical cooling off period. Interestingly enough, it doesn't require much cool-off time to save the finals. If the transmitter has a *tune* function, you can "rough in" the tuning, which is normally the period in which the transmitter tubes get their biggest test by using the "tune" function. We then finish off the job in the CW mode. SW mode is used because there is no carrier signal in the SSB modes.

Many transmitters will inject a carrier for the CW mode, and this is the function of the *CW level* control in Fig. 17-2. This will also control the output of the transmitter in the CW mode because it controls the drive level to the final amplifiers. The keying of many SSB transmitters that operate in the CW mode is done in the VOX circuit In some cases, the VOX relay is directly keyed. In others, a 1000-Hz sidetone oscillator that the operator hears in the speaker when keying the transmitter is keyed, and the 100-Hz signal is used to trip the voice-operated switch (VOX) as if it were a speech signal.

The generalized instructions for tuning up a classical transmitter are given below. In this case, assume that the transmitter contains a milliammeter that will measure either the plate or the cathode current (essentially the same) of the final amplifier tube.

- □Place the transmitter in *tune* (if available) or *CW* modes (if no tune is used).
- □Place the *tuning* control for the plate of the final amplifier either at midrange, or at the zone for the band used. Some transmitters have these zones (see Fig. 17-2) marked on the front panel. They are approximate indications, not absolute.
- □Place the *loading* control at the end of its range in which the loading capacitor is *fully meshed*. This might be marked as the minimum loading or the lightest loading on the front panel of the transmitter. The transmitter in Fig. 17-2 is set for the correct position to begin loading.
- □If there is a *drive level* or *CW level* control, then set it to either a preset point selected by the manufacturer, or to minimum.

☐Key the transmitter. In some transmitters, turning the *mode* control to tune will have done this for you. In those transmitters, you set the mode control to CW in the first step and leave the key up. Then turn the *mode* control to tune in this step. Observe which is the case in any transmitter that you are tuning.

☐Set the *meter* switch to grid and then peak up the *driver/grid tune* control. You can now adjust the *grid drive/CW level* control for a reasonable grid current, but not the full rated level. Make sure that the plate current rating of the tube is exceeded.

☐Advance the *loading* control for a small increase in plate current.

☐Use the *tuning* control to once again dip the plate current to a minimum. This minimum is higher than the previous minimum.

☐Repeat the previous two steps until there is no further increase in plate current when the *loading* control is advanced or the plate current rating of the final amplifier is exceeded.

☐Now bring the drive up or down to set the idling plate current at the level specified by the manufacturer. This is sometimes marked on the plate current meter with some special symbol.

These instructions should obtain an acceptable tune-up on most typical transmitters. Some of these transmitters, however, use an rf output meter instead of a plate milliammeter (the Health HW-series uses both). The units are proportional to rf power, but are not in watts. It is, therefore, a *relative* rf power meter. In those transmitters, use basically the same procedure but tune the controls for a *maximum* meter reading.

TUNING FM AND CB MOBILE RIGS

Many VHF FM and CB mobile rigs do not have any output power indicators or plate milliammeters. This, in fact, is almost common design in many such transmitters. For those transmitters, only an initial tune-up of internal controls is needed. Typically, an rf power meter is connected between the antenna and the transmitter. Interestingly enough, it is quite often the case to tune the radio on a *dummy load* (keeping down the interference) and then connect it to an antenna of standard impedance. In CB and 2-meter FM systems, the mobile antennas offered by most manufacturers will nicely tune to the correct frequency and will present the transmitter

Fig. 17-3. Drake antenna tuner.

with a 50-ohm impedance. In these transmitters, it is advisable to tune the internal adjustments for maximum output power.

TRANSMITTER GROUNDING

One of the major areas in which amateur radio transmitters are not correctly operated is in the matter of antenna/chassis grounding. Many rigs are either ungrounded or insufficiently grounded. This situation can result in excessive TV interference. Note that the FCC will usually be either neutral or on the radio amateur's side in a TVI/BCI dispute *if* the transmitter is operated according to "good engineering principles." These include proper grounding.

There are two rules of thumb regarding grounding of the transmitter: Make the ground wire *heavy* and *short*. It is almost standard practice in radio stations to use sheet copper in the vicintity of the transmitter and then run this to ground rods outside the building. Flashing copper that is used in roofing and available in

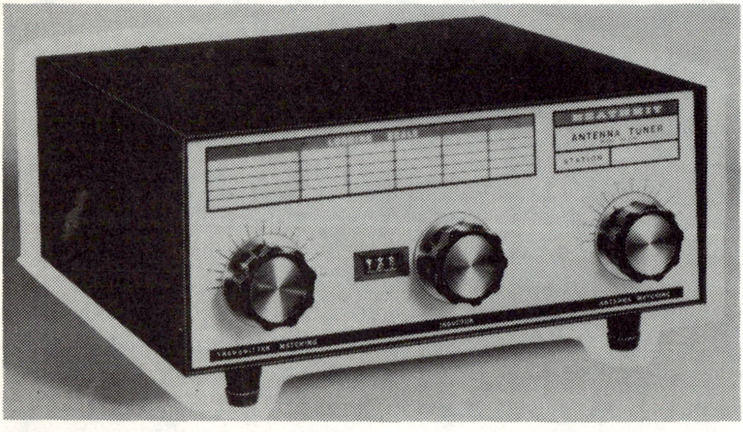

Fig. 17-4. Heath antenna tuner.

many do-it-yourself stores makes good ground straps for radio transmitters. The idea is to keep the ground connection low in DC resistance and especially low in inductive reactance. A low-impedance ground is necessary. Keep in mind that at radio frequencies, this means a short, wide ground connection. If the length of the ground wire approaches ⅛-wavelength of the operating frequency, the ground is likely insufficient. At 15-meters, ⅛-wavelength is six feet!

ANTENNA MATCHING/TUNING BOXES

The need for an antenna tuner is well known, but most amateurs assume that its principal application is to match the impe-

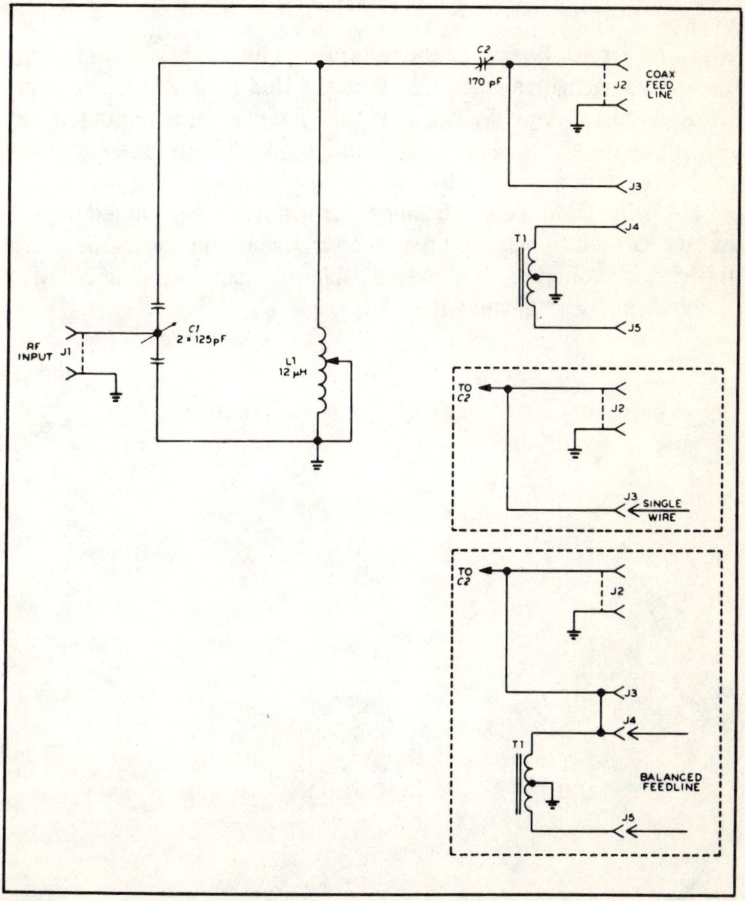

Fig. 17-5. Typical antenna tuner circuit (courtesy of Heath).

dance of some oddball antenna to the 50-ohm to 75-ohm output impedance of a typical amateur radio transmitter. This is a proper use of these devices, but it is not the *only* use.

Everybody should use an antenna tuner between the transmitter and the antenna. Some manufacturers sell tuners that are essentially coax-to-coax models that tune out some of the effects of high VSWR and *attenuate any harmonics* present in the output! This is the best reason to use a tuner, and it puts you in a better position should your transmitter interfere with someone's TV/radio receivers.

Figures 17-3 through 17-5 show two antenna tuners and a typical circuit. The tuner in Fig. 17-3 is the Drake MN-2700, while Fig. 17-4 shows the Heath SA-2040. The latter is available in kit form.

The circuit typical of these types of tuners is shown in Fig. 17-5. This particular schematic is for the Heath SA-2040. The input capacitor is a voltage divider (2 × 125 pF), while the output loading is a series capacitor. The variable inductor will both tune out some of the reactance and set the band.

A balun (*bal*anced-*un*balanced) transformer is included to permit direct feed to balanced twin-lead transmission lines. There are three modes of operation; coaxial cable to coaxial cable, single wire (long wires), and balanced feeders.

Chapter 18

Troubleshooting Transmitters

We all hope that every radio transmitter will work the first time it is turned on and will continue to work until the day it is retired honorably from service. But in the real world, we cannot really expect this to be true. Some amateur radio transmitters, especially some of the advanced transceivers on the market, should properly be serviced by the factory or a factory authorized service center. In many cases, though, you can solve the problem yourself, so the effort is worth making—provided that you don't make things worse for the trying.

THE RIGHT STEPS

If you own a piece of complex commercially made radio equipment, obtain the service manual for that equipment. In many cases, the manual is packed with the equipment, and you should keep it. Few manufacturers supply extra copies once a model is out of the line. The service manual then becomes archivel from their point of view, and they do not renew inventory when present stocks are depleted. Some companies charge a pretty penny for the manual in order to discourage people people not really qualified from attempting a repair.

Also, be prepared to buy any service extender card kit the manufacturer offers. If you own a modern transceiver, it is a good bet that the major circuits are in the form of a plug-in printed circuit boards. It is therefore almost impossible to measure voltages, inject signals, or to signal trace without these extender cards, which are plugged in in place of the actual cards, and then the actual cards are plugged into the service extenders. This will place each circuit card at a position that is easier to troubleshoot in. Service extender kits range from $10 to over $100, but are a good investment if you plan to repair anything.

Equipment made with plug-in printed circuit boards can often be repaired most economically by replacement of the defective board. This is not the elegant approach, and it offends the purist who wishes to troubleshoot to the component level. But while the purist is wrinkling his nose at your stupidity for not being able to thusly troubleshoot the rig, you are getting back on the air. It takes only a couple of days to send small packages from coast to caost, and air freight is even faster (and costlier, I might add). While the troubleshooting snob spends several nights trying to locate a leaky transistor, or open 0.001 μF capacitor, you are back on the air enjoying your radio hobby. There is much to be said for both approaches, but I suggest that you use the PCB replacement approach unless you are exceptionally qualified in the repair arts.

Learn to use the telephone. It is usually possible to talk directly to the service technician at the factory and/or importer who actually works on the equipment that you own. In some companies, one has to get mad at the sales manager before being allowed to talk directly to the service technician, but most amateur equipment companies will allow you to simply call right into the service department. A number of companies have a standing policy of trying to troubleshoot on the air and will mail out whatever board the technician feels is the problem. The usual arrangement is to allow you a credit on the returned board when received and inspected. This is a good reason not to butcher the board in any ill-advised repair attempt—the service technician would be well within rights to refuse you credit. You will then not receive the refund that you thought was due will be stuck for the full price of the new PCB.

But please, please, please keep in mind that the service technician's time is a saleable commodity for the manufacturer. Keep your conversation with the technician *short* and to the *point*. Thank him or her for the assistance by letter once the job is finished, praising the company to high heavens. If the volume of calls and the time spent on the phone become too great, company management will be forced to cancel the call-in privilege. They will keep the call-in privilege open only if they see a positive public relations return viz the cost to them.

SOLDERING

Learn to solder properly. If you think that this is overstating the problem, and begging the obvious, then you are the number-1, grade-A candidate for a quickie soldering lesson. It is especially critical on printed circuit boards, where a little too much heat or a little slip or poor technique will lift the tracks from the board, fouling up the board forever. It is well worth the low price to buy the Heath continuing education course in soldering. It's only a few dollars.

The subject of troubleshooting anything, transmitters included, is simply too broad for any reasonable treatment in one chapter. Because of this, refer to one of my other TAB books, No. 1194, *How To Troubleshoot & Repair Amateur Radio Equipment*. This is not an attempt to make you buy another book, but a blunt admission that a couple pages cannot do justice to a topic that needs an entire book.

Chapter 19

Transmitter Safety

No book on radio transmitters, especially if high-power designs are considered, is complete without a chapter on safety. Transmitters are electrical in nature and can *kill* you. I have occasionally been accused of being a little too much of an alarmist over safety, a bit too conservative. Some readers of a magazine article I wrote criticized my scare tactics on troubleshooting high-voltage power supplies. They, perhaps, failed to see a little editorial postscript from the magazine staff. The editor of the manuscript had been working on a 3000-volt power supply for his linear amplifier during the period the article was being prepared. The line touched the concrete floor and nearly killed him! Let me repeat that word for emphasis: KILLED. The kinds of electrical potentials found in radio transmitters are lethal, and they can easily kill a human being.

One night some years ago, when I was working part-time for the chief (and only) engineer of a local AM/FM broadcasting station, we brought in an electrician to work on the 440-volt three-phase panel that feeds power to the transmitter. The electrician worked the panel "live." We heard an explosion and ran down the basement stairs to find the electrician *dead*. That's right: DEAD! His skin was both red and black in places, and any further description is too hideous to write about here—and to ever forget. You might want to use my experience to become "too conservative" and live a lot longer. One should not be paralyzed by the thought of electrical shock, but it is necessary to be wary and *know* what you are doing.

THE MECHANICS OF ELECTROCUTION

There are two main causes of death by electrical shock: respiratory arrest and ventricular fibrillation of the heart. The former is not terribly frequent, because the latter would kill you first unless the path of current crosses the lungs (or a nerve thereto) and does create sufficient current density in the heart to cause ventricular fibrillation.

The heart is a muscle and operates as a pump. The muscle fibers contract in a rhythmic unison that causes blood to be pumped (see TAB book No. 930, *Servicing Medical Bioelectronic Equipment*). Ventricular fibrillation (V.Fib) is a condition in which the beating of the heart loses its synchronism; the muscle fibers begin to contract in an arrhythmic manner. V.Fib will lead to death within minutes if not corrected. A *defibrillator* is a medical instrument used to deliver a powerful (400 watt-seconds) countershock from a DC charge stored in a high-voltage capacitor to the heart. This attempts to cause all of the muscle fibers in the heart to contract simultaneously. When they relax, they are supposed to restart in their normal rhythm. It doesn't always work.

A medical lay person can support the victim of electrocution until trained medical help arrives, or the ambulance crew can get the victim to the hospital. The particular type of support is known as cardiopulmonary resuscitation (CPR) and is taught by the American Red Cross and many local hospitals. It is wise that you and as many other family members or business associates as will cooperate learn CPR. There have been innumerable instances in which a medical lay person was able to save a life because he knew CPR. It takes only a few hours to learn, but don't be so sure of yourself just because you watch *Emergency* on TV: a real, live CPR course is needed, or you can become liable for suit and may kill instead of save.

NOTES

Some practices are just accidents waiting to happen. Some things can prevent an accident. The list below is a few of these:

- ☐Install a master power switch to your workbench, any outdoor outlets, and ham shack operating desk. Instruct all members of the household where this switch is located and how to turn off the power in the event of an apparent electrocution—or even if they *think* you are electrocuted. Having the power turned off while you fall asleep during a ham contest is not too great a penalty to pay for safety, is it?

- ☐Have all members of the family or household learn CPR.
- ☐Do not operate on a concrete or dirt floor. Install insulating flooring, a rubber mat, or a wooden floor cover.
- ☐Install interlocks on all equipment chassis, doors, and access covers, and don't defeat the purpose of the interlocks.
- ☐Never work on a piece of live equipment.
- ☐Don't believe that nonsense about 115-volts AC not being lethal. Most electrocutions in the US are from 115-volt residential electricity.
- ☐Never allow temporary repairs, malfunctioning equipment, sloppy workmanship, or unrepaired appliances, equipment, cords, etc. to become hazardous. Take them out of service until repairs can be performed.
- ☐Be careful of rf burns. Never believe that stuff about rf not being able to electrocute you. Even if it is true (which I doubt), it *will* burn you very badly. Surgeons actually use high-power 1-MHz rf generators to cut tissue and cauterize cut blood vessels. RF energy will do the same to you.
- ☐Discharge and keep shorted all power-supply filter capacitors.

Chapter 20

TVI/BCI/Hi-Fi-I

One Saturday morning in 1959, I was busily enjoying a well earned sleep-in when the telephone rang. An irate neighbor *demanded* that I get off the air and stop interfering with his television reception. He was really hot! The week before, I had put up a simple 40/20/15-meter dipole, and he had seen it. Trouble was I didn't even own a transmitter. I had just passed my novice license examination and was busily waiting out the 6 to 10 weeks required to grade the exam and process the license by getting ready to go on the air. My brand new Heathkit DX-20 transmitter kit hadn't even been received in the mail yet.

Sooner or later, if you operate an amateur radio station, you will be faced with the irate neighbor problem. The interference may be real or imagined, but the result is the same: *you* get the blame. From the moment that your antenna goes up, you will automatically spring to mind whenever any neighbor's TV reception goes "bloop." You might, like one amateur was, be blamed for damage to the owner's TV set.

The best advice is to stay cool and be informed of the problem and possible solutions. Do not, under any circumstances, become irate, even if you are only matching your neighbor tit-for tat.

There are several causes of interference. It is possible that your equipment is at fault. This must be taken care of before you can be on really firm ground in any dispute. Some of the material in this chapter will help you determine whether there is a high probability that your rig is producing the problem. These are not absolute techniques, but are often good enough to resolve problems.

One word of warning: *Do not attempt to modify any equipment owned by a neighbor*! There are methods, including both addition of filtering and modification of circuits, that often help a receiver that does already not reject unwanted signals properly. Even if you are an experienced, qualified, certified, bottle-in-bond, 100-proof consumer electronics service technician, make no attempt to resolve the neighbor's problem yourself. If you do, then you will be permanently married to that set. From then on, every needed repair will automatically—in your neighbor's mind—become your responsibility. If you are *not* qualified *professionally* as a consumer electronics technician (and I don't mean to exclude electronics engineers of any type, or electronics technicians from fields other than consumer electronics), then you are just plain stupid to attempt to make modifications to the neighbor's equipment. Let the neighbor know the types of things that might be needed and then tell him to call a TV/hi-fi repairman to actually do the job.

One frequency problem with consumer electronics devices is that they are unable to reject your signal—and they are supposed to! Even some equipment by the best manufacturers is designed poorly in the interference rejection department. The owner of that $1000 color TV or $1500 hi-fi, though, sees *you doing something* that fouls up previously good operation of their expensive gear. They do not appreciate the fact that the equipment is unable to do its total job. A TV or hi-fi is not only required to respond to the desired signal, but must also be capable of rejecting the undesired signals.

TVI: TYPES OF INTERFERENCE

It is possible to diagnose the type of interference experienced by a TV receiver by noting how it affects the picture. Not all forms of video interference are caused by radio transmitters. Much interference can be traced to defective devices in the house of complainer. Furnace controllers, blower motors, defective thermostats, and elements in electrical resistance heaters, SCR lamp dimmers, vacuum cleaner or pump motors, and a host of other devices will create a snow-like interference in the picture. Such signal sources also sometimes produce a hash-like, raw sound in the audio of the TV receiver. If the interference is noted at times when you were not on the air, then it may well be that the problem is right in the home of the person who raised the complaint.

An example of genuine radio transmitter interference is shown in Fig. 20-1. This interference is characterized by horizontal or

Fig. 20-1. TVI from a CW/AM/SSB transmitter.

slanted bars in the picture that *might* move in step with the modulation of the transmitter. Note that a similar form of disruption occurs if the sound carrier of the TV is getting into the picture, and will manifest itself when the fine tuning of the TV receiver is incorrectly set. Also, be sure to not confuse this form of interference with horizontal sync problems in the TV. Of course, both of these problems become a mute issue if they exist at times when your transmitter is turned off, but try to convince your neighbor of that fact, especially if the person is someone trying to get some ham/CB operator for a free TV repair job.

Figure 20-2 shows the herringbone pattern that is typical of interference from FM transmitters. This will normally be limited to channel 6, and possibly one additional channel in the VHF band, if the source is a local FM broadcasting station. If there is a local broadcasting station, or some other station that uses FM, such as the commerical land-mobile two-way, service then the source of interference might be that transmitter. Make sure that your own FM operations are not at fault, though.

An example of co-channel interference is shown in Fig. 20-3. If your neighbor complains of seeing your picture on the TV, you might want to chuckle and investigate the possibility of co-channel interference. The channel assignments for TV stations are usually

Fig. 20-2. FM interference to a TV receiver.

made at geographical distances that limit this possibility, but atmospherics that excite your VHF DXer buddy will also cause this type of interference to VHF TV receivers. The problem is rarely chronic, however, except at locations approximately equidistant between two transmitters operating on the same channel. If the two signal strengths are approximately the same, the receiver will respond equally. The solution, however, is simple: use a high-gain directional antenna on the receiver.

There is one way to gain at least a psychological advantage over your irate neighbor—take a shotgun. Oh, not really, but the way to drive home your point that his or her equipment *might* be at fault is to keep your own equipment free of interference. If the neighbor sees similar equipment operating right in the same house with the transmitter, then he sees things your way and blames the cheap manufacturer who failed to shield and bypass his TV or hi-fi properly.

STEPS TO TAKE

The material in this section was adopted directly from an FCC publication titled *How to Identify & Resolve Radio-TV Interference Problems* (stock no. 004-000-00345-4, sold for $1.50 by the Superintendent of Documents, Government Printing Office,

Washington, D.C., 20402). It is also available from the Consumer Information Center, Pueblo, CO, under the same number at the same price. We make no apologies for using this material, as it is quite well written and presented. Have several copies of this booklet available to give out to your neighbor, especially if you plan a new installation or the purchase of a high-power linear amplifier.

Installing A High-Pass Filter

There are not set procedures for eliminating television interference—it is a matter of eliminating the most likely sources of interference at step at a time. The first step is to install an inexpensive high-pass filter on the back of your TV set. In making this installation, follow these procedures:

1. Determine the type of antenna wire that is connected to your TV set. There are two possibilities:

Coaxial Cable—a round lead-in wire which requires a filter impedance of 75 ohms (see Fig. 20-4A).

Twin Lead Wire—a flat wire which requires a filter impedance of 300 ohms (see Fig. 20-2B).

2. Purchase the filter which matches the type of antenna wire coming from your set. The impedance information mentioned above will be on the filter label. Do not use a combination of twin-lead and

Fig. 20-3. Co-channel interference to a TV receiver.

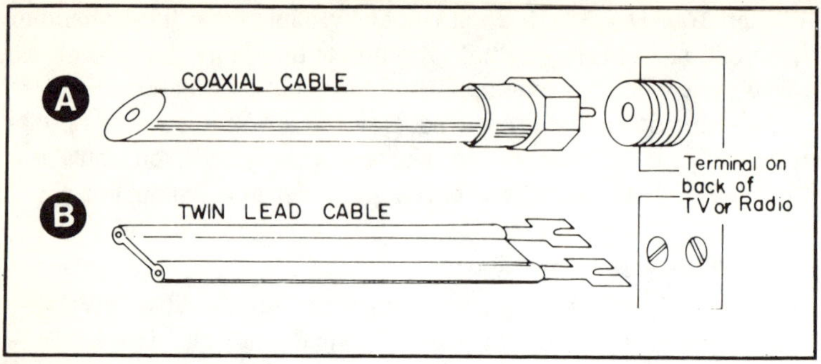

Fig. 20-4. Installing a high-pass filer with coaxial cable at A and twin-lead at B.

coaxial cable without proper matching transformers (often called baluns). Filters are available in most stores that sell or repair television sets.

3. Carefully read the instructions that are provided with the filter. You will be installing the filter on the back of your TV set, as near to the antenna terminal as possible. The antenna terminal and the filter terminal will look like either Fig. 20-4A or 20-4B depending upon the type of wire you are using—coaxial or twin-lead.

4. If you are on a cable system, you may still install the filter at the antenna terminal. However, if the interference continues, contact the cable company repair service for assistance. Do not attempt to modify the cable system yourself.

5. The following information or installing the filter should answer any additional questions you may have.

Disconnect the antenna wire (twin-lead or coaxial) from the television set antenna terminals. Connect the wire from the antenna to the input terminals of the filter. For twin-lead wire, connect a very short (1″ to 2″) "jumper" wire from the antenna input terminals of the set to the filter (see Fig. 20-5). For coaxial cable, it will be necessary to obtain a jumper cable that has the proper connectors already installed. This can be purchased at the time you buy the coaxial filter.

Be sure that in the case of twin-lead wire, the actual wires are making contact with the terminals. For coaxial cable, be sure the connector plugs are properly installed on the coaxial cable.

If you have an amplifier in your antenna system, you should have a filter installed ahead of the amplifier and another filter ahead of the TV receiver input terminals (see Fig. 20-6). If the amplifier is located close to the receiver, then install the filter before the amplifier only.

Booster amplifiers usually are located near the back of the TV set; mast mounted (outdoor) amplifiers are usually located on the antenna; and distribution amplifiers are usually located somewhere in the distribution system. If a distribution amplifier is in your antenna system, then be sure to trace the entire length of the antenna system, because amplifiers are usually in out-of-the-way places; for example, clothes closets, basements, etc.

The connecting wires between the filter and amplifier, and between the amplifier and antenna terminal, should be as short as possible.

The instructions provided with the filter you bought may call for a ground connection. The wire should be as short as possible and connected between the high-pass filter ground terminal and a metallic cold water pipe or a ground rod. Use bell wire for this connection (see Fig. 20-5). Bell wire can be obtained from most variety stores.

If installation of the filter at the TV antenna terminals does not entirely eliminate the interference, you should then contact your service representive to install a high-pass filter inside the TV set at the tuner input terminals. Internal modification to your set should be done *only* by a service representative.

Home Remedies For Resolving Electrical Interference

Electrical interference is caused by two sources: vehicle ignition systems and electrical devices. The first step in attempting to

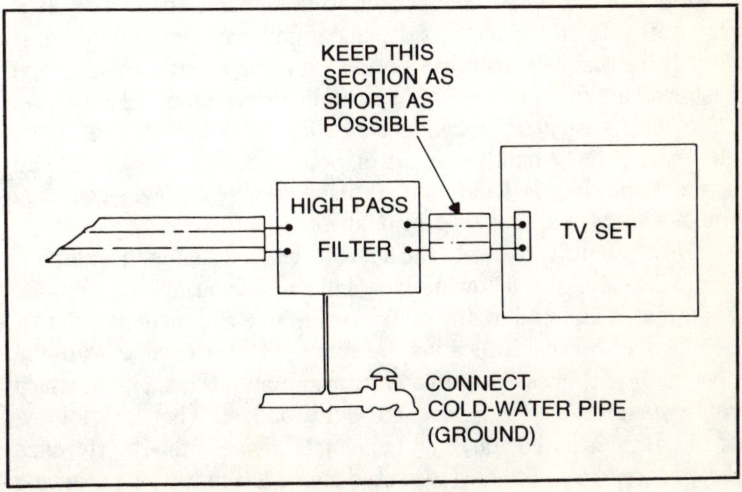

Fig. 20-5. Connecting a high-pass filter to twin-lead.

resolve electrical interference problems is to locate the source of interference.

Interference From Vehicle Ignition System. Ignition interference sounds like a "popping" noise in the sound system of your TV that rises in intensity; the "pops" occur closer and closer together as the speed of the engine speeds up. This can be caused by any vehicle ignition system, such as gasoline operated lawn mowers, snowmobiles, automobiles, etc.

If the interference is to television receivers, you may hear the same popping noise in the sound and also see "dancing dots" in the picture of the set. You may only see the interference, and not hear the "popping noise in the sound."

If your own vehicle is causing interference, you may wish to install a commerically manufactured kit in your vehicle to reduce the ignition noise. Other remedial measures include relocating your antenna, raising the antenna, and using shielded lead-in antenna wire.

Interference From Electrical Devices. Any one or more of the following electrical devices may be causing the interference you are experiencing on your television set or AM/FM radio:

Electric razor, vacuum cleaner, fan, drill, electric blankets, bake ovens, fluorescent lights, arc lights, light dimmer controls, relays, static from machinery, lightning arrestors, adding machine, cash register, circuit breakers, ultraviolet lamps, germicidal lamps, defective wiring, loose fuse, arc welder, switch contacts (such as on dishwashers and other home appliances), refrigerator, water pump, sewing machine, light blinkers (including Christmas tree light blinkers), electric heating pads, aquarium warmers, neon signs, door bell circuits/transformers, toys (such as electric trains), sign flashers, antifriction bearings, printing press static eliminators, calculators, insulators, incandescent lamp (new or old), sun lamps, electrical pole (ground wire cut or poor contact), loose electrical connection, electric fence unit, furnace controls, power company transformers, and smoke precipitators.

In attempting to locate the specific device causing the interference, consider the following suggestions. If you have a portable radio that is affected by the interference, use the radio as a detection device to assist in locating the source of interference. With the portable radio, move from room to room and determine in which room the interference appears to be the loudest. Then look for one of the devices listed above and unplug it to see if the interference disappears. If several devices listed above are in the room, unplug them, one at a time, until the interference disappears.

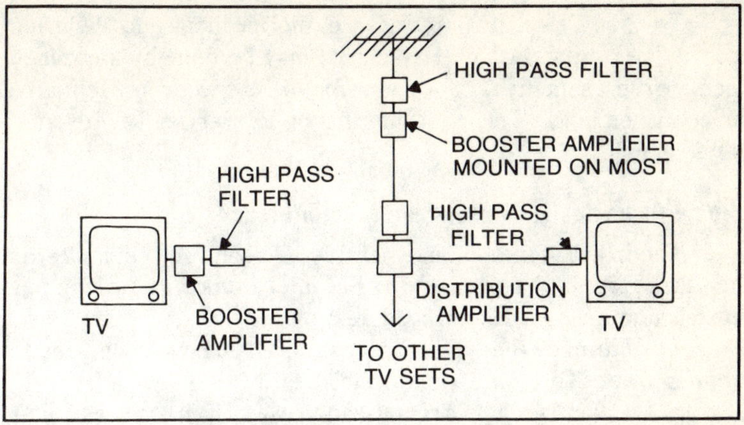

Fig. 20-6. High-pass filter installation when a filter is used.

If a portable radio is not affected, you can go to the main fuse or circuit breaker box in your home, remove one fuse at a time, or shut off one breaker at a time, and see if the interference goes away.

If it does not go away when the first fuse or circuit breaker is off, replace the fuse or turn the circuit breaker back on and continue on until the interference does disappear. When the circuit that supplies the power to the TV or radio is turned off, it will be necessary to plug that device into some other circuit to determine if the interference is being generated by a device in the same room as your TV or radio.

When the interference disappears with a fuse removed or circuit breaker off, you should go to the room supplied by that circuit and look for any of the devices listed above. If any of the listed devices are found in the room, replace the fuse of turn the circuit breaker back on. Then unplug the device suspected of causing the interference. If several devices are in the room, unplug them, one at a time.

If you are unable to locate within your own home, the device that is causing the problem, the interference may be coming from a device located in your neighbor's home. With the cooperation of your neighbor, follow the same procedures described above.

If your investigation leads you to suspect that a power line or power company equipment is the source of interference, you should contact the power company to assist you in resolving the problem.

Short duration interference, such as that from electric drills, and saws, may be very costly to attempt to eliminate; you may just want to "live with it."

To *resolve* electrical interference, modifications must be made to the interfering device. This should only be done by a qualified service representative. Information for your service representative is contained in the Technical Information for Service Representative section.

Home Remedies For Resolving FM Interference

The installation of an inexpensive FM band rejection filter is the first step to take in resolving FM interference. In making this installation, follow these procedures:

1. Determine the type of antenna wire you have connected to your TV set. There are two possibilities:

Coaxial Cable—a round lead-in wire which requires a filter impedance of 75 ohms (see Fig. 20-4A).

Twin Lead Wire—a flat wire which requires a filter impedance of 300 ohms (see Fig. 20-4B).

2. Purchase the appropriate filter, according to the type of antenna wire you have. The impedance information mentioned above will be on the filter lable. Do not use a combination of twin-lead and coaxial cable without proper matching transformers (often called baluns). Filters are available in most stores that sell or repair television sets.

3. Carefully read the instructions that are provided with the filter. You will be installing the filter on the back of your TV set, as near to the antenna terminal as possible. The antenna terminal and the filter terminal will look like either Figure 1a or 1b depending upon the type of wire you are using—coaxial cable or twin-lead wire.

4. If you are on a cable system, you may still install the same FM band rejection filter at the antenna terminal. However, if the interference continues, contact the cable company repair service for assistance. Do not attempt to modify the cable system yourself.

5. The following information on installing the filter should answer any additional questions you may have.

Disconnect the antenna wire (twin-lead or coaxial) from the television set antenna terminals.

Connect the wire from the antenna to the input terminals of the filter.

For twin-lead wire, connect a very short (1″ to 2″) "jumper" wire from the antenna input terminals of the set to the filter (see Fig. 20-5). For coaxial cable, it will be necessary to obtain a jumper cable that has the proper connectors already installed.

Be sure that in the case of twin-lead wire, the actual wires are making contact with the terminals. For coaxial cable, be sure the connector plugs are properly installed on the coaxial cable.

If you have an amplifier in your antenna system, you should have a filter installed before the amplifier and another filter ahead of the TV receiver input terminals (see Fig. 20-6). If the amplifier is located close to the receiver, then install the filter before the amplifier only.

The connecting wires between the filter and amplifier, and between the amplifier and antenna terminal, should be as short as possible.

The instructions provided with the filter you bought may call for a ground connection. The wire should be as short as possible and connected between the FM band rejection filter ground terminal and a metallic cold water pipe or a ground rod. Use bell wire for this connection (see Fig. 20-5). Bell wire can be obtained from most variety stores.

If the filter does not entirely eliminate the interference, you should call your service representative.

Identification Of And Resolving Audio Interference

Interference to audio devices, such as tape recorder, record players, electronic organs, telephones, hi-fi amplifiers, etc., is caused when the equipment responds to the transmission of a nearby radio transmitter. Audio interference (often called audio rectification) may also affect the sound (audio) portion of your TV and AM/FM radio.

When this type of interference is occuring, you will hear the voice transmissions of the radio transmitter and/or the volume level of the audio device you are using may decrease. If you have determined that this is the type of interference you are receiving, refer to the following suggested methods for eliminating audio interference.

Audio interference is a condition that usually requires internal modification of your equipment. For safety reasons, it is recommended that any modifications be made by a qualified service representative.

Due to the complexity of resolving interference to an electronic organ, again, servicing should be done only by an experienced service representative. More detailed information should be obtained from the equipment manufacturer.

For telephone interference, contact your local telephone company. They can install a 1542A or similar inductor in the telephone instrument to resolve the problem. The information provided in this bulletin applies primarily to privately-owned equipment and should not be applied to equipment owned by the telephone company. Bell system personnel can obtain additional data in Section 500-150-100 of the *Bell System Practices—Plant Series* manual.

For all other audio devices, you may wish to take the following steps before calling your service representative. Replace unshielded wire between the amplifier and speakers with shielded wire. Ground the affected equipment to a metallic cold water pipe or ground rod. A ground connection can be made with a short piece of bell wire which can be obtained at most variety stores. Do not ground AC/DC-type devices. Normally device which may safely be grounded will provide a grounding terminal. If no terminal is provided, then you should consult a qualified service representative for advice.

If the interference is not eliminated after taking these steps you must call a qualified service representative. You may also wish to discuss the matter with the operator of the radio transmitter.

Resolving Radio Transmitter Interference

There are no set procedures for eliminating television interference—it is a matter of eliminating the most likely sources of interference a step at a time. You may be required to take several steps before the interference problem is resolved. Once you have installed the filter called for, or made the adjustment that you were instructed to do, leave the modifications in place and proceed to the next step.

To begin, check to see if a high-pass filter has been installed on the TV set at the antenna terminals. If not, read the Home Remedy information beginning on page 5. If the interference is still present after the installation of a high-pass filter proceed with the following steps.

Check Radio Transmitter. Contact the operator of the radio transmitter identified as the source and, with his/her cooperation, determine if the transmitter is operating properly.

Is the transmitter properly grounded? This means a good radio frequency (**rf**) ground. A single piece of wire to a ground rod may be an open circuit to **rf**.

Are harmonics and/or spurious emissions present? Is the transmitter cabinet radiating energy?

If the transmitter is not grounded, connect the chassis to a good earth ground with large diameter wire or copper strap. This should assist in eliminating radiation of energy from the cabinet.

Next, install a low-pass filter on the transmitter antenna circuit to see if any difference occurs in the interference pattern. If a change occurs, the interference is probably caused by harmonics and/or spurious emissions from the transmitter. If no change occurs in the interference pattern, it is probably being generated at some point in the TV reception.

Check TV Reception System. Conduct a visual inspection of the TV antenna, lead-in wire, and lightning arrestors. This may reveal a source of trouble. Corroded connections or deteriorated lead-in wire could be at fault and should be repaired.

Assuming no faulty conditions are found, or if found, they are corrected, and the interference is still present, look for an amplifier in the line. Amplifiers are highly susceptible to radio frequency **(rf)** energy.

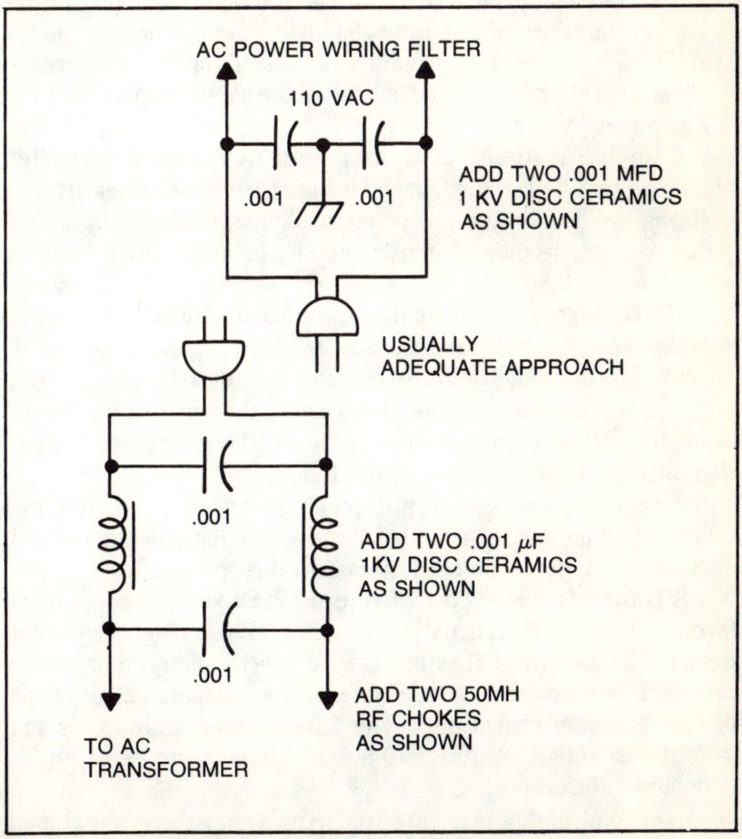

Fig. 20-7. Two methods of installing an AC power line rf filter.

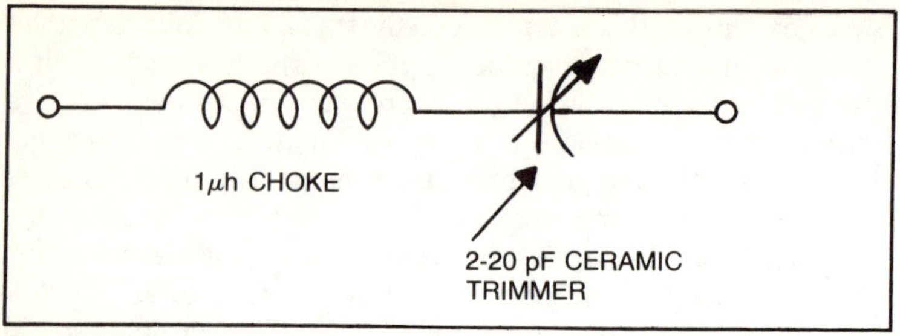

Fig. 20-8. Installation of a tuned filter across the antenna terminals of a TV set.

If an amplifier is in the system, remove it from the circuit. If you find that this eliminates the interference, reconnect the amplifier, but protect the amplifier by grounding, enclosing it in a metallic rf-proof housing and grounding the housing, or installing a high-pass filter at the input to the amplifier. If one filter improves the condition, but does not entirely eliminate the interference, install two filters in series. If no amplifier is utilized, or the interference still persists after following one or all of the above steps, check the TV receiver system.

Check TV Receiver System. An AC power line **rf** filter should be installed to determined if the **rf** from the transmitter is entering the TV via the power cord. (A line filter may be either purchased or one may be constructed by following the schematic shown in Fig. 20-7.

If no change is found with the power line filter installed, and the antenna disconnected, then the set itself is responding to the **rf** energy. The most likely internal circuit in the set to be affected by a radio transmitter is the tuner. Disconnect the antenna input lead inside the set directly at the tuner. If the interference is eliminated, then install a high-pass filter at the tuner.

If the interference is still present after installing the filter at the tuner, it will be necessary to refer to service data for the set and check each stage of the set for undesired response.

CB Interference To TV Channel 2. Second harmonic interference from a CB transmitter to channel 2 television may exist even though the transmitter meets FCC specifications for harmonic radiation. In such cases, a tuned filter across the antenna terminals of the television should help. The filter may be an inductor and capacitor in series as in Fig. 20-8. The filter should be tuned for minimum interference.

A second method is to put an open circuit, quarter-wave, tuned stub across the antenna terminals. The stub should be made of the same type of wire as the antenna input terminals of the television. The initial stub length should be 37″ for RG-59/U coax; and 48″ for 300-ohm twin-lead.

After connecting the stub, cut the unterminated end of the stub off in ⅛″ to ¼″ sections until the interference is eliminated. Refer to Fig. 20-9. For harmonics falling on other TV channels, such as channel 5, 6, or 9, length of the stub may be appropriately shortened according to the following formula.

$$\text{Length in inches} = \frac{2952V}{f}$$

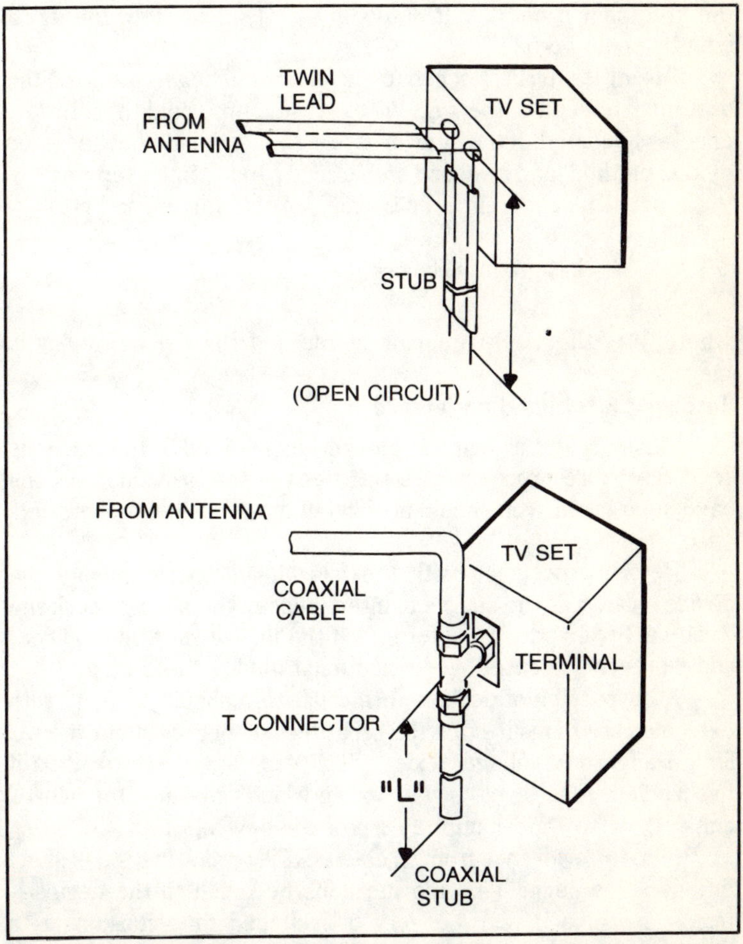

Fig. 20-9. Installation of an open circuit, quarter-wave, tuned stub for both twin-lead and coaxial cable systems.

where **v** is the velocity factor of the line and f is the frequency in megahertz.

Amateur Interference To TV Channel 2. One additional type of interference from a nearby transmitter is unique to the amateur 6-meter band—50-54 MHz. Since 6 meters is immediately adjacent to channel 2 television (54-60 MHz), interference to channel 2 may occur.

In most cases, installation of an open circuit, quarter-wave, tuned stub at the antenna terminals of the television set should be effective. It should be connected as shown in Fig. 20-9.

If RG-59/U is used as the TV lead-in wire, the initial length of the stub should be 42″. If 300-ohm twin-lead is used, the initial length should be 53″.

After the stub is attached to the television, begin cutting off the unterminated end of the stub ⅛″ to ¼″ at a time until the interference is eliminated. If the interference is reduced, but not eliminated by this method, add a second stub directly to the input terminals of the tuner. The theoretical final length of the stub should be:

$$\text{Length in inches} = \frac{2952V}{f}$$

where **V** is the Velocity factor of line and f is the frequency in megahertz.

Resolving Electrical Interference

Please read through the procedures outlined in the previous sections before proceeding. If the steps in the previous sections have been taken, you should now know the source of the interference.

Before proceeding with the following steps to modify the device located as the source of interference, you should check the local electrical codes to determine if the device may be modified, and whether a licensed electrician must modify the device.

All bypassing of devices with capacitors should be done with extreme care to insure that the capacitors do not short out the AC line. Dangerous voltages exist which can cause electrocution if mishandled. Also, avoid power wiring which can cause the full AC line voltage to appear on the case of the device.

Since interference from an electric drill or saw may be of short duration, we suggest no modifications be made to the device—mainly because it may be very difficult and time-consuming to modify the device. If, however, interference is of long duration, and you wish to take on this task, proceed as follows.

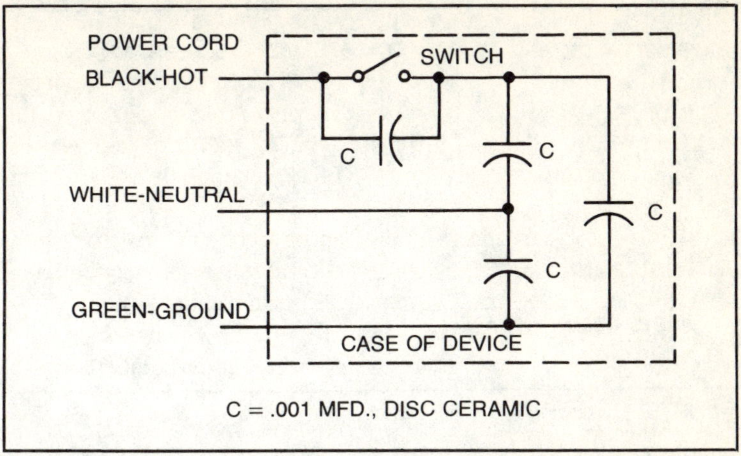

Fig. 20-10. Capacitor bypassing to remove interference from arcing.

Interference from a drill or saw is actually caused by arcing between the brushes and commutator. The interference then is transmitted through the power core. Bypassing each side of the line to ground with a capacitor, and each side to the other may be helpful. Also bypass the switch. Figure 20-10 shows the schematic involved. The bypassing should be internal to the device in question.

Electric blankets, fish tank heaters, and other thermostatically controlled appliances with worn and pitted contacts cause interference because of contact arcing of the breaker points. This can be eliminated by bypassing the contacts with a 0.001 μF capacitor or replacing the worn or pitted contacts (see Fig. 20-11).

Defective devices such as doorbell transformers should be replaced. Dimmer switches that utilize an SCR or triac can produce tremendous interference, and it is very difficult to eliminate. This is

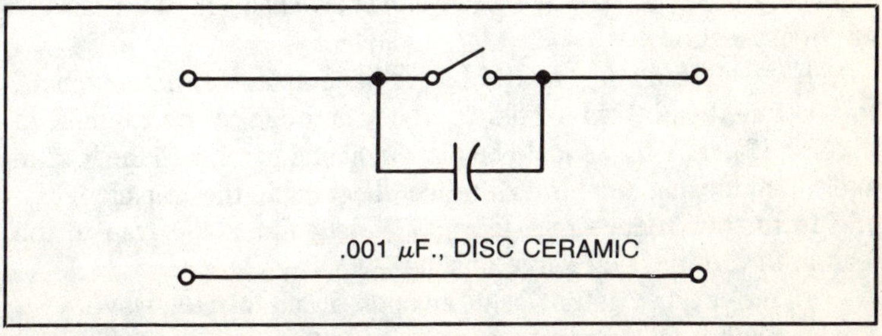

Fig. 20-11. Capacitor bypassing to remove interference from thermostatically controlled appliances.

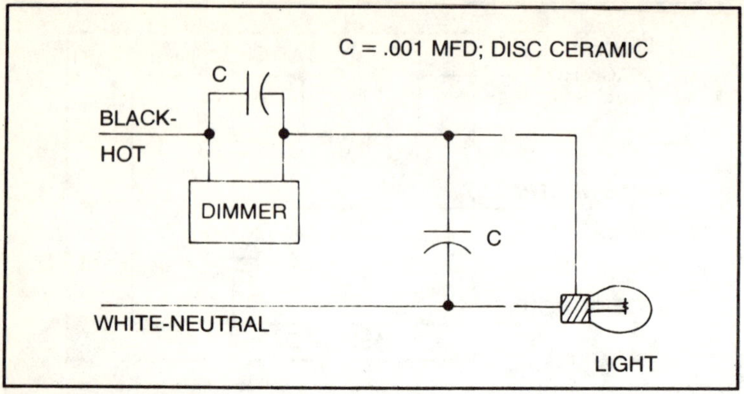

Fig. 20-12. Bypassing a dimmer switch.

due to the approximate square wave output that is produced by the switching at the SCR or triac. However, bypassing in a manner shown in Fig. 20-12 may be helpful.

Since resolving electrical interference has to proceed on a case-by-case basis, you should always consider adequately bypassing any component of the circuit that arcs or distorts the AC sine wave with ceramic condensers.

Resolving FM Interference

There are no set procedures for eliminating FM interference—it is a matter of eliminating the most likely sources of interference a step at a time. You may be required to take several steps before the interference problem is resolved. Once you have installed the filter called for, or made the adjustment that you were instructed to do, leave the modifications in place and proceed to the next step.

To begin, check to see if an FM band rejection filter has been installed on the TV set at the antenna terminals. If not, read the previous section.

If the installation of an FM band rejection filter is not effective, then a tuned stub trap should be constructed (see the example in Fig. 20-13). The trap should be placed on and parallel to the lead-in and tuned for minimum interference. Then slide the trap along the line to further reduce interference. Finally, tape the trap to the lead-in in the most effective position.

Another type of stub, called an open circuit quarter-wave type, can be made from the same type of wire as the antenna lead-in wire (see Fig. 20-9). The initial length of the stub should be 24″ for

RG-59/U coaxial cable or 29″ for 300-ohm twin-lead wire. For other cables, the initial length can be determined by the general formula:

Length in inches = (35) (Velocity factor of line)

If F-type tee connectors are not available, use BNC-type connectors.

If connecting the stub to the antenna terminals is not completely effective, connect a second stub of the same length directly to the input terminals of the tuner, inside the television set. This should eliminate the interference.

Resolving Audio Interference

Audio interference is defined as reception of radio frequency (**rf**) energy by an audio amplifier. The **rf** energy is then rectified, or more properly detected, by an electron tube, transistor, diode, poor solder joint or ground or integrated circuit, The detected signal is then treated identically as a normal audio signal appearing at the amplifier input terminals. The effects of audio interference vary with the type of modulation employed by the transmitter. The following shows expected effects:

AM—The voice or music will be heard as any normal audio signal applied to the amplifier. The voice or music may be extremely loud and slightly distorted.

SSB (single sideband)—The voice will sound practically unintelligible and garbled.

FM—Usually no sound will be heard; however, a decrease in the volume of the amplifier will be noted when the radio transmitter is on. Clicks may be heard when a two-way radio transmitter is keyed and unkeyed. A "frying" noise (such as bacon sizzling) may also be heard.

TV—Audio rectification of a TV signal will sound like a buzz. The buzz will change its sound as the television picture changes.

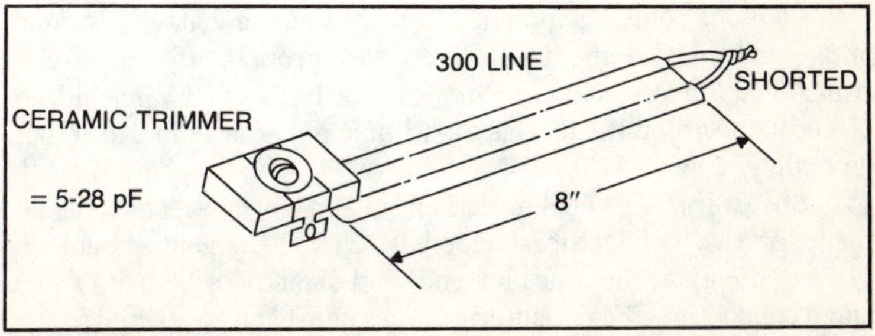

Fig. 20-13. A tuned stub trap.

In attempting to isolate where in the audio chain the rectification is taking place, check to determine if the volume control has any effect on the interference. If the volume of the interfering signal changes with a change in the volume control, then the rectification is occurring *before* the volume control. If the volume control has minimal or no effect, the rectification is occurring *after* the volume control. You should next proceed to the appropriate set of solutions. If the solutions described below do not resolve the audio interference problem, contact the manufacturer of the audio device for further assistance.

Rectification Before The Volume Control. A multiple input audio amplifier may be susceptible to audio interference on only one or some of the available inputs. Generally, low-level, high-impedance inputs, such as those in turntables, cartridges, tape heads, or microphones, are the most susceptible. If, for example, the only input affected is from a turntable, then disconnect the turntable cartridge from the amplifier at the input terminals of the amplifier.

If the interference is eliminated, then the cartridge, or wire between the cartridge and amplifier, is sensing the **rf**. Proper grounding, connections, shielding, and **rf** bypassing are the keys to solving audio rectification. Often, a "process of elimination" approach must be used.

Grounding. All grounding should be to a good earth ground such as a metallic cold water pipe or 8′ ground rod. Ground leads should be as short as possible. Remember, a DC ground may appear as an open circuit to **rf** energy. Ground leads should be of as large a diameter wire as practicable. Finally, grounding of the chassis, shields of speaker leads, and other external connections should be made to a common point to avoid ground loops. (Ground loops are circuits that form a DC ground, but contain rf circulating currents.) Figure 20-14 shows the correct and incorrect methods of grounding components.

Caution: Some equipment chassis are at line voltage potential and cannot be connected directly to ground. In these circumstances, a ceramic capacitor of 0.001 μF at 1kV should be placed in the ground lead. This capacitor appears as a short to **rf**, but an open circuit to AC.

Shielding. All speaker leads from audio equipment should be made of two conductor shielded wires. The shield should be grounded only at the amplifier end, and should not be used as an audio conductor. The two internal wires should be connected to the speaker.

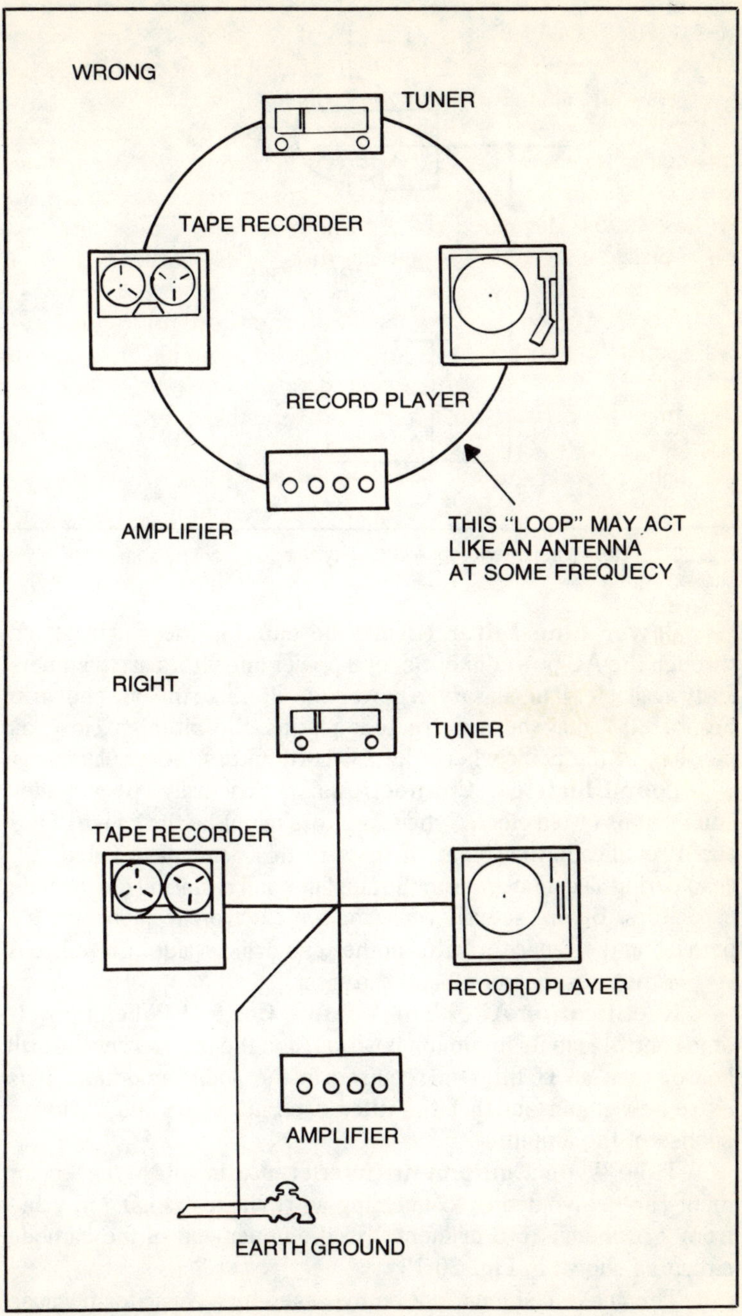

Fig. 20-14. The right and the wrong way to ground components.

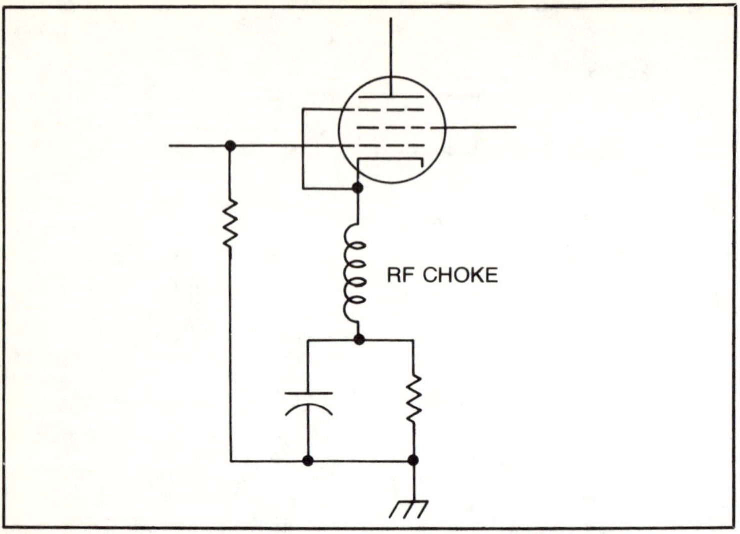

Fig. 20-15. Removing interference in tube type equipment by connecting an rf choke.

Power Line Filter. RF may be entering the audio device through the AC power line. Several power line filters are commercially available. If necessary, a power line filter like the one shown in Figure 20-7 may be constructed, placing the filter as close as possible to the point where the AC cord enters the amplifier.

Poor Electrical Connections. Occasionally, poor solder connections or old electrolytic capacitors might be the cause of the audio rectification problem. If tests to this point have failed, try resoldering all connections in the amplifier and replacing electrolytic capacitors. Before actually replacing the electrolytic capacitor, try paralleling the capacitor with another one of like value. This should reveal the presence of a band capacitor.

Rectification After The Volume Control. When the volume control is in its minimum position, and the interference is still heard, then an **rf** filter is required in the audio amplifier. It is extremely important that the filter does not affect the audio response of the amplifier.

Tube Type Equipment. Interference in tube type equipment can be avoided by connecting an rf choke (ranging in value from 2 milihenry to 5 milihenry) in the upper end of the cathode circuit as shown in Fig. 20-15.

The choke coil must *not* be bypassed by a capacitor because the DC resistance of such coil is generally quite low and the bias

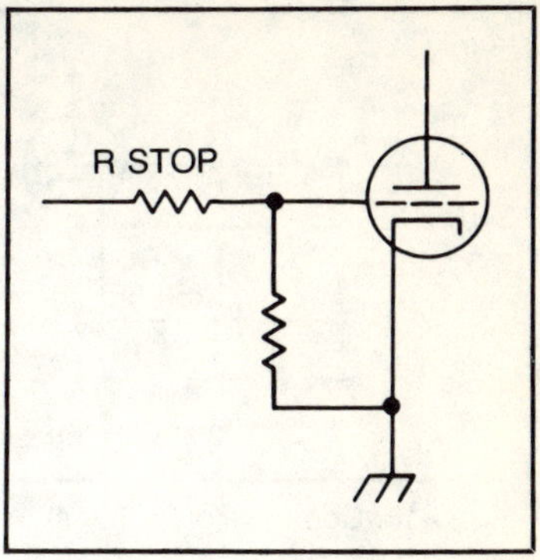

Fig. 20-16. Removing interference with a grid-stopping resistor.

voltage is not greatly affected. However, if the DC resistance does affect the bias voltage, the value of the bias resistor may be decreased to compensate for the DC resistance of the choke.

A grid-stopping or swamping resistor can also be employed. A resistor, ranging in value from 1K to 75K ohms, can be connected in series with the grid as shown in Fig. 20-16.

Capacitors, **rf** chokes, and resistors can be used in combination to make filters to eliminate the interference. For circuits such as those shown in Fig. 20-17, use a choke of 2 to 6 microhenries and a capacitor of about 10 picofarads. A combination rf filter is shown in Fig. 20-18 with the recommended values.

Transistor Equipment. Interference in transistor equipment can usually be eliminated with the use of a shunt capacitor as shown in Fig. 20-19. A resistor/capacitor combination can be used as shown in Fig. 20-20. It is important that the filter network does not affect the biasing of the transistor or the frequency response of the amplifier.

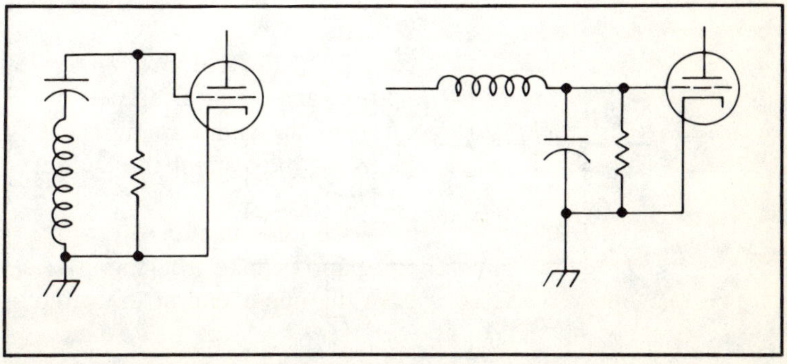

Fig. 20-17. Capacitors, rf chokes, and resistors can be combined to make interference filters.

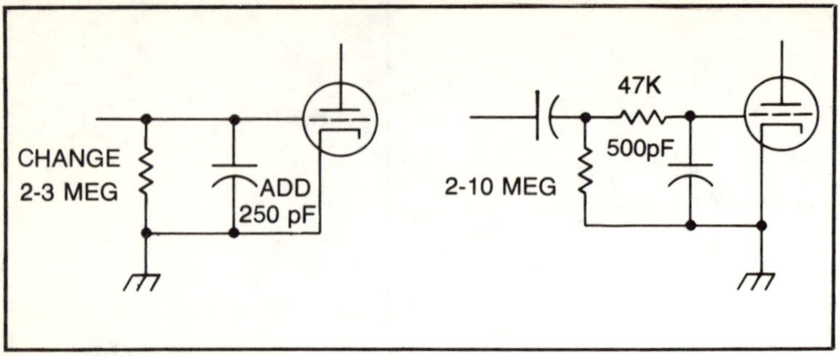

Fig. 20-18. Combination rf filters.

The values of the capacitors used are not critical, but there are some pitfalls to look out for in using capacitors. For example, ceramic caps are best, whereas paper caps do not work at radio frequencies.

Leads should be kept as short as possible. Grounds should be made directly to the emitter and not to the chassis or other grounds, since they may have more **rf** than the signal lead. If the signal increases, then a ground loop has been created, and the inductor method should be tried.

In areas of high **rf** energy, the inductor approach is more effective than the shunt capacitor. An **rf** choke can be used in series with the input and output leads of the amplifier stage since the **rf** can enter a stage through either. This method and the values are shown in Fig. 20-21.

Electronic Organs. Organ circuits can be isolated by the use of the swell pedal, band box volume, or tabs (draw bars). By

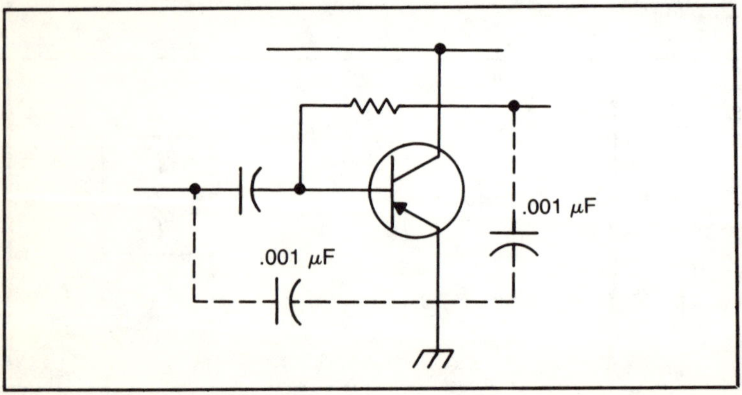

Fig. 20-19. Using a shunt capacitor to remove interference in transistor equipment.

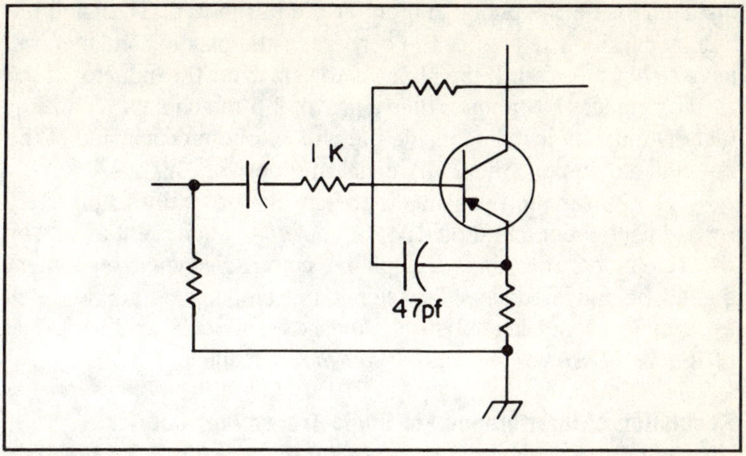

Fig. 20-20. A resistor/capacitor combination for removing interference in transistor equipment.

adjusting each one of these different controls, the effect on the interference can be noted. If the volume of the interference changes, the rf is being detected by the amplifier at a point *before* that particular control. If the volume of the interference does not change then the interference is being detected *after* that control. Using this method, the point at which the **rf** is entering the organ can be determined, and the appropriate filter, as described above, can be inserted into the circuit.

Telephones. Telephone **rf** interference can be eliminated by the use of a 1542A or similar inductor. This inductor must be

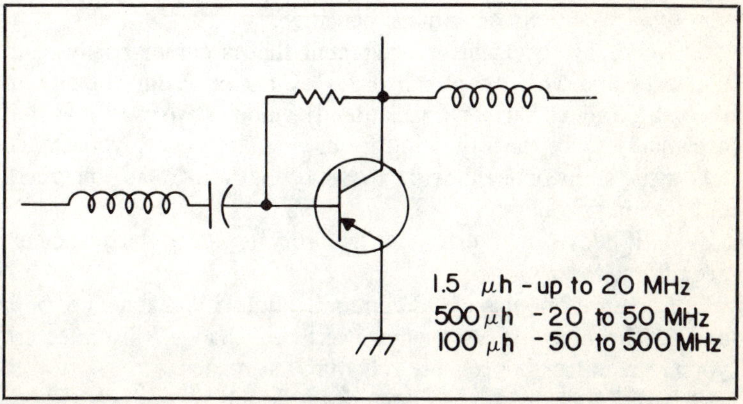

Fig. 20-21. An rf choke can be used in series with the input and output leads of an amplifier stage.

installed inside the phone and not at the baseboard. To install the inductor inside the phone, the corners of the plastic container will have to be removed. If the phone is too small for the inductor, such as the Princess telephone, then a pair of 2.5-mh chokes (75 mA or higher) must be installed inside the phone, one on each side of the line and as close to the 211A equalizing network as possible.

The information provided here applies primarily to privately owned equipment and should not be applied to equipment owned by the telephone company. Telephone company-owned equipment should be modified only by telephone company personnel. Bell system personnel can obtain additional data in Section 500-150-100 of the *Bell System Practices–Plant Series* manual.

Resolution of Interference For Radio Transmitter Operators

Although some interference problems can be attributed to television receivers, such problems can also be traced to CB and amateur radio transmitter. Therefore, upon receipt of an interference complaint from your neighbor(s), you should take all steps possible to insure that your radio transmitter is not causing the interference. Voluntary installation of a low-pass filter or other other steps might eliminate the interference and prevent you from receiving an order from the FCC to implement these measures. You are not, however, required to service or add filtering to the complainant's television, and should not take any such action without the full cooperation of your neighbors.

You are cautioned that the use of an amateur transceiver on the Citizens Band is illegal. Further, the use of external **rf** power amplifiers with CB transceivers is illegal. Both actions may subject you to FCC actions or criminal penalties.

Generally, transmitter equipment that is commercially manufactured and type-accepted by the FCC has precautions built into the set to reduce harmonic radiation. Harmonics are radiations that are multiples of the operating frequency. However, you should follow the steps outlined below to insure that your radio equipment is operating properly.

1. If television interference is occurring, note which channels are affected.

Lower harmonics of CB generally affect TV channels 2, 5, 6, and 9. Therefore, if one or more of these channels are affected, your transmitter is probably radiating harmonics.

If all TV channels are affected, the problem is more likely to be in the TV receiver.

2. If the interference is caused by harmonics, a spectrum analyzer, a calibrated field intensity meter, or frequency selective voltmeter, can be used to accurately measure harmonic and spurious radiations from your transmitter. If any lead-in devices, such as standing-wave ration (SWR) meters are used, measurements should be made with the inline device both installed and removed. This may help identify the interference and lead you to the source. These are complex measurements and should normally be made only by experienced technicians.

3. If it appears that your transmitter is at fault, you should first make sure the chassis of the set is secured to the metal case of the radio by tightening the screws holding the chassis and case together. Then assure that the case of the transmitter is grounded to a good earth ground (metallic cold water pipe or 8-foot ground rod). Solid conductor wire of at least No. 10 gauge or copper ribbon should be used as a ground lead. The lead should be as short as possible.

4. By installing one or more low-pass filters in the transmitter antenna lead, you will reduce the chances of unnecessary harmonic radiation. A low-pass filter allows frequencies up to 30 or 50 MHz, depending on brand, to pass through unattenuated to the antenna while effectively shorting out harmonic radiation. To make this test, connect the equipment as shown in Fig. 20-22 and take a power

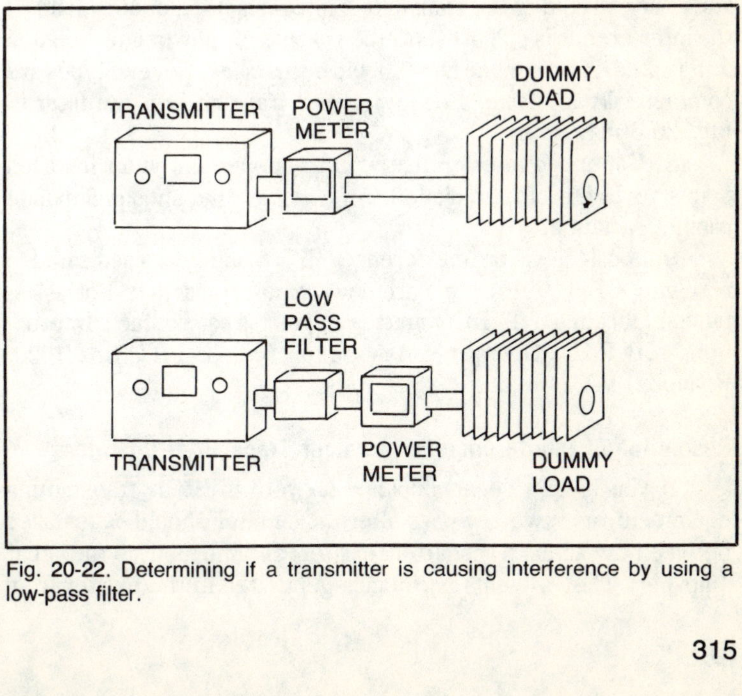

Fig. 20-22. Determining if a transmitter is causing interference by using a low-pass filter.

reading. If only a SWR bridge is available, calibrate it in the forward direction to the calibrate line in the meter. Then insert the low-pass filter and make another power measurement. Do not return the transmitter.

5. If you notice a decrease in output power on a power meter, operating to a properly matched load, with the low-pass filter installed, this is an indication that harmonic content may be present. Even though the meter reading may be lower with the filter installed, it does not mean that the transmitter absolutely has harmonic radiation. Slight detuning of the transmitter by the filter may cause a lower indication.

6. At amateur power levels, corroded metal connections in the area of the transmitting antenna may act like diodes and generate harmonics which may radiate. This type of problem can be found by vibrating suspected offenders such as galvanized downspouts, metal fences, clothes lines, etc., while viewing the affected television set. Sudden changes in the interference pattern which correspond to the vibration should be noted. This test requires on observer at the TV receiver, someone to "shake" suspicious metal objects in the area, and another person to key (but not modulate) the transmitter involved.

7. Finally, some transmitters may actually be radiating harmonic and spurious energy from their cabinet or through the power lines. Try operating the transmitter into a shielded dummy load. If the interference is still present, then cabinet or power line radiation is indicated. A power line filter should be installed. Several types are commercially available. For low power transmitters, the filter in Fig. 20-6 may be used.

8. Continued interference with the power line filter installed points toward cabinet radiation. An earth ground should eliminate cabinet radiation.

9. Local television interference (TVI) committees dedicated to resolving CB-TVI problems are now being established. For assistance in locating a TVI Committee in your area, contact: International CB Radio Operator's Association (CBA), P.O. Box 1020, Roanoke, VA 24005.

Resolution Of Interference For Amateur Transmitter Operators

If you have a linear amplifier on your amateur transmitting equipment, use two low-pass filters. One filter should be installed between the actual transmitter (exciter) and the input to the linear amplifier. This prevents harmonics generated in the exciter from reaching the linear amplifier. The second filter should be installed at the output of the linear amplifier to reduce harmonic and spurious contents.

One unique interference problem to TV channel 2 is from an amateur transmitter operating on the 6-meter band. This is due to the close proximity of the frequencies involved. You are not required to service or add filtering to the complainant's television, and should not take any such action without the full cooperation of your neighbor.

Local television interference (TVI) committees are available to assist you in resolving interference problems. Contact the nearest FCC district office (see the Appendix) or the American Radio Relay League, Newington, CT, for assistance in locating a TVI committee in your area.

Transmitter Operator Guidelines For Resolving Audio Interference

Although audio interference (often called audio rectification) is usually resolved by modification of the affected device, you as a radio operator can take certain steps to reduce the possibility of audio rectification by eliminating circulating radio frequency (**rf**) currents in ground and metal objects in the area. Your radio transmitting equipment should be effectively grounded to a metallic cold water pipe or a ground rod driven into the ground at least 8 feet. The ground lead must be at least No. 10 wire or copper ribbon. The greater the surface area of the ground lead, the more effective it will be. Also, the ground lead should be as short as possible.

You are licensed to use only the amount of power necessary to establish communications. Operating with excessive power is likely to cause audio interference problems.

If you need assistance in performing the above modifications to your equipment, you can contact the dealer or manufacturer representatives. Also, an FCC-licensed service representative may be able to assist you.

Chapter 21

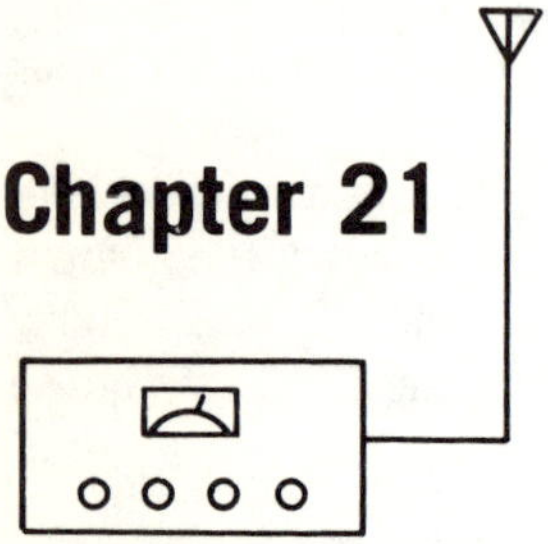

Transmitter Tests And Measurements

The responsible operation of radio transmitters sometimes requires us to be able to make certain measurements on the emissions. We must be able to be sure that the signal that we are sending out does not cause undue interference with other services and is as clean as the "state of the art" permits. In amateur radio, certain measurements are beyond us economically as the test instruments are too costly. We can make certain measurements to an acceptable degree of accuracy, however, and these can be essential when troubleshooting a problem.

In this chapter we will consider some of the most fundamental transmitter measurements: output power, modulation, and operating frequency. The emphasis will be on simple, low-cost test equipment readily available to most amateur radio operators on a borrow, if not own, basis.

TEST EQUIPMENT

Certain items of test equipment are essential to any decent attempt at measurement. Some items are "must haves" for radio amateurs, while others are highly desirable. It is necessary that an amateur with any pretentions at all of being "technical" possess at least a VOM. The old-fashioned VOM is a handy instrument around a transmitter. The high rf field produced by the transmitter often renders electronic voltmeters (VTVM, FETVM, DVM, DMM, etc.) unusable. The old standby, passive VOM, invented prior to

World War II, is still the best instrument in those situations. Imported models with decent accuracy can be had for about $30 to $50.

Don't overlook a good digital voltmeter (DVM), however. Once the costly pride of a few industrial/military laboratories, the DVM is now available for less than $100. Instruments costing less than $200 will provide accuracy and resolution that is simply nonattainable in VOMs of any grade. In addition, the input sensitivity (ohms/volt) is tremendous in these instruments. In some low-power transistor circuits, the VOM will have too low an impedance for proper operation. These will tend to turn on PN junctions and shunt some of the resistances too much for any serious measurements to be made. The digital voltmeter and the analog electronic varieties called VTVM or FETVM will not load the circuit and will not forward bias the transistor circuits at inappropriate times. There are plenty of good arguments for both types of multimeter, and you should own *both* if possible. While I prefer a digital meter (I'm a snob), the cost may make it beyond the reach of many. Other electronic meters can be bought cheap, especially used. One source might be TV/hi-fi/CB repair shops who might have an old RCA or Triplett VTVM stuck under a bench somewhere, forgotten the second day the digital wonder machine was in the place. Buy it cheap. It is quite likely that you can get both types of instrument (VOM and VTVM) for less than $50 used and not much more than $100 new. Not having these can be a pain when trying to repair something. If you can, try obtaining the high-voltage probe *for the model that you buy*, especially if you own or contemplate owning a high-power linear amplifier.

You might want to consider also an oscilloscope—and learn how to use it. The triggered sweep, dual-channel oscilloscope is now the standard in the TV and hi-fi repair business. You can sometimes pick up older recurrent sweep models for a song and a dance. There are no other instruments that are quite as useful as the oscilloscope. Of course, if you can afford some nice Tektronix or H-P 50-MHz whizzbanger, then go to it! I wish that I had one! But for us mere mortals, even a lower frequency model can be useful. Try to avoid the 500-kHz models, unless they are dirt cheap (they ought to be) and have the terminals that allow you access to the vertical deflection plates. TV servicer-grade oscilloscopes are often marked "color-TV," or something else that alludes to color servicing. This means that the scope will possess the vertical bandwidth needed for color TV work, usually speced out at 4.5 MHz or greater. If you can find some guy who turned in a 4.5-MHz scope

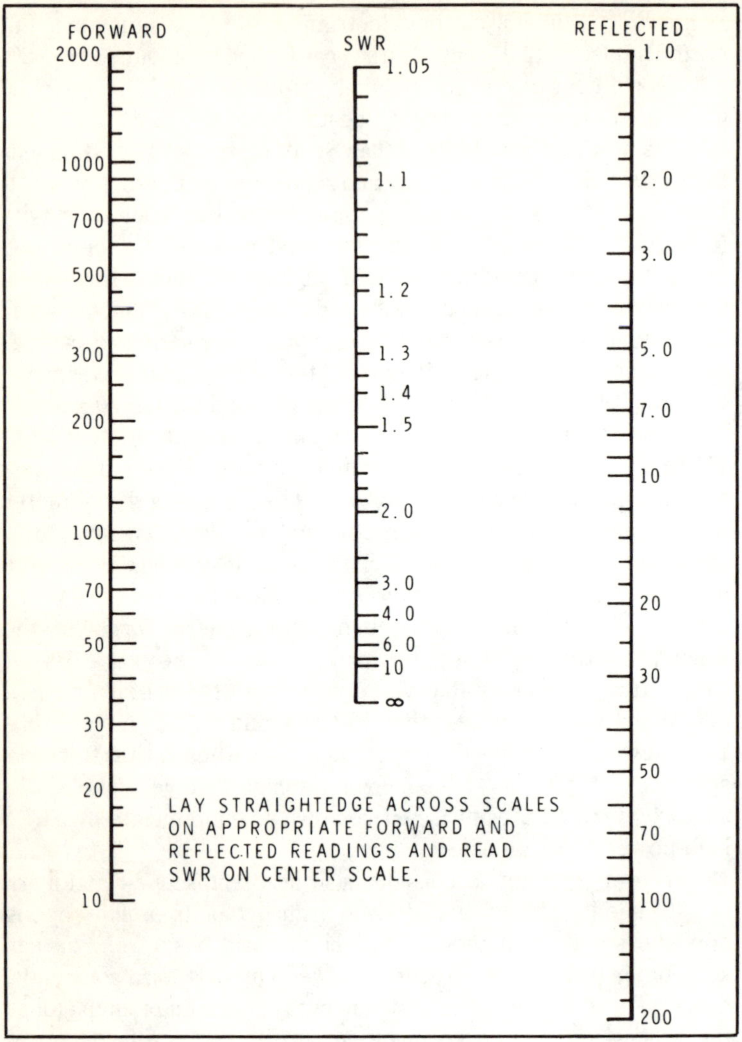

Fig. 21-1. SWR/power nomograph.

with recurrent sweep in order to ravish some nice new triggered-sweep model, then grab it as inexpensively as he will turn it loose for. If necessary, beg. Beware, though, of certain kit models other than Heath; they were dogs. RCA, Leader, B&K, Sencore, and Telequipment (a British subsidiary of Tektronix) are the names of some of the most common. I am particularly fond of B&K and Telequipment, but that is a personal opinion based as much on what I have used over the years as anything else.

Another thing that you should have is a small collection of tools—the kind needed to build typical Heathkit should suffice. Also, gather a collection of alligator clip leads, BNC/BNC path cords, UHF/UHF patch cords, and any coaxial connector adapter that you can scrounge up at the next hamfest. Adapters always come in handy someday.

RF POWER/VSWR MEASUREMENT

One of the *first* pieces of "test equipment" that any amateur should buy is a voltage standing-wave ratio (VSWR) meter. Even one of the little cheapies will do, but, at a little extra cost (not as much as you might believe), you can buy a genuine rf power meter. If you want to do it right, then go out and spend the bucks for a Bird Model 43 Thruline. If you are like the rest of us peons, though, then buy one of the less costly models. True, these devices might not deliver the precision of the Bird, but are acceptably accurate for amateur applications. Remember that the FCC only requires accurate measurements of the *DC input* power to the final amplifier (and then only in the over-900-watt range), *not* rf output. We can use the rf power information in many troubleshooting and bragging sessions, anyway. The VSWR can be computed from the rf power in those cases where the meter is not calibrated in units of VSWR. The nomograph in Fig. 21-1 shows how this is done. Locate the forward and reflected power readings on their respective scales and lay a

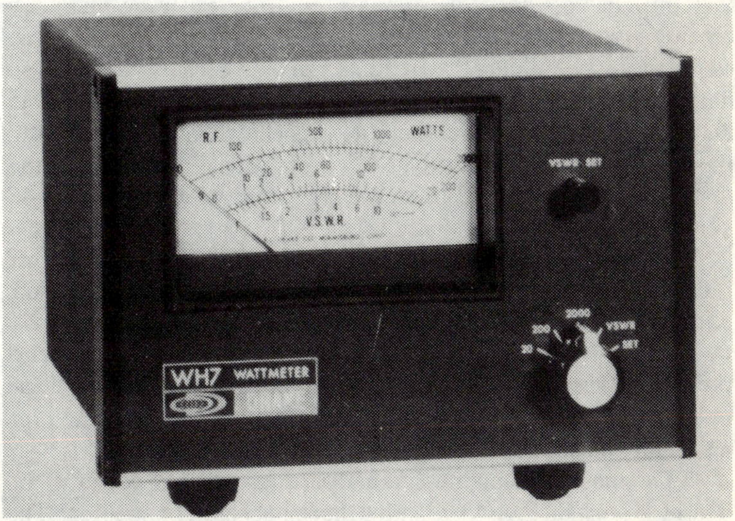

Fig. 21-2. Drake SWR/rf power meter.

Fig. 21-3. Heath SWR/rf power meter.

straightedge between them. The point where the straightedge crosses the middle scale is the VSWR for that combination of forward and reflected power.

We will not delve into the construction of VSWR and rf power meters here. This is covered in other TAB books, including one by this author. But we will show you some commercial examples. These are low-cost models by amateur equipment suppliers. Note that the cost of these instruments, and the accuracy that they obtain, is well worth the relatively low cost. While there is some justification for building the Heathkits, if only because they look so neat, there is little point in completely homebrewing such an instrument. I built a model (not the ones shown) and returned it to the factory for calibration against a professional meter. After the instrument was returned to me, it compared with Bird Model 43 very well.

The instrument in Fig. 21-2 is the Drake WH-7 meter. It will measure rf power to 2000 watts in three decade ranges (20, 200, and 2000) has a switch position for VSWR measurements.

The two Heath units shown in Figs. 21-3 and 21-4 are dual rf power meters. They each contain two meters, one for forward and one for reflected rf power. Like the Drake unit, these instruments also allow measurement of VSWR.

The most fundamental method for measuring rf power is the use of a noninductive load and a thermocouple rf ammeter. The thermocouple meter uses a low-resistance, noninductive element that is heated by the rf current in the line. A thermocouple is a device that will produce a small voltage output when a junction between two different metals is heated. The output of the thermocouple is connected a DC meter and is proportional to the current flowing in the line being measured. We can calibrate the deflection of the meter in terms of amperes. If the load is nonreactive, such as a noninductive dummy load or a perfectly resonant antenna, the power can be found from the expression $P = I^2R$. For example, if it is found that 450 mA are flowing in a 50-ohm noninductive load, the power is (0.45)(0.45)(50), or 10.1 watts.

CHECKING MODULATION IN AM TRANSMITTERS

There are several methods for checking the modulation of an AM transmitter. Many of the low-power CB "test sets" sold through mail-order and hobby retail outlets purport to test modulation. They do this by rectifying the carrier. Under no-modulation conditions, the meter is zeroed, so that the DC output of the rectifier will produce a zero reading on the meter. When the operator speaks into the microphone or a test tone is applied, the meter will indicate the peak voltage that is proportional to the percentage of modulation. This assumes that a sine wave is used, and it is really only a rough check. An oscilloscope would produce better information and would allow us to tell whether or not the modulation was distorted or whether hum and noise existed on the carrier along with the *desired* modulating signal.

Carrier shift is a phenomenon that relates to the linearity of the AM modulation process. Figure 21-5 shows a simple carrier shift

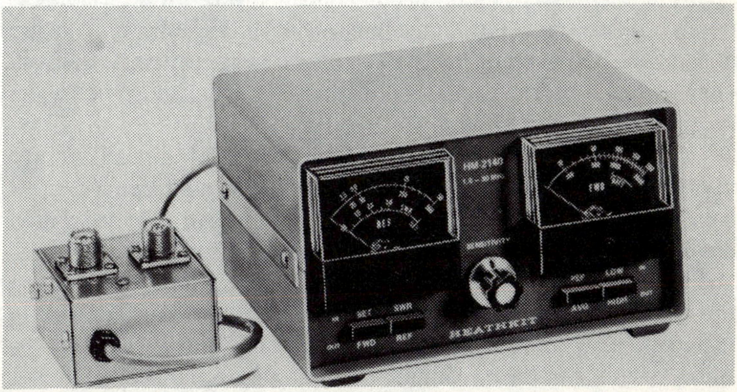

Fig. 21-4. Heath SWR/rf power meter.

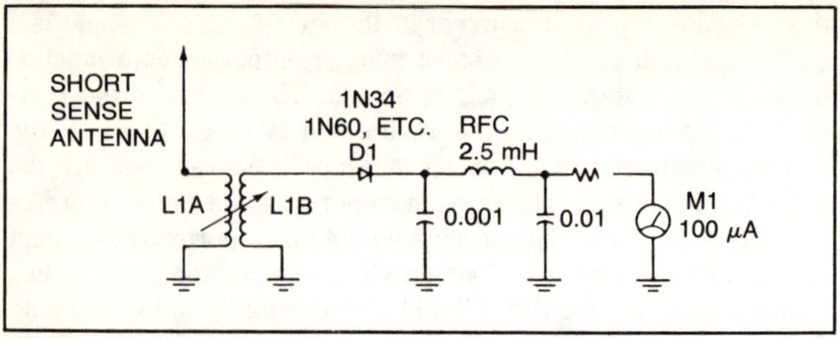

Fig. 21-5. Carrier shift meter.

indicator that will detect changes in carrier symmetry. This type of meter is used primarily in stations that always operate on the same frequency and is a monitor of the output signal.

The rf pick-up coil is adjusted to produce a convenient deflection on meter M1 (midscale seems about right). The rectified rf passes through a low-pass filter to the DC meter, M1. The component values in this filter are selected to attenuate rf, but to not produce any smoothing (like a power-supply filter) action at audio frequencies.

If the modulation envelope emitted by the AM transmitter is symmetrical, then the positive and negative peaks will have equal value. All analog meters provide some damping action due to their own inertia, and this will integrate the audio frequency to read out an average value. But if the modulator becomes nonlinear, the modulation envelope will develop assymmetry, and this will change the average value of the rectifier output, producing a change in the deflection of the meter pointer.

The carrier shift meter is not useful for measuring the percentage of modulation or indicating much about the extent of modulation impurities. It is useful, though, as a monitoring technique that will keep a broadcaster out of trouble with the FCC field engineer when he comes snooping around.

The best way to check the emission of the AM transmitter is to use an oscilloscope. If the oscilloscope has sufficient vertical bandwidth, then it can be used directly to monitor the modulation envelope. If the oscilloscope is located in the field of the antenna, and the transmitter is powerful enough, then it is possible to obtain the needed waveform simply by connecting a short piece of wire to the vertical input of the oscilloscope. The pattern will not be stationary unless the transmitter is being modulated by a single-tone sine wave. In that case, it will prove possible to lock the time base of the oscilloscope onto the modulating frequency and obtain a pattern that resembles Fig. 13-2. On lower power transmitters, some people will make a *gimmick*. This consists of a short piece of hook-up wire connected to the center conductor of a piece of coaxial cable. The coax is connected to the scope input, and the wire is dangled inside the transmitter near the final tubing coil. The gimmick will pick up sufficient rf to display a pattern on the CRT screen of the oscilloscope. This procedure is *dangerous as the dicken's*—you could get *killed* trying that if you aren't careful and don't really know what you're doing! A better procedure is to sample the output. Most scopes with any degree of vertical sensitivity at will require only a breath of hot air on the input in order to produce some form of readible display. We can give the input a real hurricane-force signal is we use an appropriate signal pickup. But first, a warning:

> Never connect the output of the transmitter directly to the input of an oscilloscope, even with a dummy load! Only oscilloscopes that are *specifically monitoring* scopes for radio transmitters allow direct connection. These models contain a built-in pickup such as we describe next.

Figure 21-6 shows two types of rf pickups that can be used to sample the rf waveform for application to the input of oscilloscopes,

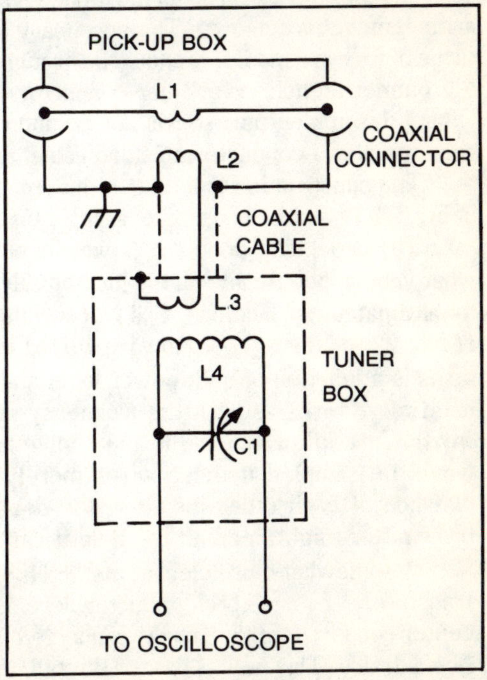

Fig. 21-6. Oscilloscope/counter signal pickup.

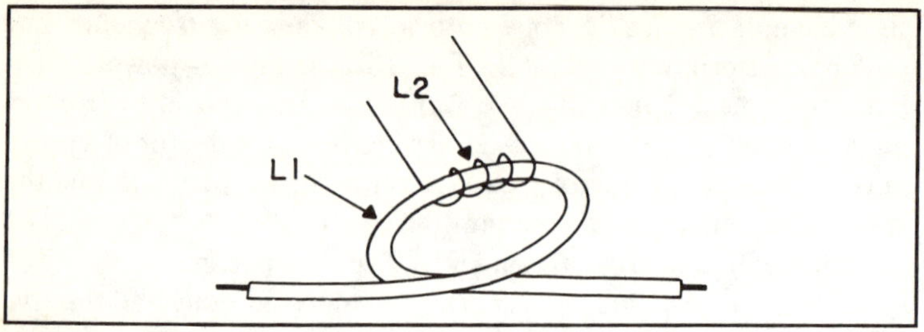

Fig. 21-7. Oscilloscope/counter gimmick.

digital frequency counters, and so forth. Both of these methods are capable of producing very high signal levels—more than adequate for most measurement purposes—without danger to the measurer or its operator.

The pick-up box is a small shielded aluminum box that has a pair of coaxial connectors on the ends. The box may be made of a Bud sheet metal enclosure, or one of the more substantial Bud or Pomona die-cast aluminum boxes. Note that the Pomona type of box can be purchased with the connectors already in place.

Coil L1 is the coaxial cable center conductor, extended. This "coil" might be either a straight run or a single-turn loop, and both arrangements are shown. The secondary coil, L2, will be several turns of wire around L1. It should be brought out of the box through a grommetted hole or another connector. Many people prefer to "float" this link output, so will not ground one side of L2. But this requires a two-conductor shielded cable instead of mere coax.

The construction of L1/L2 with a single-loop of wire is shown in Fig. 21-7. Coil L1 consists of a single turn of No. 16 through No. 10 wire, depending upon the power level of the transmitter and what you happen to have in the junk box. No. 14 wire should work for all amateur applications. Coil L2 consists of from one to 10 turns of No. 22 to No. 26 hook-up wire wound over L1. The number of turns is a function of both power level and frequency. You would need more turns as both the frequency and the power level *decrease*. In a kilowatt pick-up unit, for example, adequate signal would be sampled if these wires merely came within shouting distance of each other inside of the box. But a QRP 80-meter transmitter might need all the help it can get.

A somewhat more elegant method is shown in Fig. 21-8. This transformer uses a toroid inductor core as the pick-up unit. The center conductor of the coax is connected to a piece of No. 12/No. 14 solid wire. This wire is passed through the center of a 3/16-inch black rubber grommet that is cemented into the center hole of the toroid core. This is L1. Select a core that is rated for the frequency range of interest and that has an inside diameter that will allow the grommet to be squished into the hole without too much deformation. The grommet will probably stay without any cement, but if it needs help, then a little dap 'ill do ya—use super glue if you lack patience and epoxy if you're cool. Coil L2 is approximately 25 to 100 turns of No. 26/No. 28 enamelled or formvar coated wire for 1-MHz to 10-MHz operation and fewer turns (down to about five) for the rest of the HF spectrum. Again, the power level of the transmitter and the frequency of operation make the difference. It might be wise to try it with, say 20 turns, and see if it works. If not, add more turns. Do not cement the grommet into place, however, until you are satisfied that the pick-up unit works properly.

Both types of pick-up units should keep L1 and L2 mounted away from the sides of the aluminum box. That is why I prefer the toroidal method of Fig. 21-8 over the other. A toroid naturally keeps the magnetic field closer to home, so it is less affected by nearby objects. In addition, the construction of the toroidal pick-up unit allows the thing to be placed almost dead-center in the box, along the line of centers between the center pins of the two coaxial connectors.

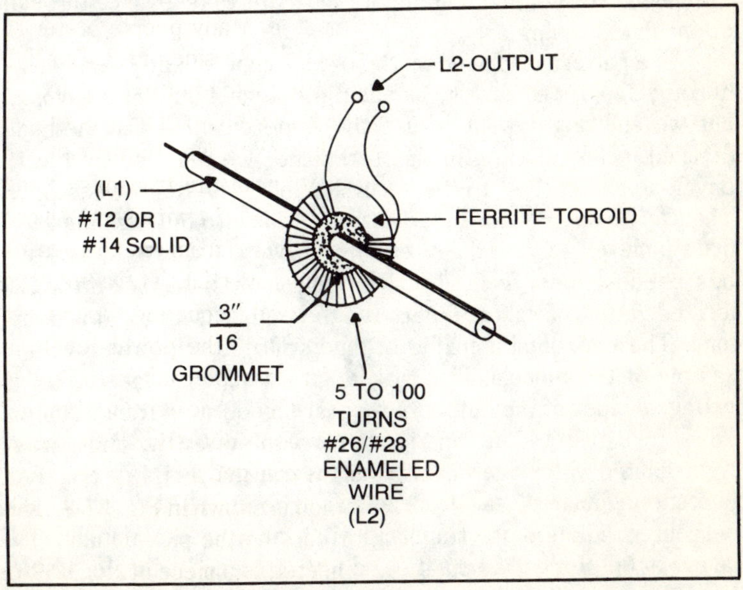

Fig. 21-8. Another type of pickup.

The rf pick-up unit may produce sufficient output signal by itself to drive the oscilloscope. This is especially true if the operating frequency of the radio transmitter is well within the vertical amplifier bandwidth of the oscilloscope and the vertical deflection sensitivity goes down to the low millivolts range. When these factors are not met, however, we must help things along with the LC tank circuit shown in the tuning box. This tank circuit can be connected directly to the vertical deflection plates of the oscilloscope to produce a usable deflection on the CRT. Some older scopes that lacked something in the vertical bandwidth department made the deflection plates available on the rear or side panels of the instrument. You could disconnect the vertical amplifier from the deflection plates by removing 1, 2 or 4 shorting bars from their sockets. This would allow you to connect the tank circuit directly to the plates. In most cases, the plates available to the outside world are AC-coupled through a capacitor. But this was not true in all cases—those pins could be very hot *electrically*! There could also be a leaky capacitor in the circuit. Most of the older scopes that offered this capability made extensive use of paper dielectric capacitors, and this means a high probability of leakage currents and a nasty shock for you. The answer, of course, is to make all connections with the scope turned off and the plates initially grounded. You could then tune the capacitor to resonance through an insulated tuning shaft.

The pattern displayed on the oscilloscope will be the modulation envelope. We can tell a lot about the signal from this envelope, but we still might want to use the trapezoidal pattern method, detailed in Fig. 21-9A. In this technique, we must sense the rf envelope and apply it to the vertical amplifier of the oscilloscope. This can be done with either a link on or close to the final amplifier plate tank, or a gimmick. Alternatively, we could use the pick-up box just described. As in all such measurements, the pick-up box is located in the coaxial cable between the transmitter output and the load. The horizontal amplifier of the oscilloscope is driven with a sample of the modulating signal; in this case, it is taken from a voltage divider at the output of the plate modulation transformer.

The patterns shown in Fig. 21-9B result when the transmitter is modulated with a sine wave. When the transmitter is less than 100 percent modulated, the trapezoid is truncated. At exactly 100 percent modulation, the trapezoid will form a perfect triangle, and at over-100 percent modulation, a foreshortened triangle. Using the designations shown in the less-than-100-percent modulation

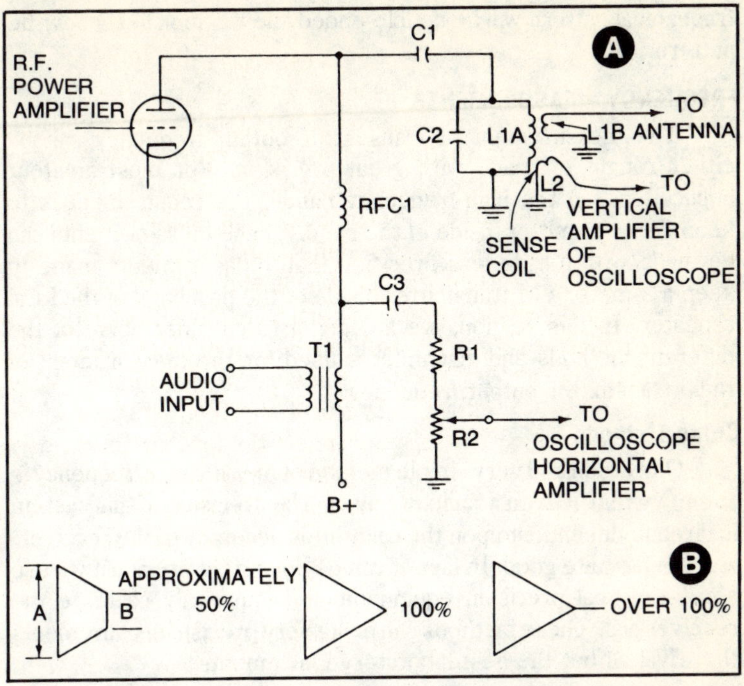

Fig. 21-9. Measuring modulation by trapezoidal method at A with waveforms at B.

example, the percentage of modulation can be measured using the expression:

$$\% \text{ Mod} = \frac{A - B}{A + B} \times 100\%$$

☐ **Example:**

Using the circuit of Fig. 21-9A, it is found that the trapozoidal pattern has 2 cm for the B side and 3 cm for the A side. What is the percentage of modulation?

$$\begin{aligned} \%\text{MOD} &= (A-B)\ (100\%)/(A+B) \\ &= (3-2)\ (100\%)/(3+2) \\ &= (1)\ (100\%)/(5) \\ &= 20 \text{ percent} \end{aligned}$$

Single-sideband transmitters, which are basically amplitude-modulated transmitters in which the carrier and one sideband are suppressed, are tested using a two-tone test signal. The input to the modulator is a pair of sine waves, spaced a little apart. The trapezoidal pattern will be double-ended and will look like a bow-tie pattern.

FREQUENCY MEASUREMENTS

The measurement of transmitter output frequency can be either rough, or made with great precision. For most amateur applications, in the high-frequency bands, the requirement is to keep the transmitter inside of the band. These measurements can be made with a little less precision than measurements made to keep a 2-meter FM transmitter inside of the passband of the local repeater. In this section, we are going to consider some of the different methods and instruments used for the measurement of radio transmitter output frequency.

Crude Methods

One crude and very simple method of measuring a frequency is to simply tune it in on a radio receiver. The precision of this system is directly dependent upon the calibration accuracy of the receiver, and can be quite good. In fact, it can be downright precise if we use some external precision equipment to temporarily calibrate the receiver dial. These methods can make very precise measurements that rival all but the best laboratory instruments.

Most crude frequency measurement techniques fall into the *absorption* classification. Figure 21-10A shows a simple absorption wavemeter. It consists of a low-current lamp connected in series with a fixed inductor and a variable capacitor. The inductor is placed in close proximity with the final amplifier tank circuit, so that energy from the tank is induced into the wavemeter coil. When the capacitor is tuned to resonate the coil with the frequency of the tank circuit, the circulating tank currents will cause the lamp to light up. Peak the intensity of the lamp and then read the frequency from a calibrated dial on the capacitor shaft.

Another type of absorption wavemeter is shown in Fig. 21-10B. In this case, the signal is picked up by a small sense antenna connected to the LC tank circuit. Some energy from the tank is coupled through capacitor C1, which is connected to an impedance matching tap on L1. The sample of the rf signal passing through C1 is rectified by D1 and applied to the meter. The capacitor can be tuned for a maximum reading on the meter. Then the frequency can be read on the calibrated dial connected to the capacitor shaft. This circuit can be used at a short distance from the transmitter, so it does not suffer from the danger of accidental contact with the final amplifier tank circuit.

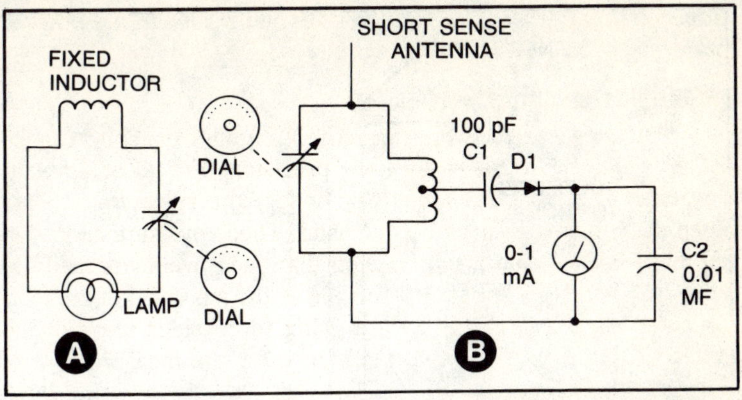

Fig. 21-10. Absorption wavemeters. A simple one at A and a more complex one at B.

The dip meter is also a means of making absorption frequency measurements. This instrument was once known as a *grid dip oscillator* because vacuum tubes were used. The grid current would suddenly drop if the oscillator tank coil was closely coupled to an external tank circuit that was tuned to the oscillator frequency. The grid current was monitored by a microammeter, and this allowed the operator to find the resonant point. But the dipper only finds the frequency of unenergized tank circuits. If the meter is replaced with a set of earphones, however, the dipper can be used as an *oscillating detector*. When the dip oscillator frequency approaches the frequency of the transmitter, an audio beat note is heard in the earphones. Adjust the oscillator frequency for zero beat and then read the frequency from the dial of the dip oscillator.

The last crude method which we will examine is the *slotted line* or *Lecher wires*. These are variations of the same basic idea. Lecher wires are shown in Fig. 21-11. These wires form a parallel transmission line with a shorting bar between the elements. A calibrated meter stick alongside the wires is used to measure the distance of the shorting bar from the feedpoint. The lecher wires should be made from No. 12 or larger wires or small diameter copper tubes.

The slotted line is a coaxial version of the Lecher wires. The outer conductor is slotted to admit a diode detector. In this case, do not short the line, but look for the amplitude of the rf signal in the line.

The Lecher wires shown in Fig. 21-11 use an rf voltmeter across the feedpoint. The wires form a shorted stub. A quarter-wavelength shorted stub exhibits a high impedance at the resonance

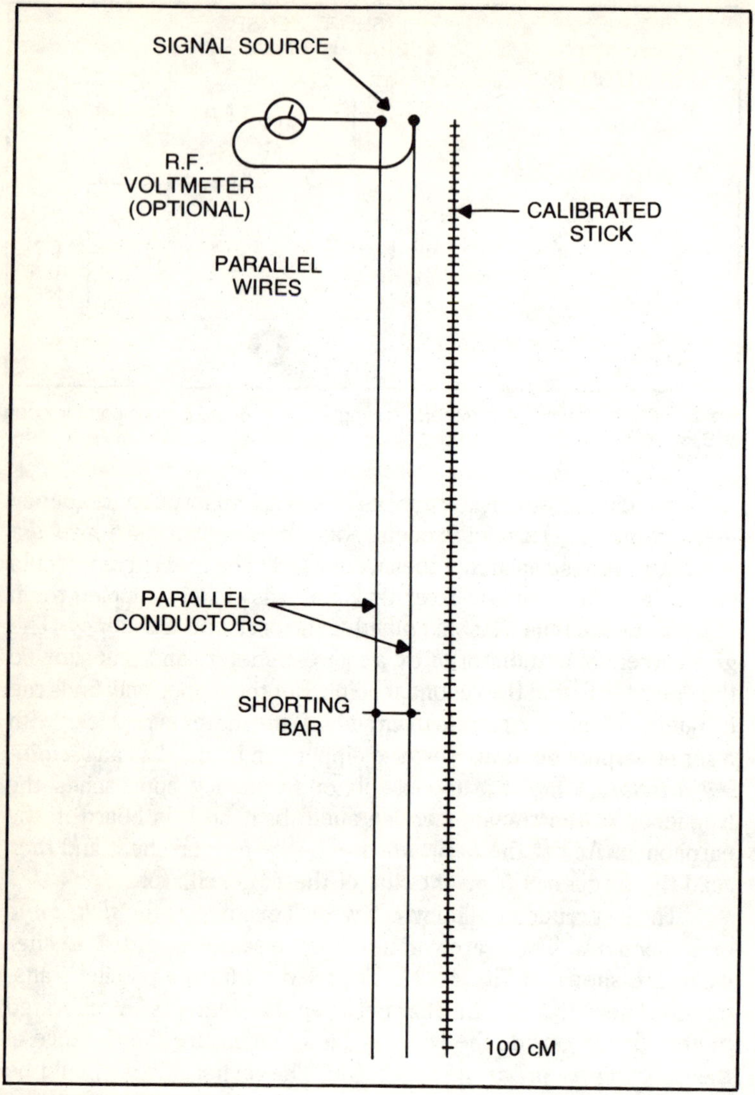

Fig. 21-11. Lecher wires.

frequency. This frequency can be measured by moving the shorting bar until a peak is found in the rf voltage across the input. The relationship is the well known:

$$F_{MHz} = \frac{k(300)}{L}$$

where F is the frequency in megahertz, L is the length in meters at which the peaks occur, and k is a constant, which depends upon whether the 1st, 2nd, 3rd or 4th peak is used. If the first peak is used, then k = ¼. If the second peak is used, then k = ½, and so on.

Comparison Methods

A comparison method is one in which some external frequency of known frequency and high precision is compared with the unknown. There are several such methods, all lumped under the rubric of *heterodyne* methods. Figure 12-12 shows the basics of a heterodyne frequency meter. The block diagram of this type of instrument also demonstrates the elements of the comparison method of frequency measurement. In this instrument, there are three basic sections: a mixer circuit, an output section (in this case, earphones), and a precision oscillator. The signal to be measured and the oscillator signal are mixed together to form an audio beat note. The frequency of the oscillator is adjusted until the zero beat is heard. The frequency of the signal can be then read from the dial on the oscillator tuning control.

Most commercial heterodyne frequency meters of the older design use a limited range oscillator with a micrometer frequency readout. The instrument requires a calibration booklet to interpolate the correct frequency from the micrometer dial setting. This was necessary because it was easier to make high-precision VFO circuits over a limited range, especially because a linear frequency readout characteristic on the dial was desired. Today, though, it is possible using PLL techniques to make wide-range oscillators that have the precision of the crystal reference of the PLL.

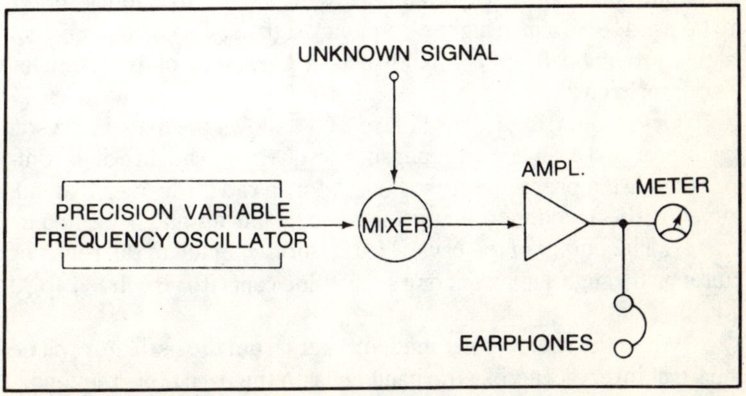

Fig. 21-12. Heterodyne frequency meter.

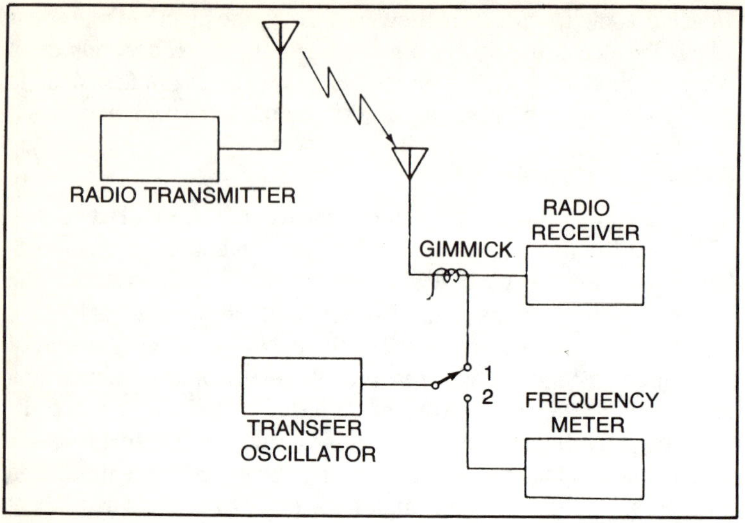

Fig. 21-13. Transfer oscillator.

Transfer oscillator. Figure 21-13 shows the transfer oscillator technique. This is one of the heterodyne systems, and it uses the radio receiver as the indicator. The transfer oscillator is connected through a gimmick—a couple turns of wire around the antenna lead—to the receiver. The receiver is tuned to the unknown signal, and then the transfer oscillator is tuned until it is in zero beat with the unknown signal. We can then measure the frequency of the transfer oscillator on an external frequency meter.

Now just why would anyone want to use the transfer oscillator if they had a frequency meter? Simple! Because it allows a low-frequency precision oscillator and a low-frequency frequency-meter to be used to measure higher frequency signals. The frequency that actually beats with the unknown is a *harmonic* of the transfer oscillator signal.

Crystal calibrators. Figure 21-14 shows the use of a crystal calibrator and a receiver to measure frequency. The principle components are a precision crystal oscillator, a radio receiver that will tune to the frequency being measured, and an audio frequency meter. The output of the crystal oscillator is coupled to the receiver antenna through a gimmick or a small value capacitor, such as 7 to 33 pF.

When the receiver is tuned, markers from the oscillator will be heard at intervals across the band equal to the oscillator frequency. For example, if a 100-kHz crystal is used, the markers will appear

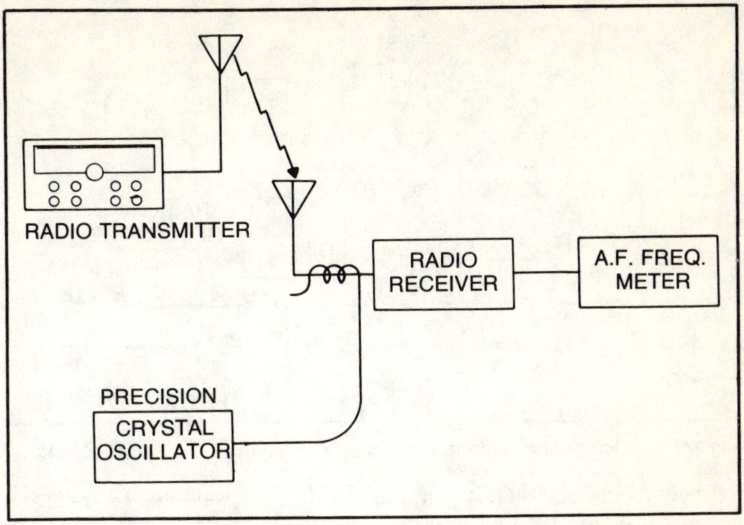

Fig. 21-14. Crystal oscillator.

every 100 kHz across the band. Similarly, a 5-kHz crystal oscillator produces markers every 5 kHz across the band.

Figure 21-15 shows a hypothetical situation in which the frequency of a signal in the 40-r ıeter amateur band is being measured. The unknown signal exists at 7183.56 kHz. The procedure is as follows:

☐ Apply a 100-kHz signal to the antenna inputs and adjust the receiver to one of the markers near the unknown frequency. If the dial cursor of the receiver is movable, then adjust it to lay over the

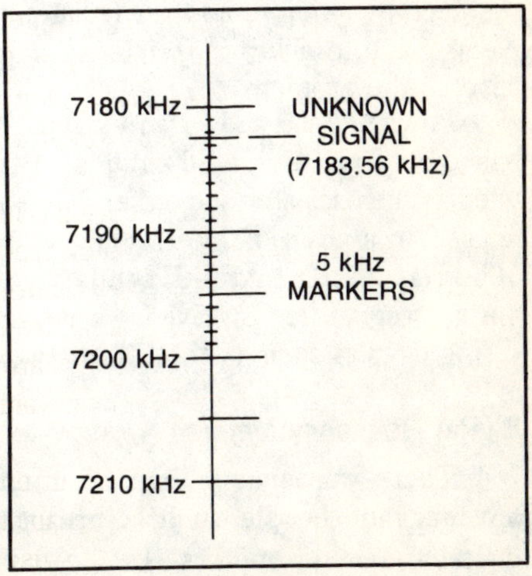

Fig. 21-15. Marker oscillator.

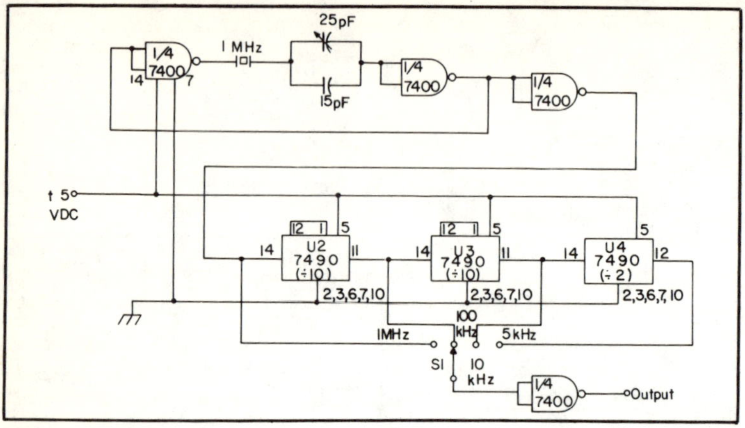

Fig. 21-16. Using a digital frequency counter without a direct connection.

correct frequency marking on the dial. In this example, 7200 kHz is the closest 100 kHz point to the unknown frequency.

- ☐ Tune the receiver to the unknown frequency.
- ☐ Apply a 10-kHz signal to the receiver antenna terminals. This will place markers every 10 kHz across the band. In this case, the 7180-kHz marker will beat with the unknown signal to produce a 3560-Hz beat note.
- ☐ Measure the frequency of the audio beat note and add this frequency to 7180 kHz.

In many cases, the frequency of the unknown signal will lay midway between two marker points. This will produce two beat notes of approximately equal frequencies. This problem can be resolved by going to a lower frequency marker, (such as 5 kHz). Keep going to a lower frequency marker until it is possible to resolve the problem.

Figure 21-16 shows an example of a simple crystal marker based on TTL integrated circuits. A type 7400 NAND gate IC is used as the oscillator and buffer stages. The output of the buffer is applied to a chain of 7490 frequency divider stages that produce 100-kHz, 10-kHz, and 5-kHz outputs. Note that you can make other output frequencies by dividing any of these outputs by either 10 (using circuits such as U2/U3) or two (U4).

Digital Frequency Meters

There was once a time when any book aimed primarily at amateur radio people would be presumptuous for even mentioning digital frequency meters. In the early '60s, the typical digital counter cost not less than 10 times what most transmitters cost! It was simply too expensive for the typical amateur to afford. Today, IC digital frequency counters are so inexpensive that many amateurs have them. In fact, some receivers and transmitters use digital counters in place of the analog dial.

Figures 21-17 and 21-18 show two methods for using the digital counter to make a transmitter frequency measurement. The method in Fig. 21-17 is illegal in all bands except the citizens band and possibly the amateur bands. In even these bands, however, it is discourteous to the other users unless the measurement is incidental to a genuine use of the transmitter. I have used a digital frequency counter next to the rig on occasion, but note that this only works if you can adjust the trigger control on the counter to eliminate the effects of modulation on the frequency counter; otherwise, chaotic readings result.

The technique simply requires a short whip antenna connected to the input of the frequency couter. In most cases, the little slip-up or rubber-duckie antennas used on portable hand-held units will work nicely. They will generally have a connector which is compatible with the input connector on the counter.

A better approach is shown in Fig. 21-18. In this case, a tee connector samples the rf signal and passes it to the counter input

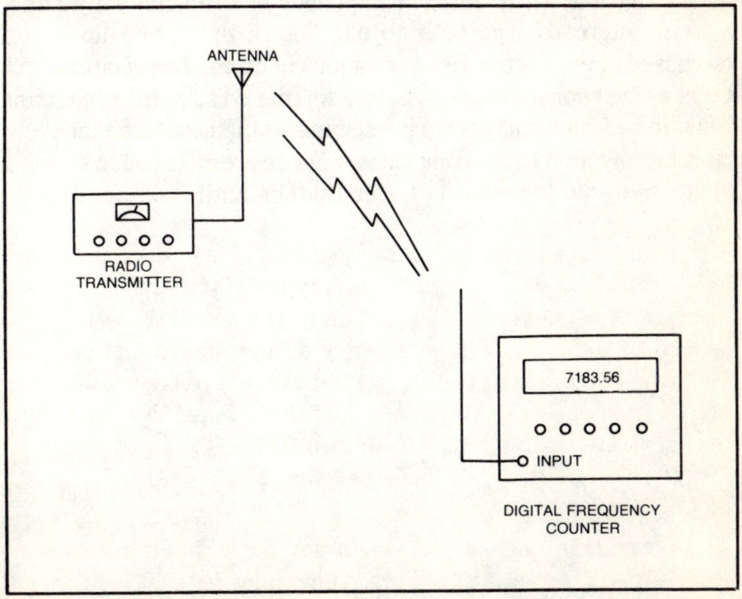

Fig. 21-17. Using a digital frequency counter with a direct connection.

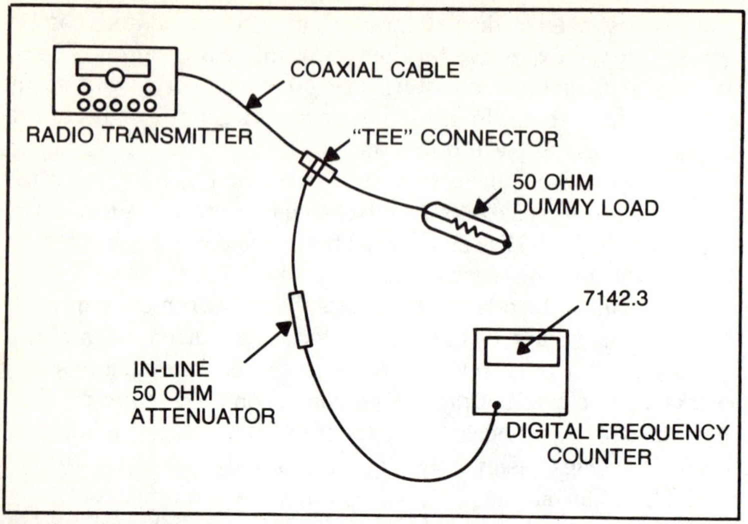

Fig. 21-17. Using a digital frequency counter with a direct connection.

through an *in-line attenuator*. This is absolutely necessary! Do not ever connect the output of a transmitter directly to a counter, unless instructed to do so by the counter manufacturer. The attenuator must have a loss sufficient to drop the high-power transmitter signal down to a level compatible with the counter input.

An alternate scheme is to use one of the rf pick-up boxes discussed in the section on modulation checking. The rf output will be at a level consistent with safety for the counter input circuits. This, in fact, is usually what is used inside the counters that allow direct connection to the transmitter. These are intended to be used in-line between the transmitter and the load.

Chapter 22

How to Make the Antenna Think It's a Loudspeaker

How much is *high* power? That somewhat rhetorical question has many answers, as do most rhetorical questions. But, really, what do *you* consider high power? To the citizens band operator limited to 5 watts, the illegal 100-watt linear amplifier is deliciously high powered! But to the amateur radio operator, who is allowed to run up to 1000 watts DC input to the final amplifier stage, 100 watts is considered an exciter. In fact, 100-watt class transmitters were sold for years under the term "exciter," a practice that presumed that you would rush out and buy a linear amplifier to follow it. The 5 watts of the CB operator is considered by amateurs little more than a *preamplifier* for the oscillator signal.

THE BROADCASTER

But let's consider someone else's point of view. The broadcast engineer considers the 1000-watt power level of the radio amateur to be on the low-power end of the definition spectrum. When researching this book, I called on a professional broadcast engineer in order to obtain some information. When I referred to his AM transmitter as "high power," he snickered. When I unabashedly changed my tune and called it "medium power," he laughed out loud. To the broadcast engineer, the idea of a 1-gallon rig being called either high or medium is just plain ridiculous. Transmitters in the AM broadcast band can be licensed for up to 50,000 watts of rf output power. To them, a 50-gallon rig is high power. Anything under 5 gallons is strictly *low* power. To the BC engineer, the amateur 100-watt transmitter is as ridiculous as the CB 5-watt transmitter seems to radio amateurs.

The domestic broadcast engineer can't talk too much once he looks at his international shortwave cousin. Some of the transmitters make the 50 gallons of the AM clear-channel station look like a preamplifier. In fact, one manufacturer of custom megawatt transmitters uses the same box that is the 50,000-watt model as the *driver* for the megawatt final amplifier. What a preamp! As you can see, it is probably easier to skin an amoeba than to pin down some power definitions.

The use of radio broadcasting on the shortwave bands was prematurely buried by some who looked too seriously at satellite broadcasting possibilities. The problem of cost to the user seems to be a factor. We can assume that a lot of people in foreign countries have broadcast receivers capable of operation on the shortwave bands, but few have 10-meter microwave dishes and low-noise microwave GaSFET receiver preamplifiers needed for direct reception of the satellite. We can therefore obtain radio entry into those countries without the consent of the government.

The furor in the international shortwave bands is a not-so-mute testimony to the power of the radio signal. Almost all national governments operate large shortwave transmitters for broadcasting their view of reality to countries who might not share that view. Christian organizations make good use of the shortwave spectrum. One of the earliest—possibly *the* earliest—was HCJB (*H*eralding *C*hrist *J*esus *B*lessings), which is located in Quito, Ecuador. It has maintained a strong shortwave voice in the Andes for several decades. Others, including the Family Radio stration WYFR in Massachusetts and Florida, and Trans World Radio on Bonair Island in the Carribean, also have an evangelical religious flavor.

But the power sweepstakes keep forcing the required power levels even higher. In the '50s, a powerhouse shortwave station was 250,000 watts, with most being 100,000 watts or less. But as bands became more crowded, stations increased their power levels to keep pace with the interference from like minded stations. The United States International Communications Agency operates a multi-megawatt transmitter on the coastal plains of North Carolina, near Greenville. This mammouth transmitter sends out the US position in 37 languages, without having to rely on having some friendly host government grant us space for "transmittering." The electrical bill for such a place must be enough to snap your mind clean out of its socket.

Fig. 22-1. 1000-watt AM transmitter (courtesy of WABS).

DIFFERENT CLASSES

Figures 22-1 through 22-3 show several different power classes of radio transmitter. For other examples, see Chapter 1. The transmitter in Fig. 22-1 is a 1000-watt AM broadcast model manufactured by Sparta. This transmitter is in daily service at an AM broadcasting station. Our second example is the 25,000-watt FM transmitter shown in Fig. 22-2. This transmitter was made by

Rockwell/Collins, one of the earliest top-grade transmitter/receiver manufacturers in the United States, dating back to the '30s with their amateur line.

Figure 22-3 shows a 50,000-watt international shortwave broadcasting transmitter manufactured by Brown Boveri in Switzerland (they *do* make more than watches, you know!). How would you like to load *that* one up on 20-meters?

Transmission lines for transmitters such as this are a little bit, shall we say, "heavy duty" Parallel feeder transmission lines for the amateur service consist of a couple No. 12 wires a couple inches apart. A transmission line for this power level looks like a pair of irrigation pipes a couple feet apart. In fact, they probably *are* irrigation pipes! Coaxial cable for this level looks like a heating duct for a large house.

Can a high power transmitter make an antenna tower think that it's a loudspeaker? Yes, it can. If you doubt it, then ask the guard who was on duty at the Voice of America transmitter near Cincinatti, Ohio, one night during World War II. The guard was making his rounds, when all of a sudden he seemed surrounded by sound—very loud sound—and it seemed to be coming from everywhere. He

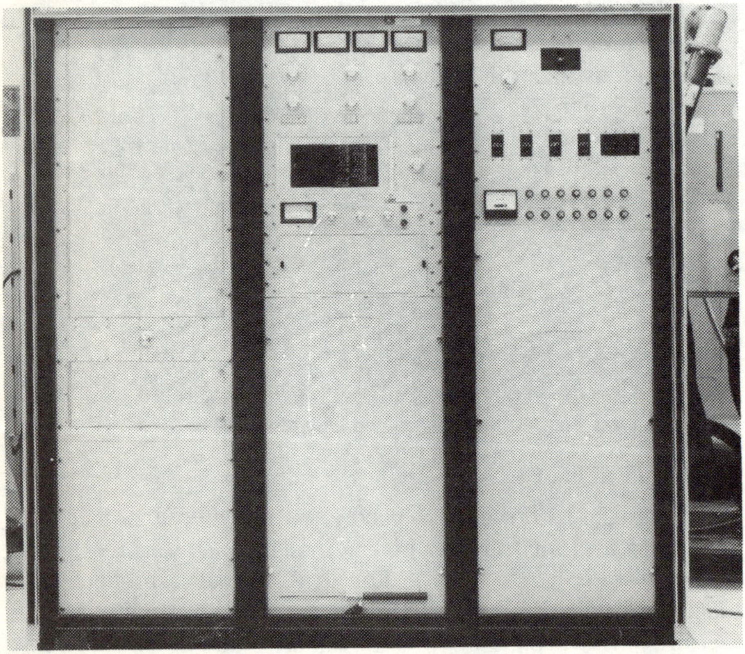

Fig. 22-2. Collins 25-kW FM rig (courtesy of Collins/Rockwell).

Fig. 22-3. 500,000-watt shortwave rig (courtesy of Brown-Boveri of Switzerland).

was frightened so much that he frantically drew his .38 Special revolver and promptly shot himself in the foot.

The problem was traced to rusty joints in the antenna tower construction. These joints acted like diode rectifiers, which rectified the rf signal. The signal was modulated, so a large audio current began to course up and down the tower. The audio currrent vibrated the tower and made it think it was a loudspeaker. The poor guard probably didn't understand all that "detection" talk. All he knew was that his foot got shot, and no Purple Heart for all the pain.

Appendix

FCC Field Offices

Radio transmitters in the United States and all territories are regulated and licensed by the Federal Communications Commission. Any questions regarding licensing, operating technical requirements, or any other matter regarding the use of radio transmitters should be directed to the nearest FCC field office. The addresses of most FCC offices follow.

ALASKA, Anchorage
U.S. Post Office Building Room G63
4th & G Street, P.O. Box 644
Anchorage, AL 99510
Phone: Area Code 907-265-5201

CALIFORNIA, Los Angeles
3711 Long Beach Blvd.
Suite 501
Long Beach, CA 90807
Phone: Area Code 213-426-4451

CALIFORNIA, San Diego
Fox Theatre Building
1245 Seventh Avenue
San Diego, CA 92101
Phone: Area Code 714-293-5460

CALIFORNIA, San Francisco
323A Customhouse
555 Battery Street
San Francisco, CA 94111
Phone: Area Code 415-556-7700

COLORADO, Denver
Suite 2925 Executive Tower
1405 Curtis Street
Denver, CO 80202
Phone: Area Code 303-837-4054

DISTRICT OF COLUMBIA (WASHINGTON, D.C.)
1919 M Street N.W. Room 411
Washington, D.C. 20554
Phone: Area Code 202-632-8834

FLORIDA, Miami
919 Federal Building
51 S.W. First Avenue
Miami, FL 33130
Phone: Area Code 305-350-5541

FLORIDA, Tampa
Barnett Office Bldg., Rm 809
1000 Ashley Drive
Tampa, FL 33602
Phone: Area Code 813-228-2872

GEORGIA, Atlanta
Room 440, Massell Bldg.
1365 Peachtree St. N.E.
Atlanta, GA 30309
Phone: Area Code 404-881-3084

GEORGIA, Savannah
238 Federal Office Bldg. and Courthouse
125 Bull Street, P.O. Box 8004
Savannah, GA 31402
Phone: Area Code 912-232-4321 ext. 320

HAWAII, Honolulu
502 Federal Building
P.O. Box 1021
335 Merchant Street
Honolulu, HI 96808
Phone: Area Code 808-546-5640

ILLINOIS, Chicago
3935 Federal Building
230 South Dearborn Street
Chicago, IL 60604
Phone: Area Code 312-353-0195

LOUISIANA, New Orleans
829 F. Edward Hebert Federal Bldg.
600 South Street
New Orleans, LA 70130
Phone: Area Code 504-589-2094

MARYLAND, Baltimore
George M. Fallon Federal Building
Room 823 31 Hopkins Plaza
Baltimore, MD 21201
Phone: Area Code 301-962-2728

MASSACHUSETTS, Boston
1600 Customhouse
165 State Street
Boston, MA 02109
Phone: Area Code 617-223-6609

MICHIGAN, Detroit
1054 Federal Building
231 W. Lafayette Street
Detroit, MI 48226
Phone: Area Code 313-226-6078

MINNESOTA, St. Paul
691 Federal Building
316 N. Robert Street
St. Paul, MN 55101
Phone: Area Code 612-725-7810

MISSOURI, Kansas City
1703 Federal Building
601 East 12th Street
Kansas City, MO 64106
Phone: Area Code 816-374-6155

NEW YORK, Buffalo
1307 Federal Building
111 W. Huron Street at Delaware Ave.
Buffalo, NY 14202
Phone: Area Code 716-842-3216

NEW YORK, New York
201 Varick St.
New York, NY 10014
Phone: Area Code 212-620-3437

OREGON, Portland
1782 Federal Office Bldg.
1220 S.W. 3rd Avenue.
Portland, OR 97204
Phone: Area Code 503-221-3098

PENNSYLVANIA, Philadelphia
11425 James A. Byrne Federal Courthouse
601 Market Street
Philadelphia, PA 19106
Phone: Area Code 215-597-4411

PUERTO RICO, San Juan
U.S. Post Office and Courthouse
Room 323 P.O. Box 2987
San Juan, PR 00903
Phone: Area Code 809-753-4567

TEXAS, Beaumont
Room 323, Federal Building
300 Willow Street
Beaumont, TX 77701
Phone: Area Code 713-838-0271

TEXAS, Dallas
Earle Cabell Federal Bldg.
Room 13E7, 1100 Commerce Street
Dallas, TX 75242
Phone: Area Code 214-749-1719

TEXAS, Houston
5356 Federal Building
515 Rusk Avenue
Houston, TX 77002
Phone: Area Code 713-226-5624

VIRGINIA, Norfolk
Military Circle
870 North Military Highway
Norfolk, VA 23502
Phone: Area Code 804-461-6472

WASHINGTON, Seattle
3265 Federal Bldg.
915 Second Avenue
Seattle, WA 98174
Phone: Area Code 206-442-7653

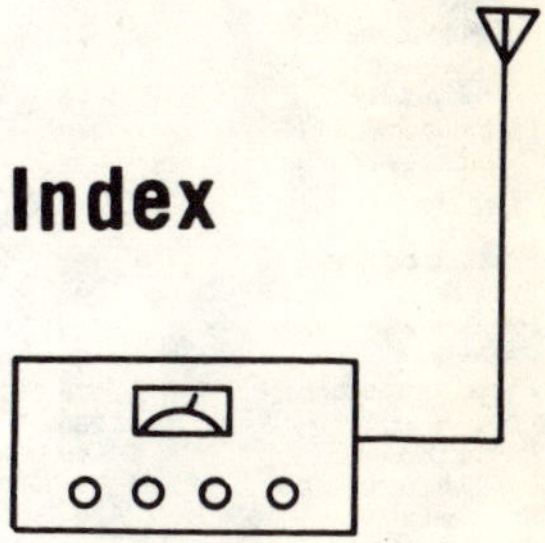

Index